5th International Conference on Man Machine Systems (ICoMMS 2019)

IOP Conference Series: Materials Science and Engineering
Volume 705

Pulau Pinang, Malaysia
26 - 27 August 2019

ISBN: 978-1-7138-1035-3
ISSN: 1757-8981

Printed from e-media with permission by:

Curran Associates, Inc.
57 Morehouse Lane
Red Hook, NY 12571

Some format issues inherent in the e-media version may also appear in this print version.

Institute of Physics
Dirac House, Temple Back
Bristol BS1 6BE UK

Phone: 44 1 17 929 7481
Fax: 44 1 17 920 0979

techtracking@iop.org

Additional copies of this publication are available from:

Curran Associates, Inc.
57 Morehouse Lane
Red Hook, NY 12571 USA
Phone: 845-758-0400
Fax: 845-758-2633
Email: curran@proceedings.com
Web: www.proceedings.com

TABLE OF CONTENTS

IOP Conf. Series: Materials Science and Engineering **705** (2019) 011001 doi:10.1088/1757-899X/705/1/011001

Preface

With great honour, we would like to welcome all delegates to our 2019 International Conference on Man Machine Systems (ICoMMS). It is a great pleasure and privilage for the Mechatronic Engineering Programme, under the School of Mechatronic Engineering, Universiti Malaysia Perlis (UniMAP) to organize this conference.

ICoMMS conference covers a wide range of current research areas all of which are important elements in the related engineering fields. This also reflects the diverse research interests conducted under the School. This conference gathers experts from different countries to discuss and present the state of the art and new research findings and perspectives. We are sure all participants will use this conference as a platform to present their work and address current problems in this challenging but promising field and subsequently create new avenues for the commercialization of the research outcomes.

We would like to express our sincere gratitude to all reviewers, both local and international, for their time and effort in reviewing the papers. Their feedbacks to the authors have been excellent and useful to ensure the high quality of papers for the conference. Our sincere thank also goes to all sponsors for their generosity. Last but not least, we very much appreciate and thank UniMAP and all members of the organizing committee for their hard work, dedication, cooperation and suggestions in ensuring the success of this conference.

Thank you very much and hope to see you again in our next conference.

Prof. Dr. Abdul Hamid Adom
Dr. Norasmadi Abdul Rahim
Dr. Lam Chee Kiang
Dr. Eng Swee Kheng
Dr. Kamarulzaman Kamaruddin

Editors
2019 International Conference on Man Machine Systems (ICoMMS)

IOP Conf. Series: Materials Science and Engineering **705** (2019) 011002 doi:10.1088/1757-899X/705/1/011002

Peer review statement

All papers published in this volume of *IOP Conference Series: Materials Science and Engineering* have been peer reviewed through processes administered by the proceedings Editors. Reviews were conducted by expert referees to the professional and scientific standards expected of a proceedings journal published by IOP Publishing.

IOP Conf. Series: Materials Science and Engineering **705** (2019) 012001 doi:10.1088/1757-899X/705/1/012001

A Review on Moisture Measurement Technique in Agricultural Silos

M S M Abdullah[1], M H F Rahiman[1,a], A Zakaria[1], L M Kamarudin[1] and L Mohamed

[1]Centre of Excellence for Advanced Sensor Technology (CEASTech), Universiti Malaysia Perlis, Pauh Putra Campus, 02600 Arau, Perlis, Malaysia.

Email: hafiz@unimap.edu.my

Abstract. Moisture content (MC) is that the weight of water contained in paddy and other grans expressed in percent. Moisture content is typically mentioned the wet basis which means the whole weight of the grain including the water (MCwb). Accurate MC testing is very important in managing and selling paddy and rice because depending on the aim rice has completely different ideal MC. Inaccurate MC measurements lead to lower head rice once milled at wrong moisture content, additional drying cost and loss of quality if paddy is dried too much, spoilage if the grain is just too wet in storage, additional drying cost and harvesting loss if paddy is harvested wetter than necessary and Weight loss (loss in profit) if grain is sold too dry. This paper intends to review the available techniques of moisture content measurements, its principles and sources of inaccuracies. Relevant properties of cereal grain also are studied with their impact on moisture content, temperature, density and humidity. There are several ways of grain moisture detection such as drying technique, resistance technique, capacitance technique, microwave technique, neutron technique, V/F Conversion, Ftir spectroscopic technique and Microcontroller technique. Current MC of grains are based- on sampling and individual single point measurement, so far, there is no available overall measurement of MC in silo, but by using tomography method can measurement MC in whole silo and investigate the efficiency and accuracy of localization of moisture distributions sensing in agricultural silo by using Gaussian algorithm.

1. Introduction

Moisture content (MC) typically refers to the wet basis, which suggests the entire weight of the grain, together with the water (MCwb). accurate mc testing is incredibly necessary in managing and marketing paddy and rice as a result of counting on the aim rice has totally completely different ideal MC. Inaccurate mc measurements cause additional drying value and harvest loss if paddy is harvested wetter than necessary, spoilage if the grain is simply too wet is storage, additional is drying worth and loss of quality, if paddy is dried too, so much, lower head rice once milled at wrong moisture content and weight loss (loss in profit) if grain is sold too dry. MC of grain can measure by using either primary method, based on weight measurements like the (a) oven method or (b) infrared moisture balance or secondary method, using an electronic instrument that uses electrical characteristics of the grain. As moisture content measurements are done manually and this is the biggest problem in grain storage due to inefficient and not well distributed in moisture content reading and lead to false measurement, in turn, damaging the grain due to excessive moisture. To maintain grain quality throughout storage, grain

IOP Conf. Series: Materials Science and Engineering **705** (2019) 012001 doi:10.1088/1757-899X/705/1/012001

should be protected against the expansion and reproduction of fungi, insects and mites[1]. Storage temperatures below 15C will forestall insect development and mite and fungi growth[2]. temperature in storage is one in all the most necessary factors that verify grain storage quality[3].

The MC of stored rice grain changes according to the weather which affects MC in the silo with poor aeration, this could lead to moisture build-up within the storage overtime and the moisture might be unevenly distributed. Paddy drying technique embrace ancient and mechanical systems with variable technological quality and capacities for either farm or industrial level, High MC level throughout storage will lead to grain discoloration, increase the likelihood of attack from pests and encourage the development of molds [4]. Table 1 show the moisture content for safe storage.

Table1. Grain should be dried to a specific MC counting on storage period[5].

Required MC for safe level storage	Storage period	Potential problems
9% or less	<1 year	Loss of germination
12% or less	Storage of farmer's seeds	Loss of germination
13% or less	8-12 months	Insect damage
14% or less	Weeks to a few months	Moisture adsorption, discoloration, respiration loss insect damage, Molds

Ecological conditions like uneven temperature dissemination and wet wind current will make higher wet substance create over the keep grains. The RH inside the air shut a grain test at a particular temperature (T) relies upon the wet substance of the grains [6]. it's along these lines important to appropriately live and the executives the natural conditions for finding out the standard of the grains, keep up an upper hand inside the nourishment technique exchange, and certification sustenance security[7].

MC measurements are done manually, and this is the biggest problem in grain storage due to inefficient and not well distributed in MC reading and lead to false measurement, in turn, damaging the grain due to excessive moisture. Thus, this study aims to investigate the efficiency and accuracy of localization of moisture distributions sensing in agricultural silo using tomography approach. The advantages of tomography lower cost, safety, easy handling, and high speed performance are important reasons why this type of tomography is chosen particularly within the method industry and medical area[8]. The correlation between multiple RF attenuation information can be gathered and analyzed using tomography method to locate the moisture distribution in the silo. This would benefit the agricultural sector in securing the food storage for the country and reduces damages of grain in the silos.

2. Methods of Measurement

There are several ways of grain moisture detection, as well as drying technique, resistance technique, capacitance technique, microwave technique, Electrical moisture meter technique, Wi-Wheat technique, Smart detector Network System technique, FTIR spectroscopic technique, Microwave technique, Microcontroller technique, neutron technique, V/F Conversion and Radio-frequency (RF) technique.

2.1. Capacitive moisture detection technique

Developed quickly however the measuring value isn't only sensitive to temperature, however additionally terribly sensitive to the grain flow rate and grain compactness within the drier and lots of different factors in grain moisture measuring, And this technique is limited to the detector re-calibration once very long time use[9][10].

IOP Conf. Series: Materials Science and Engineering **705** (2019) 012001 doi:10.1088/1757-899X/705/1/012001

2.2. *Wi-Wheat method*

Presented a non-destructive and economic wheat moisture detection system with artefact wireless local area network. The experiment validates the practicability of wheat moisture detection by victimization channel state information (CSI) amplitude and phase difference data. Data pre-processing, feature extraction and support vector machine (SVM) classification are enforced for CSI process module, However Wi Wheat achieved higher classification accuracy for line of sight and non- line of sight eventualities[11].

2.3. *DeepWMD technique*

A profound LSTM organize fundamentally based framework for multi-class wheat wet location. They utilized two Wi-Fi devices(5 GHz), while one is prepared in light of the fact that the transmitter and furthermore the inverse as the collector, framework misused CSI [12] abundancy and stage refinement information from exchange products Wi-Fi gadgets for wheat MC detection, System accomplished high-exactness multi-class wheat wetness recognition in numerous indoor stockpiling environment [12], however the measuring device is complicated, high in price, unavailable to measure the whole silo, unavailable outdoor and need repeat steps more than just once.

2.4. *Smart detector Network System supported ZigBee Technology*

To observe Grain Depot is planned to monitor grains in storage depots. The harm of grains is controlled by observation the parameters like temperature, humidity and light that influence the storage of grains. The optimum temperature for the insects is 25°C to 32°C, Grain harm by fungi once the grain is keep at MC content below 14% and also the temperature is less than 20°C and higher than 40°C respectively, accurate measurements for parameters like, temperature, humidity and light[13], however this method available for various laboratory experiments and affected by surrounding environment.

2.5. *FTIR spectroscopic technique (Fourier Transform Is Required)*

An accurate, efficient and non- destructive techniques which might be used for multi-component analysis. These techniques can even be used for continuous on-line measurements. FTIR needs no external calibration, operation is incredibly easy and this method has most optical throughput with accrued sensitivity[14], However, technique will fail due to very localized and not represent whole grain in the silo.

2.6. *Microcontroller technique*

Based system for safe grain stockpiling in an exceedingly silo system which will monitor temperature and RH, live MC non-damagingly, and the management stockpiling condition to a protected level, system utilized Arduino software programming. The created CS was made out of Mega2560 (as the microcontroller), SHT21 sensors (temperature and RH), GSM SIM900A module, and data feller safeguard (V1.0 Deep-Robot). These deliberate parameters were wont to initiate and off the fan and warmer to oversee or keep up the capacity condition to a protected dimension, however framework could keep up and manage the capacity surroundings to safe dimension conditions [15][17], however, the temperature was very important for storage however, RH was thus high that the grain MC is laid low with the relative humidity (RH) of the storage surroundings.

2.7. *Microwave method*

Microwave wetness *meters* are developed specifically for oat grains. Okabe et al portray a microwave meter to be utilized with rice and wheat, operational at 9.4 GHz and upheld the lessening of the microwave motion by the grain test, that was set among sending and receiving horn antennas. The according precision was ±0.5% dampness content over the range from 10 to 30th. A microwave wetness meter for wheat, rye, grain, and oats, operational on the construction of a 9.35 gigahertz wave engendering through the grain and intended for nonstop wetness perception applications. Microwave movement at frequencies 9 GHz give preferred standpoint to solid assurance of grain test wetness content in either group or on-line application [18]. At these frequencies, varieties among grain tests in material properties on account of not meddle with solid discovery of the amount of water inside the grain

[19]. estimating with microwave thunderous cavities of the resounding recurrence move and changes in hole transmission losses due the presence of a part inside the pit will offer size free and shape-autonomous estimation of the moisture content of single kernels[16][17].

2.8. Microwave moisture meter method

For estimate the relationships between wetness density inside the grain and so the attenuation and section shift, A model grain wetness meter consisting of a 10.5 GHz oscillator, patch array antennas, isolator, detector and electronic voltmeter provided a constant of determination, microwave is extremely effective for activity the moisture content of grain and will be pliant for fast, accurate, non-destructive on-line wet measurements on flowing grain [22]. However, the measurement device is complex and high in worth. However, it is affected by heat.

2.9. Relative humidity (RH) technique

May be measured using electronic meters, or perhaps a lot of inexpensively using indicator paper that changes colour response to RF. This method provides a spreadsheet to convert RF to MC for several common seeds and commodities. This technique may be used in the sector for quickly estimating seed or commodity moisture content to determine whether any drying is required or is suitable for storage, milling or different uses. Although the colour scale can distinguish only differences of 10 to 20% in RH, this is sufficient to estimate grain moisture content to at intervals or so a 1% within the range from 40 to 60% RH. At the extremes (80% RH), the conditions are clearly excellent or very poor for storage, therefore precision in conversion to moisture content is less necessary in those RH regions[18].

2.10. Radio-frequency (RF) technique

Basically, principles are represented for practice (RF) (including high frequencies and microwaves) material properties, or permittivity, of grain for detecting status material properties, or permittivity, of grain for detecting status through their relationship with wet substance. Radio frequencies are useful in quick and non-ruinous detecting of status content gratitude to high relationships that exist at any recurrence between the material properties of grain thus the amount of water blessing at interims the grain. The greater part of those dielectric-type status meters uses frequencies at interims the rang between with respect to 1-20 MHz. Another methodology. All through that weakening and area move were thought of as elements of status content, grain mass thickness, and temperature, uncovered that temperature compensated, and thickness free status content judgments were gotten with 15.2 GHz estimations on shelled corn. [23].

2.11. A density-independent

Perform of the granular material permittivity offers guarantee for an all-inclusive institutionalization for detecting condition substance of granular or particulate materials. The procedures of thickness autonomous condition detecting by high-recurrence and microwave estimations give guarantee to advancement of instruments for brilliant use in perception MC of grain and completely very surprising materials on-line for strategy the board and the board.[19].

3. Design and Tomography method

Based on the principle of RF sensor, the diagram of detection circuit can be described as: RF sensor get the received signal strength indication each sensor by using 20 ESP8266 sensor then send data to cloud. It is shown in Fig.1.

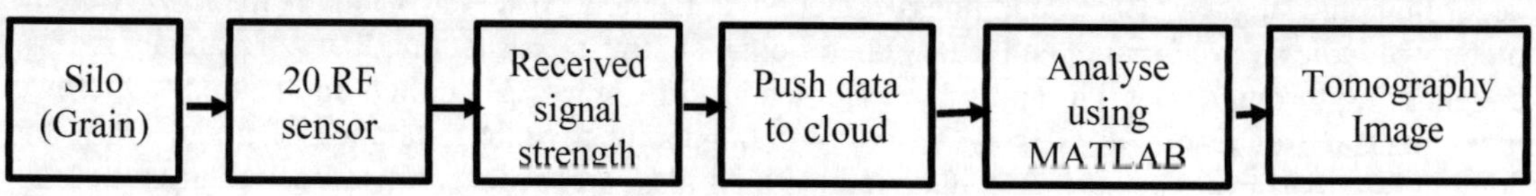

Fig.1 The design diagram of system

The hardware of system is designed based on RF sensor, system has 20 ESP8266 sensor each sensor get data from 19 sensor then push it to cloud this part goes with all sensor after all sensor push data to cloud MATLAB analyse data to get tomography image.

Current MC of grains are based- on sampling and individual single point measurement, so far, there is no available overall measurement of MC in silo. The MC of stored rice grain changes according to the weather which affects MC in the silo. With poor aeration, this could lead to moisture build-up within the silo overtime and the moisture might be unevenly distributed. In this condition, conventional measurement of MC will fail due to very localized and not represent whole grain in the silo. RF sensor network signals are known to have some level of sensitivity towards moisture. This characteristic can be exploited to locate the source of the moisture in the form of tomogram (moisture distribution map). To reconstruct the map, some information is required to be collected form the RF sensor network and suitable reconstruction can be studied. This project will focus on modelling RF sensor network communication to act as the sensor for sensing the moisture content in rice grain. A novel tomography project will be proposed to cater for the uneven moisture distribution in the silo. This tomography approach (RF sensor network) is possibly giving a full spectrum of MC reading inside the silo. System cover all silo by using mesh network each sensor become sender and receiver to get data in every place inside silo. By using gaussian algorithm to analyses data to localization of moisture distribution and get several image reconstruction algorithms for mapping the moisture in the silos. In Linear Back Projection algorithm (LBP), the concentration profile is generated by combining the projection information from every device with its computed sensitivity maps, the image reconstructed by LBP lands up in blurring the image that ends up in high AE in each live taken, This blurring image is due to the character of back-projection technique [20].

4. Comparison of measurement of moisture content (MC) strategies

Several methods have reported the values of the parameters used in drying their material likewise because the accuracy of the moisture content determination based on a typical test. These methods are given in Table.2 . As can be seen, the electromagnetic Imaging technique has been used for an oversized of grains, with a large range of sample masses, applied, drying durations and initial MC.

Table.2 Measurement the moisture content by several method.

No	Methods of measurement	Setup of the system	Description
1	Electromagnetic Imaging	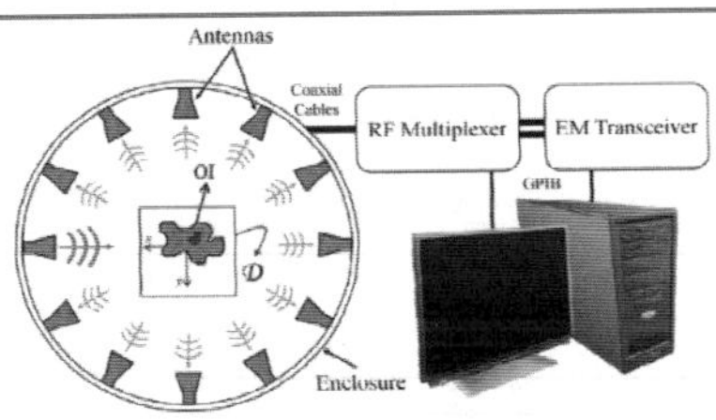	The measured data collected by an electromagnetic imaging system are used as the input to a picture reconstruction algorithm[21].
2	Microwave Transmission at 10.5 GHz	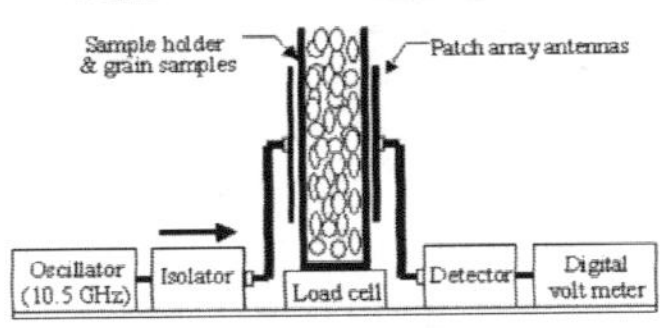	System carried out to work out the optimum dimensions of a single patch antenna for wetness measure with antenna style [22].
3	Microwave Dielectric at 5 to 15 GHz		Dielectric properties of cereal grains were measured at 23°C over a variety of wet contents and over microwave frequencies from 5 to 15 GHz [17].

IOP Conf. Series: Materials Science and Engineering **705** (2019) 012001 doi:10.1088/1757-899X/705/1/012001

4	Electromagnetic Imaging	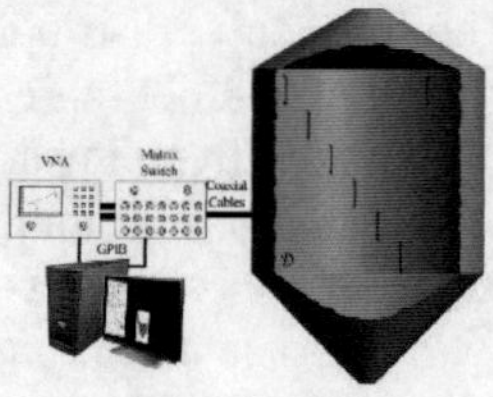	The setup is comprised of twenty four flux transmitter/ receiver probes mounted on the inner walls of the bin. The system quantitatively pictures the insulator properties of the grain used as indirect indicators of wetness content, temperature, insect infestation[23].
5	Commodity WiFi		Wi-Wheat setup with off-the-peg laptops and native space network cards. the instance consists of associate horsepower ProBook 4411s PC with 2.1GHz Intel (R) Pentium 2 central process unit and 2GB RAM as receiver, and a Sony PCG-6S1T PC as a transmitter[12].
6	Relative Humidity and Temperature sensors	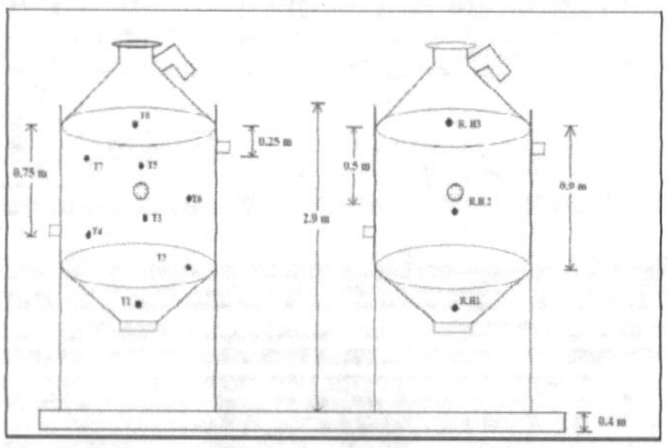	A data lumberman model TC-800D (Make: Ametropias, Mumbai) was record the temperature among the silo. The instrument had eight thermometer sensors K of a length of fifteen cm and a wire of length three m [24].

5. Conclusion

Apparently, a scientific study has not been conducted over associate outsized vary of condition in grains by drying to the vary varied grain wetness meters and diverse characteristics of grain that contribute to variability in insulator properties. to boot, there ought to be plenty of study conducted on the however the physical properties of grains influence the last word condition content throughout acquisition, storage and methodology. the foremost effective insulator properties and best frequency ranges ought to be designated whereas springing up with the instrument for fast condition determination. the correct choice and use of circuits are necessary on springing up with the sample holder capability for various commodities. the other necessary wishes for grain condition instrument embody the great issue concerning use, accuracy, consistency, precision, irresponsibleness and lightweight thus one will simply carry the instrument from heap to heap throughout acquisition season. Finally, there's another disadvantage that deals with the indirect techniques for activity of condition in grains that's that the "standard wetness determination method" to that all the measured electrical information should be referred. Typically, RF sensor network technique is to be considered because is to investigate the efficiency and accuracy of localization of moisture distributions sensing in agricultural silo using tomography approach analyse data by gaussian algorithm. The correlation between multiple RF sensor network attenuation information can be gathered and analysed using tomography method to locate the moisture distribution in the silo. This would benefit the agricultural sector in securing the food storage for the country and reduces damages of grain in the silos and can measurement of moisture content in whole silo.

Acknowledgement

This research was financed by the Ministry of Education Malaysia under Transdisciplinary Research Grant Scheme (Ref: TRGS/1/2018/UNIMAP/02/4). The authors would also like to acknowledge Universiti Malaysia Perlis (UniMAP) for the facilities and technical support.

References

[1] Sun D W and Woods J L 1997 Deep-bed simulation of the cooling of stored grain with ambient air: A test bed for ventilation control strategies *J. Stored Prod. Res.* **33** 299–312

[2] Sun D W and Woods J L 1994 Low Temperature Moisture Transfer Characteristics of Barley: Thin-Layer Models and Equilibrium Isotherms *J. Agric. Eng. Res.* **59** 273–83

[3] Jia C, Sun D W and Cao C 2001 Computer simulation of temperature changes in a wheat storage bin *J. Stored Prod. Res.* **37** 165–77

[4] Anon Drying - IRRI Rice Knowledge Bank

[5] Weinberg Z G, Yan Y, Chen Y, Finkelman S, Ashbell G and Navarro S 2008 The effect of moisture level on high-moisture maize (Zea mays L.) under hermetic storage conditions-in vitro studies *[1] Z. G. Weinberg, Y. Yan, Y. Chen, S. Finkelman, G. Ashbell, S. Navarro, "The Eff. moisture Lev. high-moisture maize (Zea mays L.) under Hermet. storage Cond. Vitr. Stud. J. Stored Prod. Res., vol. 44, no. 2, pp. 136–144, 2008.* **44** 136–44

[6] Onibonoje M O and Kehinde L O A Wireless Sensor Network for Controlling the Effect of the Moisture Content in Stored Maize Grains *Int. J. Eng. Res. Technol.* **Vol. 4** 2278–0181

[7] Onibonoje M O and Jubril A 2012 Determination of Bulk Grains Moisture Content *Ife J. Technol.* **21** 55–9

[8] Zarina S, Muji M, Abdul R, Hafiz M and Rahiman F 2013 Optical tomography : Image improvement using mixed projection of parallel and fan beam modes *Measurement* **46** 1970–8

[9] Liu Z, Wu Z, Zhang Z, Wu W and Li H 2015 Research on online moisture detector in grain drying process based on V/F conversion *Math. Probl. Eng.* **2015**

[10] Oommen T V 2002 On-line moisture sensing in transformers *IEEE ABB Transnission Technol. Inst.* 236–40

[11] Yang W, Wang X, Song A and Mao S 2018 Wi-Wheat: Contact-Free Wheat Moisture Detection with Commodity WiFi *IEEE Int. Conf. Commun.* **2018-May** 1–6

[12] Yang W, Wang X, Cao S, Wang H and Mao S 2018 Multi-class wheat moisture detection with 5GHz Wi-Fi: A deep LSTM approach *Proc. - Int. Conf. Comput. Commun. Networks, ICCCN* **2018-July** 1–9
[13] Kumar S, Hiremath V and K R 2012 Smart Sensor Network System based on ZigBee Technology to Monitor Grain Depot *Int. J. Comput. Appl.* **50** 32–6
[14] Thakur R 2015 Determination of Grain Moisture Content Using Ftir Spectroscopy ME (Electrical Engineering) NITTTR , Chandigarh (INDIA) Assistant Professor Electrical Engineering Department , 34–5
[15] Dal-uyen D, Yaptenco K, Peralta E and Suministrado D 2019 Microcontroller-based control system for safe grain storage in silo *IOP Conf. Ser. Earth Environ. Sci.* **230** 012020
[16] Anwar S I 2010 Determination of moisture content of bagasse of jaggery unit using microwave oven *J. Eng. Sci. Technol.* **5** 472–8
[17] Trabelsi S and Nelson S O 2012 Microwave Dielectric Properties of Cereal Grains *Am. Soc. Agric. Biol. Eng.* **55** 1989–96
[18] Bradford K J, Dahal P and Bello P 2016 Using Relative Humidity Indicator Paper to Measure Seed and Commodity Moisture Contents *Ael* **1** 0
[19] Nelson S O, Trabelsi S and Kraszewski A W 2001 RF sensing of grain and seed moisture content *IEEE Sens. J.* **1** 119–26
[20] Hafiz M, Rahiman F, Rahim R A and Rahim H A 2012 Image reconstruction algorithms for ultrasonic tomography *Int. J. Innov. Comput. Inf. Control* **8** 527–38
[21] Asefi M, Jeffrey I, LoVetri J, Gilmore C, Card P and Paliwal J 2015 Grain bin monitoring via electromagnetic imaging *Comput. Electron. Agric.* **119** 133–41
[22] Kim K, Daejeon Y and Kim M S 2005 Moisture Measurement in Grain by Free-Space Microwave Transmission at 10 . 5 GHz **0300**
[23] Asefi M, Gilmore C, Jeffrey I, LoVetri J and Paliwal J 2017 Detection and continuous monitoring of localised high-moisture regions in a full-scale grain storage bin using electromagnetic imaging *Biosyst. Eng.* **163** 37–49
[24] Sawant A A, Patil S C, Kalse S B and Thakor N J 2012 Effect of temperature, relative humidity and moisture content on germination percentage of wheat stored in different storage structures *Agric. Eng. Int. CIGR J.* **14** 110–8

A Review on Frequency Selection in Grain Moisture Content Detection

N A M Ramli[1], M H F Rahiman[1, a], L M Kamarudin[2], A Zakaria[1] and L Mohamed[1]

[1]Centre of Excellence for Advanced Sensor Technology (CEASTech), Universiti Malaysia Perlis, Pauh Putra Campus, 02600 Arau, Perlis, Malaysia.
[2]School of Computer and Communication Engineering, Universiti Malaysia Perlis, Pauh Putra Campus, 02600 Arau, Perlis, Malaysia.

Email: hafiz@unimap.edu.my

Abstract. A brief review on the dielectric properties of grain and their use in sensing moisture content is presented. Dielectric constant and loss factor were defined generally, and their correlation with grain permittivity from historical studies is presented. References are cited on dielectric measurement techniques using various frequency ranges from radio frequency through microwave frequencies. The described techniques including from the past decades until current studies.

1. Introduction

Grains such as wheat and paddy represent the majority of worldwide consumed daily food. However, as the grain is naturally hygroscopic, the grain's quality post-harvesting revolves around their level of moisture content. Moisture content generally referred to the wet basis, meaning the total weight of the grain including the water (MCwb) expressed in percentage. According to [1][2][3], moisture content is the essential parameter that impacts the quality and longevity of the granular material to be stored without degrading before processing for commercial purpose. It is also a key factor that influences the selling price because the quality of dry grain is more valuable, thus extra drying costs for safe storage must be considered. Even so, for other grain processing such as flour, packaging foods, and animal feeds, achieving desired moisture content is necessary in obtaining high quality products. In short, accurate moisture content sensing is important because different purpose in managing and marketing the grain required different ideal moisture content.

The grain moisture content generally being measured by using primary and secondary method. The primary method [4][5] is sensing the absolute moisture content in grain through oven drying process. The grain is heated over a certain duration under specified temperatures, and the weight loss in grain is used to calculate the moisture content. Weight measurement constantly reliable and high precision; thus, this method is referred as the reference method. Since the reference method involves tedious procedures, time consuming and not suitable for online and on-site measurement, a secondary method which is simpler, low cost, rapid and non-destructive have been developed. This method works on the principle of sensing the clear relationship between moisture content and the dielectric properties of the grain [6][7][8]. Over the past decades, [9–13] presented that the grain dielectric properties is a function of moisture content, applied frequency, temperature of the grain and the material density. Their influence therefore is significant for moisture level-sensing applications in agricultural industry as well as in dielectric heating processes.

In agricultural industry, dielectric properties, or permittivity affect the absorption and dissipation of electromagnetic energy when they are subjected to radio frequency (RF) or microwave (MW) heating and practically expressed as the complex permittivity relative to free space, $\varepsilon = \varepsilon' - j\varepsilon'' = |\varepsilon|e^{-j\delta}$ where δ is the loss angle of the dielectric. The real part, ε' is the dielectric constant associated with the material's ability to store the applied electrical energy while the imaginary part, ε'' known as dielectric loss factor represents the dissipation of applied electrical energy in the form of heat [14]. The conductivity of the dielectric in S/m is $\sigma = \omega\varepsilon_o\varepsilon''$ where $\omega = 2\pi f$ is the angular frequency with f in Hz and ε_o is free space permittivity, $\varepsilon_o = 8.854 \times 10^{-12}$ F/m [6].

2. Historical Perspective

Appropriate moisture content sensing techniques for any particular application mainly depend on the frequency of interest and the nature of the dielectric material to be measured [15]. Different kinds of instruments have been used for measurement, but any instrument that provides reliable determinations of the required electrical parameters involving unknown material in the interest frequency range should be considered. In this section, previous works on the moisture detection over the past decades in low, medium and high frequency were briefly reviewed.

2.1. Frequency Range 1 – 50 MHz

The first recorded quantitative data about the dielectric properties of grain was reported by [16] using frequency range 1 – 50 MHz. It was presented that permittivity of grain is clearly correlated with the moisture content, as illustrated in Figure 1 and 2, which therefore can be used for moisture sensing. Diverse kinds of grain tested by [16] has shown that the dielectric constant, ε' either remains constant or decreases as frequency increased of all moisture content percentage while the loss factor, ε'' varies with frequency accordingly. The changes in dielectric properties due to the variations in moisture content are greater at lower frequencies.

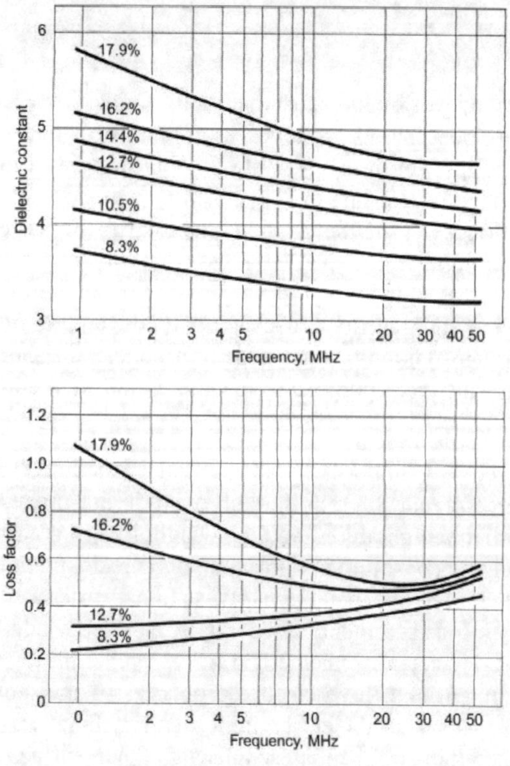

Figure 1. The permittivity of 'Nebred' hard red winter wheat at 24°C. Test weight: 768 kg/m³ (59.7 lb/bu) [16].

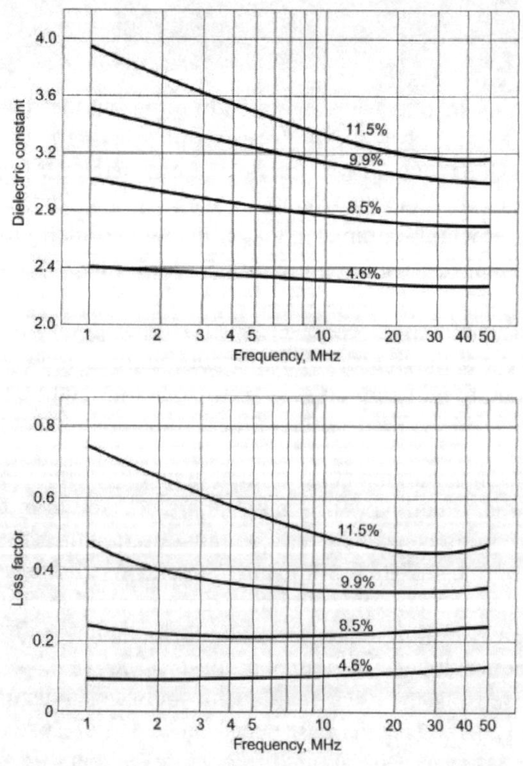

Figure 2. The permittivity of 'Hawkeye' soybeans at 24°C. Test weight: 738 kg/m³ (57.3 lb/bu) [16].

IOP Conf. Series: Materials Science and Engineering **705** (2019) 012002 doi:10.1088/1757-899X/705/1/012002

2.2. Frequency Range 50 – 500 MHz

According to [17], a slightly similar method was developed principally for measurement the grain and seed dielectric properties. A coaxial-line sample holder with open-circuit terminations and with sample holders exhibited as transmission-line sections were used for similar measurements with bridges and admittance meter at frequency 50 – 250 MHz and 200 – 500 MHz respectively.

By using bridges meter, the loss factor for lossy materials is not determined accurately. [18] has reported that accuracy of permittivity for liquid materials is about 2%, but as the interest frequency increases, the accuracy decreases. However, as the losses of grain and seed samples were high, this method is useful to be used in studying the frequency dependency of the dielectric properties. The measurement is taken on hard red winter wheat sample using admittance meter [19] also gave out the same conclusion. Both techniques in this frequency range were suitable for dielectric sensing on grain.

2.3. Microwave Frequency

In previous work, while measuring the permittivity of static grain samples for moisture detection, the dependency of permittivity on temperature and bulk density must also be thought out. However, later, it was recognized that using microwave frequency, the sensing of grain moisture content had the capability of providing a measurement independent of bulk density by single-frequency measurements of attenuation (decrease in amplitude) and phase shift (delay in phase) or by using density-independent functions of dielectric constant and loss factor [20][21]. This finding provides the possibility for on-line and real-time monitoring for grain moisture content sensing.

At lower frequency, [19] stated that the precision of conductance measurement is about 5%. At 2% alteration in conductivity value, less than 0.1% dielectric constant being changed and about 2% changes in the loss factor. However, [22] reviewed that at frequency higher than 1 GHz, materials' conductivity would not be affected. Besides that, in [11] which measured moisture content at 16.8 GHz, they concluded that using higher frequency, the characteristic of equipment can be smaller, and the thickness of material layer should be thinner than the required at lower frequencies. Therefore, better sensitivity to moisture variation can be allocated along with the layer of the material.

2.4. Wide Band Frequency, 250 Hz – 10 GHz

In Figure 3, the dependence of hard red winter wheat permittivity on moisture content in the frequency range 250 Hz to 10 GHz is illustrated [12]. The plotted graph determined that the dielectric constant, ε′ continuously decreases as the frequency increased, but for loss factor, ε″ varies differently as the frequency increased while both values constantly depending on the moisture content of the wheat.

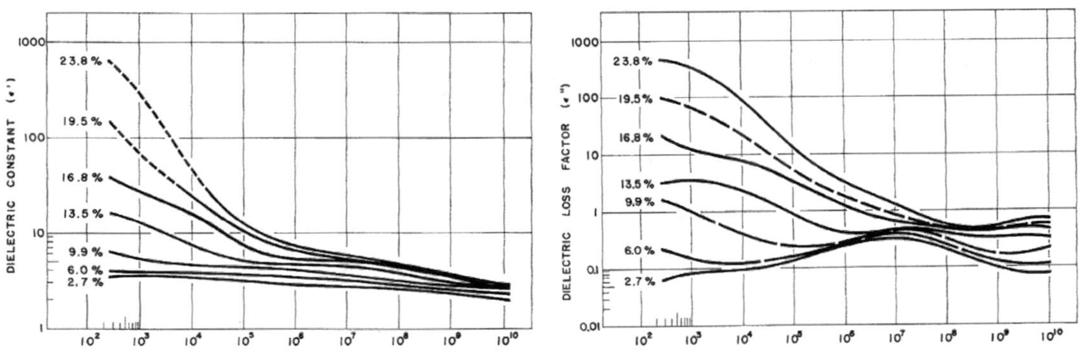

Figure 3. Correlation of dielectric constant and loss factor of hard red winter wheat with frequency from 250 Hz to 10 GHz at 24°C [12]

3. Current Frequency Selection

Based on the diverse grain moisture range, and the continuous reliance between grain dielectric properties and frequency from Figure 3, instrument designers and researchers had a wide range of frequency selection in set up the grain moisture meter. Commonly, the frequency was selected based on the tradition, familiarity with the design as well as the components cost. Table 1 briefly reviews the current frequency selected for grain moisture content detection over the past three years.

Table 1. Frequency selection for grain moisture content detection

No.	Frequency Selection (MHz)	Setup of the system	Description
1	3000 to 4500	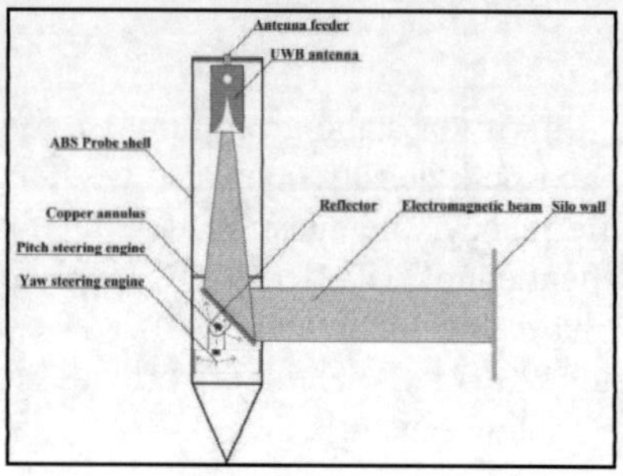	The system is proposed by [23] to detect the moisture distribution of the grain in silo storage. The beam of far-field microwave was able to scan the whole horizontal plane [24]. The results obtained shows that the moisture distribution can accurately be calculated with 0.5% error.
2	78 and 93	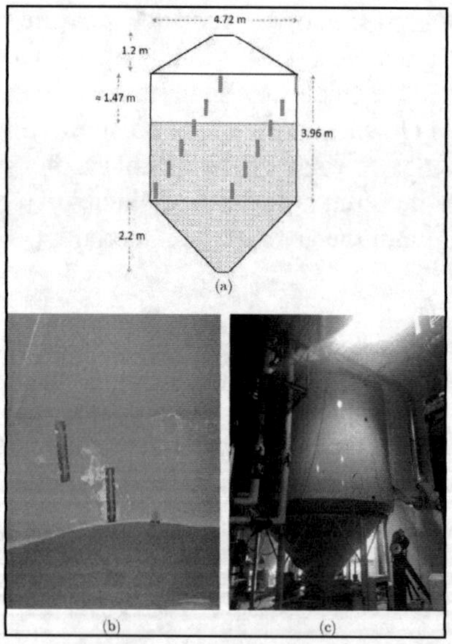	[25] presented a system capable in sensing spoiled grain region inside an industrial-scale grain bin. Both frequencies selected are tested suitable based on the modelling error where the difference between the real scale system and simulated model are reduced. The spoilage region can be clearly distinguished even for the region up to 0.24% of total grain volume. The contrast image and knowledge of the spoiled grain inside the bin later being studied using inversion algorithm [26].
3	5800	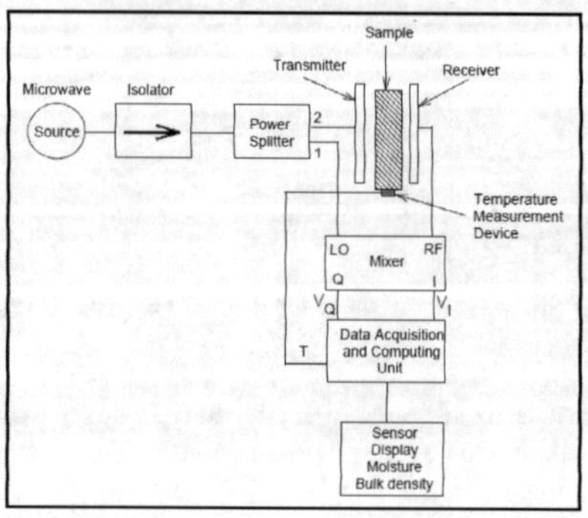	The microwave meter built by [27] is used for instantaneous shelled peanut kernel moisture content detection. The operated signal on 5800 MHz using free-space transmission measurement principles determined the standard error of performance of 0.53% compared to the oven-drying method, whereas official moisture meter error performance was 0.87%.

4	433		The capacitive measuring device designed by [28] was suitable to test on-line corn grain moisture content. The selected frequency of 433 MHz has ensured the real-time, accurate, and stable data communication between the device and the computer.
5	100	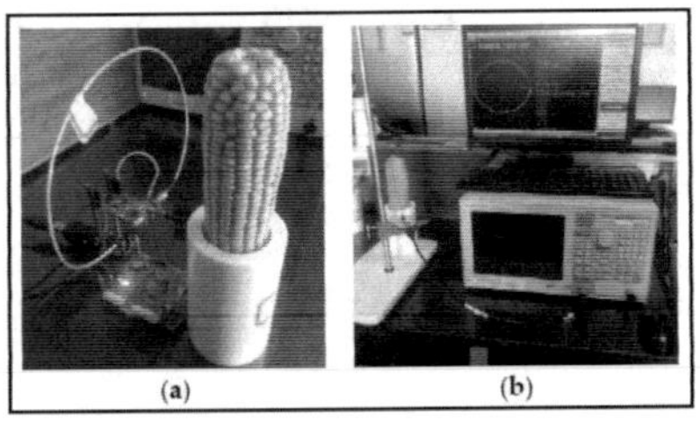 (a) (b) 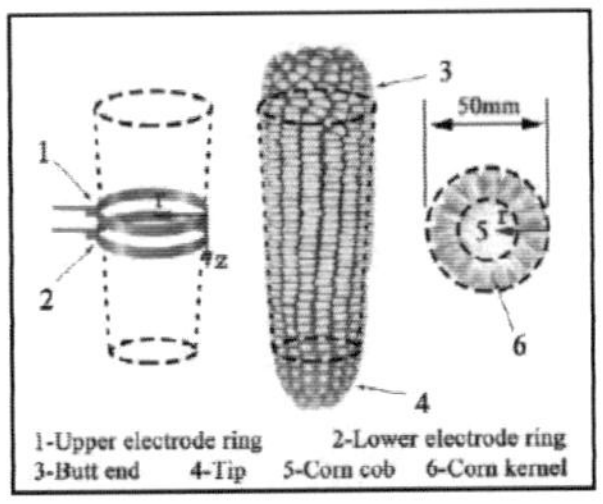	[29] presented a non-invasive in situ system to measure corn kernel moisture content. At 100 MHz, the effect of conductivity has been reduced. They reviewed that permittivity of dielectric materials under electric field is affected by the ionic polarization, thus conductivity is one of the factors to be considered in measurement moisture content. The results presented that the signal variated less than 1% if the conductivity change less than 1 mS/cm. By using RF frequency (100 MHz), the conductivity effect has been reduced.
6	1000 to 10,500	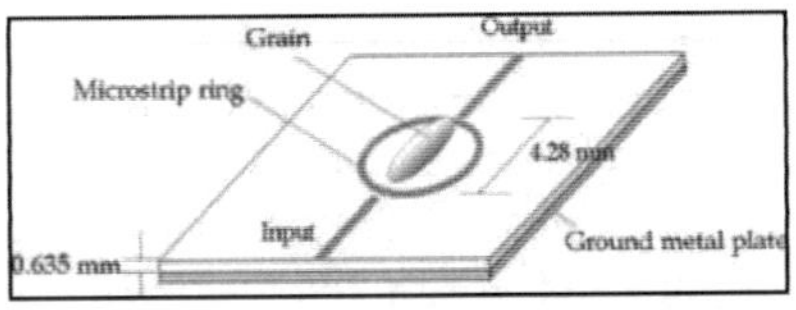 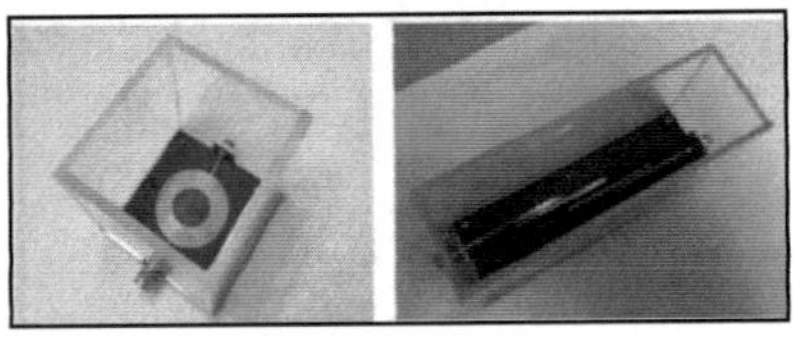 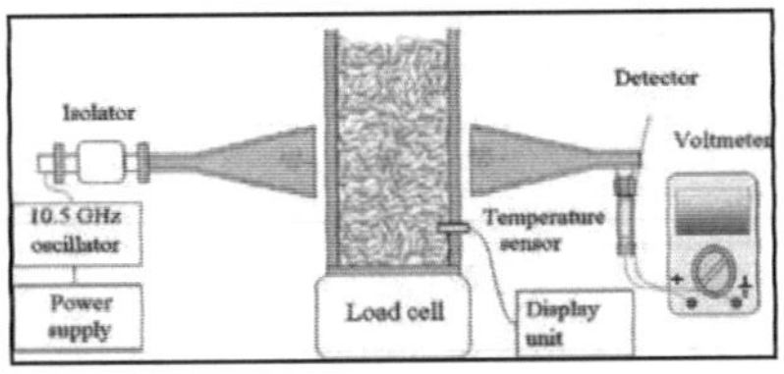	All microwave systems mentioned in [30] were able to predict the moisture content in single and bulk rice grain. For granular material, the water molecules polarization is sensitive and showed a significant response when exposed to microwaves. As for moisture sensing in broken rice, the sensor used microwave up to 13.5 GHz because the short wavelength is required to enhance the air gap sensitivity and reduce the penetration energy.
8	915 and 2450	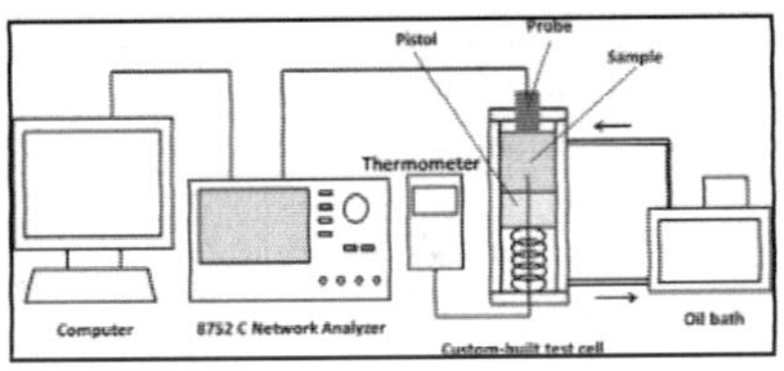	The setup shown in [31] is used to measure the dielectric properties of rice grain (RG) and rice flour gel (RFG). At the same value of conductivity for both materials, the penetration depth of 915MHz is deeper compared to 2450 MHz. It is also reported that, at higher frequency (more than 1000 MHz), microwave measurement is unaffected by conductivity of materials.

7	10 to 3000	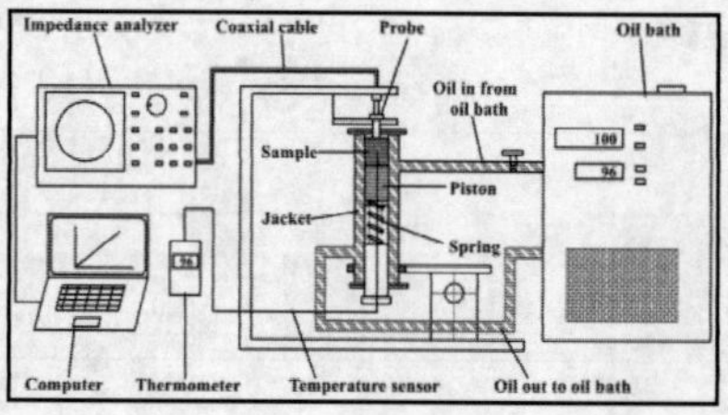	The setup shown in [32] is used to established advance pasteurization treatments using RF (10-300 MHz) and MW (300-3000 MHz) energy. Based on the results, the detection of almond kernels moisture content within RF ranges showed a significant effect on the frequency-dependent dielectric properties. At higher moisture content, the effect of RF (10-200 MHz) on dielectric properties also was illustrated greater. They also developed quadratic polynomial equations to analyse the relationship between dielectric properties and frequency at 27, 40, 915, and 2450 MHz. The penetration depth at RF ranges was larger compared to those at MW frequencies at similar moisture content and temperatures.

4. Conclusion

Determination of dielectric properties on agricultural products are prominent mainly because of their usefulness in correlation with the material moisture content. Familiarity between the variation of both dielectric constant and loss factor with variables such as moisture content, frequency, temperature, and bulk density is vital in selecting the most suitable techniques for the measurement system. In selecting the most suitable frequency range for the moisture detection system, both radio frequency and microwave offer different advantages depending on the parameters for example moisture content range, temperature, material conductivity and size of the tested system. The ability to monitor online and on-site grain moisture content will provide significant information for better handling and storage management which later improvise the quality of commercial products.

Acknowledgment

This research was fully financed by the Ministry of Education Malaysia under Transdisciplinary Research Grant Scheme (Ref: TRGS/1/2018/UNIMAP/02/4). The authors gratefully thank Universiti Malaysia Perlis (UniMAP), Malaysia for the facilities and technical support.

Reference

[1] D Saida Naik and M B Chetti 2017 (PDF) Influence of Packaging and Storage Conditions on the Moisture Content and its Effect on Fungal Load of Paddy *Res. Pap. Res. J. Agric. Sci.* 370–4

[2] Simonyan K J, Yiljep Y D, Oyatoyan O B, Bawa G S, Programme I, Extension N A, Services L, Engineering A and Programme T 2009 Effect of Moisture Content on Some Physical Properties of Paddy Grains *Agric. Eng. Int. CIGR Ejournal* **XI** Manuscript 1279

[3] Karimi M, Kheiralipour K, Tabatabaeefar A, Khoubakht G M, Naderi M and Heidarbeigi K 2009 The effect of moisture content on physical properties of wheat *Pakistan J. Nutr.* **8** 90–5

[4] Ahn J Y, Kil D Y, Kong C and Kim B G 2014 Comparison of oven-drying methods for determination of moisture content in feed ingredients *Asian-Australasian J. Anim. Sci.* **27** 1615–22

[5] Klomklao P, Kuntinugunetanon S and Wongkokua W 2017 Moisture content measurement in paddy *J. Phys. Conf. Ser.* **901**

[6] Nelson. S.O 2008 Dielectric properties of agricultural products and some applications *Res. Agric. Eng.* **2008** 104–12

[7] Stuart O Nelson and Samir Trabelsi Nelson 2010 Grain and Seed Moisture and Density Measurement through Sensing of Dielectric Properties *2010 Pittsburgh, Pennsylvania, June 20 - June 23, 2010*

[8] S. O. Nelson and S. Trabelsi 2012 A Century of Grain and Seed Moisture Measurement by Sensing Electrical Properties *Trans. ASABE* **55** 629–36

[9] Lawrence K C and Nelson S O 1993 Radio-frequency Density-independent Moisture Determination in Wheat *Trans. ASAE* **36** 477–83

[10] Sang Ha Noh and Stuart O. Nelson 2013 Dielectric Properties of Rice at Frequencies from 50 HZ to 12 GHZ *Trans. ASAE* **32** 0991–8

[11] S. O. Nelson and A. W. Kraszewski 2013 Simple Grain Moisture Content Determination By Microwave Measurements *Trans. ASAE* **33** 1303–5

[12] Nelson S O and Stetson L E 1976 Frequency and moisture dependence of the dielectric properties of hard red winter wheat *J. Agric. Eng. Res.* **21** 181–92

[13] Nelson S O, Trabelsi S, Member S and Kraszewski A W 2001 RF Sensing of Grain and Seed Moisture Content **1** 119–26

[14] S. O. Nelson, Trabelsi S 2002 Sensing Grain Moisture Content Through Dielectric Properties *IEEE* 320–3

[15] Nelson S O 1981 Review of Factors Influencing the Dielectic Properties of Cereal Grains *Cereal Chemestry* **58** 487–92

[16] S. O. Nelson 2013 Dielectric Properties of Grain and Seed in the 1 to 50-MC Range *Trans. ASAE* **8** 0038–48

[17] Nelson S O 2010 Fundamentals of dielectric properties measurements and agricultural applications *J. Microw. Power Electromagn. Energy* **44** 98–113

[18] J. L. Jorgensen, A. R. Edison and S. O. Nelson and L. E. Stetson 2013 A Bridge Method for Dielectric Measurements of Grain and Seed in the 50- to 250-MHz Range *Trans. ASAE* **13** 0018–20

[19] Stetson L E and Nelson S O 1970 A Method for Determining Dielectric Properties of Grain and Seed in the 200- to 500-MHz Range 491–5

[20] Nelson S O, Kraszewski A W, Trabelsi S and Lawrence K C 2003 Using cereal grain permittivity for quality sensing by moisture determination **49** 237–42

[21] S. O. Nelson, S. Trabelsi and A. W. Kraszewski 2013 Advances in Sensing Grain Moisture Content By Microwave Measurements *Trans. ASAE* **41** 483–7

[22] Pelletier M G, Wanjura J D and Holt G A 2016 Microwave moisture sensing of seedcotton: Part 1: Seedcotton microwave material properties *Sensors (Switzerland)* **16**

[23] Zhang C, Zhou X, Shi Z and Wang L 2018 A novel method for measuring the moisture distribution of grain in the silo based on microwave image technology *Int. Conf. Adv. Mechatron. Syst. ICAMechS* **2017-Decem** 157–62

[24] Muji S Z M, Rahim R A, Rahiman M H F, Tukiran Z, Ayob N M N, Mohamad E J and Puspanathan M J 2013 Optical tomography: Image improvement using mixed projection of parallel and fan beam modes *Meas. J. Int. Meas. Confed.* **46** 1970–8

[25] Gilmore C, Asefi M, Paliwal J and LoVetri J 2017 Industrial scale electromagnetic grain bin monitoring *Comput. Electron. Agric.* **136** 210–20

[26] Rahiman M H F, Rahim R A, Rahim H A, Muji S Z M and Mohamad E J 2012 Ultrasonic tomography - image reconstruction algorithms *Int. J. Innov. Comput. Inf. Control* **8** 527–38

[27] Trabelsi S, Lewis M A and Nelson S O 2016 Microwave moisture meter for in-shell peanut kernels *Food Control* **66** 283–90

[28] Mai Z, Li C and Wang B 2016 Design and Test of Grain Moisture Online Measuring System Based on Floating Ground Capacitance 821–30

[29] Zhang H L, Ma Q, Fan L F, Zhao P F, Wang J X, Zhang X D, Zhu D H, Huang L, Zhao D J and Wang Z Y 2016 Nondestructive in situ measurement method for kernel moisture content in corn ear *Sensors (Switzerland)* **16** 1–15

[30] You L L, Yue C S, Mun H K, Lee C Y and You K Y 2017 Physical and Chemical Characterization of Rice Using Microwave and Laboratory Methods *Rice - Technol. Prod.*

[31] Auksornsri T, Tang J, Tang Z, Lin H and Songsermpong S 2018 Dielectric properties of rice model food systems relevant to microwave sterilization process *Innov. Food Sci. Emerg. Technol.* **45** 98–105
[32] Zhang S, Kou X, Wang S, Li R and Ling B 2017 Dielectric properties of almond kernels associated with radio frequency and microwave pasteurization *Sci. Rep.* **7** 1–10

IOP Conf. Series: Materials Science and Engineering **705** (2019) 012003 doi:10.1088/1757-899X/705/1/012003

A Review on Service Robots: Mechanical Design and Localization System

M. Q. Bakri[1*], A. H. Ismail[1], M.S.M. Hashim[1] and M. J. A. Safar[1]

[1]School of Mechatronic Engineering, Universiti Malaysia Perlis (UniMAP), Perlis, Malaysia.
[*]mohdkhai95@ymail.com

Abstract. Service robot is becoming more popular in recent study. It has opened up various opportunity to solve problems as well as the challenges need to be studied. While there are many successfully developed service robot out there, the general design principle with respect to the application need a description with solid justification. This review paper seeks and compare the design principle, specification, and hardware selection of several successfully developed service robot such as Sacarino and Blackbot. Based on the study, the most important design principles include simple mobile platform, modular structure, and aesthetic design.

1. Introduction

Service robot can be defined as the robot that perform predefined tasks originally done by human in a specific area or field. However, less expectation were to be set on these robot to accomplish the tasks as perfect as human did, at least it can achieve some degrees of acceptable results [1]. One cannot doubt that service robot is getting popular nowadays either to the researcher, industry, or commercial. The emergence of Industrial Revolution 4.0 (IR4.0) has deem the importance of service robots particularly in automation process, medical services, and many other fields. Furthermore, the affordability of high processing computer nowadays enable these robots to be brought to the consumer as part of their living [2].

Researcher and developer are mostly interested in designing the service robot for daily life assistance such as elderly care. This is because studies showed that aging population has become the common problem nowadays due to higher aging population and declining birth rate in recent years. It is expected that robot technology have an important role for the sustainable health of the society and overcome the manpower shortage in the medical industry. These kind of task require the robot at least to be able to know the surrounding conditions, recognize objects, and avoid obstacles [3]. Researcher in [4] carried out a study from a focus-group about the system requirement of a service robot for elderly care which shows that the focus-group prefer a robot with simple user interface, require no prior knowledge to be operated, and easy to be understood.

Apart from that, a robotic company Savioke has created the first hotel receptionist robot Botlr that has been introduced in Aloft Hotel in Cupertino, California. The robot can do the hotel management tasks such as greeting people and deliver items to the specific hotel room. The robot has a storage compartment that can store items like toiletries, newspapers, and water bottles to be delivered to the guest room. According to the survey, 56% of respondent prefer to have a service robotic in the hotel [5].

The rest of the article is organized as follows: Section 2 described the service robot design layout, hardware selection, and service robot comparison. Section 3 introduces the important design components of service robot, and section 4 describe the technological aspect related to service robot which currently gain the interest among researchers.

IOP Conf. Series: Materials Science and Engineering **705** (2019) 012003 doi:10.1088/1757-899X/705/1/012003

2. Design Layout and Consideration

When it comes to design, there are several layout and consideration that has been proposed to design a service robot. This aspect is important as the general reference for the early development phase of a service robot so that it gives the idea of the starting point.

The works in [6] developed a restaurant service mobile robot and described the five layers structure of the robot. The first layer consists of drive and castor wheels, rechargeable power supply, and infrared navigation sensor. The second layer has the electronic driver and I/O board as well as the emergency stop mechanism. The third layer is where the PC based controller is placed. It is connected to the electronic driver and I/O board via RS232. The fourth layer is the storage compartment. This is where the dishes are placed for transportation within the restaurant. The fifth layer is for user interface utilities such as touch screen panel, speaker, and microphone for voice command. Other than that, the robot also have wireless network module and laser positioning system.

A receptionist robot codenamed Xinxin is well explained in [7]. The robot is designed to be modular and the author described the hardware design of the robot into six module which are mobile module, laser scanning module, somatosensory module, arm module, display module, and voice module. The mobile module is based on the commercial robot named Kobuki. The battery has been customized with 12-cell 18650 Lithium ion which enable the robot to work for 6 hours on a single charge. Rplidar A1 is used as the laser scanning module. In the other hand, Microsoft Kinect camera is used as the somatosensory module which detect the human gesture and movement. The arm module consists of several motors that provide extra degree of freedom instead of the mobile platform. 7-inch touch-screen device is used as the display module for this robot. Lastly, the voice module developed based on iFlyTek Speech Lab System that provide many features related to voice command and synthesis.

Other than that, researcher in [8] developed 'BlackBot' a receptionist robot. In general, this robot consist of two modules joined by a pan-tilt mechanism. The modules are the body module with mobile platform and the head module with LCD display for user interface. The mobile platform has three-wheeled mobile with differential drive mechanism powered by two DC motors. In the other hand, the control system architecture of the robot consists of 4 layers which are the hardware, software framework, application program interface, and service application. To be more specific, the author described the design specification of BlackBot such as the height is 1600mm, weight is 30kg, maximum travel speed is 0.25m/s, material used is aluminum frame, DC motors rated power is 35W each, and the processor used is Core i5 with 4GB RAM.

The works in [9] describe the design structure of 'Sacarino' the hotel receptionist service robot into three levels. The first level includes the hardware, mechanical structure, sensors, and actuators. The second level is the robot's control architecture which include the control algorithm, software, modelling, and programming. The third level is the application in which the robot is given the specific services or tasks. According to the author, development of a service robot should consider the application level in the first place. This to ensure the robot able to provide the service successfully with minimal failure. In the hardware level, there are two main parts which are the mobile base for robot to maneuver and the anthropomorphic upper body for user interface and interaction with human. The mobile base has synchro-drive mechanism which comprises of two motors, one for turns and the other for traction.

In addition, researcher in [10] described the design philosophy of service robot and identified the requirements for future robots such as friendly and cooperative interface, simple structure, function that are easy for humans to understand, and gentle appearance in harmony with surroundings.

A service robot that is designed for household tasks should have enough workspace area for it to get access to the objects and furniture just like human do [4]. However, in terms of the robot size, the limitation is the typical door sizes because the robot is needed to move from one point to another inside the house. The robot weight is hard to compromise because it often depends on battery capacity, platform stability, payload, and manipulators. In most cases, the robot operates autonomously alongside human, thus safety is the main criteria for service robot. Furthermore, according to interviews with market player it was found that a service robot should be kept low cost and affordable in order to be accepted by the consumers [4]. Due to this limitation factor, a design architecture of minimal cost such as selecting the differential drive platform with an arm manipulator as the basic design has been introduced. The dimension factor makes the robot to be designed so that it can pass through smallest door in the house/building and even maneuver across the corridors. The stability of the robot is kept to the maximum by placing the lead batteries on the lowest point possible to lower the center of gravity, thus making the robot safer to work alongside human

According to researcher in [5], the design process of a service robot begins with selecting the material. The chosen material depends on its weight and motor torque. The motor torque in the other hand must follow the design principle of minimizing the cost, thus cheaper motor need to be selected. This in turns determine the size, weight, and travel speed of the robot.

Last but not least, researcher in [11] developed MILO a personal service robot for general application such as housekeeping, mobile tele-conferencing, and tele-presence at home. The author pointed out five considered criterions in the development of MILO. The first one is safety since MILO is intended to work and interact closely to people. This include sensors redundancy and intelligent algorithm control algorithm. The second criterion is low cost hardware and component. This can be achieved by simplifying the mechanical structure and materials for fabrication. Next, the third criterion is form factor of the design. Since MILO is intended to work alongside human, the shape should not be threatening. The fourth consideration criterion is expandability, easy for maintenance, and flexibility. The mechanical design is made modular while the electronics design allows extension with other module. The fifth criterion is simple and efficient user interface. Table 1 summarizes the important aspect in service robot by comparing the service robot in [6], Xinxin, Blackbot, and Sacarino.

Table 1. Summary of service robot design layout.

Name	Unnamed [6]	Xinxin [7]	Blackbot [8]	Sacarino [9]
Picture	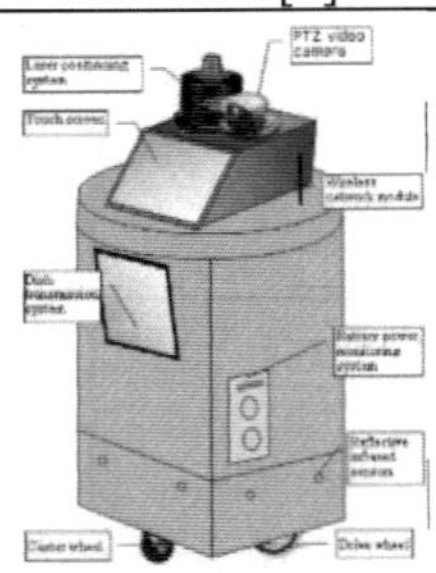			
Application	Restaurant waiter	Hotel receptionist	Hotel receptionist	Hotel receptionist
mobile platform	Differential drive	Differential drive	Differential drive	Synchro drive
Design layout	Five layer structure	Six modular	Two modular	Three level architecture
User interface	LCD screen	LCD screen on chest	LCD screen head and facial expression	LCD screen on chest

3. Design Components

3.1 Mobile Platform Mechanism

There are a few options of mobile platform mechanism that are typically used in service robot. The three most popular used mechanism are differential drive, omni-directional with mecanum wheel, and synchro-drive. Each mechanism has its own advantages and disadvantages.

Differential drive is the most widely used mobile platform mechanism. This is due to low cost, simplicity, and easy to implement with minimal complexity in control system [12]. This mechanism is powered by at least two independent motor which drive the wheels. Fuzzy PID can be used as the controller for differential drive wheel which able to produce sophisticated trajectories even though non-identical motors were used [13]. Other than that, a steering wheel can also be added to the differential drive platform to provide extra degree of freedom and extend the ability to manoeuvre making this mechanism expandable [14].

For the omni-directional with mecanum wheel mechanism, at least two wheels are needed to move the platform. It is quite efficient mechanism to move a robot platform since no need for a stopping for making a turn or curvature [15]. The most important component is the wheel itself. It is a specially designed wheel which comprises of small roller along the main wheel oriented co-axial with the main wheel shaft. The small roller enable the wheel to skid sideways [16]. With tangential superimposition, the platform with mecanum wheel can move in any direction without changing the orientation with respect to the global frame [17] [18]. Other than that, this wheel provide more advantage in narrow space maneuvering [19]. Reseracher in [19] choose the omni-directional with four mecanum wheel mechanism in the design of state-of-the-art wheel chair. This new configuration gives extra ability to maneuver for the wheel chair user. Low cost microcontroller also used to control the whole system which proves that this mechanism can be controlled with lower processing system. Reseracher in [15] enhanced the mecanum wheel by using odometry sensors to measure the wheel speed more accurately. It consist of two dummy wheel with encoder placed at the center of the mobile platform that are perpendicular to each other.

Synchro-drive mechanism is also a type of omni-directional but there is no need for a specially designed wheel like mecanum wheel. It uses the normal wheel, but there is a specially designed steering mechanism for each wheel. In synchro-drive, all wheels are steered simultaneously so that they all facing the same orientation and direction. In common configuration, the synchro-drive consist of one motor for steering and the motors on each wheel for driving. All of the wheel have vertical steering shaft and are connected either via the gears or chain drive for a synchronize rotation in the vertical axis. This mechanism has two degree of freedom and the platform will always stay on the same orientation at all time. In the other hand, this mechanism is holonomic and can move in any direction. However, the wheel need to stop before changing the direction of travel [20] [21].

Table 2. Summary of mobile mechanism

Mobile Mechanism	Advantage	Disadvantage
Differential drive	– Simple – Low cost	– Cannot maneuver in sideways
Mecanum wheel drive	– High maneuver ability – 3 DOF	– High cost mecanum wheel
Synchro drive	– Can maneuver in tight space	– Complex steering mechanism

3.2 Upper Body

On the upper body of service robot, there are several components need to be selected and decided with respect to the robot application. This include manipulator, storage compartment, and user interface screen.

Researcher in [22] developed the 6 degree-of-freedom (DOF) arm on both side of the service robot. Besides that, the robot is also equipped with a human like head. Both the arm and the head are designed so that the service robot appear to be human like which is one of the attraction factor. Other than that, a touch screen is mounted on the chest of the robot to provide information as well as the user interface.

A robot explained in [23] is known as Care-O-bot and it has 28-DOF. This service robot has single arm manipulator and tray compartment for storage. The manipulator itself is a 7-DOF lightweight arm. The workspace of the manipulator is designed so that it can reach the floor space. The tray compartment is used to place the items such as food and medicine for the patient. Due to it focused application, this robot has no anthropomorphic head and upper body. The overall structure is designed to be able to support human thus making the stability is an important criteria instead of appearance.

In the other hand, researcher in [24] stated that service robot with dual arm is more practical than the single arm, albeit its subjectivity to pre-determined tasks. For a subjective task application such as to grasp object, it is better that the service robot has dual arm. Hence, dual arm manipulator which has different structure and functionality between each other has been developed to cover different operation tasks for elderly care. The whole upper body structure of the robot is kept simple without anthropomorphic appearance.

4. Positioning of a Service Robot

As it opens various application opportunity, service robot also require technological improvements in many aspects especially the software and control algorithm such as collision avoidance, path-planning, localization, positioning, and Human-Robot Interaction behaviour. Recent studies show high focus and interest in these aspect.

A positioning system for service robot which is based on dual vision sensor is presented in [25]. The proposed system consist of two vision sensor which is placed on the robot and on the room to construct the mapping of the room. This system automatically making the service robot smart in terms of positioning, navigation, obstacle avoidance, and human-robot interaction. The system use V-SLAM technique to run the algorithm. It compares the predictive information from its depth visual information to form navigation and avoidance. The robot is no longer depends on guide line to manoeuvre instead it navigate through the dual vision system and able to avoid obstacle. These enhancement makes the robot more efficient and improve the speed in completing the tasks.

The works in [26] in the other hand developed a positioning system by using wireless fingerprinting technique. This technique involve two stages which are offline and online. During offline stage, the RSS information is collected and recorded. Next, during online stage, the unknown RSS information acquired is compared with the recorded information in the database by using specific matching algorithm. Hence, the closest value is selected as the positioning results. This method have several advantage compared to SLAM because it utilizes the existing wireless infrastructure in the room or building without the need to implement a new system, therefore cost facilitative.

5. Conclusion

As a conclusion, service robot is a promising technology because it offer many possibilities in terms of application, research, and enhancement. Since the development process and parameter of service robot is very subjective with respect to its application, the design consideration and principle from the previous researcher is summarized in this paper. The design parameters and factors has been justified with further comparing the design of successfully developed service robot. Next, the critical components of service robot discussed with details along with several option available to be selected. The selection of the design component is important to determine the functionality of the robot and ensure the efficiency. Last but not least, the enhancement aspect of service robot is highlighted by summarizing the previous work from researchers including service robot navigation, positioning, obstacle avoidance, and Human-Robot interaction.

6. Acknowledgement

The authors would like to acknowledge the support from the Fundamental Research Grant Scheme (FRGS) under a grant number of FRGS/1/2018/ICT05/UNIMAP/02/4 from the Ministry of Education Malaysia.

7. References

[1] Van Osch M, Bera D, Van Hee K, Koks Y and Zeegers H 2014 Tele-operated service robots: ROSE *Autom. Constr.* 39 152–60

[2] Dobrev Y, Gulden P and Vossiek M 2018 An Indoor Positioning System Based on Wireless Range and Angle Measurements Assisted by Multi-Modal Sensor Fusion for Service Robot Applications *IEEE Access* 6 69036–52

[3] Pyo Y, Nakashima K, Kuwahata S, Kurazume R, Tsuji T, Morooka K and Hasegawa T 2014

Service robot system with an informationally structured environment *Rob. Auton. Syst.* 74
[4] Hendrich N, Bistry H and Zhang J 2015 Architecture and Software Design for a Service Robot in an Elderly-Care Scenario *Engineering* 1 027–35
[5] Ashhad B K, Bristy I S, Khatun H, Kalam K and Zahir E 2015 Design and Implementation of an Interactive Service Robot *J. Comput. Commun.* 3 121–5
[6] Tzou J H and Su K L 2008 The development of the restaurant service mobile robot with a laser positioning system *Proc. 27th Chinese Control Conf. CCC* 662–6
[7] Zhaohui Z, Mei X, Bian X, Cai H and Ti J 2017 Development of an intelligent interaction service robot using ROS *Proc. 2016 IEEE Adv. Inf. Manag. Commun. Electron. Autom. Control Conf. IMCEC 2016* 1738–42
[8] Kantharak K, Sombooncha C, Trong N and Thinh N T 2017 Design and Development of Service Robot based Human - Robot Interaction (HRI) 293–6
[9] Pinillos R, Marcos S, Feliz R, Zalama E and Gómez-García-Bermejo J 2016 Long-term assessment of a service robot in a hotel environment *Rob. Auton. Syst.* 79 40–57
[10] Kawamura K, Pack R T, Bishay M and Iskarous M 1996 Autonomous Systems Design philosophy for service robots *Rob. Auton. Syst.* 18 109–16
[11] Salemi B, Reis J, Saifhashemi A and Nikgohar F 2005 MILO: Personal robot platform *2005 IEEE/RSJ Int. Conf. Intell. Robot. Syst. IROS* 3833–8
[12] Yazdjerdi P and Meskin N 2018 Fault tolerant control of differential drive mobile robots using sliding mode controller *2017 5th Int. Conf. Control. Instrumentation, Autom. ICCIA 2017* 2018–Janua 270–4
[13] Heikkinen J, Minav T and Stotckaia A D 2017 Self-Tuning parameter fuzzy PID controller for autonomous differential drive mobile robot *Proc. 2017 20th IEEE Int. Conf. Soft Comput. Meas. SCM 2017* 382–5
[14] Prayudhi L H, Widyotriatmo A and Hong K S 2015 Wall following control algorithm for a car-like wheeled-mobile robot with differential-wheels drive *ICCAS 2015 - 2015 15th Int. Conf. Control. Autom. Syst. Proc.* 779–83
[15] Jain P 2014 Odometry and motion planning for omni drive robots *Proc. Int. Conf. Innov. Appl. Comput. Intell. Power, Energy Control. with Their Impact Humanit. CIPECH 2014* 164–8
[16] Padgett S T and Browne A F 2017 Vector-based robot obstacle avoidance using LIDAR and mecanum drive *Conf. Proc. - IEEE SOUTHEASTCON* 1–5
[17] Birari A, Kharat A, Joshi P, Pakhare R, Datar U and Khotre V 2016 Velocity control of omni drive robot using PID controller and dual feedback *2016 IEEE 1st Int. Conf. Control. Meas. Instrumentation, C. 2016* 295–9
[18] Mohamed T Bin, Karim N A, Ibrahim N, Suleiman R F R, Anwar M, Rashid M F B A and Idris M I Bin 2017 Development of mobile robot drive system using mecanum wheels *2016 Int. Conf. Adv. Electr. Electron. Syst. Eng. ICAEES 2016* 582–5
[19] Kundu A S, Mazumder O, Chattaraj R and Bhaumik S 2014 Door negotiation of a omni robot platform using depth map based navigation in dynamic environment *2014 7th Int. Conf. Contemp. Comput. IC3 2014* 176–81
[20] Tatar M O, Haiduc F, Szalontai A and Pop A 2016 Design and development of the synchronous mobile robots *2016 20th IEEE Int. Conf. Autom. Qual. Testing, Robot. AQTR 2016 - Proc.* 6–11
[21] Schwartz M, Hwang S, Lee Y, Won J, Kim S and Park J 2015 Aesthetic design and development of humanoid legged robot *IEEE-RAS Int. Conf. Humanoid Robot.* 2015–Febru 13–9
[22] Kuang Z, Yang Y, Hau T and He Y 2018 Research on Biomimetic Coordination Action of Service Robot Based on Stereo Vision *Proc. 2018 2nd IEEE Adv. Inf. Manag. Commun. Electron. Autom. Control Conf. IMCEC 2018* 769–73
[23] Reiser U, Connette C, Fischer J, Kubacki J, Bubeck A, Weisshardt F, Jacobs T, Parlitz C, Hagele M and Verl A 2009 Care-O-bot® 3 - Creating a product vision for service robot applications by integrating design and technology *2009 IEEE/RSJ Int. Conf. Intell. Robot. Syst. IROS 2009* 1992–8
[24] Zhihua Z, Jianyong Z, Yang L and Gang L 2009 Development of a movable service robot

with double working arms for the elderly and the disabled *Proc. - 2009 Int. Conf. Electron. Comput. Technol. ICECT 2009* 636–40

[25] Guo J, Xiao X, Pan P and Luo X 2017 A Design of Multi-vision Localization and Navigation Service Robot System 787–90

[26] Ismail A H, Kitagawa H, Tasaki R and Terashima K 2017 WiFi RSS fingerprint database construction for mobile robot indoor positioning system *2016 IEEE Int. Conf. Syst. Man, Cybern. SMC 2016 - Conf. Proc.* 1561–6

IOP Conf. Series: Materials Science and Engineering **705** (2019) 012004 doi:10.1088/1757-899X/705/1/012004

Point Cloud Data Segmentation Using RANSAC and Localization

S. N. Mohd Isa[1, a], S. A. Abdul Shukor[1, b], N. A. Rahim[1], I. Maarof [2], Z. R. Yahya [3], A. Zakaria[1], A. H. Abdullah[1], R. Wong[4]

[1]School of Mechatronic Engineering, Universiti Malaysia Perlis, 02600 Arau, Perlis, Malaysia

[2]Faculty of Architecture, Planning and Surveying, UiTM, 02600 Arau, Perlis, Malaysia

[3]Institute of Engineering Mathematics, Universiti Malaysia Perlis, 02600 Arau, Perlis, Malaysia

[4]Geodelta System Sdn. Bhd., 22, Jalan SS20/11, Damansara Utama, 47400 Petaling Jaya, Selangor, Malaysia

[a] sitinurhashima.isa@gmail.com

[b] shazmin@unimap.edu.my

Abstract. In this paper, we present 3D point cloud data segmentation from a Terrestrial Laser Scanner (TLS) using RANSAC and localization approaches. Our work is using the real world data acquired from outdoor and indoor scene of Tawau Bell Tower (also known as the Belfry), one of the heritage building located in Sabah. We adapt the methods to segment ground and non-ground data for data reduction. Based on the non-ground information and geometrical constraint, the data managed to show the building structure. The aim is to use and handle a smaller set of data rather than using the entire point cloud data in processing the region of interest. This will also help to reduce processing time as point cloud data from a TLS is usually large in size. The ground plane was fitted with RANSAC algorithm restricted by distance to describe the geometrical of the plane. As a result, the point cloud data was reduced to the environment around Bell Tower and able to highlight the region of interest. Building segmentation is then retrieved from the reference centre location to form a bounding box and data were organized using kd-tree to denote the Bell Tower structure. The framework is tested for outdoor and indoor scan of the Bell Tower and able to segment the data appropriately.

1. Introduction

At present, documentation of heritage buildings are moving towards on the usage of laser scanner technology for cultural preservation [1]. Terrestrial Laser Scanner (TLS) is preferred because of its capability of acquiring accurate data in a short time [2][3]. TLS captures the object and the scene information in millions 3D point cloud that able to ensure the details of the building surface and its surrounding are collected. However, this will leads to a large file size to be managed. Hence, this has attracted the interest among researchers in processing and analysing the data.

Apart from that, the challenges in point cloud data processing could also be started as early as the data acquisition process. TLS acquires data at same platform and different terrain height can cause some difficulty to differentiate ground and non-ground elements [4]. The system too could not automatically capture targeted object only as it will collects data from any physical surfaces of the entire scene. By segmenting ground and non-ground, this will help to differentiate the data into regions and give topology simplification to the data [5]. Thus, the data will be smaller in size and will be easier to be managed for further processing, plus at a reduced processing time. In this paper, we adapt data reduction approach to segment it into ground and non-ground. Random Sample Consensus (RANSAC) approach is used for segmenting ground plane. Geometrical constraint representing non-ground data consists of building and vegetative is used to describe the object of interest retrieved from the location of the reference centre that forms the bounding box. The point cloud was organized using kd-tree data structure. This method is applied in the real world scene for outdoor and indoor scan of Tawau Bell Tower.

2. Related work

Researchers use geometric adaptation as the approach in segmenting object of interest from terrain. RANSAC is one of the robust method used in geometrical fitting [6]. The method adapts mathematical concept based on model shape such as plane, cylinder or sphere. The algorithm randomly select model data points according to geometrical shape. In contrast, other methods like 3D Hough transform requires sensitive parameters such as discrete axis value and threshold for segmentation which difficult to determine because of its dependency on point cloud characteristic [7]. Furthermore, small value parameters are required by 3D Hough transform for good plane detection thus it needs more processing time and memory. Hence, RANSAC provide advantage in negligible processing time over large data size by randomly search that will reduce iterations [8].

Other researchers used hybrid model fitting method to improve the efficiency of RANSAC for segmentation. A study by B.Oehler et al. stated that by combining RANSAC and 3D Hough transform for planar segmentation will give good efficiency. They used Hough transform as a coarse solution and then make it finer with RANSAC to fit the plane. Although their method provide large possible plane surface, it is only limited for indoor scene [9]. Meanwhile, T. Czerniawski et al. extracted plane of building using extended RANSAC by concentrating in region of interest. Their approach able to select feature extraction for further classification process [10]. F. Poux et al. deal with image and point cloud data for their study. They used RANSAC to detect shape of tessera for cultural heritage preservation. The algorithm managed to extract details of tessera with best fit of bounding box [11]. From here, it can be seen that RANSAC is useful for data segmentation. Therefore, our study utilizes RANSAC approach for outdoor and indoor 3D point cloud data segmentation. As the data size will be reduced, working with small data will make it easier for segmenting object of interest from the scene.

3. Datasets

Point cloud data used in this research is from a heritage building environment. The data was acquired using a TLS and the data collection process is explained in details in this section.

3.1. Study area

Tawau Bell Tower or also known as the Belfry is located in Tawau and was constructed as the evidence of previous British government before Sabah became part of Malaysia. The historic building is almost 100 years old and it is the oldest monument that was safe during the Second World War. The Bell Tower was built by prison labours to remember the resolution undersigned during First World War when Japan became cronies of England. This building was refurbished by Rotary Club of Tawau and passed to Jabatan Muzium Sabah on 2006 [12].

3.2. Point cloud data acquisition

The point cloud data representing the building is provided by Geodelta Systems Sdn. Bhd. acquired using Leica HDS6000 TLS. The TLS is capable to scan up to 500,000 points per second and this system is equipped with a NIKON D80, 10.2 Megapixel camera. They located the scanner into 10 stations respectively to obtain the whole Bell Tower building data. Figure 1(a) shows the location of the TLS from station one to station ten (St1-St10), while Figure 1(b) and 1(c) represent St1 and St10 point cloud data respectively. In this paper, we process the first station (St1) for evaluating the effect of different radius of obstacle and object of interest as well as sensor location to the segmentation approach.

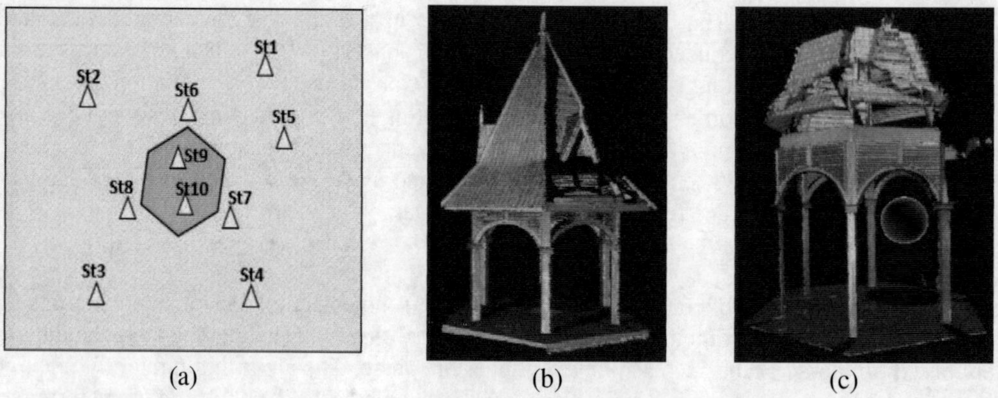

Figure 1. Bell Tower data: (a) Location of scan station, (b) station one (St1) and (c) station ten (St10) point cloud data.

The raw point cloud data are available in .zfs, which is in binary file format consist of XYZ coordinate intensity and colour information. In order to process the data, it needs to be converted to a non-proprietary format like ASCII. Autodesk ReCap and CloudCompare software were used in this case to obtain the .ply file format and the conversion process can be seen in Figure 2. Then the data processing is done by using MATLAB R2018b and 64 Bits Windows 10 with Intel Core i5-4200U (1.60GHz) 12 GB RAM.

Figure 2. File format conversion (binary to ASCII).

4. Methodology

There are three stages involved in this point cloud data segmentation using RANSAC for ground plane fitting and localization for nearest neighbour in radius to find the object of interest. Firstly, the region of interest was selected to highlight the environment around object of focus. Secondly, segmentation of ground plane was used to show information for ground surface and non-ground objects in the scene. Lastly, the non-ground objects were segmented according to their locations.

As the normal directions of the plane are pointing in vertical direction, thus it can be calculated using equation 1:

$$aX + bY + cZ + d = 0 \tag{1}$$

where *X*, *Y*, and *Z* represents 3D point in the space, while *a*, *b*, and *c* are the normal vector for the respective axes and *d* is the normal vector passing through origin.

The distance from the point to the plane, *D* can be found using equation (2):

$$D = \frac{aX+bY+cZ+d}{\sqrt{a^2+b^2+c^2}} \tag{2}$$

The number of RANSAC iteration, *N* can be calculated using equation (3):

$$N = \frac{\log(1\text{-}p)}{\log(1\text{-}(1\text{-}e)^S)} \tag{3}$$

where p is the probability of choosing a good sample, e is the probability of choosing outliers with 99 of desired confidence percentage and s is the initial number of points to fit the model, where in this case, s=3.

Algorithm 1 shows the pseudo-code of the overall work, where the input data are 3D point cloud, *pc*, *d*, *v*, *s*, *r1* and *r2* respectively, and their description are discussed in Section 4.1 to 4.3.

Algorithm 1: Pseudo-code of building segmentation using RANSAC and localization approach

```
pc= {Xlimit; Ylimit; Zlimit};
for maxDistance = d; reference vector = v;
fit plane outliers (pc, d, v);
then label inPlanePointIndices = greenIdx;
else select outliers = pcWithoudGround;
end for;
for sensor location = s; radius = r1;
find neighbours in radius, nearIndices (s, r1, pcWithoutGround);
then label nearPointIndices = RedIdx;
end for;
while radius = r2 < r1;
find neighbours in radius nearIndices = (pcWithoutGound; s, r2);
then buildingPointIndices = outliers (nearIndices);
select pcBuilding = (pcWithoutGround; nearIndices);
do bounding box for pcBuilding then label BlackIdx;
end; Return for sequence pc
```

4.1. Region of interest selection

This work started off with the selection of the region of interest from raw point cloud, *pc* in order to emphasis on the main information to be delivered from the scene. This method is aimed to manage big data focused on region of interest rather than processing the data of the entire scene. The data from CloudCompare software (in .ply file) was loaded into array structure and it stored object in 3D point cloud properties as location (X, Y, Z), intensity value (I) and RGB colour. The displayed region of point cloud was set by limiting the range for each axis (Xlimit, Ylimit, Zlimit) using properties in location. We reduce the raw data of every scan stations to region of interest by 16m x16m for x-axis and y-axis respectively, based on the region of Bell Tower location. No limitation set for the z-axis as the captured data are from panorama view and to allow free constraint for the horizontal scene that represents the whole building. Figure 3(a) and 3(b) show actual raw data and its region of interest for St1.

4.2. Segmentation of ground plane

The ground plane was then fitted using RANSAC algorithm. Therefore all inlier points are assumed as the ground plane. The points that did not belong to the inliers of ground plane are considered to be the outlier points which were then removed. The ground was segmented according to the outliers point set at maximum distance, *d* by 20cm above the ground plane constrain from reference vector, *v* at (0, 0, 1). Figure 3(c) shows the resulting data with the ground plane removed for St1.

(a) (b) (c)

Figure 3. (a) Raw data from St1, (b) region of interest, (c) region of interest without ground.

4.3. Segmentation of non-ground objects

From the scene shown in Figure 3(c), it can be seen that the Bell Tower are surrounded with obstacles due to overlapping from nearby tree. Therefore, a radius of obstacle, *r1* is created to note the boundary to the Bell Tower. The obstacle point was retrieved from the sensor location, *s* of reference point at centre of the Bell Tower. The radius of obstacle should be larger than radius object of interest ($r2 < r1$) in this approach. The nearest neighbour points are considered outliers at given radius within the non-ground points. We denote the obstacle with red colour.

The centre point of Bell Tower can be estimated from the location of 3D point cloud plot in Cartesian plane. This reference point of sensor location was used to track the Bell Tower. The algorithm finds the nearest neighbour point to the sensor location as the object of interest, thus smaller radius was set to encompass Bell Tower out of the obstacle radius. The point cloud data was arranged using kd-tree structure and bounding box to represent the Bell Tower.

5. Results and discussion

Point cloud data captured from outdoor scanning are more challenging compared to indoor scanning in terms of size data capture by TLS [13] . Some researchers also stated that outdoor scans required more attention as it involve interferences such as light reflection [14]. Evaluations on parameters of radius and sensor location are discussed in following section.

5.1. Effect of localization and parameter of radius

The localization of the object of interest with respect to the reference centre is subject to the data acquisition by the TLS at different platform. Changes in reference centre of sensor location and radius will produce oversegmentation and undersegmentation based on experiment using St1. Figure 4 shows the effect of changing sensor location and radius of obstacle and radius. The best segmentation using this approach can be seen in Figure 4(a). With respect to radius of Bell Tower, it can be seen that the black colour over segment the leaves from the tree that overlapping on the roof of the Bell Tower shown in Figure 4(b). Meanwhile, undersegmentation can be observed in Figure 4(c), as the parameter of the same reference centre and smaller parameter for radius. Here, the black colour is under segment the leaves. This scenario usually happen for case dealing with trees as an obstacles [15]. Therefore, reference centre of object detection and radius set for algorithm are in relationship with each other to determine the success of the segmentation. Overall, our approach has able to segment the ground data and the object of interest.

(a) Reference centre (-5,5,0), radius Bell tower = 8m, radius obstacle = 6m. **(b)** Reference centre (-7,4,0), radius Bell tower = 8m, radius obstacle = 6m. **(c)** Reference centre (-5,5,0), radius Bell tower = 7m, radius obstacle =5m.

Figure 4. Segmentation of St1 scan station (a) Well-segment, MSE = 0.402m, (b) oversegmentation, MSE = 0.482m and (c) undersegmentation, MSE = 0.501m.

5.2. Evaluation of segmentation approach

Researchers used data management using region of interest for point cloud processing and object detection from big data in application of culture heritage [16]. Data acquisition using TLS has to be done in many scans to collect the complete object in the scene. In this work, data from St1 to St10 are utilized for evaluation on the effect of radius object interest and radius of obstacle for segmentation performance of ground and non-ground with respect to mean square error, MSE as shown in Table 1. The distance of inliers to fits the ground for outdoor scan station (St1- St5) give more error, 0.0468m due to higher altimetric variations compare to indoor scan station, 0.0374m. As the exterior scans contain lower number of building point cause difficulty to precisely cite the building shapes [17]. But, from results shows in Figure 5, this method has managed to segment ground plane, Bell tower and trees in green, black and blue colour respectively. Red colour near the Bell Tower shows obstacle that surrounded the object of interest (i.e. the tower).

Table 1. Mean error for St1 to St10 with radius of obstacle, *r1*, radius of object interest, *r2* and sensor location, *s* respectively.

Scan station	Number of points	Radius of obtacle,*r1*	Radius of object interest,*r2*	Sensor location, *s*	Mean square error, MSE (m)
St1	937985	8	6	(-5, 5, 0)	0.0402
St2	1044604	8	6	(-5, 5, 0)	0.0557
St3	1257631	8	6	(-3, 3, 0)	0.0599
St4	1025035	8	6	(-1, -5, 0)	0.0416
St5	1139172	8	6	(-4, 6, 0)	0.0367
		Average of mean square error for outdoor scan			0.0468
St6	2458917	7	5	(0, 2, 0)	0.0336
St7	2470413	7	4	(0, 2, 0)	0.0373
St8	2368642	7	4	(0, 3, 0)	0.0467
St9	3716494	6	4	(1, -1, 0)	0.0355
St10	3752373	7	5	(1, -0.5, 0)	0.0338
		Average of mean square error for indoor scan			0.0374

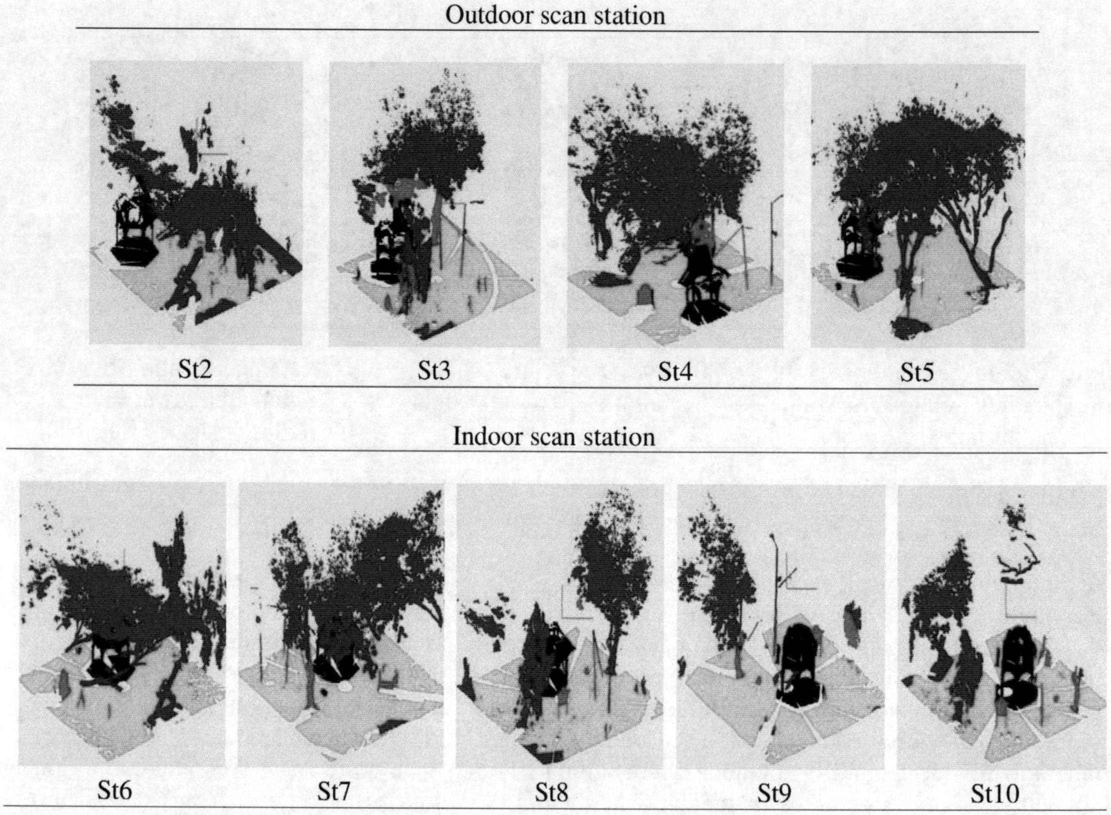

Figure 5. Segmentation of Bell Tower (black), ground plane (green), trees (blue) and other obstacles (red) for outdoor and indoor scans station.

6. Conclusion and future work

In this research, point cloud data of Tawau Bell Tower or the Belfry is treated as the object of the interest to study the usage of RANSAC and localization in segmenting ground and non-ground data. The ground plane and non-ground objects were able to be segmented using RANSAC approach. Data reduction implemented by limiting the range of each axis has managed it to only highlighting region contains information of the building and nearby entities. Moreover, we apply reference point location for bounding box to trace the Bell Tower as the extended approach to RANSAC algorithm. In conclusion, our method has able to segment ground plane and the object of interest.

Some future work can further be studied to improve the algorithms where overlapping area between object and obstacle required detail assessment for actual structure. Moreover, other geometrical constrain can help in defining the boundary of the object of interest. Extended data structure also can be utilised to speed up processing time.

7. Acknowledgment

The authors would like to acknowledge the support from the Fundamental Research Grant Scheme (FRGS) under a grant number of FRGS/1/2017/ICT04/UNIMAP/02/1 from the Ministry of Education Malaysia.

8. References

[1] Chiabrando F, Lo Turco M and Rinaudo F 2017 *ISPRS Int. Arch. Photogramm. Remote Sens. Spat. Inf. Sci.* **XLII-2/W5** 605

[2] Baik A 2017 *Digit. Appl. Archaeol. Cult. Herit.* **4** 1

[3] Klapa P, Mitka B and Zygmunt M 2017 *IOP Conf. Ser. Earth Environ. Sci.* **95** 1755
[4] Jaboyedoff M, Oppikofer T, Abella A, Derron M, Loye A, Metzger R and Pedrazzini A 2010 *Nat Hazards* **61** 8
[5] Wang L and Zhang Y 2016 *LiDAR Ground Filtering Algorithm for Urban Areas Using Scan Line Based Segmentation* p4
[6] Nguyen A. and Le B 2013 *IEEE Conf. Robot. Autom. Mechatronics RAM - Proc* 229
[7] Tarsha-Kurdi F, Landes T and Grussenmeyer P 2007 *ISPRS Work. Laser Scanning 2007 SilviLaser 2007* **34** 412
[8] Aitelkadi K, Tahiri D, Simonetto E, Sebari I and Polidori L 2013 *ISPRS Ann. Photogramm. Remote Sens. Spat. Inf. Sci.* **24** 3
[9] Oehler B, Stueckler J, Welle J, Schulz D and Behnke S 2011 *Proc. 4th Int.Conf. on Intelligent Robotics and Applications (Aachen:Germany)* p10
[10] Czerniawski T, Sankaran B, Nahangi M, Haas C and Leite F 2018 *Autom. Constr.* **88** 46
[11] Poux F, Neuville R and Billen R 2017 *ISPRS Ann. Photogramm. Remote Sens. Spat. Inf. Sci.* **4** 208
[12] Ali A M 2016. *J. Built Environ. Technol. Eng.* **1** 326
[13] Theiler P W, Wegner J D and Schindler K 2013 *ISPRS Ann. Photogramm. Remote Sens. Spat. Inf. Sci.* **2** 287
[14] Hackel T, Savinov N, Ladicky L,Wegner Jan D, Schindler K and Pollefeys M 2017 *ISPRS Ann. Photogramm. Remote Sens. Spat. Inf. Sci.* **4** 92
[15] Alonzo M, Bookhagen B and Roberts D A 2014 *Remote Sens. Environ.* **148** 77
[16] Shanoer M M and Abed F M 2017 *Egypt. J. Remote Sens. Sp. Sci.*1
[17] Florent L and Mallet C 2012 *Int. J. Comput. Vis.* **1** 79

IOP Conf. Series: Materials Science and Engineering **705** (2019) 012005 doi:10.1088/1757-899X/705/1/012005

Face recognition using machine vision for Security system

F S A Saad[1], M V Lee[1] and S Basah[1]
[1]School of Mechatronic Engineering, Universiti Malaysia Perlis (UniMAP)

Abstract. The aim of this project is to design and develop a face recognition based security system using machine vision. In this paper, a face recognition system using Principal Component Analysis (PCA) with Euclidean distance classifier is proposed. PCA is selected as it is less sensitive to noise and interference, besides reducing the number of variables in face recognition which reduce computational time and increase accuracy. A number of experiments were done to evaluate the performance of the face recognition system. A training database of 20 students is created and used in this project. Overall, the face recognition system on static face images has 86% success rate and 14% error rate which is mainly due to the variation of pose orientation. It is also found that the threshold value is an important factor for the performance of face recognition. The distance of the person to the camera play a role as the further the person is away from the camera, the more blur the face image captured. As for the anti-spoofing approach, the proposed detection of eye blinking is successfully tested with training eyes open and eyes close images.

1. Introduction

Face recognition technology is the machine vision system that emulates the capabilities of human eyes to detect and recognise faces. Due to its nonintrusive nature, face recognition can easily be executed where it works with human face that is the most obvious individual identifier without the need of participation of that person [1]. The input image characteristics of a person's face image will be analysed and necessary facial features are taken out from the face and retained in a database which will then be used as a comparison when a user stands before the camera [2]. This paper aim is to design and develop a security system using machine vision for face recognition using IoT. The best feature extraction and classification approach is chosen to develop a more robust system which yields more accurate results, meet the standard of security systems, besides being spoof-proof to take care of our privacy and safeguard our assets.

2. Methodology

2.1. Introduction

The security system developed consists of several components, which is face database component, motion detection component, face detection component, face recognition component, IoT and the anti-spoofing approach.

2.2. Facial database

A face database is required for comparison and scoring during the recognition process. For this project, 20 samples are taken for the creation of face database. 5 images are taken from each sample with 3 frontal face images of different facial expression and 2 more side face images. All 100 face images are pre-processed on MATLAB, where it is converted from RGB to grayscale and then resized into resolution of 112 X 92 pixels.

2.3. *Face detection component*

The initial step of human face recognition system is face detection. The face detection algorithm used is the Viola-Jones Algorithm. Viola-Jones combined many weak classifiers to create a strong classifier. Four main concepts are introduced in the Viola-Jones algorithm, which are the Haar-like features, the integral image, the use of AdaBoost in machine training and a cascade of classifiers.

2.3.1. *Haar-like features*

At this stage, the Haar feature is found by calculating the difference of the pixels on the black portion to the pixels on the white portion [3]. There are few types of Haar feature, for instance the two rectangle, three triangle and four triangle Haar features.

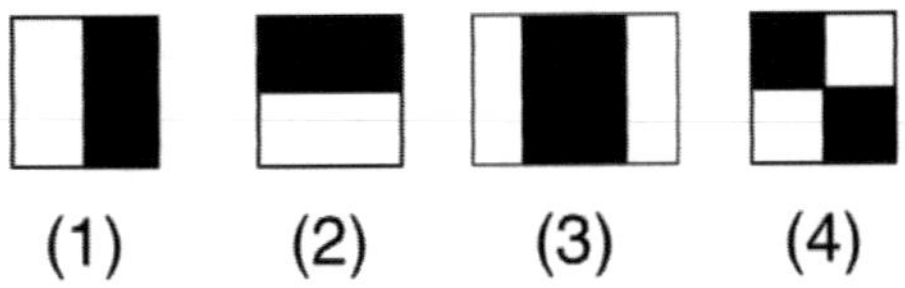

Figure 1. The type of Haar features

2.3.2. *Integral image*

Next step is turning input image into integral image. Integral image is the intermediate form of the rectangular features of an image [4].

2.3.3. *AdaBoost machine learning*

AdaBoost is a machine learning boosting algorithm where large very number of features is reduced and a small set of features is selected. Adaboost develops the classifier that self learns and adapts the threshold value for selecting the features from the overall features to build a strong classifier [5].

2.3.4. *Cascading*

In the cascading step, the face area of the image is determined. The classifier uses the form of a degenerate decision tree, the window is either rejected or passed and this process is repeated to get accurate face detection. All the intermediate classifiers need to be passed. The weight given by AdaBoost determines the order on the cascade. The aim is to obtain the least false negative and to get low rates of error as set by AdaBoost thresholds value in default. Eliminating the undesired regions is the key to enhance the probability for the correct face detection.

2.4. *Face recognition component*

In the facial recognition component, the input face image is tested with the training image in the database. Both training image and test image goes through the principle component analysis (PCA) feature extraction, which is then classified using the Euclidean distance classifier.

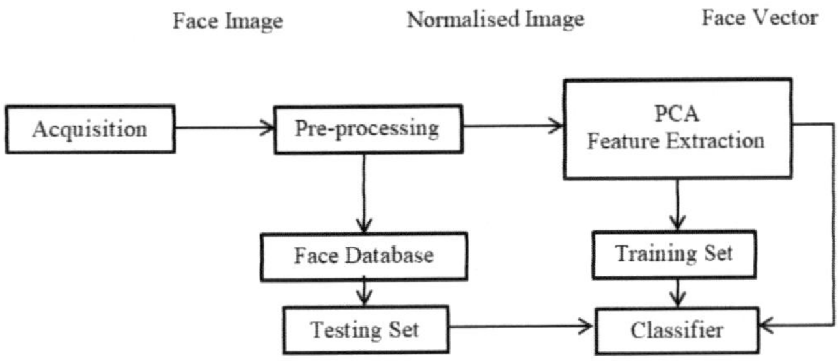

Figure 2. The general idea of the face recognition system [6]

2.4.1. Principal Component Analysis (PCA)

For feature extraction, PCA is chosen because it is good at reduction of dimension where data is compressed. By representing high dimension data of face images with lower dimension data, it reduces the complication and intricacy of sorting the images in groups. PCA defines original data with calculated eigenvectors and eigenvalues when it is projected onto a lower dimensional feature space. PCA captures the most distinctive data component from the face image which helps in maximising between-class data separation [7]. The detailed procedure of PCA algorithm is described in Figure 3.

Step1: A set of M images (I_1, I_2, I_3,…., I_M) with size N*N can be represented by column or row vector of size N_2.

Step2: The average (μ) of the training set image is defined by

$$\mu = \frac{1}{M}\sum_{n=1}^{M} I_n$$

Step3: Each trainee image differs from the average image by vector (ϕ)

$$\phi_i = I_i - \mu$$

Step4: Total Covariance Matrix is calculated from ϕ as follows:

$$C = \frac{1}{M}\sum_{n=1}^{m} \phi_n \phi_n^T$$

$$= AA^T, \text{ where } A= [\phi_1, \phi_2, \phi_3, \ldots, \phi_n]$$

Step5: Calculate the eigenvalues λ_k and eigenvectors U_k of the covariance matrix C.

Step6: To classify an image, it can be projected into feature space. Calculate the vectors of weights $\Omega^T = [\omega_1, \omega_2, \ldots, \omega'_M]$

Where,

$$\omega_k = u_k^T(1-\mu), k = 1,2,\ldots,M'$$

Where M' represents not the total eigenfaces, but the ones with greater value.

Figure 3. The detailed procedure of PCA algorithm

2.4.2. Classification of face images

In this project, classification is performed by comparing the projection vectors of the training face images with the projection vector of the input face image based on the Euclidean distance classifiers between the faces classes and the input face image. The classifier computes the square root of distance between the coordinates of two objects.

2.4.3. Anti-spoofing approach

The anti-spoofing approach proposed in this system is by detecting blinking eyes. Eye blinking is detected from analysing one by one each sequence image to classify the state of either opened eye or closed eye. The eye area is detected by Viola-Jones method, while the blinking of eyes are detected using Histograms of Oriented Gradients (HOG) as features and Support Vector Machines (SVM) as binary classifiers.

3. Results and discussion

3.1.1. Static test

Test 1 is frontal face with normal expression, Test 2 is frontal face with smile expression, Test 3 is frontal face with other expression, Test 4 is left-sided face and Test 5 is right-sided face. The recognition rate is 86% with 14% error rate which is mainly caused by pose variation of face images. Side face image has parts of the face occluded affecting the feature extraction thus affecting the recognition process. Those highlighted are the false positive points, where the faces recognised is not that of the test face image but other identity.

Table 1. Recognition result on static face images with variation of expression and pose

Test	Test 1		Test 2		Test3		Test4		Test5	
Image	Min. Dist.	T/F	Min. Dist.	T/F	Min. Dist.	T/F	Min. Dist.	T/F	Min. Dist.	T/F
S1	2.06E+13	TRUE	3.14E+13	TRUE	1.98E+13	TRUE	2.42E+14	TRUE	1.94E+14	TRUE
S2	8.54E+13	TRUE	4.41E+13	TRUE	4.33E+13	TRUE	3.18E+14	TRUE	4.23E+14	TRUE
S3	6.42E+13	TRUE	5.39E+13	TRUE	3.21E+14	TRUE	2.03E+14	TRUE	1.02E+15	FALSE
S4	2.90E+14	TRUE	2.24E+14	TRUE	2.17E+14	TRUE	2.96E+14	TRUE	1.62E+14	TRUE
S5	2.13E+14	TRUE	7.19E+14	FALSE	3.94E+14	TRUE	1.91E+14	TRUE	3.18E+14	TRUE
S6	3.37E+14	TRUE	1.22E+14	TRUE	2.34E+13	TRUE	7.37E+14	FALSE	6.53E+14	FALSE
S7	5.67E+13	TRUE	4.70E+13	TRUE	9.74E+13	TRUE	9.48E+14	TRUE	8.15E+14	FALSE
S8	2.98E+14	TRUE	2.32E+14	TRUE	2.57E+14	TRUE	1.56E+14	TRUE	3.27E+14	TRUE
S9	2.76E+13	TRUE	2.45E+13	TRUE	2.49E+14	TRUE	5.40E+14	FALSE	8.24E+14	TRUE
S10	4.24E+14	TRUE	3.34E+14	TRUE	2.69E+14	TRUE	5.14E+14	TRUE	3.71E+14	TRUE
S11	9.14E+13	TRUE	7.14E+13	TRUE	2.32E+14	TRUE	2.10E+14	FALSE	8.63E+14	TRUE
S12	8.48E+13	TRUE	8.90E+13	TRUE	7.46E+13	TRUE	4.16E+14	TRUE	3.39E+14	TRUE
S13	2.96E+14	TRUE	2.48E+14	TRUE	1.28E+14	TRUE	3.09E+14	TRUE	9.74E+14	FALSE
S14	2.20E+14	TRUE	1.82E+14	TRUE	2.11E+14	TRUE	5.76E+14	TRUE	9.82E+14	FALSE
S15	5.08E+14	TRUE	1.25E+13	TRUE	2.09E+13	TRUE	6.87E+14	TRUE	1.12E+15	FALSE
S16	8.14E+13	TRUE	7.40E+13	TRUE	3.54E+14	TRUE	4.92E+14	FALSE	5.56E+14	FALSE
S17	1.74E+14	TRUE	2.60E+14	TRUE	1.61E+14	TRUE	9.90E+14	TRUE	1.42E+15	TRUE
S18	3.62E+14	TRUE	4.37E+14	TRUE	3.58E+14	TRUE	6.34E+14	TRUE	4.82E+14	TRUE
S19	3.74E+14	TRUE	6.18E+14	TRUE	3.07E+14	TRUE	4.86E+14	TRUE	3.79E+14	FALSE
S20	1.80E+14	TRUE	1.74E+14	TRUE	1.29E+14	TRUE	2.56E+14	TRUE	6.87E+14	FALSE

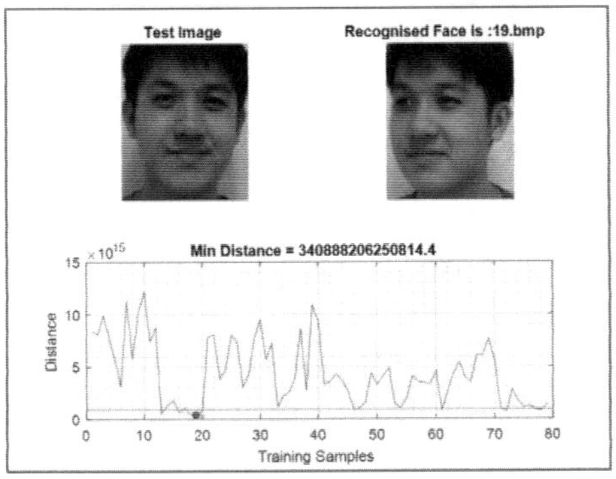

Figure 3. Successfully recognised face

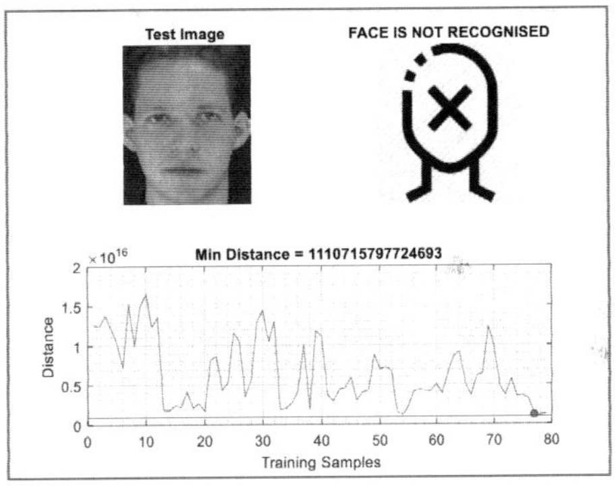

Figure 5. Unregistered face which is not recognised

3.1.2. Real-time test

The distance of the camera to the person testing the system is being considered in the real-time test besides having the variables of facial expression and pose variation as in the static test. The distance between the camera to the person tested in real-time test is 90cm, 120cm and 150cm as shown in figure 6. When the distance of the person to the camera is further, the face image captured is blurer. Thus, the principal components extracted from blur face image is not accurate when compared to those of training images in the database. Besides, the anti-spoofing approach is successfullly tested in real-time as shown in figure 7.

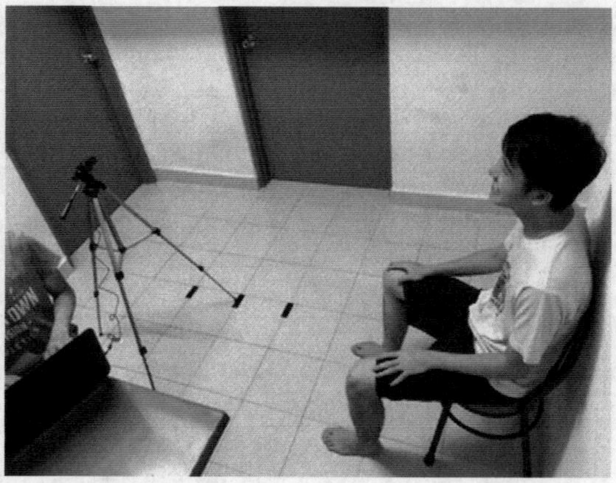

Figure 6. The implementation of the real-time test at 90cm, 120cm and 150cm

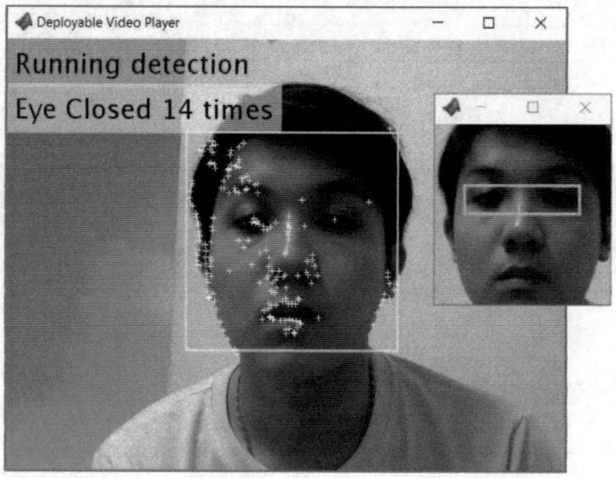

Figure 7. The successfully tested anti-spoofing approach

4. Conclusion

In conclusion, face recognition can be applied for security purpose, such as access control, immigration verification and even criminal identification. In this paper, a security system using machine vision for face recognition is developed using Viola-Jones method to detect face, Principal Component Analysis (PCA) for feature extraction and Euclidean distance classifier. The integration of Internet of Things (IoT) with the system is also proposed. The algorithm performs relatively well in terms recognition rate and computational time under controlled environment. The face recognition can be further improved by increasing the training image and using 3D image processing where the risk of spoofing is hugely reduced.

Acknowledgement

First and foremost, special appreciation goes to my student Mar Vin for his assistance and constant support. His dedicated involvement throughout the whole project contributed to the success of this

IOP Conf. Series: Materials Science and Engineering **705** (2019) 012005 doi:10.1088/1757-899X/705/1/012005

research and accomplishment of this paper. To those who indirectly contributed in this research, your kindness means a lot to me. Thank you very much.

References

[1] K. Dharavath, F. A. Talukdar, and R. H. Laskar, Study on Biometric Authentication Systems , Challenges and Future Trends. (IEEE Int. Conf. Comput. Intell. Comput. Res. image).
[2] N. Bakshi and V. Prabhu, Face recognition system for access control using principal component analysis, (ICCT 2017 - Int. Conf. Intell. Commun. Comput. Tech).
[3] I. Gusti Ngurah Made Kris Raya, A. N. Jati, and R. E. Saputra, Analysis realization of Viola-Jones method for face detection on CCTV camera based on embedded system, (Proc. 2017 Int. Conf. Robot. Biomimetics, Intell. Comput. Syst. Robionetics 2017),
[4] R. B. B and M. Sharmila Kumari, (Comparative Study on Face Detection Algorithms),
[5] M. Ayi, A. K. Ganti, M. Adimulam, and B. Karthik, Interfacing of MATLAB with Arduino for face detection and tracking algorithm using serial communication, (Proc. Int. Conf. Inven. Comput. Informatics).
[6] M. Saraswathi and S. Sivakumari, Evaluation of PCA and LDA techniques for Face recognition using ORL face database, (Int. J. Comput. Sci. Inf. Technol).
[7] K. S. Sodhi and M. Lal, Face Recognition Using PCA , LDA and Various Distance Classifiers, (J. Glob. Res. Comput. Sci).

F-ready: A support device that allows disabled people to play the guitar

Y Nishinohira[1, 2], M Ikeda[3], C Oshima [2], H Matsui[1], and K Nakayama [2]

[1] Graduate school of Engineering, Mie University, 1577 Kurimamachiyatyo, Mie, 5148507, Japan
[2] Graduate school of Science and Engineering, Saga University, 1 Honjo, Saga, 8408502, Japan
[3] Graduate School of International Social Development, Nihon Fukushi University, Mihamatyo Okuda, Aichi, 4703233, Japan

yukiko@robot.mach.mie-u.ac.jp

Abstract. Although many people with disabilities have a strong desire to play a musical instrument, there are a few support devices that allow them to play it. We propose four considerations that have been identified for the development of such a support device and developed "F-ready," allowing people with disabilities in their upper limbs to play the guitar. We demonstrate the results of a trial in which a male with a cervical spinal cord injury practiced playing the guitar with F-ready for five weeks. He could make sounds with the guitar using F-ready, even from his first day of practice. Then, he could simultaneously engage in three movements: moving his left hand to the appropriate position on the strings, pressing the F-ready switch, and strumming the strings with his right hand. After only five weeks, he performed a musical piece at a research meeting on the harmonica and guitar using F-ready. In addition, he faced several tasks: deciding on an optimum posture, how to deal with arm pain, and learning the song's chord progression. The experts solved these tasks together. The experiment demonstrated that F-ready supports players' ability to play the guitar at will and enthusiastically.

1. Introduction

In this paper, we indicate four considerations that were identified for the development of a support device for people with upper limb disabilities who want to play the guitar. Then, we show a support device, "F-ready [1][2][3]," which has been developed based on the four considerations and present data on a trial conducted with a male with cervical spinal cord injury who practiced playing the guitar with F-ready for five weeks.

Many people with disabilities have a strong desire to play musical instruments [4]. They often play these instruments as a part of their physical and psychological rehabilitation efforts [4]. There are many instruments and devices to assist in the playing of musical instruments, and new musical instruments have been developed for people with disabilities. Cymis is a musical scoring program and includes several ways to play the instrument. For example, users can touch the music notes displayed on a monitor [5], use expiratory pressure [6], various switches [7], and other means for making music; these are each effective and highly accessible. Users who have worked with Cymis since the beginning of its

IOP Conf. Series: Materials Science and Engineering **705** (2019) 012006 doi:10.1088/1757-899X/705/1/012006

development have used it to play in some European orchestras and play the Cymis. Freqtric Drums are another interface for detecting human skin contact and its intensity; by touching others, a user can create sounds that simulate drums and other percussion instruments [8]. Kyoraku is a device developed to help people with cervical spinal cord damage play the guitar [9]. The guitar used in the Bunne method, on which a user can sound the four chords of D, A, G, and E by moving a red lever, is also well known [10].

Despite the advanced mentioned above, it has not yet been not discussed how the development these support devices and musical instruments for the people with disabilities can allow them to practice whenever they choose. We know that the people with disabilities need specialized input devices, such as single switches and sensors that require only gestures to use them [11]. However, people with disabilities often cannot play a musical instrument without such modifications or equipment. Moreover, many musical therapists also do not use electronic music technology and/or clinical tools with their clients all of the time, because the therapists select what they see as the appropriate tools for each client [11]. If considerations for the development of these support device for people with disabilities are established, they come to be used well, by not only by people with disabilities, but also by music therapists.

With the above in mind, we propose four considerations for the development of a support device that people with disabilities in their upper limbs can use to practice a musical instrument at will. In Section 2, we introduce F-ready [1][2][3], which we have developed, and we propose four considerations. In Section 3, we provide details of a trial conducted with a male a cervical spinal cord injury who practiced playing the guitar with F-ready for five weeks. The results and a discussion of the trial are shown in Sections 4 and 5, respectively. We conclude the paper in Section 6.

2. Support device "F-ready"

F-ready [1][2][3] is a support device for people with disabilities in the upper limbs to play the guitar.

Figure 1 shows F-ready as a support device. Figure 2 shows the total system used to support people with disabilities in their use of F-ready. Many experts, such as physical and occupational therapists, were involved in the development of F-ready and identified four considerations, as shown in Sec. 2.1.

2.1. Considerations for the development of F-ready

In this section, we explain the four considerations identified by experts for the development of F-ready. A "user" refers to "a person with a disability in the upper limbs."

1. A user can play the guitar immediately using only the slightest power from his or her fingertips.

   A related study found that "Kyoraku [9]" is a self-help device developed to help people with cervical spinal cord damage play the guitar. It is attached to a player's hand, to help him or her press down on the guitar strings using the principle of leverage. As such, it is difficult for many disabled people to use. People with disabilities in their upper limbs, including those with a cervical spinal cord injury at the C3 level, cannot use Kyoraku. In response, F-ready was developed so that a user can play the guitar using only the power of a fingertip, like using a computer mouse, and he or she can press the strings easily.

2. A user can select and press a part of the string (position) and strum the strings by him/herself.

   We have observed that a person with a disability may try to play the guitar as much as possible, but he or she will still require someone else for support in the performance. However, it is important for the user to select and press a part of the string by him/herself. In a related study, the guitar's swing bar, which is one of "Bunne" musical instruments [10], allows the user to play by him/herself, but only to produce four chords. By contrast, F-ready allows a user to press anywhere on the string, from the first fret through the 11th, to play freely, if the user fixes the guitar's tuning to one chord. We expect that users will be able to play a favorite piece and realize that his/her musical skills have improved. Then, the motivation to play the guitar will continue for a long time.

3. The operations needed to play a guitar are unchanged, regardless of whether an assistance device is used or not.

In the case of Kyoraku [9], the playing method differs, based on whether or not the device is in use; hence, Kyoraku requires the mastery of a different performance style to control the pressing of the strings. Further, the method cannot be used with a guitar without the Kyoraku device. Assistance devices of this kind require the user to master the device before they can, eventually, start to practice the instrument. As a result, the user cannot and does not acquire the skill of playing the instrument at all. Moreover, if a user wants to play the instrument without the device, because they have recovered their physical function, he or she has must resume practicing once more as a novice. In contrast, if the physical movement of playing an instrument with an assistance device is the same as that of playing without it, the user might feel able to practice playing for longer periods of time and, ultimately, become attached to the musical instrument itself. If someone then wants to play the instrument without the device, he/she will not need to modify the movements they have already mastered.

4. The assistance device should let the user's hands operate simultaneously, so that playing the guitar has a rehabilitative effect.

When a user plays the guitar with or without F-ready, he/she must coordinate the following three movements (α, β, γ): move the left (right) hand to the appropriate position on the strings (α), press (and keep pressing) the position or the F-ready switch at that position (β), and strum the string(s) with the right (left) hand (γ). Although the user may, at first, perform these three movements one by one, they should eventually occur in cooperation, following repeated practice. Thus, playing the guitar is expected to have a rehabilitative effect for the upper limbs.

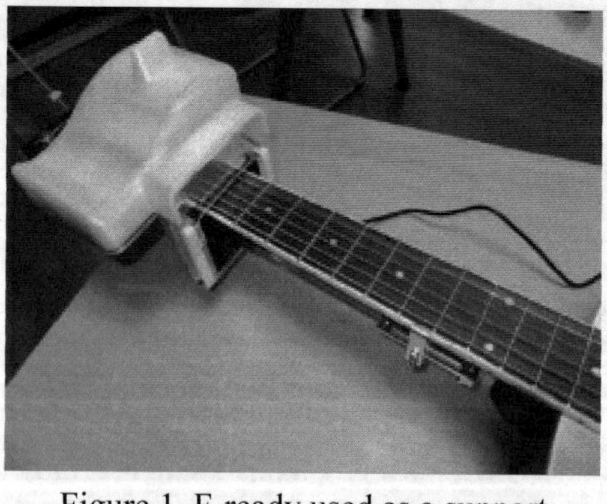

Figure 1. F-ready used as a support device for playing the guitar.

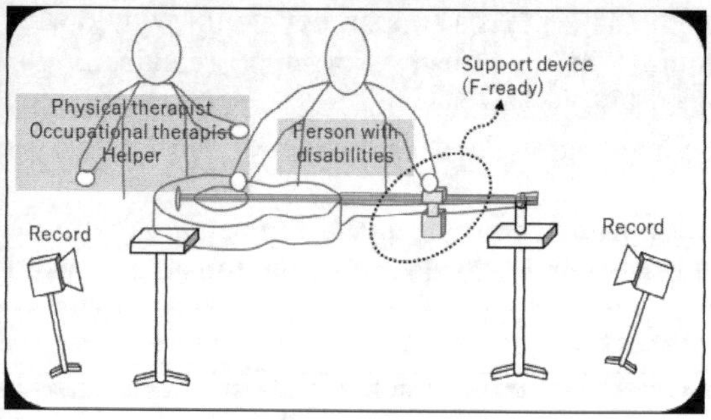

Figure 2. Total support system that allows experts to check the performance of a person with disabilities.

2.2. *Support system*

It is important to construct a support system (in collaboration with experts) for F-ready's users. As shown in Figure 2, the user's performance is recorded with video cameras during his/her practice sessions. Then, experts (physical, occupational, and music therapists; caregivers; and so on) who understand the user's daily life discuss the his or her physical state and performance by watching the recording. Physical movements are complex, so we should break them down to determine what is currently possible and what abilities may or may not be recoverable through rehabilitation. In this way, it becomes clearer whether the user's movements require no assistance, what levels of assistance need to be adjusted in response to changes in his/her physical state, or whether permanent assistance is required.

The experts' advise the user and his or her caregiver on how to practice, how to move the upper limbs, and how to assist in the performance based on the goals for the performance and the user's rehabilitation.

IOP Conf. Series: Materials Science and Engineering **705** (2019) 012006 doi:10.1088/1757-899X/705/1/012006

3. Experiment

A disabled male with a cervical spinal cord injury used F-ready to practice the guitar for five weeks. We made sure that F-ready meets the four specifications shown in Section 2 and that people with disabilities are able to play whenever they choose, using F-ready.

3.1. Subject information

The subject was a 30-year-old male who, at 20 years of age, injured his cervical spinal cord in a car accident. He has paralysis in his upper limbs, below the chest. He has a C3-C4 cervical spinal cord injury, but can lift his elbows and move both hands, but he cannot move the fingertips on his right hand. He cannot move his fingers independently.

3.2. Experimental conditions

The conditions of use for the device and the recording method are shown below.

(a) Use a typical, unmodified acoustic guitar.

(b) As shown in Figure 3, the guitar body was placed face up, and the neck was placed on another desk.

(c) As shown in Figure 4, before he began practicing the guitar, the first author explained to the subject how to use the device, and a physiotherapist was brought in to adjust the placement of the instrument.

(d) The practice period for the trial detailed in this paper was five weeks. The subject decided on the number of exercises and the practice time devoted to each exercise.

(e) The first and second authors supported and advised the user's practice from perspectives of physics and music.

(f) To demonstrate the usefulness of the support device during a research meeting, the subject played a two-minute-long musical piece on the guitar using the support device

Figure 3. The guitar torso is placed face up.

Figure 4. Before starting to practice the guitar, some experts had a meeting at the user of F-ready.

4. Results

The user practiced the guitar 26 times in five weeks. We call the practice, from the first though ninth sessions "Phase 1," the 10th to 15th sessions "Phase 2," and the 26th session "the final practice day." Then, he played the guitar during a research meeting.

In these results, we focus on the following three operations, α, β, and γ (see Section 2.1).

4.1. Phase 1

The user could move the F-ready's input device to the appropriate position on the strings, even on the first day. However, since each movement was not one to which he was accustomed in everyday life, he had a lot of muscle tension in his arms and shoulders. He considered the optimum posture to use that would allow him to play the guitar most easily. Then, he decided on the position that put the guitar's body on his knees and the neck on a desk, as shown in Figure 3.

He could strum the strings while pressing them using the F-ready switch. However, he operated α, β, and γ independently during Phase 1. He took a little time to perform each operation, because he had to visually confirm the movements of his hands.

4.2. Phase 2

The user's performance style became established, and he was able to perform each operation without excessive muscle tension.

Although he could continue playing the guitar without pausing for longer lengths of time, he began to feel pain in his arms, and his fatigue was growing.

He became used to each operation, α, β and γ. In particular, during operation γ, his need for visual confirmation was reduced. By contrast, he did not try to perform operations α and β without visual confirmation, as in Phase 1. However, it seemed that he could perform all operations at once, although it still took time to make a sound.

He started practicing a portion of a song. He played the melody on the harmonica and using the guitar as an accompaniment. He played the guitar at a very slow tempo.

4.3. Final practice day

The user could perform operations α, β, and γ simultaneously. Then, he could move the F-ready's input device to the correct positions on the strings while playing the piece at the intended tempo. He became able to play the harmonica with his guitar as an accompaniment for 24 measures. He could play more than 18 tones on the guitar while the playing the piece.

4.4. Performance at a research meeting

Five weeks after beginning his guitar practice, the user played the harmonica, with his F-read-equipped guitar as accompaniment, at a research meeting to demonstrate the usefulness of the F-ready device. He could play one chorus of the song in about two minutes with fluency and musicality. The audience applauded.

5. Discussion

In this section, we discuss whether F-ready follows the four considerations (see Section 2.1), based on the process experienced by the user who participated in the trial outlined in this paper.

He could play the guitar by himself using F-ready from the first day of practice, even though he cannot move his hands freely, due to a C3-C4 cervical spinal cord injury. This result shows that F-ready follows Consideration 1 (A user can play the guitar immediately using only the slightest power from his or her fingertips) and Consideration 2 (A user can select and press a part of the string (position) and strum the strings by him/herself). However, his performance style was different from that of our assumptions. It was not easy for him to press the F-ready's input device with his fingertips; therefore, he used his arm to press it. He could draw his arm towards his body by pulling his shoulder up and while bending his chest backward. Then, he could catch the input device. In contrast, he could return his arm to the original position by relaxing his shoulder. However, it was somewhat difficult for him to hold the input device, even though the input device has a rest attached. His left hand was sometimes lost its grip on the input device. Therefore, we should consider changing and enlarging the attached rest, so that the user's hand does not slip.

If F-ready is removed, the user can play the guitar in the same way he plays it with F-ready. He will strum the strings with his right hand by pressing them with his left hand from above, even without F-ready. He does not need to modify the movement that he has already mastered. Namely, F-ready follows Consideration 3 (The operations needed to play a guitar are unchanged, regardless of whether an assistance device is used or not). However, he did not remove F-ready during this trial, because the physical function of his upper limbs did not recover enough to play the guitar without F-ready. Because he has injured his cervical spinal cord, he cannot directly press the strings with sufficient power, even if he were to practice for a long time. However, he can move his left hand from side to side on the strings and strum them using his right hand.

Although he could perform each operation (α, β, and γ) independently, he became able to perform these operations simultaneously and smoothly. He practiced the guitar not for rehabilitation purposes, but also for his own enjoyment. However, we have considered the rehabilitation effect of moving the left hand from side to side on the strings and pressing the input device using the upper limbs. Therefore, F-ready follows Consideration 4 (The assistance device should let the user's hands operate simultaneously, so that playing the guitar has a rehabilitative effect).

We consider that F-ready allows people with a disability in the upper limbs to practice the guitar whenever they choose, because F-ready follows the four considerations.

Despite the above successes, the user had some tasks to tackle other than making a sound with the guitar for the five weeks. He had to consider the optimum posture that would allow him to play the guitar most easily. Then, his arm began to hurt. He needed to learn a chord progression. The physical and musical therapists addressed these tasks together. Meanwhile, these matters show that F-ready supports users with disabilities in their upper limbs in practicing the guitar at will and enthusiastically.

6. Conclusion

F-ready was developed as a support device for people with disabilities in their upper limbs who want to play the guitar based on four considerations: (1) play the guitar immediately with the support device, (2) play by oneself, (3) the performance styles playing with/without the support device are the same, and (4) playing the guitar has a rehabilitative effect.

In this paper, a male with a disability in his upper limbs practiced the guitar with F-ready for five weeks to examine whether F-ready follows the four considerations. He could play the guitar by himself with F-ready even on the first day of practice. However, he performed operations that are necessary for playing the guitar independently. At first, he moved the F-ready input device to an appropriate position on the strings by his left hand. Then, he pressed the input device, to press the strings using his left hand. Finally, he strummed the strings using his right hand. After two weeks, he became able to perform these operations simultaneously and smoothly. Although he could not play the guitar without F-ready during this five-week trial, he would be able to play the guitar using the same performance style, even if F-ready were removed. Furthermore, because he moved his left hand from side to side on the strings and pulled his shoulder and arm up to move his hands repeatedly, playing the guitar can be considered a part of his rehabilitation.

In contrast to the above, the user developed pain in his arms after practicing for a long time. Moreover, he wanted to find a chord progression that would be easier for him. Therefore, the second author, who is a physical therapist, and the first author, a music expert, advised him. In this way, we should provide not only just support for the device for the people with disabilities, but also a support system that experts can use to evaluate a person's current state and physical condition, sot hat they may advise them appropriately.

In the near future, we will conduct further experiments that some neuromuscular patients try to practice playing the guitar with F-ready for a long period. Moreover, we will develop a system that allows experts to review practice videos in a short time.

7. Ethical considerations

This study was conducted in line with the Helsinki Declaration of Ethics and was approved by the Ethics Committee of Mie University. We also obtained the permission of the corresponding individual regarding this work's publication.

Acknowledgment

This work was supported by JSPS KAKEN Grant Number JP 18K18642.

References

[1] Kumagai Y, Kato N, Uto Y, Takamura Y 2016 *Developing assist device for playing guitar*, International Conference on Engineering Technology (ICEC 2016)

[2] Nishinohira Y, Matsui H, Oshima C, Nakayama K 2017 *A Performance Support Device for People with Motor Neurologic Deficit to Practice the Guitar with Enjoyment Continually*. IPSJ SIG Technical Report, 2017-AAC-5(9). (in Japanese)

[3] Oshima C, Nishinohira Y, Ito Y, Ikeda M 2018 *Toward a Musical Instrument Which Even Performers with Disabilities Can Enjoy Practice Continually -Discuss a Way of Supporting for the Performer's Physical Movement with Assist Devices*, IPSJ SIG Technical Report, 2018-AAC-8(9) (in Japanese).

[4] Schneider S, Schonle PW., Altenmuller E, Munte TF 2007 *Using musical instruments to improve motor skill recovery following a stroke*. J Neurol. 254(10), pp. 1339–1346

[5] Akazawa K, Kawai T, Okuno R, Masuko T, Nishida H 2012 *Novel electronic musical instrument for persons with cerebral palsy to play and enjoy together.* Proc. 9th Intl Conf. Disability, Virtual Reality & Associated Technologies. pp. 419–422

[6] Akazawa K, Okuno R, Kajiyama S, Kitamura T, Kawai T, Matono H, Nishida H, Ichie M, Aoki T, Masuko T 2012 *Development of breath pressure controlled electronic musical instrument for persons with motor dysfunction to play and enjoy.* JSMBE Trans. 50(6), pp. 629–636 (in Japanese).

[7] Ichinose T, Takehara N, Matsumoto K, Aoki T, Yoshizato T, Okuno T, Watabe S, Sato K, Masuko T, Akazawa K 2015 *Applying a novel electronic musical instrument and kinect in music therapy for children with autism spectrum disorders.* Proc. of World Congress on Education. pp. 90–93

[8] Baba T, Ushiama T, Tomimatsu K 2007 *Freqtric drums: a musical instrument that uses skin contact as an interface.* In Proc. of the 7th international conference on New interfaces for musical expression (pp. 386-387). ACM

[9] LUMINOUS JAPAN Official page<http://www.luminous.co.jp/kyouraku.html> [accessed on April 30, 2019] (in Japanese)

[10] Bunne Japan Official page <http://www.bunnemusic.jp/instruments/ > [accessed on April 30, 2019] (in Japanese)

[11] Magee W L 2006 *Electronic technologies in clinical music therapy* A survey of practice and attitudes. Technology and Disability, 18(3), 139-146

Drag Assessment of Vertical Lift Propeller in Forward Flight for Electric Fixed-Wing VTOL Unmanned Aerial Vehicle

Z. Sahwee[a], N. L. Mohd Kamal[a], S. Abdul Hamid[a], and N. Norhashim[a], N. Lott[a], M. H. Mohd Asri[a]

[a]Universiti Kuala Lumpur, Malaysian Institute of Aviation Technology, 43800, Dengkil, Selangor, Malaysia

Abstract. A fixed-wing UAV that is capable of vertical takeoff/landing is a hybrid aerial vehicle that could take off as helicopter and then transition into conventional airplane for forward flight. There are several powertrain configurations for the VTOL aircraft, one of them is called Separate Lift Thrust (SLT) configuration where the forward flight uses different powertrain compared to the hover phase. Without complex mechanism to store the inactive hover powertrain, the hover powertrain components added a significant amount of aerodynamic drag during forward flight. This paper presents the assessment of the drag caused by the inactive propellers during the forward cruising flight phase. For this research, 26 propeller samples with diameter from 5 inch to 9 inch were used with various configuration and materials. They were then tested in wind tunnel facility and the resultant drag was measured. The results from the wind tunnel test shows that for lowest drag penalty, smaller propeller diameter with low pitch blade provide the lowest drag. Interestingly, from the results, it shows that there is a flight speed that can provide an optimum drag from the chosen propeller. A selection of optimum motor and battery can be made in the future based on result presented in this paper to further improve the performance of the UAV.

1. Introduction

Typically, it is desired that airplanes can take off and land at fairly low speeds, so that they can use smaller and shorter airfields. The initial climb and the final descent are the most dangerous segments of a flight where most accidents occurs [1]. Various control schemes were studied to improve the reliability of flight [2]. One of the methods is by having Vertical take-off and landing capability which greatly reduces this risk because the plane is accelerated and decelerated while it is high in the air by adding margin for error and recovery. The VTOL configuration also bridges the performance gap between traditional helicopter and conventional aircraft [3]. Based on recent and projected advancements in battery specific energy, these electric VTOL (eVTOL) aircraft are anticipated to become feasible in the near term for short flights of 20-50 miles [4].

In Vertical Take-Off and Landing (VTOL) aircraft, the vertical lifting propulsion system used by Separate Lift Thrust (SLT) configuration contributes to an additional drag to the airplane during forward cruising mode. While in hovering mode, the drag penalty in not significant due to the short duration of the hovering phase. Most of the flight phase of the VTOL aircraft is in forward cruising mode. Thus, it imposed a significant increase of drag while in forward flight. The ability to characterized the drag introduced from the installation of VTOL components and structure is important in order to design an aerodynamically efficient aircraft. The VTOL structure and component adds to the total parasitic drag of the aircraft. The additional VTOL components used are arms, electric propulsion system and the propellers. Each of this components adds different drag value to the aircraft. While it seems that making smaller and shorter parts is the better option in reducing drag, there will be a point of conflict where the smaller propulsion system size would not be able to lift the aircraft.

Challenges in developing VTOL aircraft lies within its flight control, stability and endurance which is of prime area of research presently in this field [5]. As an example, study by Aoki and Muraoka [6] explores the design of forward wingtip to reduce induced drag of a VTOL aircraft. Another study by Footohi et.al [7] found that, for a VTOL aircraft, a sweptback tailless design provides better performance in comparison to conventional fixed-wing aircraft with a tail.

To increase performance of an aircraft, its drag needs to be reduced. Drag is categorized into two groups which are Parasitic drag and Induced drag [8]. Parasitic Drag is a combination of Form Drag, Interference Drag and Skin Friction Drag. In terms of reducing drag of the VTOL components and systems, various approaches have been recommended. For example, the parasitic drag of propeller in horizontal flight can be reduced by using folded propeller. From the study conducted by Goetten et. al [9], it is found out that individual aircraft design needs specific countermeasure to reduce its drag. Thus, study by Stahl [10] try to reduce drag contributed by the exposed VTOL motors to increase the aircraft performance. Some other researcher such as [11] use the deflected slipstream technology to improve the VTOL efficiency while study by Teo, [12] explorers on propeller blade design to be used specifically for VTOL application.

Thus the objectives of this research are to investigate the drag contributed by the vertical lifting propeller with various diameter, pitch and airspeed.

2. Drag of VTOL Aircraft

It is important that the drag force on the aircraft is minimized to ensure high efficiency of the aircraft. The drag force, D on an aircraft is given by:

$$D = 0.5\, S\, \rho\, C_D\, V^2 \qquad (1)$$

where S is the area of the wing, ρ is the air density, C_D is the coefficient of the drag and V is the aircraft velocity. In addition to the drag induced by the wings of the aircraft, a VTOL aircraft also suffers from drag due to the vertical lifting propulsion system therefore the drag coefficient consists of summation of the following 2 components:

$$C_D = C_{D,i} + C_{D,O} \qquad (2)$$

where $C_{D,i}$ is the lift-induced drag coefficient while $C_{D,O}$ is the zero-lift induced drag coefficient which is the parasitic drag. Induced drag is affected by the geometry of the wing and is given by [10]:

$$C_{D,i} = C_L^2 / (\pi\, \Lambda\, \epsilon) \qquad (3)$$

where C_L is the lift coefficient, Λ is the wing's aspect ratio and ϵ is the Oswald efficiency number. If the wingspan is given by d and the area of the wing is S, the wing's aspect ratio, can be computed from the following [13] :

$$\Lambda = d^2 / S \qquad (4)$$

This equation shows that the induced drag increases with increasing wing area. Furthermore, the hover rotors of the VTOL aircraft introduce additional drag in which is categorized as a parasitic drag and relates to the viscous effects of the aircraft. This parasitic drag comprises of pressure drag, friction drag, interference drag and rotor shaft drag [10].

3. Methodology

3.1. Propeller selection

There are many propellers available off the shelf components which have various size, pitch and shape. As the purpose of the vertical propeller is to provide lift in hovering phase only, a high static thrust type of propeller which have low pitch were chosen. Propeller size from 5 inch to 9 inches was chosen due to the suitability of the chosen airframe. Table 1 shows the selection of propeller used in this research. There are 26 propellers with various diameter, pitch, number of blade and material.

Table 1. Propellers sample data.

Prop number	Diameter (inch)	Pitch (degree)	Number of blade	Material
1	5.0	3.0	2	EP
2	5.0	4.5	3	Nylon
3	5.0	4.5	2	Carbon
4	5.0	4.5	2	Nylon
5	5.5	5.0	2	EP
6	5.0	5.0	2	EP
7	5.0	5.0	2	Carbon
8	6.0	5.0	2	Carbon
9	6.0	4.0	2	Carbon
10	6.0	4.5	2	Nylon
11	6.0	4.5	2	Nylon
12	7.0	4.5	2	EP
13	8.0	4.5	2	Nylon

Prop number	Diameter (inch)	Pitch (degree)	Number of blade	Material
14	8.0	4.5	2	Carbon fibre
15	8.0	4.5	2	Carbon
16	8.0	6.0	2	EP
17	8.0	6.0	2	EP
18	9.0	4.5	2	Carbon
19	9.0	4.5	2	Carbon fibre
20	9.0	6.0	2	EP
21	9.0	6.0	2	EP
22	9.0	4.7	2	Nylon
23	9.0	4.5	2	Carbon fibre
24	9.0	4.7	2	Carbon
25	9.0	5.0	2	Carbon
26	9.0	4.5	2	Nylon

Propeller is measured by their diameter and pitch. The pitch is the theoretical distance travelled, by the prop, in one revolution. It is also the steepness of the angle of the blades. A low-pitch propeller will make less thrust and generally be more efficient. A high-pitch prop will make more thrust and generally draw more current from the battery.

Figure 1 shows the propeller sample used for this research. In general, the diameter of the propeller influences the amount of thrust generated. For a similar static thrust, a larger diameter prop rotates at slower speed thus generates less noise.

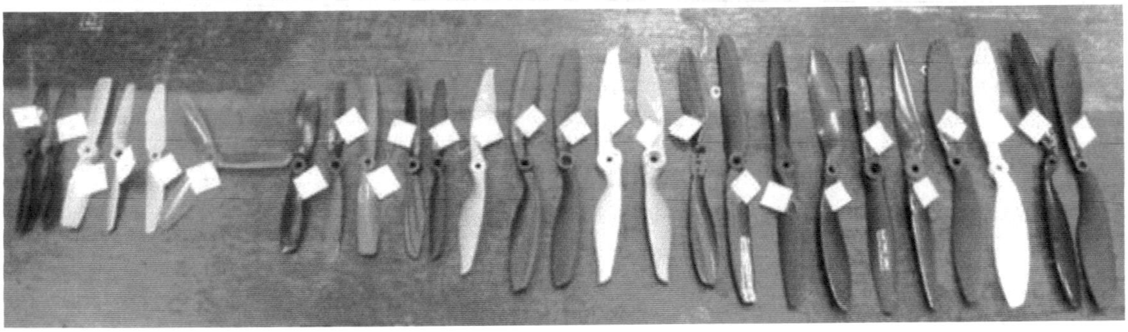

Figure 1. Propeller samples used for the wind tunnel test.

3.2 Wind tunnel setup and measurement

The wind tunnel experiment was conducted in the open circuit subsonic wind tunnel available in UniKL MIAT. This wind tunnel can operate at maximum airspeed of 40m/s. Figure 2 shows the subsonic wind tunnel and its control panel for the experiment.

Figure 2. Subsonic wind tunnel and its control panel.

The test section of the wind tunnel measures of 30cm x 30cm x 30cm as shown in Figure 3. The propeller was mounted horizontally as they would be mounted on the VTOL UAV. Since each of the chosen propellers have different pitch and diameter, measurements were taken by replacing the propellers attached to the motor inside the wind tunnel test section. The assembly of the motor and propeller are mounted on the force balance equipment within the test section. This force balance equipment which are attached to a load sensor enables the drag to be measured in Newton (N).

Figure 3. Subsonic wind tunnel test section and test setup arrangement for propeller.

IOP Conf. Series: Materials Science and Engineering **705** (2019) 012007 doi:10.1088/1757-899X/705/1/012007

4. Results

Each of the propeller sample was tested by varying the airspeed and the resultant drag value was recoded. The airspeed used were 10m/s, 12m/s, 14m/s, 16m/s, 18m/s and 20m/s. These airspeeds were chosen because the UAV is intended to be flown within these airspeed constraints. The propeller samples were arranged such that the propeller diameter is increased from 5 inch for sample 1 to 9 inch for sample 26. Figure 4 shows the result from the test. It can be seen from Figure 4 that the amount of drag produced is directly proportional to the propeller diameter size. A bigger diameter propeller has higher drag compared to smaller diameter propeller. This is due to the bigger frontal area exposed to the wind. The experimental data also coincide with Equation (1) where as airspeed is increased, the drag will also increase proportionally. As a result, an optimal selection for propeller diameter is chosen with the smallest possible diameter that could provide an adequate thrust for the UAV.

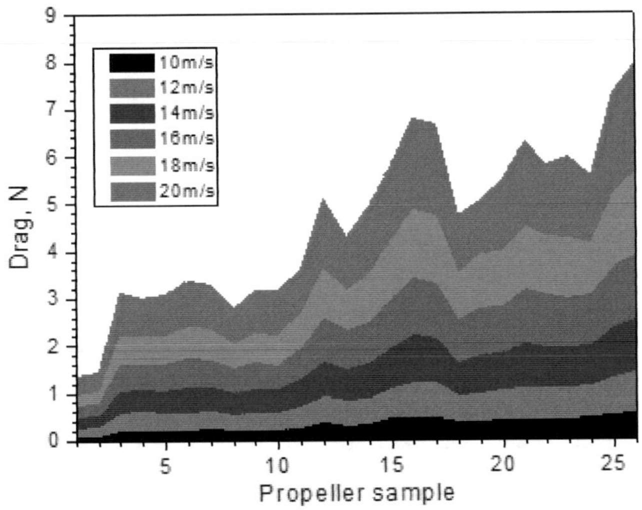

Figure 4. Propeller sample and its horizontal drag with different airspeed.

There are three factors that influence the drag produced by the placing the propeller horizontally facing the wind, which are propeller diameter, airspeed and blade pitch of the propeller. The first factor, as the propeller diameter increases, the amount area exposed to the wind is also increases thus increasing the drag. Figure 5 indicates an increase in drag as the propeller diameter increases.

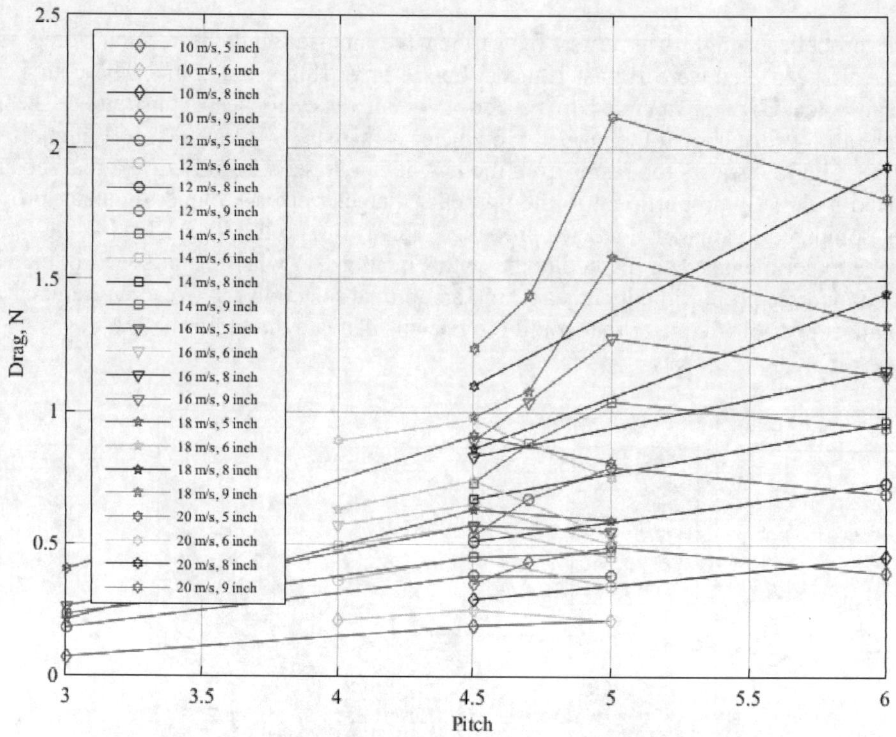

Figure 5. Comparison between the propeller sizes and blade pitch.

The second factor that affects the amount of drag is the airspeed. As shown in equation (1), the drag increase as a factor of velocity squared. As shown in Figure 5, each propeller produced an increasing drag values as the airspeed increased. Furthermore, as the blade pitch of the similar diameter increases, the area that is exposed to the incoming air is also increase, thus increasing the drag. For example, a 5 inch propeller, with pitch of 3, 4.5 and 5, as the value of pitch increase, the resultant drag also increase.

A detailed graphs showing the relationship between blade pitch propeller diameter and airspeed for each similar propeller diameter is shown in Figure 6. It shows an interesting result that as the airspeed is increased, the intermediate blade pitch has higher drag compared to higher blade pitch. For example, a 5 inch propeller tested in 20 m/s airspeed, a propeller with blade pitch of 5 has lower drag compared to a propeller that has blade pitch of 4.5. This trend also can be seen from propeller with 6 inch and 9 inch. The same condition can also be observed for 9 inch propeller as the pitch change from 5 to 6, the amount of drag is reduced nearly half. This finding requires further analysis since the drag should increase linearly with increasing pitch. Thus, details study is planned for future works.

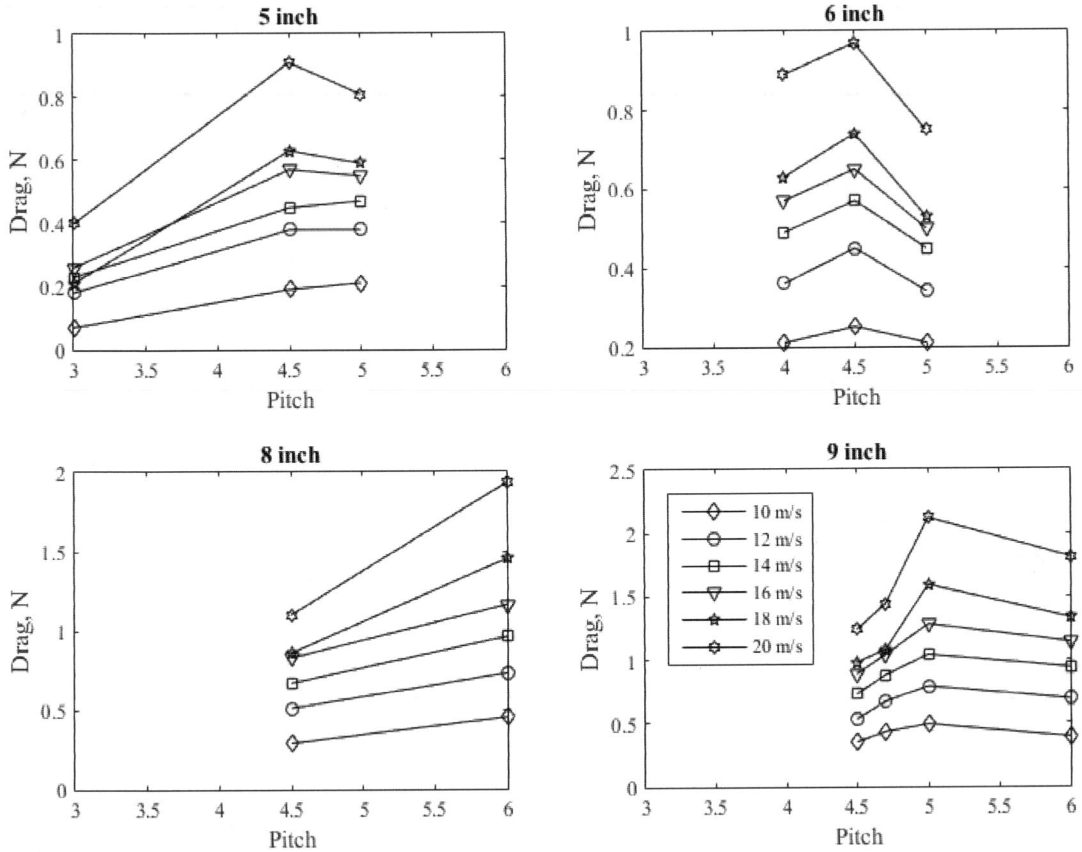

5. Conclusion

In contrast to other studies in which the air flow is in line with propeller shaft, this paper presents the drag produced by placing the propeller horizontally. This is to replicate the condition of a SLT VTOL aircraft where the propeller is exposed to the incoming air during forward cruising phase. For these vertical propellers, three factors affect the amount of drag produced which are propeller diameter, blade pitch and airspeed. 26 propeller samples were tested in wind tunnel facilities and data for each propeller was analyzed. From the data gathered, the least drag from all the tested propellers was those that are smallest in diameter with lowest pitch blade. As the airspeed plays an important factor in drag produced, lower forward speed is desirable to reduce the drag form the propeller. The result from these analysis is important for making a decision for optimum selection of motor and battery for the VTOL propulsion system.

6. References

[1] Aircraft Owners and Pilots Association 2010 22nd Joseph Nall Report: General Aviation Accidents in 2010 52

[2] Sahwee Z, Mahmood A S, Rahman N A and Sahari K S M 2017 Flight testing of control allocation for elevator actuator fault in small UAV *2016 International Conference on Control, Automation and Information Sciences, ICCAIS 2016*

[3] Abd Rahman Y A, Hajibeigy M T, Al-Obaidi A S M and Cheah K H 2018 Design and

IOP Conf. Series: Materials Science and Engineering **705** (2019) 012007 doi:10.1088/1757-899X/705/1/012007

Fabrication of Small Vertical-Take-Off-Landing Unmanned Aerial Vehicle *MATEC Web Conf.* **152** 02023

[4] Warren M M, Garbo A, Herniczek M T K, Hamilton T K and German B J 2019 Effects of Range Requirements and Battery Technology on Electric VTOL Sizing and Operational Performance 1–15

[5] Intwala A 2015 A Review on Vertical Take Off and Landing (VTOL) Vehicles *Innov. Res. Adv. Eng.* **2** 186–91

[6] Aoki Y and Muraoka K 2019 Forward Wing Tip Design for Induced Drag Reduction of Quad Tilt Wing VTOL Un-Manned Plane in Cruise *AIAA Scitech 2019 Forum* 1–10

[7] Footohi P, Bouskela A and Shkarayev S V. 2019 Aerodynamic Design of Long-Range VTOL UAV *AIAA Scitech 2019 Forum* 1–17

[8] Federal Aviation Administration 2000 Pilot's handbook of Aeronautical Knowledge: Aerodynamics of Flight

[9] Goetten F, Havermann M, Braun C, Gómez F and Bil C 2018 On the Applicability of Empirical Drag Estimation Methods for Unmanned Air Vehicle Design *2018 Aviat. Technol. Integr. Oper. Conf.*

[10] Stahl P, Rössler C and Hornung M 2018 Benefit Analysis and System Design Considerations for Drag Reduction of Inactive Hover Rotors on Electric Fixed-Wing VTOL Vehicles *2018 Aviat. Technol. Integr. Oper. Conf.* 1–9

[11] Cui G-P, Feng L-H and Wang W-J 2019 Aerodynamic Characteristic of Deflected Slipstream Aimed at Vertical Takeoff and Landing *J. Aircr.* 1–9

[12] Teo B S, Weigl J D and Henz M 2016 Test and development of efficient blades for use in VTOL aircraft *2016 11th Int. Conf. Ecol. Veh. Renew. Energies, EVER 2016*

[13] Vestlund O 2018 *Aerodynamics and Structure of a Large UAV* (Mälardalen University School)

Minimization Response Time Task scheduling Algorithm

Maria Maqsood[1], Saima Anwar Lashari[2], Murtaja Ali Saare[3], Sari Ali Sari[4], Yaqdhan Mahmood Hussein[5], Hatem Oday Hatem[6]

[1]Department of Computer Science, University of Lahore, Sargodha Campus, Pakistan.

[2]College of Computing and Informatics, Saudi Electronic University, Dammam, Kingdom of Saudi Arabia.

[3]Human-Centered Computing Research Lab, School of Computing, Universiti Utara Malaysia, Sintok, 06010, Kedah, Malaysia

[4]Faculty of Computer Science and Information Technology Universiti Tun Hussein Onn Malaysia,86400 Parit Raja, Batu Pahat, Johor, Malaysia.

[5,6]Communication Engineering Department, Faculty of Electrical Engineering, University Technology Malaysia, 81310 Johor Bahru, Johor, Malaysia

s.lashari@seu.edu.sa

Abstract. Cloud computing is a type of computing that base on shared computer resources for delivering information technology services. Distributed Computational system has been considered as best systematic model for sharing resources. Minimization of response time is the requirement in the distributed computational system. Previous Algorithm belongs to the optimization in which every new Virtual Machine reaching as a request to the cloud computing manager. This creates disturbance in the processing during life cycle of others Virtual Machines. For overcome this problem, there is a proposed an algorithm for minimization response time in distributed computational system by using the multi agents cloud structure. The overall system is architecture as a distributed over multi layered model. This leads to a situation that demands high care in allocating a Virtual Machine (VM) to a server machine in order to improve allocation time and efficiency of resources. Focus of this work remains to decrease time of allocation to optimize response time using a Virtual Machine (VM) placement and allocation algorithm that accounts Virtual. Machines (VM), their types and available resources among the data center server machines. Presented algorithm is implemented by using cloud simulator tool, through this tool, the calculated results of presented algorithm have been quite better with respect to traditional algorithm, and the difference presented with graphs for the both algorithms. Results support proposed methodology.

1. Introduction

Cloud computing is a type of computing technique which plays an important role for increasing the efficiency of data sharing and storage capacity for the business in the world. Cloud computing improves the capabilities of services like as Iaas (Infrastructure as a service), Paas (Plateform as a service) and Saas (Software as a service) but there are many security and management problems that are faces by Cloud computing [1]. With increasing advancement of information technology, the

Computer has become the essential utility of human life [2]. Cloud computing is introducing the capability for data access, interconnection between resources and exchangeable data with different resources through cloud computing [3]. For organizations and enterprises data sharing is available within low cost and with high capacity that is provided by new technology of cloud computing. Cloud computing is open base environment that provide the facility of sharing in which main aspect of security issues during sharing of resources [4].In past years many methodologies are presented for achieve the high quality of service from server to client in the cloud infrastructure with fast response time in different ways. Such as traditional methods is bin package algorithm for placement of VM to host. Earlier algorithms belong to the optimization of every new virtual machine reaching as a request to the cloud computing manager. This disturbs the processing during the life cycle of a virtual machine. Traditional polices has not been contained such method that accounts time optimization in placement policies of VM. This situation needs improvement of design towards efficient methodologies that can lead to a dynamic computing technological improvement. A standard cloud terminology adopted by Cloud simulator tool which is used for executed with different specifications of virtual machine with modified algorithm. Figure 1. presents the cloud computing infrastructure.

Figure 1 Cloud Computing

2. Cloud Services Models

There are three types of cloud computing service models.

1. Software as a service (SaaS)
2. Platform as a service (PaaS)
3. Infrastructure as a service (IaaS)

2.2.1 Software as a Service (SaaS)

User uses the services of cloud computing by executing their software and application properly with its subscription [5]. The other option is that user accesses the services of cloud computing through Software as a Service (SaaS) without not knowing about the architecture and backend of the applications and software's that run behind the frontend. User does not concern with the installation and maintains of application [6]. The main advantage and rapidly increasing use of Software as a Service (SaaS) is its scalability, easy configuration and compatibility because there is no need user to concern with the hardware and software of applications. If we compare with traditional system the user faces many problems of configurations and there is no level of scalability, compatibility and accessibility [7]. Example of Software as a Service (SaaS) providers are Google Docs, Microsoft Office 36, Amazon etc [8].

2.2.2 Platform as a Service (PaaS)

The Configurations, implementation and maintenance of applications, Platform as a service (PaaS) is an essential part in the cloud computing for application developers to maintain their application .Service providers need Platform as a service (PaaS) to maintain the whole services of cloud computing so there is necessary to implement the Platform as a service (PaaS) for successfully execution of services for cloud computing [9]. Service provider provide platform which enable the user to access the software ,operating system, servers ,development techniques and the database but all of this there is no too used to need of concern of user about the hardware of system like as processing unit , memory and storage etc. Platform as a service (PaaS) allow the application developers to configure and implementation of applications on cloud .The differentiation between SaaS and PaaS is that Platform as a service PaaS provides full life cycle for implementation of software and user can easily implemented the software directly on cloud while by using the SaaS, user can just host the executed applications [10].

2.2.3 Infrastructure as a Service (IaaS)

The third layer of service model is Infrastructure as a service (IaaS). IaaS deal with hardware resources and infrastructure such as storage, processing unit and connectivity of computer in the network. For execution of application there is need of infrastructure that is provided by IaaS. Infrastructure as a service (IaaS) is completely new concept of virtualization that only provides an infrastructure for application execution [11]. Infrastructure as a service (IaaS) work like a middle layer between two services which act as a physical or a virtualization machine for provide the infrastructure to help the user for execution the application. Infrastructure as a service (IaaS) enable the user to execute the application without taking the responsibility of implementation, administration and maintains. Example includes Amazon EC2 of IaaS [12].

3. Related Work

There is critical issue in cloud computing of management of resources and for the fast response time the placement of multiple virtual machines. For gaining the better results, many efforts and techniques are applying for solving this problem. Virtual machines work on the physical layer which is share available resources for the better productivity and effectiveness of computing resources. For solving the virtual machines placement problem the approached which has been modeled as Constraint Satisfaction Problem to increases the effectiveness of computing resources. With solving the virtual machine placement problem, the main focus on decreasing the cost of using the Computing resources [13]. These approaches focus is not on performance of network and effectiveness of computing resources that has a great impact on the whole effectiveness and efficiency of cloud architecture layers. For improving the response time and quality of service there is necessary to manage the computing resources and virtual machines in best way. In the case of migration of virtual machines few approaches [14][15][16][17] are declaring for gaining the effectiveness and efficiency of performance from one physical machine to another. Some approaches which focus on the performance of layers and response time of cloud for clients with the high quality of service. However, the existing approach is heavily focused on increasing the response time from cloud to client with managing the quality of service in better way. The drawback is that migration of virtual machines and their placement taking time has affected the response time to the front end and also effect on the quality of service. In the existing approach the algorithm is managing the placement of virtual machine in such a way one new appointed virtual machine does not disturb the life cycle of already working virtual machine. As the result the response time is increased from lower layer to upper layer and also it has good effect on the quality of service.

4. Traditional Scheduling Algorithm

A Virtual Machine allocation policy performs as an outline for provisioning process. The detail of the algorithm is given in the following lines. This is a one dimensional policy that accounts one Virtual Machines specification given in the Virtual Machine creation demand sent by the user to Data Center Manager. Data Center Manager further enlists the host that fulfills the demand of mentioned policy. This policy further tries to create a Virtual Machine after it finds a proper host. Output of this algorithm is either true or false depending upon availability of resources.

```
Step1:   If (Not VM)
Step2:          Do
Step3:   For each (host_i in Hosts);
Step4:Search_Host(Demanded_Processing_cores);
Step5:   End For;
Step6:   Results_list = Index_of (host i, j, k......);
```

IOP Conf. Series: Materials Science and Engineering **705** (2019) 012008 doi:10.1088/1757-899X/705/1/012008

Step7: Counter = Counter +1;
Step8: If (Resuls_list)
Step9: Modify:
Step10:VM_Table (VM);
Step11:Set_Used_Processing_Cores(demanded_processing_cores);
Step12:Set_free_Processing_Cores(demanded_processing_cores);
Step13:Acknowledge_Flag = True;
Step14: End if;
Step15:Else host=host (where host index is minimum);
Step16:While(Results_flag=Fasle && Counter<demanded_processing_cores);
Step17:Return Acknowledge_Flag

5. Proposed Response Time Minimization Algorithm

Step1: If (!VM_Created)
Step2: Do: Send_Vm_Specs(DCM);
Step3: For Each(host i in Hosts)
Step4: If(Host i>= Processing_Cores_Req && Host i >= Req_Mips)
Step5:Calcualte_remaingin_capcaity(host i)
Step6:Calculate_Serving_ratio(host i) //after current VM
Step7: If(remaining capcaity>serving ratio)
Step8:Index=host_index$_i$
Step9:Remaining_capacity=serving_ratio;
Step10: End If
Step11: End if
Step12: End for
Step13: If (! (index)
Step14: Break;
Step15: End if
Step16: Else
Step17: Host_list=index
Step18:Acknowledge_Flag=create_VM
Step19:Try_counter++;
Step20:If(Acknowledge_flag)
Step21: Update:
Step22:Vm_table(VM$_i$)
Step23:Sheduler_Queue(VM$_i$)
Step24:Used_processors(Required_VM_specs);
Step25:Free_Cores(Used_processors)
Step26:Acknowledge_Flag=true
Step27: Break;
Step28: Else
Step29:Free_cores(main_value)
Step30:Used_processors(main_value)
Step31:Remaining_capacity(main_value)
Step32: End if
Step33:While(Result&&try_counter<free_cores)
Step34:End if
Step35:Return Acknowledge_flag

6. Experimental Setup

CloudSim is a simulator that supports simulation of cloud data center using java based classes. It supports two decision based function, one works as coordinator of data center that is only one for each data center. And the second one represent host. Second agent function is associated exactly one with each host. Agent layers and their communication can be ordered and maintain but it increases the complexity of simulation. To keep the simulation simple, two layers architecture is used for the purpose of simulation. CloudSim allows the pooling of resources by using intelligent allocation at the time of Virtual Machine creation.

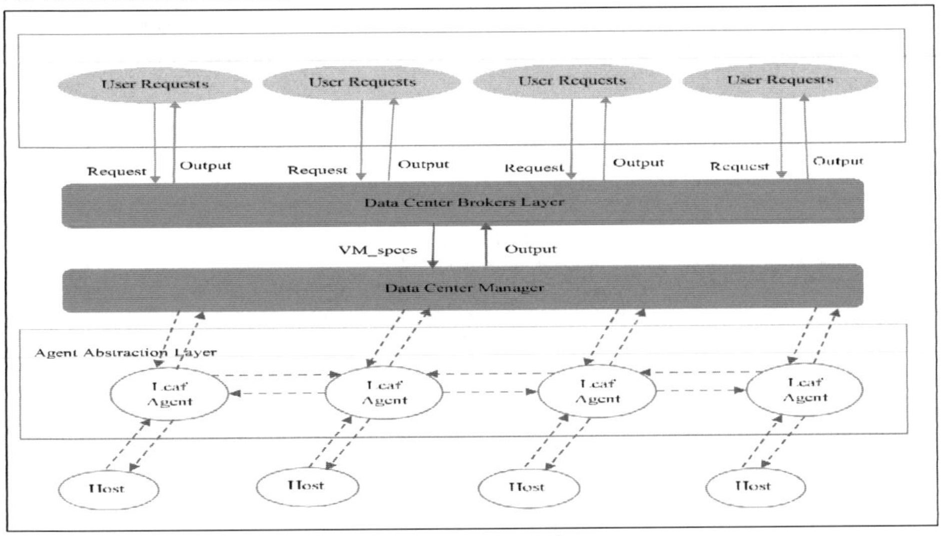

Figure 2. Data Center Configurations

In figure 2, the distributed Computational system show with multi agents dividing them into layers. For the purpose of simulation, the configurations specification that is used in implementation show in the table 1 with quantity.

Table 1. Data center configuration for simulation purpose

Configurations	Details
No of Data Centers	1
No of hosts	5
Cores on each host:	1(total 5 cores)
Storage capacity: 1 Tb	(100 GB=100000 MB)
Bandwidth:	1 mbps =8000 kbits/sec
No of Broker instances:	1
VMs	10
Tasks per VM	50
Instruction length	50000
Input size	300 kb
Output file size	400 kb
clock Speed	1000 mips per core
Storage per VM	20 GB
RAM per VM	512MB
CPU core per VM	1 virtual core
MIPS Per VM	1000 mips
Bandwidth per VM	1000 kb/s

Table 2. Specifications of Different Virtual Machine categories

Spec	VM 1	VM 2
Storage	512 mb	1024 mb
Ram	256 mb	512 mb
Cores	1	1
Mips	1000	1000
Bandwidth	512	512

IOP Conf. Series: Materials Science and Engineering **705** (2019) 012008 doi:10.1088/1757-899X/705/1/012008

7. Results

Results are observed in two different ways. Impact of both algorithm (default & modified) are noted down in two different observed variable. Increasing number of VMs has been observed for their impact on Allocation time & resources pooling. The observations are given in the form of graphs. Figure 3. is a graph which shows the difference of time allocation between default VM placement algorithm and modified algorithm.

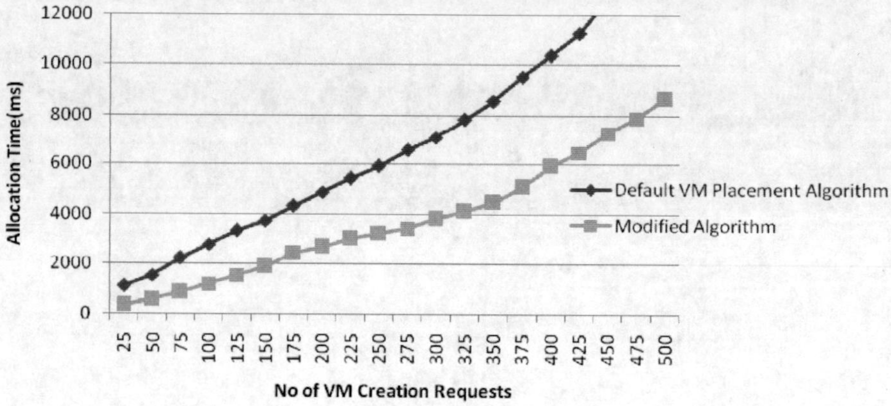

Figure 3. Time allocation difference between default algorithm and modified algorithm.

Figure 3. Graph of time allocation between default (traditional) Virtual Machine placement algorithm and modified algorithm. With respect to response Time, the modified algorithm has been quickly response to the maximum Virtual Machines requests. During the second scenario, different kinds of Virtual Machines (with different specs) are used to test the effect of placement method on allocation time. Four different kinds of specification are defined for Virtual Machines (show in table 6.2) and then their observations are made. Data center configurations are kept same as in the above example.

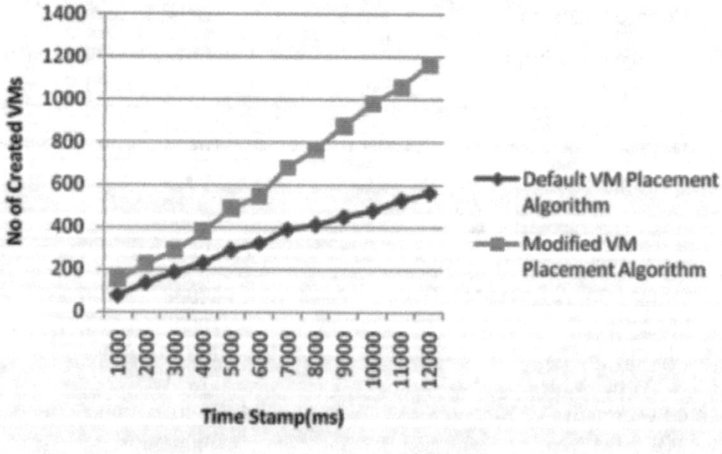

Figure 4. Throughput Comparison

According to the observed results, when number of Virtual Machines requests crossed over 225, the difference between allocation times began to clear. In case of default (traditional) algorithm, the allocation time started rising abruptly as number of demanded resources bypassed number of available resources. On the other hand, agent based scenario with modified algorithm smoothly increased its allocation time with increasing numbers of virtual machines requests. The overall ratio with increasing the number of virtual machines with respect to time is less than the traditional algorithm that is show in figure 4.

In the case of maximum virtual machine request, the failures of request have been happened in both traditional and modified algorithm but with quite difference.

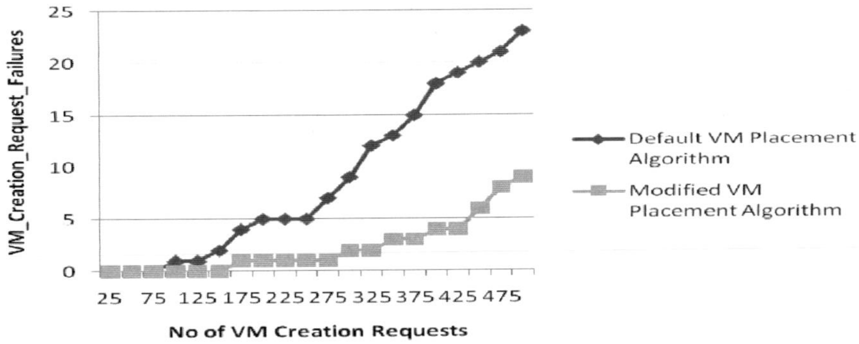

Figure 5. Ratios of Virtual Machine Request Failures

Figure 5 shows the ratio of Virtual Machines Failures Request. Modified Placement Algorithm is quite less ratio of failures of Virtual Machines Request than traditional Algorithm.

8. Time Complexity of Algorithm

The algorithm's complexity is dependent on number of Virtual Machine creation requests and the time needed by the creation attempts, time needed for communication and scheduling time. So the Complexity is described as

$$F(n)= T_c*n+T_{ca}+T_{com}+T_{sch}$$

Time Complexity = Time of creation *num of virtual machine + time of communication +time of scheduling

$$\Rightarrow \quad F(n)=O(n)$$

All of these are time constants therefore we can say that time complexity of this algorithm is linear.

9. Results of Observations

Different kinds of observations that can be made from simulation include response time and turnaround time. Response time depends on number of factors that include time to send request, number of tries, creation time of VM, time to send acknowledgement. Therefore mathematical notation of these two observations is given in the following.

$$Response\_Time=T_{req}+Counter_{tries}*T_{vm\_cr}+T_{ack}$$

The second observation that can be made is turnaround time. This is time is dependent on following factors.

$$T_{ta}=T_{Response\_Time}+T_{cl\_res}+T_{ex}+T_{sr}+T_{dr}+T_{des\_vm}$$

Here T_{ta}= turnaround time

$T_{Response\_Time}$=VM request response when the VM is properly created

T_{cl_res}= Response Time to cloudlets

T_{ex}= Execution time for cloudlets

T_{sr}=Time to request VM destruction

T_{rr}= Time to send back results

T_{dr}=time to destroy a VM Request

T_{des_vm}=time to destroy a VM.

10. Conclusion and Future Direction

While, in the improvement of response time depends on the interaction with cloud manager layer to the physical layer where multiple virtual machine share computing resources. At every client request, Cloud manger find new virtual machine according to the requirement of client request. Thus, it is necessary to control and manage the placement of virtual machine so that improve the response time and quality of service by selecting virtual machine according to the requirement in the short time. Some Virtual Machines placement approaches that work for distributed layer system in the cloud environment but for every new request arrival, cloud manger find the new virtual machine with considering the requirement of requesting task that disturb the life cycle of busy virtual machines. To address this issue, this paper present a Virtual Machines placement approach focus on the best and in short time selection of Virtual Machines by managing their placement in a way that new virtual machine does not disturb life cycle of working virtual machine.in addition, our approach has also considered the communication between distributed layers in the cloud architecture. Contract net protocol use for communication purpose in the distributed layers. Our simulation on cloud sim tool that proved purposing approach has been improved the task completion time. Results are showing in graphical representation with comparison traditional approach. This study can be further extended by using hybrid algorithm to participate the Virtual Machines migration and data centre joining aspects by considering energy saving.

References

[1] Xue, C.T.S. and Xin, F.T.W., 2016. Benefits and challenges of the adoption of cloud computing in business. *International Journal on Cloud Computing: Services and Architecture (IJCCSA)*, *6*(6).

[2] M. A. Saare and A. Hussain., 2018. "Successful Aging among Older Adults Using Computers: a Systematic Review," *Int. J. Eng. Technol.*, vol. 7, no. 20, pp. 50–54.

[3] Buyya, R., Yeo, C.S., Venugopal, S., Broberg, J. and Brandic, I., 2009. Cloud computing and emerging IT platforms: Vision, hype, and reality for delivering computing as the 5th utility. *Future Generation computer systems*, *25*(6), pp.599-616.

[4] P.Dhir and S. Garg., 2017."Survey on Cloud Computing and Data Masking Techniques", International Journal of Innovation & Advancement in Computer Science 6, no. 4, pp.1-7.

[5] D. Assante, M. Castro,I .Hamburg and S. Martin., 2016."The use of cloud computing in SMEs", Procedia computer science 83, pp.7-12.

[6] K. K. M. Kumar., 2014. "Software as a service for efficient cloud computing", International Journal of Research in Engineering and Technology, pp. 178-181.

[7] M. Al-Roomi,S. Al-Ebrahim, S. Buqrais, and I. Ahmad,. 2013 "Cloud computing pricing models: a survey." International Journal of Grid and Distributed Computing 6, no. 5, pp. 93-106.

[8] Q.Zhang, L. Cheng and R. Boutaba., 2010. "Cloud computing: state-of-the-art and research challenges", Journal of internet services and applications 1, no. 1, pp.7-18.

[9] T.Devi, and R. Ganesan., 2015. "Platform-as-a-Service (PaaS): model and security issues", Indonesian Journal of Electrical Engineering and Computer Science 15, no. 1, pp. 151-161.

[10] S. Kumar, and R.H. Goudar., 2012. "Cloud computing-research issues, challenges, architecture, platforms and applications: A survey", International Journal of Future Computer and Communication 1, no. 4, pp. 356-360.

[11] G. Kulkarni, R.Sutar and J.Gambhir, "Cloud Computing- Infrastructure as Service – Amazon EC2", International Journal of Engineering Research and Applications 2(1), pp. 117-125, 2012.

[12] S.R.Aniruddha and D.N. Chaudhari., 2013. "Cloud Computing: Instructure as a Service", International Journal of Inventive Engineering and Sciences, 1(3), pp. 1-7.

[13] J-L. Kim, and R. D. Ellis., 2005. A Framework for Integration Model of Resource-Constrained Scheduling using Genetic Algorithms. Simulation Conference, Proceedings of the Winter, 2119 – 2126, Orlando, Florida,USA, (December).

[14] S. K. Bose, and S. Sundarrajan., 2009. Optimizing Migration of VirtualMachines across Data-Centers. Parallel Processing Workshops ,ICPPW '09. International Conference.

[15] T. Hirofuchi, H. Ogawa, H. Nakada, S. Itoh, and S. Sekiguchi., 2008 AStorage Access Mechanism for Wide-Area Live Migration of Virtual Machines. In Summer United Workshops on Parallel, Distributed and Cooperative Processing, pages 19–24.

[16] Y. Luo, B. Zhang, X. Wang, Z. Wang, Y. Sun, and H. Chen., 2008. Live and incremental whole system mıgratıon of vırtual machıne usıng blockbıtmap.In IEEE Internatıonal Conference on Cluster Computing.
[17] T. Maoz, A. Barak, and L. Amar., 2008. Combining Virtual Machine migration with process migration for HPC on multi-clusters and Grids. Cluster Computing,IEEE International Conference.

IOP Conf. Series: Materials Science and Engineering **705** (2019) 012009 doi:10.1088/1757-899X/705/1/012009

Cellular Network and its Relevance for Unmanned Aerial Vehicle Application in Malaysia

N. L. Mohd Kamal, Z. Sahwee, S. Abdul Hamid, N. Norhashim and N. Lott

Universiti Kuala Lumpur, Malaysian Institute of Aviation Technology, 43800 Dengkil, Selangor, Malaysia

E-mail: nadhiyaliyanamk@unikl.edu.my

Abstract. Range of operation is one of the main limitation in the operation of Unmanned Aerial Vehicle (UAV). The ability to exploit currently deployed data-based cellular network for UAV application enables better utilization of UAV for various applications. Understanding the requirements of UAV application allows the identification of challenges to obtain coverage offered by terrestrial-based cellular networks. This paper outlines the challenges and identifies the relevance of cellular based UAV application in the context of Malaysia as well as its potential application. The main challenges highlighted include the design of cellular network which is optimized for terrestrial users. Additionally, in the context of Malaysia, the tropical weather imposes a further challenge to the reception of reliable radio signals. It has been identified that provided that these challenges can be overcome, the cellular based UAV can benefit the agriculture, security and utility sector in Malaysia.

1. Introduction

Previously, electric-powered UAV are limited to short distance flight due to the limitation on its battery. Recent improvement on battery technology has enhanced its flight range and duration. Today, with an optimized propulsion system, flight range of over 100 km can be achieved. With the extended range capability, the next bottleneck for flight range is data communication range between the ground control station and the flight vehicle. Current consumer system available on the market however, with its actual communication range is limited by transmitter radiated power and is not more than 15 km [1]. Various solutions have been explored such as increasing transmitter power, improving antenna design and using satellite as communication system.

Currently, UAV fly utilizing a direct radio link with the pilot ground station. The reliability in unmanned system is very important [2] and this is potentially achieved by the increased coverage of cellular network. It has been made possible for a remotely situated pilot at an operation centre to launch, fly and land a drone that has cellular network coverage, including in urban areas where structures can obstruct radio communications. This telecommunication network is able to transmit control and telemetry data of the UAV thus allowing an enhanced line-of-sight (LOS) flight [3].

The advantage of using the cellular network as a high bandwidth data-link provides virtually unlimited range which increase the operational range for many applications such as surveillance, delivery, search and rescue and emergency response flights in both urban and rural areas.

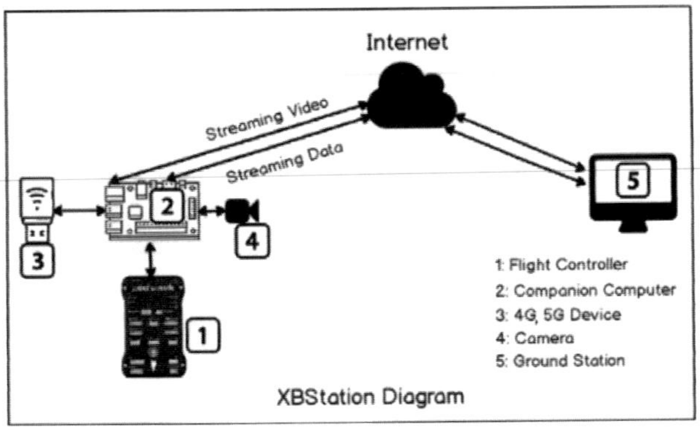

Figure 1. Cellular based communication system offerred by XBStation.

Today's cellular networks are, however, not designed for aerial coverage, and deployments are primarily optimized to provide good service for terrestrial users [4]. Another limitation of this method however, is the cell tower coverage. For this technology to be successfully being used in Malaysia, the area coverage and the limitation due to geographical feature needs to be further investigated.

The communication between the ground station and the flight vehicle is best described by Figure 1 which shows the data connectivity by using system offered by XBStation [5].

The organization of this paper are as follow. Section II discuss the needs and requirements for UAV application. Section III describes in detail the challenges in propagation environment imposed on cellular based UAV. Section IV address the potential application of cellular based UAV in Malaysia. Lastly, Section V presents the conclusions of this paper.

2. Needs and Requirement for UAV Application

2.1. Operational Height

Flight operation is based on the mission. Legally it is below 400 ft, unless special permission is granted by Civil Aviation Authority of Malaysia (CAAM). For civilian used in mapping and remote sensing, this height depends on the camera specification and the resolution of the image.

2.2. Collision Avoidance

Once the UAV can fly beyond line-of-sight, there is a safety requirement to ensure it does not collide with another flying vehicle either manned or unmanned.

2.3. Signal Coverage

The flight mission sometimes require the airborne aircraft to fly in remote areas. Since the operation of UAV over cellular network depends highly on data-based network, coverage from 3G and beyond network is important as these network provides packet-based services [6]. This however might be a challenge in remote areas whereby deployment of 3G (and beyond [1]) networks are absent. Figure 2 illustrates the coverage for 4G network in Peninsular Malaysia and East Malaysia. These data are obtained from users' devices using an app and represent a maximum coverage of 50 meters for uplink and 200 meters for downlink. Based on the coverage map, Malaysia currently shows their readiness in providing 4G network for users [7, 8] although it might be limited to urban areas as illustrated in Figure 2 [9].

[1] Third generation of mobile network and onwards.

IOP Conf. Series: Materials Science and Engineering **705** (2019) 012009 doi:10.1088/1757-899X/705/1/012009

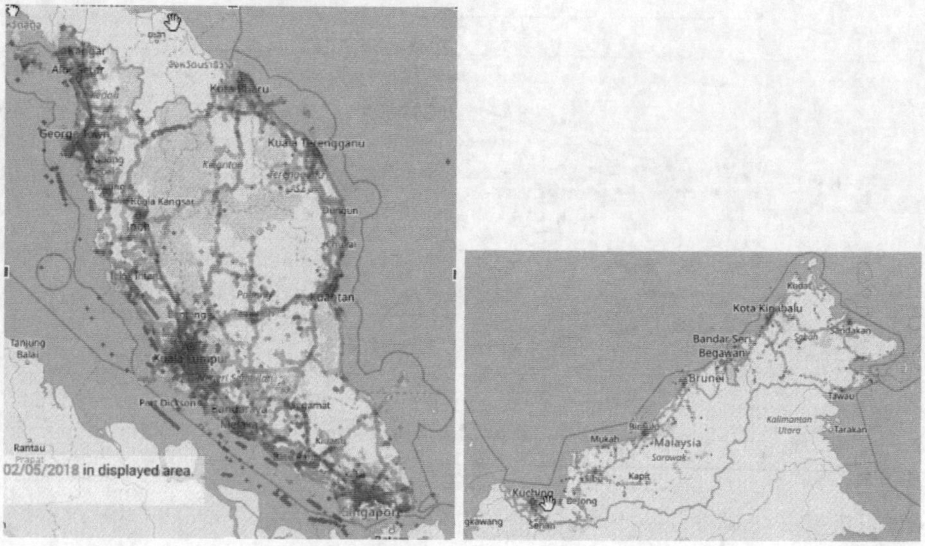

Figure 2. Map showing 4G connection in Malaysia (information updated as of 23 March 2019) [9].

Legend: (M) Maxis (C) Celcom (D) DiGi (U) U Mobile (U) Unifi

Region	Download Speed: 4G	Download Speed: 3G	Download Speed: Overall	Upload Speed: 4G	Latency: 4G	Latency: 3G	Availability: 4G
Ipoh	M	M C D U	M	M U	M	M C	M C
Johor Baharu	M	M C	M	M	M	M	C
Kuala Lumpur	M	C	M	M C	M	C	C
Malacca	M	U	M	U	M U	U	C
Seberang Perai and Penang Island	M	C	M	M	M U	M C	C

Figure 3. Regional Performance [8].

Figure 3 shows the regional performance of five largest operators in each category by OpenSignal measures. In OpenSignal's regional analysis of Malaysia's biggest cities, they examined seven core metrics for Malaysia's five largest operators. Many of the results reflected the national results with Maxis winning 4G download speed and overall download speed awards in all five areas [8].

3. Challenges in Propagation Environment

3.1. Antenna Tilt Angle

The antenna at the base station (BS) is often tilt downwards for optimum coverage of terrestrial users [10]. Generally, as show in Figure 4 (a), a larger coverage area among terrestrial users can be achieved if the tilt angle of the antenna is small (shown by θ_1). Conversely, increasing the downwards tilt angle of the antenna (shown by θ_2) results in smaller coverage area (smaller cell size). The decision on the size of the cell depends on the demand of users wanting to connect

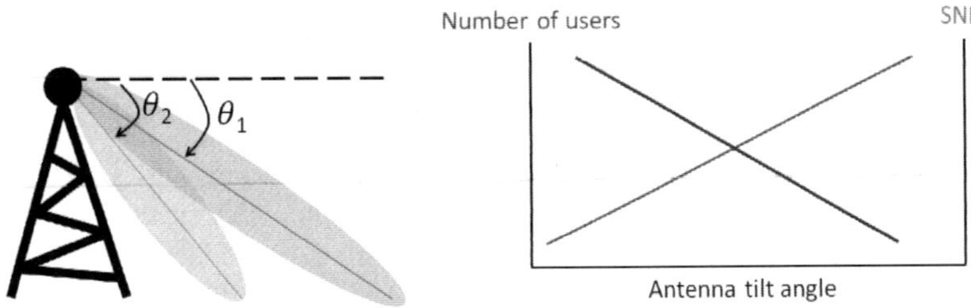

Figure 4. (a) Antenna with different tilt angle. (b) Relation between SNR of terrestrial users and number of users against the antenna downwards tilt angle.

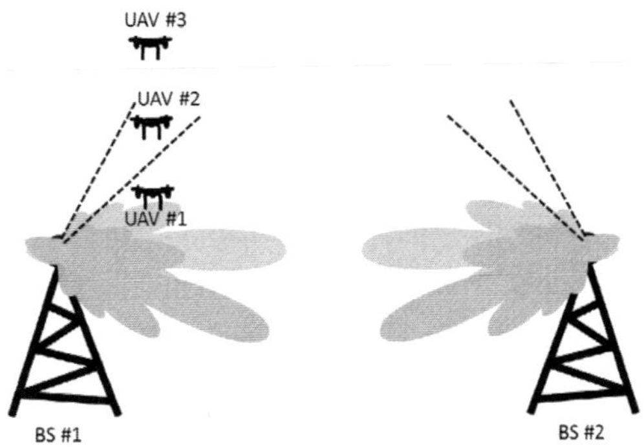

Figure 5. UAV coverage affected by tilt angle.

to the BS as well as the signal-to-noise-ratio (SNR) threshold that the users can tolerate.

Based on Figure 4 (a), a larger downwards tilt of the antenna results in larger SNR at the cell edge (relative to the case where antenna's tilt angle is smaller) at the expense of reduced coverage area hence smaller number of users are able to connect to the BS (assuming a uniform user distribution with fix density). On the other hand, a larger coverage area can be achieved with a smaller downwards tilt angle at the expense of reduced SNR at the cell edge.

Based on the standard 3GPP antenna radiation pattern in a 3 sector cell, the antenna gain is given by [11]:

$$A(\theta) = -\min\left[12\left(\theta/\theta_{3dB}\right), A_m\right] \tag{1}$$

Where, $-180 \leq \theta \leq 180$ is defined as the angle between the direction of interest and the bore sight of the antenna, θ_{3dB} is the 3 dB beamwidth in degrees, and A_m is the maximum attenuation. This shows that the greater the angle between direction of interest and the bore sight of the antenna (direction with strongest radiation), the smaller is the gain. Hence not only the users at the edge are affected by the tilt angle, but users closer to the BS are also affected by different configuration of the antenna tilt angle. The general relation between tilt angle of the antenna and the received signal strength (represented by SNR) as well as the number of users connected to the BS is illustrated in Figure 4 (b). This figure shows that a larger downwards tilt angle of the antenna results in larger SNR being established but at the expense of reduced number of users obtaining coverage. This is because as the antenna is tilt further downwards, the radiation from the antenna is imposed onto a smaller area.

Consider the illustration of 3GPP antenna radiation pattern in Figure 5 [12], UAV network at high altitudes could potentially be connected to the BS through the side lobes of the antenna.

For a given horizontal distance with respect to the BS, the higher the altitude of the UAV, the smaller the antenna gain at the UAV (UAV2 has smaller received signal strength than UAV1 from the blue-shaded antenna radiation pattern). If the antenna is tilt further downwards, the UAV reception will be worsen and eventually disconnected from the BS if the received signal strength is below the specified threshold. For a given horizontal position, unlike the performance of the terrestrial users, the performance of aerial users degrades with increasing tilt angle.

The distribution of users also affects the decision on cell size and hence the tilt angle of the antenna. Smaller cell sizes such as Pico cellular network are often adopted to accommodate high number of users. In such dense cellular network, the UAV may receive connection not from its nearest BS but from a BS further away (UAV3 may receive reception from BS 2 in Figure 5).

The work in [4] investigated the performance of cellular based UAV network and considered practical interference mitigation techniques in their evaluation. However, their investigation is based on a typical rural area network. 4G networks in Malaysia illustrated in Figure 2 are mostly deployed in urban or sub urban areas. Since different environment would result in different types of cellular network deployment (which varies in cell sizes), the network performance might too differ. Therefore this opens for further investigation for cellular based UAV in urban or sub urban areas.

The investigation on UAV coverage performance particularly in important areas requiring UAV application in Malaysia would provide useful insight on the robustness of cellular based UAV network in the country.

3.2. Propagation Environment

Al-Hourani and Gomez [13] identified that several challenges exist in utilization of current cellular networks for UAV application, as they are purely designed to cater for terrestrial users. Flying UAV might face a large amount of neighboring interference as well as a different coverage behavior. Moreover, Qazi et al. [14] identified the shadowing loss and multi-path propagation as challenge for wireless signal propagation. When the receiver and transmitter have no clear line of sight, such as when wireless signal has to pass through obstacles (such as walls) it is referred to as shadowing loss. If the wireless signals sees intermittent reception of Line of Sight (LOS) signal and reflected versions of signals from other obstacles it is referred to as multipath propagation or fading.

3.3. Weather Conditions

Malaysia is known as a tropical region which involves two different peak seasons; a dry and wet season which are distinguished based on the average rainfall. The dry season refer to low rainfall rate with an average rainfall total of about 50 mm/hr [15, 16] whilst the wet season experience high rainfall rate. The normal characteristic of the rainfall in tropical region, including Malaysia is heavy with the average of 200 mm/hr and large raindrop sizes [17, 18]. Rain contributes to the attenuation in propagating radio wave through absorption and scattering process. To implement the 4G network with the UAV technology, several factor related to the effects of rain on the performance of 4G network need to be considered. This include rainfall rate under different operating frequency and effective path length. The other factor which needs to be taken into consideration is the type of polarisation which is horizontal or vertical polarization.

The 1st factor which is rain rate attribute to the attenuation of the propagated signal during rainy period. The propagation of signals from the source to the destination is effected by the rainfall in term of the signal can be absorbed, scattered and diffracted by raindrops. As the rain rate increases, raindrop size increases resulting in significant attenuation to the propagated signal. This lead to the communication link becomes disrupted and the probability of having connection lost is high and become more severe in tropical regions which experience high rainfall rates.

IOP Conf. Series: Materials Science and Engineering **705** (2019) 012009 doi:10.1088/1757-899X/705/1/012009

The 2nd factor is the frequency operation which affects the attenuation level due to rainfall. The higher frequency affects attenuation the most because the wavelengths become smaller and approach the size of the raindrop. When operate at frequency above the 10GHz, rain can give severe effect to attenuation. The width size of the raindrop in average is 1.67mm [19] for example, if high wavelength used such as 100 GHz, the raindrop size could be 3 mm, thus the attenuation of propagation signal increases due to the interchange of energy happened between the propagated wave and the raindrops.

The 3rd factor is effective path length which is the length between the source and destination that contribute to attenuation and disturbance of the propagation of waves during rainfall [19]. The effective path length is calculated based on the rainfall distribution as rainfall is not evenly distributed along the radio path length. Thus, the accurate rain attenuation is calculated as a function of the effective path length. The attenuation increases as the effective path length increases, more area is covered by rainfall between source and destination which increases the attenuation [20]. This is concluded that the total rain attenuation is directly proportional to the effective communication path length. Due to this, during heavy rainfall, only small coverage can be provided with short communication length. The path length between the source and destination should be taken into consideration when designing the 4G channel systems to reduce the effect of rain on the wave propagation.

The 4th factor is the type of polarization which affects the propagation of electromagnetic waves. The effect of the polarization has been studied by other research group such as Shresthal and Choi [21] which found that horizontal polarization resulted in 12.5 dB more compared to vertical polarization with the same frequency band of 18 GHz, the same path length, the same rain rate, and utilizing the same system. Thorvaldsen and Henne [22] also found that more attenuation is resulted in horizontal polarization with 26 GHz frequency band. The horizontal polarization provides more attenuation compared to the vertical polarization based on the higher rainfall rate and higher frequency bands. Due to these factors, the communication distance become short during rainfall and becomes severe in tropical regions that considered as heavy rainfall. As a result, the 4G coverage system will be very small compared to the cellular systems and need numerous small base stations for the respected area.

The path loss and signal coverage at certain location need to be determined first before setting up the wireless network for communication. Rain attenuation cause propagation loss, degrades the system performance and limits the usage of higher frequencies for communication system. The rain attenuation is calculated based on the amount of attenuation that occurs due to rain per unit distance. Specific rain attenuation is calculated based on the rain rate level and incident electromagnetic waves at specific location. For mathematical equation, rain attenuation, A is calculated using the following equation:

$$A = aR^b \tag{2}$$

Where a and b denote the regression factors that are subject to several factors such as operating frequency and radio wave polarization. Polarization type also gives effect to specific attenuation due to non-spherical nature of the raindrops, where normally vertical polarization attenuation level is less than horizontal waves. Value a and b can be accessed from ITU [10]. R is the rain rate in mm/hr exceeded for 0.01% of an average year which can be obtained from the meteorological data.

Other factor need to be considered for rain attenuation is effective path length, L_{eff}. The effective path length can be calculated using the following equation:

$$L_{eff} = rL \tag{3}$$

Where L_{eff} refer to effective path length, r refer to path reduction factor or distance factor introduced by ITU [11]. L refers to actual path length between source and destination. The

IOP Conf. Series: Materials Science and Engineering **705** (2019) 012009 doi:10.1088/1757-899X/705/1/012009

model proposed by ITU provides rain attenuation estimation which is widely accepted around the world. However, the work performed in [18] demonstrate that such model does not fit well in tropical region. This section emphasizes that rainfall rate is directly proportional to rain rate, operating frequency, effective path length, and horizontal polarization which result in increasing path loss, limiting the coverage area and degrading the system performance such as reliability and stability. Thus, the 4G system require a lot of small base stations to cover the affected area.

4. Potential Application in Malaysia

Cellular based UAV would be useful for border security. The discovery of human trafficking camps and graves near Wang Kelian in May 2015 was a big blow to the national boarder security. The government has set up a royal commission of inquiry (RCI) in January 2019 to investigate the matter [23]. Malaysia shares land border with three neighbouring countries namely Thailand (646.5 km), Indonesia (2,019.5 km) and Brunei (481.3 km). UAV can be used to patrol these borderlines for the benefit of all related countries. In May 2007, the Malaysian government conveyed that UAV will be used to conduct surveillance on the countrys land and sea borders [24]. In United States, the U.S Customs and Border Protection (CBP) operated the Predator B drone since 2003 to relay information on suspected border crossings [25]. In September 2017 the organisation started using smaller hand-launched drones and in summer 2016 uses small drones with face recognition feature [26]. These examples show that cellular-based UAV plays an important role to enhance security and hence need to be explored.

Other than that, cellular based UAV can be utilized for agriculture activities. Malaysia is actively involved in agriculture activities. The country is the world's number two in palm oil producer and exporter behind Indonesia. The Malaysian Palm Oil Council in its official website stated 4.49 million hectares of land in the country is under palm oil cultivation [27]. The country is also the worlds top 10 rubber producer to date. Although the focus on rubber is not as extensive as many years before there are still slightly over one million hectares of land cultivated with this crop [28]. With such land coverage, potential of UAV use can be very significant in many ways. Noting that both industries are still relying on hard labour, UAV can be utilised in various farm management applications.

Challenges may incur to implement cellular based UAV for agriculture as these agriculture areas are most likely located in rural areas. Based on Section II, 4G coverage in Malaysia is easily available in urban areas as compared to rural areas. With regards to the current deployment of 4G network in Malaysia presented in Section II, it would be desirable to focus the application of cellular based UAV in urban areas. Inspection of utilities such as power grid lines, railway track and river is important for maintenance purposes to avoid or detect any potential damage to these utilities. Some utilities however are not easily accessible directly for inspection hence the use of cellular based UAV would be useful. In 2017, it has been reported that Tenaga Nasional Berhad, TNB utilized drones for monitoring the power grid replacing the use of helicopter in the past which is costly [29]. Such inspection can be further enhanced if inspections using drones are not limited to a certain range between the drone and the controller which is possible with cellular based UAV.

5. Conclusions

Large range of UAV operation can be achieved with the aid of cellular network coverage whereby the information to control the UAV can be made remotely via the wireless network. The increased range of operation of the UAV has made it a desirable option for many applications given that the requirements for UAV operation are fulfilled. Some of the requirements are the need to operate the UAV over a minimum height, avoidance of collision with other flying vehicle and the presence of a reliable data-based coverage. In terms of the capability of providing sufficient network coverage, it has been identified that Malaysia has demonstrate its readiness in providing 4G network although it may be limited to urban areas. In order to utilize cellular based UAV, some challenges must be acknowledge and addressed which includes compensating the network coverage of cellular network which are designed to optimized terrestrial users in the first place. Additionally, since Malaysia is a tropical region, challenges also arise in combating signal attenuation due to rainfall. Given that these challenges are carefully taken into account, the cellular based UAV would be a promising option to facilitate various sectors in Malaysia which include agriculture, security and utilities maintenance.

References

[1] D. Brodnevs, *Development of a flexible software solution for controlling unmanned air vehicles via the internet* (Transport and Aerospace Engineering vol. 6, pp. 37–43) 2018
[2] Z. Sahwee, N. A. Rahman, and K. S. M. Sahari *Experimental evaluation of data fusion algorithm for residual generation in detecting uav servo actuator fault* (Int. J. Micro Air Veh, vol. 7, no. 2, p. 133145) June 2015
[3] A. Fotouhi, H. Qiang, M. Ding, M. Hassan, L. G. Giordano, A. Garcia-Rodriguez, and J. Yuan *Survey on UAV cellular communications: Practical aspects, standardization advancements, regulation, and security challenges* (IEEE Communications Surveys Tutorials, pp. 1–1) 2019
[4] H. C. Nguyen, R. Amorim, J. Wigard, I. Z. Kovcs, T. B. Srensen, and P. E. Mogensen *How to ensure reliable connectivity for aerial vehicles over cellular networks* (IEEE Access, vol. 6, pp. 12 304–12 317) 2018
[5] *3g system overview* [Online]. Available: http://xb-uav.com/getting-started/overview
[6] N. L. Mohd Kamal, K. W. Sowerby, and M. J. Neve *On the influence of the propagation environment on throughput performance in indoor wireless network* (Wireless Networks) Sept 2018
[7] N. F. B. A. Aziz and N. A. M. *Ali 4g coverage in Malaysia* (International Journal of Science and Research) Jan 2015
[8] *State of mobile networks: Malaysia - opensignal* [Online].
Available: https://www.opensignal.com/reports/2018/10/malaysia/state-of-the-mobile-network
[9] *2g / 3g / 4g coverage map, Malaysia* [Online]. Available: https://www.nperf.com
[10] M. M. Azari, F. Rosas, A. Chiumento, and S. Pollin *Coexistence of terrestrial and aerial users in cellular networks* (IEEE Globecom Workshops, pp. 1–6) Dec 2017
[11] *3rd generation partnership project; technical specification group radio access network; spatial channel model for multiple input multiple output (MIMO) simulations (Release 10)* (3GPP TR 25.996 V10.0.0 (2011-03))
[12] A. Rahmati, Y. Yapici, N. Rupasinghe, I. Guvenc, H. Dai, and A. Bhuyan, *Energy Efficiency of RSMA and NOMA in Cellular-Connected mmWave UAV Networks* 2019
[13] A. A. and G. K. *Modeling cellular-to-uav path-loss for suburban environments* (IEEE Wireless Communication Letter, vol. 7, no. 1, pp. 82–85) Feb 2018
[14] S. Qazi, A. S. Siddiqui, and A. I. *Wagan UAV based real time video surveillance over 4g LTE* (International Conference Open Source System Technology Proceeding, pp. 141–145) Dec 2015
[15] D. H. Janzen, *Tropical dry forests: The most endangered major tropical ecosystem* (Wilson, E.O. and Peters, F.M., Eds., Biodiversity, National Academy Press, Washington DC, pp. 130–144) Jan 1988
[16] P. W. Richards *The tropical rain forest: An ecological study* (Cambridge University Press, Cambridge)
[17] *Characteristics of precipitation for propagation modelling* (P Series, Radiowave propagation, Recommendation ITU-R)
[18] I. Shayea, T. Abd. Rahman, M. Hadri Azmi, and M. R. Islam Real *Measurement study for rain rate and rain attenuation conducted over 26 GHz microwave 5G link system in Malaysia* (IEEE Access, vol. 6, pp. 19 044–19 064) 2018
[19] A. Tokay, P. G. Bashor, E. Habib, and T. Kasparis, *Raindrop size distribution measurements in tropical cyclones* (Monthly Weather Review) 2008
[20] P. A. Watson *Survey of measurements of attenuation by rain and other hydrometeors* (Proceedings of the Institution of Electrical Engineers, vol. 123, no. 9, pp. 863–871) Sept 1976
[21] S. S. abd D. Y. Choi *Rain attenuation over terrestrial microwave links in south korea* (IET Microwaves, Antennas and Propagation, vol. 11, pp. 1031–1039) 2017
[22] P. Thorvaldsen and I. Henne *Outdoor transmission measurement at 26 GHz; results of a 4 years trial in prague* (1st URSI Atlantic Radio Science Conference (URSI AT-RASC), May 2015, pp. 1–9) 2015
[23] *RCI to probe mass graves in wang kelian - nation.* [Online]. Available:
https://www.thestar.com.my/news/nation/2019/01/28/rci-to-probe-mass-graves-in-wang-kelian-humantrafficking-camps-to-be-investigated-too/
[24] *Army set to use made-in-malaysia drones - nation.* [Online]. Available:
https://www.thestar.com.my/news/nation/2007/05/16/najib-army-set-to-use-madeinmalaysia-drones/

IOP Conf. Series: Materials Science and Engineering **705** (2019) 012009 doi:10.1088/1757-899X/705/1/012009

[25] C. C. H. abd J. Gertler *Homeland security: Unmanned aerial vehicles and border surveillance* (CRS Report for Congress) July 2010

[26] D. J. Bier and M. Feeney *Drones on the border - efficacy and privacy implications* (Immigration Research and Policy Brief) May 2018

[27] *The oil palm tree.* [Online]. Available: http://www.mpoc.org.my

[28] *Natural rubber statistics 2016.* [Online]. Available: http://www.lgm.gov.my/nrstat/nrstats.pdf

[29] *Exclusive: Why Malaysia uses drones to monitor power lines.* [Online]. Available: https://govinsider.asia/smart-gov/malaysia-tnb-power-lines-drones-fauzan-mohamad/

IOP Conf. Series: Materials Science and Engineering **705** (2019) 012010 doi:10.1088/1757-899X/705/1/012010

Assessment of Muscles Fatigue Based on Surface EMG Signals Using Machine Learning and Statistical Approaches: A Review

Hayder A. Yousif [1,2*], Ammar Zakaria[1], Norasmadi Abdul Rahim[3], Ahmad Faizal Bin Salleh[3], Mustafa Mahmood[1,2], Khudhur A. Alfarhan[1], Latifah Munirah Kamarudin[1], Syed Muhammad Mamduh[1], Ali Majid Hasan[2], and Moaid K. Hussain[2]

[1]Centre of Excellence for Advanced Sensor Technology (CEASTech), School of Mechatronic Engineering, Universiti Malaysia Perlis (UniMAP), Pauh Putra Campus, 02600 Arau, Perlis, Malaysia.

[2]Al-Hussain University College, Iraq.

[3]School of Mechatronic Engineering, Universiti Malaysia Perlis (UniMAP), Pauh Putra Campus, 02600 Arau, Perlis, Malaysia.

Corresponding author: hayderabdulazeez@huciraq.edu.iq

Abstract. Muscle fatigue is described by the decline in muscle maximum force during contraction. The fatigue occurs in the nervous or muscle fibre cells. The nerves produce a high-frequency signal to gain the maximum contraction, but it cannot sustain the high frequency signal for a long time, and that leads to a decline in muscle force. The surface Electromyography (EMG) is the dominant method to detect muscle fatigue because the EMG signals give more information about the muscle's activities. This review discussed the EMG signal processing and the methods of detection muscles fatigue with three domains (time domain, frequency domain, and time-frequency domain) based on EMG signals that are collected from the muscles during dynamic and static movements.

Keywords: muscles fatigue, EMG, dynamic and static movement

1. Introduction

Muscle fatigue is described by the decline in muscle maximum force during contraction [1]. Acute fatigue affects our ability to lift or move. Many types of research have carried out a further study to detect and study muscle fatigue. Currently, there are many detection methods that are applied to muscle signals in order to detect fatigue [2]. However, the surface electromyography (sEMG) is the main method to record and study the muscles functions, by recording the electrical signal of the muscles. There are also many factors that can contribute to fatigue [3], like muscle fiber composition, regulation of Ionic in the bloodstream, energy supply, neural factors, and many other factors. The studies have indicated that muscle fatigue is related to the incidence of musculoskeletal injuries occurring in sports training and competition [4]. Nyland [5] made the comment that fatigue during running might be related to lower extremity injury. Recently, evidence has shown that muscle activation patterns and kinematics are altered due to fatigue. Hence, it may raise the possibility of the occurrence of an injury to both the muscles as well as the bones injury [6]. Pinniger [7] found that after a specific hamstring fatigue exercise, the kinematic data of sprinting were changed. The kinematics could display alteration owing to the functional changes which occur in order to prevent or minimize the level of unpleasantness and the occurrence of fatigue [8]. Rodacki [9] echoed that the regulation of movement of the individual portions is controlled and ultimately decided by the force of muscle, which when sufficient fatigue has occurred may display certain changes. They also concluded that this phenomenon gave a clue as to the available of a supporting methodology to negate the losses occurred during the process of the force-development characteristics which occurs within the muscles owing to fatigue. In this paper, we will discuss the muscles fatigue during dynamic and static movements, EMG signal processing, methods of detection muscles fatigue.

2. Muscles Fatigue Measurements

The disparity in the opinion of the mechanism of muscle fatigue can be partially elaborated with the use of various models of exercise, procedures, and processes done applied to quantify fatiguing of muscles. With an objective, a quantifiable and continuous method to assess muscle fatigue, this will amplify the knowledge of ergonomics, work, and injury resulting from work physiology [4]. The single most relevant determination of fatigue is done through the measurement of force or power measurement, which is produced during the course of a voluntary effort of maximum intensity, maximal voluntary contractions (MVCs) test. In general, when the subject performs the task of interest or the fatigue task continuously, at the pre-, post- and/or the interim time point, brief MVC tests will be conducted to register the drop of maximal force output from particular muscle. This will quantify the muscle fatigue pattern in relation to the task performed. The force output decline rate measured in these MVCs tests will indicate the muscle fatigue pattern. The foundation of similar tests which assists in the direct evaluation includes force measurement equipment. Although the drop of maximal voluntary contractions output directly indicates the presence of muscle fatigue. *However,* the surface electrodes have the ability to record up the superficial muscle layer's electrical transmissions, after which it magnified and ultimately the determination of the signal power spectrum is done when the response produced in sEMG.

3. Signal Processing

Electromyogram signals are gradually becoming more essential in various fields of application, such as in prosthesis devices, human to machine interactions, clinical/biomedical, and rehabilitation devices [10]. But, distorted EMG signals possess a major challenge in further expanding the performance applications mentioned above [11]. The EMG signals that are collected from the muscles by the electrodes will have noise, where the noise has a role in the hampering of the recording signals. For this reason, the EMG signal must be filtered in a proper manner to remove any noise [12]. The frequency of noise contaminating the EMG signal may be low or high. Low-frequency noises usually come from an amplifier's direct current offsets. Normally this noise can be avoided with the use of a high pass filter, while the high-frequency noises arise from nerve conduction. The interference of high frequency comes from computers, radio broadcasts and can be removed by using a low pass filter. The high pass filter is used to eliminate the low ranged frequencies that result at the electrical signal collection. The frequencies that are passed with a filter transmission as a band are known as a passband. The frequencies that cannot be passed with a filter are known as stop band [12]. The concept of a low pass filter is opposite to that of high pass filters. This means the frequencies exceeding the cut off value are removed and those below it will be transmitted. The EMG transmission needs to remove the high frequencies and low frequencies, while a specific band of frequencies must be transmitted forward, this is done by using a specific filter called a bandpass filter. This bandpass filter permits for specific bands to be transmitted according to the range that is determined by the user, however, the bandpass filter is very suitable for EMG signals to take a specific range of frequencies that is perfect for the type of analyzing [13]. Figure 1 shows the general procedures of EMG signal processing.

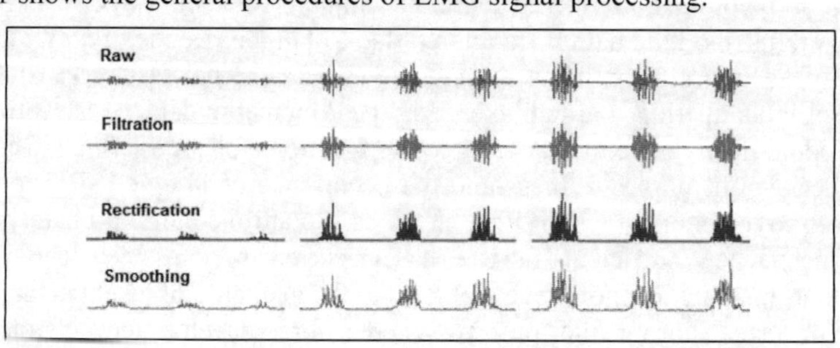

Figure1. EMG signal processing procedures included filtering, rectifying, and smoothing [14].

4. Feature Extraction

Feature extraction is a method to extract useful information from EMG signals during fatigue condition is simply explained as the diminishing of the capacity to generate force within the neuromuscular system. Generally, muscle fatigue is recognized by the EMG signal as long as there is a mutual decline in parameter examined in the frequency domain and time-frequency domain, and an increment in the parameter evaluated in the time domain [15], as shown in Figure 2. The type of EMG features can be divided into three domains: time domain, frequency domain, and time-frequency domain.

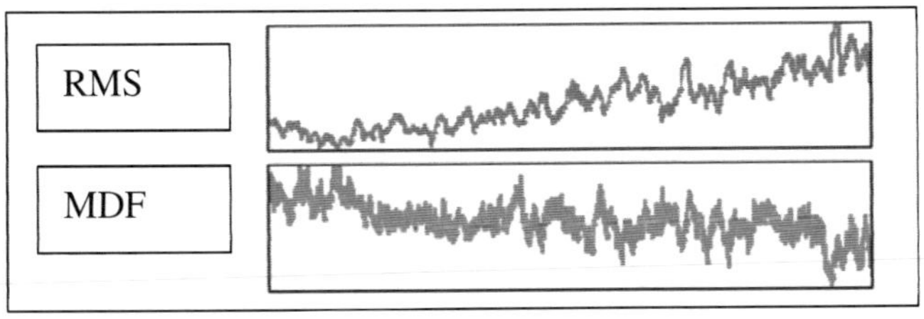

Figure 2. The increment in the pattern of RMS and decline in the pattern of MDF features for the EMG signal during muscle fatigue [24].

4.1. Time Domain Features

The signal recorded in the time domain is filtered to eliminate and minimize the noise, crosstalk, and line noise. The fatigue is related to the increment of the EMG amplitude [16]. The muscle fatigue detection in the time domain can be estimated by extracting features like root mean square (RMS) and mean absolute value (MAV). The detection of fatigue by observing the amplitude of the EMG signal is rarely used, and it will be more accurate if it is combined with other methods such as spectral analysis [17] to get better detection and more reliability. Later was discovered that the connection occurring between signal amplitude and force generated within the muscle are different among fatigue protocol [18]. However, there is a method that use multiple time windows (MTW) to get over this issue, by using different functions like Multiple hamming windows (MHW), multiple Slepian windows (MSW) and Multiple trapezoidal windows (MTRW) [19]. Thus, the signal processing to be combined with another feature extraction for better detection.

4.2. Frequency Domain Features

The shift towards lower frequencies of the EMG frequency spectrum calculated by fast Fourier transform (FFT) method [20]. Where the most common procedure is to monitor the relative changes in the mean (MNF) and median power frequencies (MDF) and to relate these measures to the initial value or non-fatigue state mean and median power frequencies. Kaljumae [21] used the phenomena of the shift of sEMG MNF and MDF towards lower frequencies under the isometric condition to document the change of fatigability of vastus lateralis (VL) and vastus medialis (VM) after 10 weeks of bicycle-ergometer training program in male subjects. Gerdle and Fugl-Meyer [22] had 10 healthy women perform repetitive maximal isokinetic plantar flexion. The sEMG MNF measured on the soleus, gastrocnemius medialis (GM) muscle along with the gastrocnemius lateralis (GL) were decreased in parallel with the work output within the first 50 contractions which was later superseded by constant levels in the resultant 50 contractions. So [23] examined the surface sEMG of rectus femoris(RF) and biceps brachii (BB) collected from 18 athletes when they performed maximal extension of the knee and flexion of the elbow exercises. The calculated MPF by FFT was found to reduce along with the reduction of work output and they suggested that an approximate reduction of 37.2% in relative MPF might be handy in performing the role of a parameter for assessing the level of muscle fatigue.

However, the MDF and MNF are very important to detect the muscles fatigue based on the power spectrum, but it will be more reliability and clarity if it combined with the time domain features to get more information about the function of the muscles.

4.3. Time Frequency Domain Features

Muscle fatigue is related to both EMG signal characteristics each in a deferent way, that leads to the Joint analysis of EMG spectrum and amplitude (JASA). This analysis has 4 situations to determine force and fatigue, first when both amplitude and spectrum increases that's mean force increase, second when both amplitude and spectrum decreases that's mean force decrease, third when amplitude increase and spectrum decrease that's mean fatigue, fourth when amplitude decrease and spectrum increased that's mean recovery [4]. Time-frequency methodologies are utilized to extensively study signals in which there is considerable variation of the frequency content with the change in time, which is similar to the situation seen in muscle fatigue. Short-Time Fourier Spectrum (STFT) is considered as a relatively common time-frequency distribution. However, the time-frequency functions can be utilized in the analysis of sEMG signals arising during a localized fatiguing of a muscle. Research conducted by Davies and Reismann clearly states that the STFT has the ability to most accurately depict spectrum compression which occurs during the time of muscle fatigue. In a study conducted on the relationship which exists between STFT and continuous wavelet transforms (CWT) to analyze EMG signals from the back and hip muscles during fatiguing isometric contractions, it was discovered that both methods ultimately give out the same output on EMG spectral variables [2]. The discrete wavelet transform (DWT) has been used to indicate the low-level muscles fatigue of brachioradialis (BRD), biceps brachii long head (BBL), and biceps brachii short head (BBS) when the subjects performing force exertion [25]. The muscle fatigue is seen in children suffering from cerebral palsy [26]. The muscle fatigue that occurs to PC gamers after continuous playing [27], and in helicopter pilots who are suffering from lower back pain [28]. Multiple classical, as well as newly designed modern methods of signal processing modalities, are used [17] these include RMS, zero-crossing rate (ZCR), averaged instantaneous frequency, wavelet analysis, fractal analysis, and both MNF and MDF. The instantaneous mean and median frequency (IMNF and IMDF) are added so as to complete the need [29] through the utilization of time-frequency or time-scale approaches, such as STFT [17], Wigner distribution (WD), Choi-Williams distribution (CWD) [30], time-varying autoregressive approach (TVAR) [31], and CWT [32].

5. Methods of Detection Muscles Fatigue

There are many non-invasive methods to detect muscle fatigue, the main methods are surface Electromyography (sEMG) and Mechanomyography (MMG). The EMG records the electrical activity signal from the muscle, while the MMG record the mechanical activity of the muscle [33]. There are many other methods but less use in clinical or research in fatigue like sonomyography (SMG) that use ultrasounds to detect fatigue and controlling prosthesis [34], near-infrared spectroscopy (NIRS) that use near-IR light to measure the hemoglobin absorption properties, and Acoustic myograph (AMG) that record the sound of muscles it's particular application of MMG. Each method tries to record and analyze one or multi symptoms, signals, and characteristics of muscle. However, the better method for detection muscles fatigue is surface EMG. There are many researchers collected the EMG signals from subject's muscle like BF, Medial Hamstrings (MH), GM, RF, Tibialis Anterior (TA), GL, Medial Gastrocnemius (GMS), semimembranosus (SEMBS), VM, GA, BB, Triceps Brachii (TB), and VL) to use it as indicator for muscles fatigue index based on using the machine learning (regression) or statistical approach (ANOVA test), and some of them based on the pattern of the features to indicate the muscles fatigue as shown in Table 1.

IOP Conf. Series: Materials Science and Engineering **705** (2019) 012010 doi:10.1088/1757-899X/705/1/012010

Table 1. Literature review of the methods for detection muscles fatigue during contraction based on EMG signals.

References	Muscles	Contraction Protocol	Analysis Methods
Marco [36]	RF, BF, GM	Cycling 30 minutes constant	IMNF
Christin [37]	GM, BF, VL, RF, TA, GA	Incremental running test on a treadmill	iEMG
Kumar [38]	RF, GL, GM, VL, VM	Cycling for prolonged constant	RMS, MF
Bing Yu [39]	SEMBS, BF	Running over ground with maximal speed.	EMG Peak, ANOVA
Takayuki [40]	BB, TB	Utilized the dumbbell as a burden	FFT, MPF
Kenichi [41]	Long Head BB	Exercise for 8 minutes during low-level isometric contraction.	recurrence quantification analysis
Andrzej [42]	RF, BF	Run 400m on tartan athletic track with a different intensity.	MPF, Linear Regression
Ahamed [43]	Middle BB	Eccentric and concentric contractions	RMS
Tatsushi [44]	RF, TA, BF, GM	Cycling with 70 RPM, 100 watts	RAW EMG and statistical
Rubana [45]	Right RF	During walking.	DWT
Kiran [46]	TB	Dumb-bell curl exercise.	FFT & Spectral Density
Crozara [47]	RF, VL, BF, GL	Incremental running on a treadmill	RMS, Linear Regression
Maner [48]	RF, VL, VM	5km running on variables surface	iARV, iMAV, iRMS, WL, IMNF, IMDF
Ridzuan [49]	GA	Running on a treadmill for 30min	MDF, Linear regression
Mastaler [50]	RF, BF, TA, GAS	Running, 200m/outdoor and 400m/ treadmill.	MPF, Linear Regression

6. Discussion

The sEMG is the most dominant method used for muscle function and fatigue detection, where there are many papers describing the fatigue and detecting it. However, the most used analyzing methods are RMS, MNF, MPF, WL, MDF, iMAV, iRMS, IMNF, and IMDF and then applied the statistical analysis or machine learning (ANOVA, Regression Line) based on the linear regression slope values where these values describe the muscles fatigue index as shown in Figure 3. In the time domain the fatigue is related to increment of the EMG amplitude [16], in the frequency domain the shift towards lower frequencies, and in time-frequency domain when the amplitude increase and spectrum decrease that's mean fatigue [4]. Indeed, the better methods that used for detection muscles fatigue are MDF and MNF were based on the power spectrum analysis of the EMG signals that result from the FFT, because the spectral analysis of the data be more reliable and give more information about the muscles functions compared with the other methods. But it will be more reliability and clarity if it combined with the time domain features to get more information about the function of the muscles and to avoid losing the information. However, the most researches detect fatigue in GM, RF, BF, GMS, GL, VL, and VM muscles because they are easy to implement the electrodes on it and they are superficial muscles, also can control it efficiently. The sEMG can detect fatigue during dynamic and static contraction.

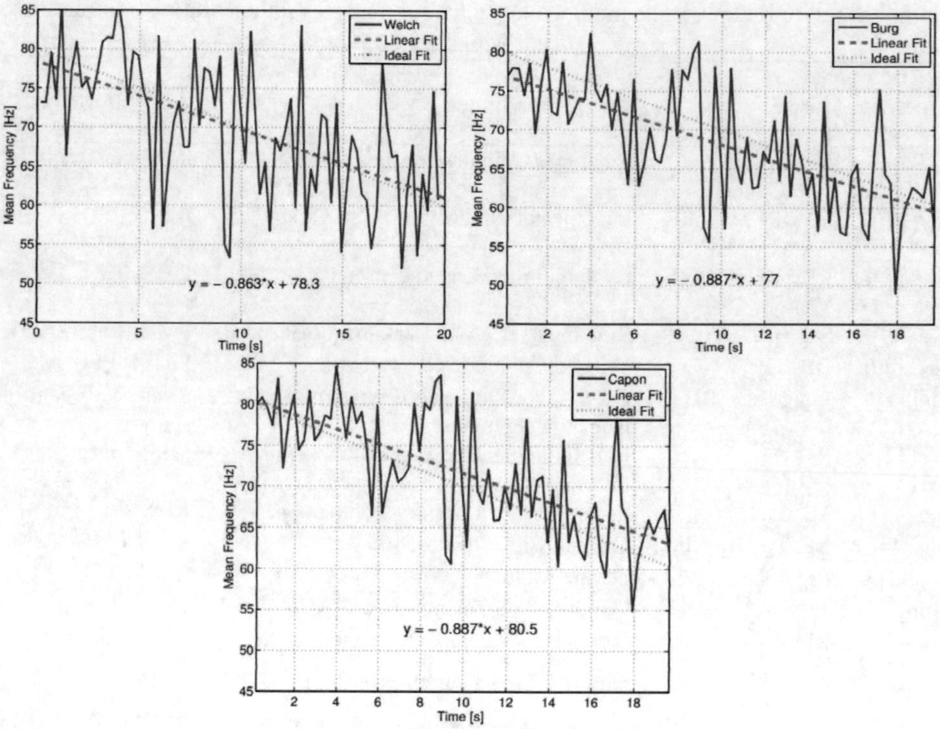

Figure 3. Muscle fatigue detection based on frequency decline with applying linear regression [35].

7. Conclusion

Muscle force production involves a sequence of events, extending from cortical excitation to motor unit activation to excitation–contraction coupling, and ultimately leading to muscle activation. Changes at any level in this pathway, including changes in the nervous, ion, vascular, and energy systems, impair force generation and contribute to the development of muscle fatigue. However, the current states for fatigue detection are still in the research area, therefore must be real application are found for fatigue. Even though the sEMG electrodes and MMG transducers are not meant to be for long time monitoring. Also, very little real time myography fatigue detection is done, there should be some trials in fatigue detection and classification using Neural Network for fast accurate fatigue detection. We believe that the combination of two methods (features) will provide more useful information, compared with the information obtained from uni-method. Therefore, the combination of multi features should be considered to develop and improve the performance of the methods that related with the muscles fatigue assessment.

References

1. González-Izal M, Malanda A, Navarro-Amézqueta I, Gorostiaga EM, Mallor F, Ibañez J, et al. EMG spectral indices and muscle power fatigue during dynamic contractions. J Electromyogr Kinesiol [Internet]. 2010;20(2):233–40.
2. Al-Mulla MR, Sepulveda F, Colley M. A review of non-invasive techniques to detect and predict localised muscle fatigue. Sensors. 2011;11(4):3545–94.
3. Bogdanis GC. Effects of physical activity and inactivity on muscle fatigue. Front Physiol. 2012;3 MAY(May):1–16.
4. Jonkers I, Nuyens G, Seghers J, Nuttin M, Spaepen A. Muscular effort in multiple sclerosis patients during powered wheelchair manoeuvres. Clin Biomech [Internet]. 2004;19(9):929–38.
5. Nyland A, Shapiro R, Stine RI, Horn TS, Ireland ML. Relationship of Fatigued Run and Rapid Stop to

Ground Reaction Forces , Lower Extremity Kinematics , and Muscle Activation. J Orthop Sport Phys Ther. 1994;20(3).
6. Brereton LC, McGill SM. Effects of physical fatigue and cognitive challenges on the potential for low back injury. Hum Mov Sci. 1999;18(6):839–57.
7. Pinniger GJON, Steele JR, Groeller H. Does fatigue induced by repeated dynamic efforts affect hamstring muscle function? Med Sci Sport Exerc [Internet]. 2000;32(3):647–53.
8. Van Dieën, J. H., Toussaint, H. M., Maurice, C., & Mientjes M. Fatigue-related changes in the coordination of lifting and their effect on low back load. J Mot Behav. 2005;28(4):304–314.
9. Rodacki ALF, Fowler NE, Bennett SJ, Bota J. Multi-segment coordination : fatigue effects. Med Sci Sport Exerc. 2001;33(5):1157–67.
10. Chowdhury RH, Reaz MBI, Alauddin M, Mohd B, Bakar AAA, Chellappan K, et al. Surface Electromyography Signal Processing and Classification Techniques. Sensors. 2013;12431–66.
11. Reaz MBI, Hussain MS. Techniques of EMG signal analysis : detection , processing , classification and applications (Correction). Biol Proced Online. 2006;8(April 2005):2006.
12. Kumar JS, Kannan. Human Hand Prosthesis Based On Surface EMG Signals for Lower Arm Amputees. Int J Emerg Technol Adv Eng. 2013;3(4).
13. De Luca CJ, Donald Gilmore L, Kuznetsov M, Roy SH. Filtering the surface EMG signal: Movement artifact and baseline noise contamination. J Biomech [Internet]. 2010;43(8):1573–9.
14. By Leandro Ricardo Altimari, José Luiz Dantas, Marcelo Bigliassi, Thiago Ferreira Dias Kanthack AC de M and TA. Influence of Different Strategies of Treatment Muscle Contraction and Relaxation Phases on EMG Signal Processing and Analysis During Cyclic Exercise. Comput Intell Electromyogr Anal Perspect Curr Appl Futur Challenges IntechOpen. 2012;
15. Roman-liu D, Tokarski T, Wo K. Quantitative assessment of upper limb muscle fatigue depending on the conditions of repetitive task load. 2004;14:671–82.
16. Kallenberg, L. A., Schulte, E., Disselhorst-Klug, C., & Hermens HJ. Myoelectric manifestations of fatigue at low contraction levels in subjects with and without chronic pain. J Electromyogr Kinesiol. 2007;17(3):264–274.
17. Cifrek M, Medved V, Tonković S, Ostojić S. Surface EMG based muscle fatigue evaluation in biomechanics. Clin Biomech. 2009;24(4):327–40.
18. Dideriksen, J. L., Farina, D., Baekgaard, M., & Enoka RM. An integrative model of motor unit activity during sustained submaximal contractions. Am J Physiol Circ Physiol. 2010;
19. Venugopal G, Navaneethakrishna M, Ramakrishnan S. Extraction and analysis of multiple time window features associated with muscle fatigue conditions using sEMG signals. Expert Syst Appl [Internet]. 2014;41(6):2652–9.
20. Petrofsky, J. S., Glaser, R. M., Phillips, C. A., Lind, A. R., & Williams C. Evaluation of the amplitude and frequency components of the surface EMG as an index of muscle fatigue. Ergonomics. 2000;25(3):213–223.
21. Kaljumäe, U., Hänninen, O., & Airaksinen O. Knee extensor fatigability and strength after bicycle ergometer training. rchives Phys Med Rehabil. 2003;75(5):564–7.
22. Gerdle, B., & Fugl Meyer AR. Is the mean power frequency shift of the EMG a selective indicator of fatigue of the fast twitch motor units? Acta Physiol Scand. 1992;145(2):129–138.
23. So, R., Chan, K. M., & Siu O. EMG power frequency spectrum shifts during repeated isokinetic knee and arm movements. Res Q Exerc Sport. 2002;73(1):98–106.
24. Guo W, Sheng X, Zhu X. Assessment of muscle fatigue by simultaneous sEMG and NIRS: From the perspective of electrophysiology and hemodynamics. In: International IEEE/EMBS Conference on Neural Engineering, NER. 2017. p. 33–6.
25. Zhang G, Morin E, Zhang Y, Etemad SA. Non-invasive detection of low-level muscle fatigue using surface EMG with wavelet decomposition. In: Conference proceedings : . Annual International Conference of the IEEE Engineering in Medicine and Biology Society IEEE Engineering in Medicine and Biology Society Annual Conference. 2018. p. 5648–51.
26. Leunkeu, A. N., Keefer, D. J., Imed, M., & Ahmaidi S (2010). Electromyographic (EMG) analysis of quadriceps muscle fatigue in children with cerebral palsy during a sustained isometric contraction. J Child Neurol. 2010;25(3):287–93.
27. Oskoei MA, Member S, Hu H, Member S. Support Vector Machine-Based Classification Scheme for Myoelectric Control Applied to Upper Limb. 2008;55(8):1956–65.
28. Balasubramanian, V., Dutt, A., & Rai S. Analysis of muscle fatigue in helicopter pilots. Appl Ergon.

2011;42 (6):913–918.

29. Roy, S. H., Bonato, P., & Knaflitz M. EMG assessment of back muscle function during cyclical lifting. J Electromyogr Kinesiol. 1998;8(4):233–245.
30. Knaflitz M, Bonato P. Time-frequency methods applied to muscle fatigue assessment during dynamic contractions. 1999;9:337–50.
31. Zhang, L., Xiong, G., Liu, H., Zou, H., & Guo W. Time-frequency representation based on time-varying autoregressive model with applications to non-stationary rotor vibration analysis. Sadhana. 2010;35(2):215–232.
32. Karlsson S, Member A, Yu J, Akay M, Member S. Time-Frequency Analysis of Myoelectric Signals During Dynamic Contractions : A Comparative Study. 2000;47(2):228–38.
33. Sarillee M, Hariharan M, Anas MN, Omar MI, Oung AMNQW. Non-invasive Techniques to Assess Muscle Fatigue using Biosensors : A Review. 2014;187–92.
34. Shi J, Chang Q, Zheng Y-P. Feasibility of controlling prosthetic hand using sonomyography signal in real time: Preliminary study. J Rehabil Res Dev [Internet]. 2010;47(2):87. 35. Alty SR, Georgakis A. Mean Frequency estimation of surface EMG signals using filterbank methods. Eur Signal Process Conf. 2011;(Eusipco):1387–90.
35. Alty SR, Georgakis A. Mean frequency estimation of surface EMG signals using filterbank methods. In2011 19th European Signal Processing Conference 2011 (pp. 1387-1390). IEEE.
36. Knaflitz, M., & Molinari F. Assessment of muscle fatigue during biking. In: IEEE Transactions on neural systems and rehabilitation engineering. 2003. p. 17–23.
37. Hanon C, Thépaut-Mathieu C, Vandewalle H. Determination of muscular fatigue in elite runners. Eur J Appl Physiol. 2005;94(1–2):118–25.
38. Singh, V. P., Kumar, D. K., Polus, B., & Fraser S. Strategies to identify changes in SEMG due to muscle fatigue during cycling. J Med Eng Technol. 2007;31(2):144–151.
39. Yu B, Queen RM, Abbey AN, Liu Y, Moorman CT, Garrett WE. Hamstring muscle kinematics and activation during overground sprinting. J Biomech. 2008;41(15):3121–6.
40. Sakurai, T., Toda, M., Sakurazawa, S., Akita, J., Kondo, K., & Nakamura Y. Detection of muscle fatigue by the surface electromyogram and its application. In: In 2010 IEEE/ACIS 9th International Conference on Computer and Information Science. 2010.
41. Ito, K., & Hotta Y. EMG-based detection of muscle fatigue during low-level isometric contraction by recurrence quantification analysis and monopolar configuration. In: In 2012 Annual International Conference of the IEEE Engineering in Medicine and Biology Society. 2012. p. (4237-4241).
42. Mastalerz A, Gwarek L, Sadowski J, Szczepański T. The influence of the run intensity on bioelectrical activity of selected human leg muscles. Acta Bioeng Biomech. 2012;14(2):101–7.
43. Ahamed, N., Sundaraj, K., Ahmad, R. B., Rahman, M., Islam, A., & Ali A. Non-invasive electromyography-based fatigue detection and performance analysis on m. biceps brachii muscle. In: In 2012 IEEE International Conference on Control System, Computing and Engineering. 2012. p. 302–6.
44. Tokuyasu, T., Kushizaki, S., Matsumoto, S., & Kitawaki T. Development of automatic positioning system for bicycle saddle based on lower limb's EMG signals during pedaling motion. In: 2013 IEEE 6th International Workshop on Computational Intelligence and Applications (IWCIA). 2013. p. (pp. 27-32).
45. Chowdhury, R. H., Reaz, M. B. I., & Ali MAM. Determination of muscle fatigue in SEMG signal using empirical mode decomposition. In: In 2014 IEEE Conference on Biomedical Engineering and Sciences (IECBES). 2014. p. (pp. 932-937).
46. Marri, K., Jose, J., Karthick, P. A., & Ramakrishnan S. Analysis of fatigue conditions in triceps brachii muscle using sEMG signals and spectral correlation density function. In: In 2014 International Conference on Informatics, Electronics & Vision (ICIEV). 2014. p. (pp. 1-4).
47. Crozara LF, Castro A, De Almeida Neto AF, Laroche DP, Cardozo AC, Gonçalves M. Utility of electromyographic fatigue threshold during treadmill running. Muscle and Nerve. 2015;52(6):1030–9.
48. Manero RBR, Shafti A, Michael B, Grewal J, Fernandez JLR, Althoefer K, et al. Wearable embroidered muscle activity sensing device for the human upper leg. Proc Annu Int Conf IEEE Eng Med Biol Soc EMBS. 2016;2016–Octob:6062–5.
49. Ridzuan N, Azaman A, Soeed K, Zulkapri I, Aa W. Muscle Fatigue Evaluation using Non-invasive Infrared Thermography Technique With Assisted Electromyography : A Preliminary Study. 2017;251–4.
50. Mastalerz A, Urbanik C, Iwańska D, Tabor P, Karczewska-Lindinger M, Grabowska O. P 034 - the effect of treadmill geometry on muscle fatigue. Gait Posture. 2018;65(xxxx):293–4.

IOP Conf. Series: Materials Science and Engineering **705** (2019) 012011 doi:10.1088/1757-899X/705/1/012011

Development of CAN Bus Converter for On Board Diagnostic (OBD-II) System

M A C Din, M T A Rahman*, H A Munir, A Rahman and A F A Hamid

School of Mechatronic Engineering, Universiti Malaysia Perlis (UniMAP), Kampus Pauh Putra, 02600 Arau, Perlis, Malaysia

*Corresponding author: tasyrif@unimap.edu.my

Abstract. CAN-bus technology is adapted in various vehicles to reduce the wiring harness and controls the vehicle exploitation physical system. In this paper, Control Area Network (CAN)-bus protocol is enforced employing a Single Board Computer (SBC) and access with each other by set a sender and receiver to get the data transmission. A special CAN-bus application layer protocol is intended for the high reliable and high period management network of the vehicle system. However not all microcontroller that used for Computer module in vehicle network had CAN-bus communication port. Thus the CAN-bus Converter is used as tool for converting data from a computer into an understandable language. The CAN-bus converter can also be used in various ways to obtain the final results of the data sent to it. Although different methods may be used, the most appropriate method cannot be identified. This project is therefore designed to find and evaluate which method is suitable for obtaining accurate data results and for avoiding abnormal data and data transfer speed.

1. Introduction

The Controller Area Network (CAN) was developed in the early 1980s by Bosch as a serial bus communication protocol. It sets the standard for efficient and reliable communication in real - time applications between sensor, actuator, controller and other nodes [1]. On-board diagnostics (OBD-II) is an automotive term that refers to the ability of a vehicle to diagnose itself and report. OBD systems provide access to the status of different vehicle subsystems to the vehicle owner or repair technician. OBD System. OBD is a computer-based system originally designed to reduce emissions by monitoring the performance of major components such as the engine, catalytic converter, particulate filter, oxygen sensor, emission control system, fuel system, and exhaust gas recirculation (EGR) [2]. In some cases, some of the microcontroller cannot communicate with can bus due to the communication converter protocol. It means that communication converter protocol consists of variety of protocol such as serial, serial peripheral interface (SPI) and inter-integrated circuit (I2C).

Therefore, this study will analyse the data packet transferring over conversion of difference communication protocol. Another case is mostly the converter of integrated circuit (IC) come in the form of chips and not fully support the protocol to communicate, so that the project needs to develop electronic circuit for data protocol conversion. Since this project will be implemented at the car, there are lots of method that can be used to get the data from electronic control unit (ECU) to on board diagnostic (OBD-II). Hence, this project needs to determine best protocol conversion method. This experiment focuses on communication protocol which is to develop can bus converter that can communicate with communication converter that consist of Serial and Serial Peripheral Interface

IOP Conf. Series: Materials Science and Engineering **705** (2019) 012011 doi:10.1088/1757-899X/705/1/012011

(SPI). This study will used two types of communication converter which are MCP2515 and CAN-bus shield V2 (MCP2551) to test the performance on latency and data loss of this project and determine which type of communication converter provide more efficient result.

2. CAN-bus protocol framework

The frame is the data link layer transmission unit [3]. In the CAN protocol, it is consists of four different types of frames [4-5]. The data frame and remote frame send messages during the transmission of error frames and overloading frames. The data frame is the format of the standard message. As shown in Fig. 1 Start-of-Frame (SOF's) indicates the frame start. The field of arbitration contains an identifier and Remote transmission request (RTR) bit indicating whether the message is a remote application [6]. The control field specifies the length of data while the current content is stored in the data field [7]. A Cyclic redundancy check (CRC) sequence is included in the CRC field to verify the integrity of the message received [8]. The Acknowledge slot (ACK) field acknowledges that at least one of the nodes successfully receives the transmitted message.

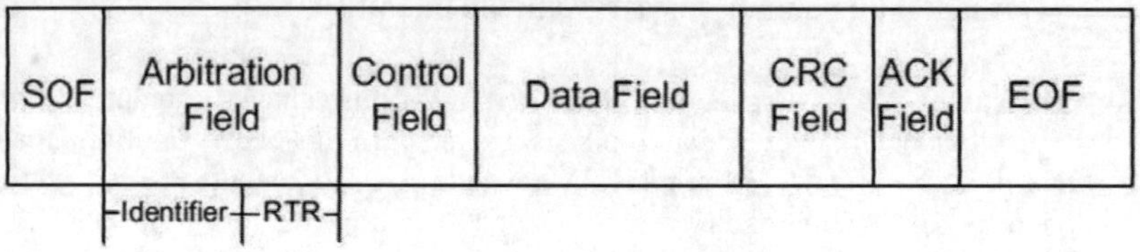

Figure 1. Frame Format.

The End-of-Frame (EOF) marks the end of the seven-bit frame. The remote frame is a message sent by any node to request a different node to send the identical identifier to a data frame [9]. Data frame format differs in that the RTR bit marks 0 and the data field is empty.

3. Experiment Setup

To test which CAN-bus communication method is better, a hardware system needs to be setup as shown Figure 2. Where the OBD-II port in vehicle is connected to CAN-bus converter module and Microcontroller Arduino Uno. Then for the firmware to test performance of CAN-bus communication with different method, algorithm flow as displayed in Figure 6 is needed based on four stages to communicate the data between OBD-II and Arduino Uno. Stage 1 focuses in examining three bitrate and design electronic circuit for data transferring from ECU to sensor node device. Then, for the second stage, mathematical algorithm for converting different base number in data packet is applied. Next, the latency and data loss is analysed based on two different CAN-bus converter. Lastly, from the analysis of the result, high performance communication device for connecting ECU to sensor node is determined.

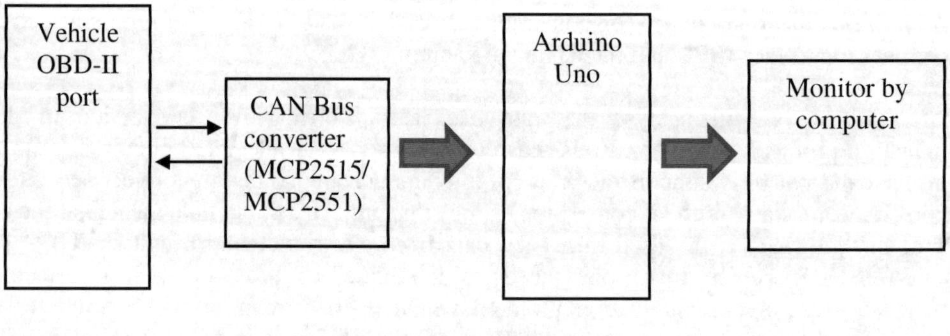

Figure 2. Communication node hardware setup that will receive data from communication node.

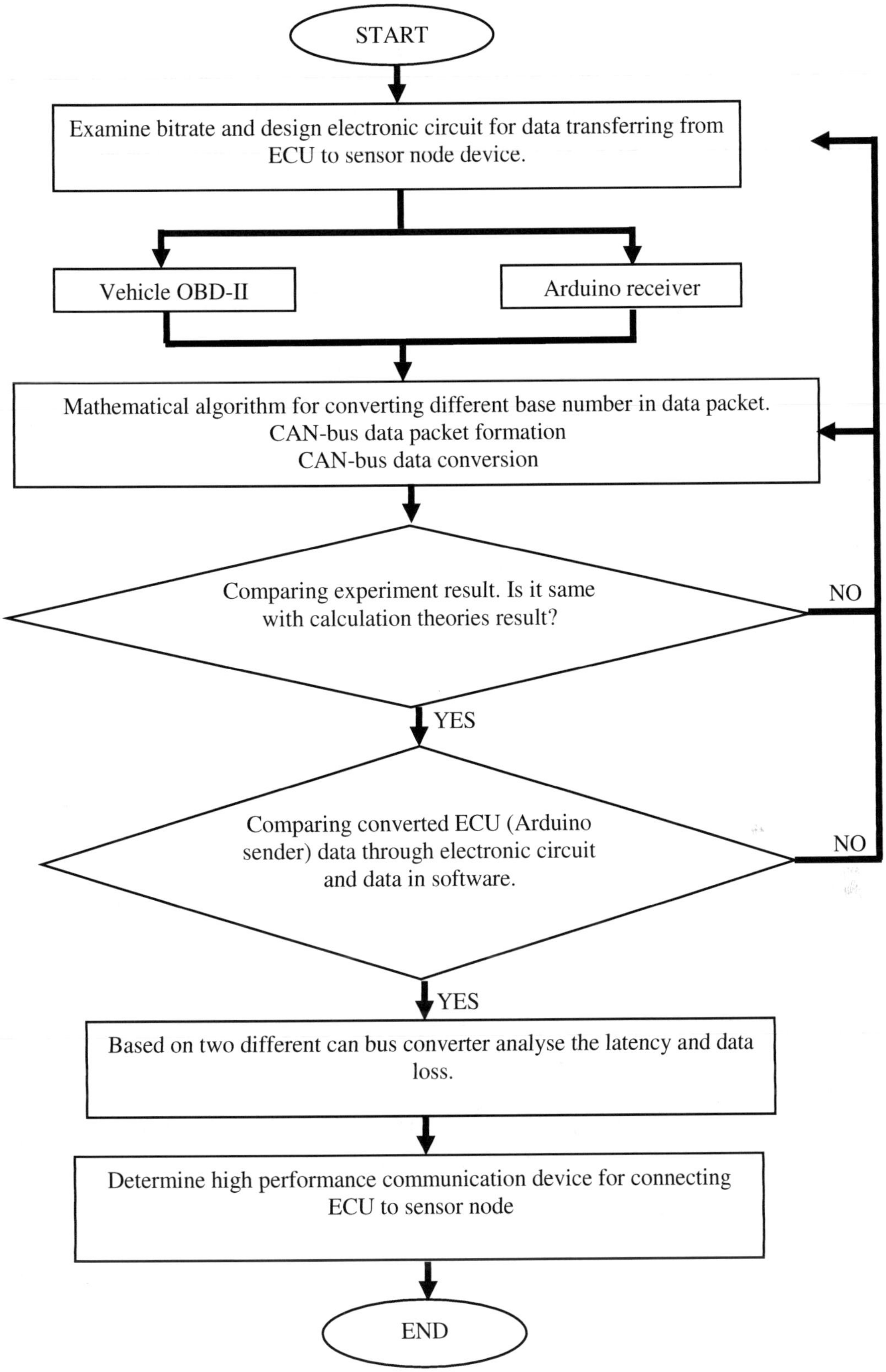

Figure 3. Flowchart.

4. Experiment Result and Discussion

Based on 20 samples of data, the latency result from communication converter experiment is separate to three different baud rate 250kbps, 500kbps and 1000kbps as in Figure 4. The result shows decreasing trend of latency corresponding to the increasing of baud rate.

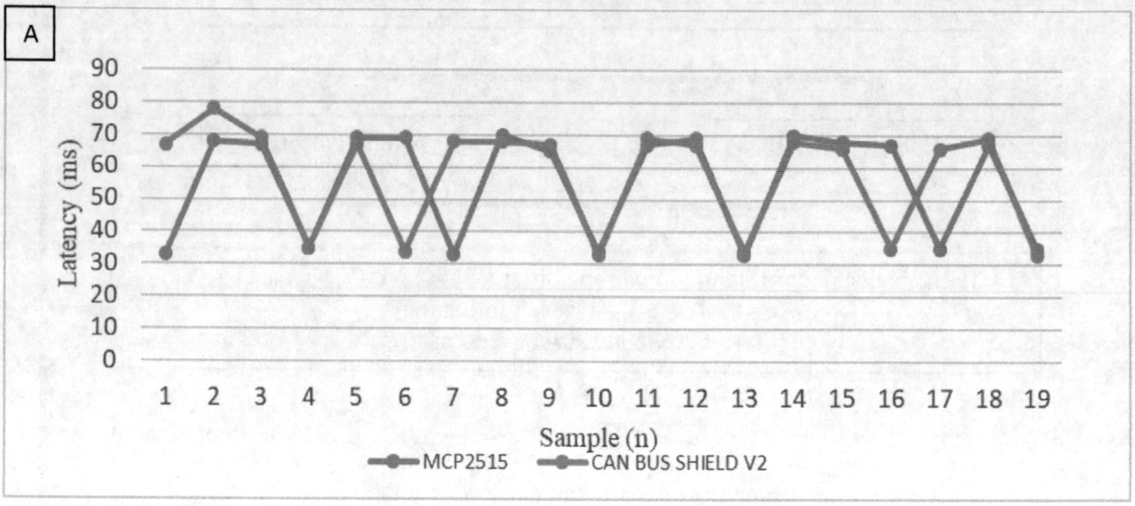

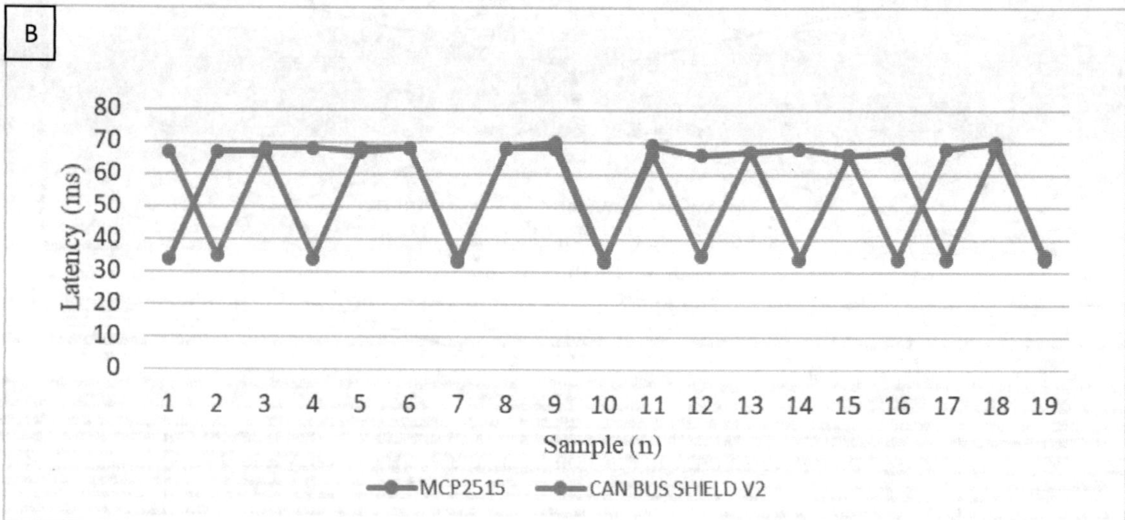

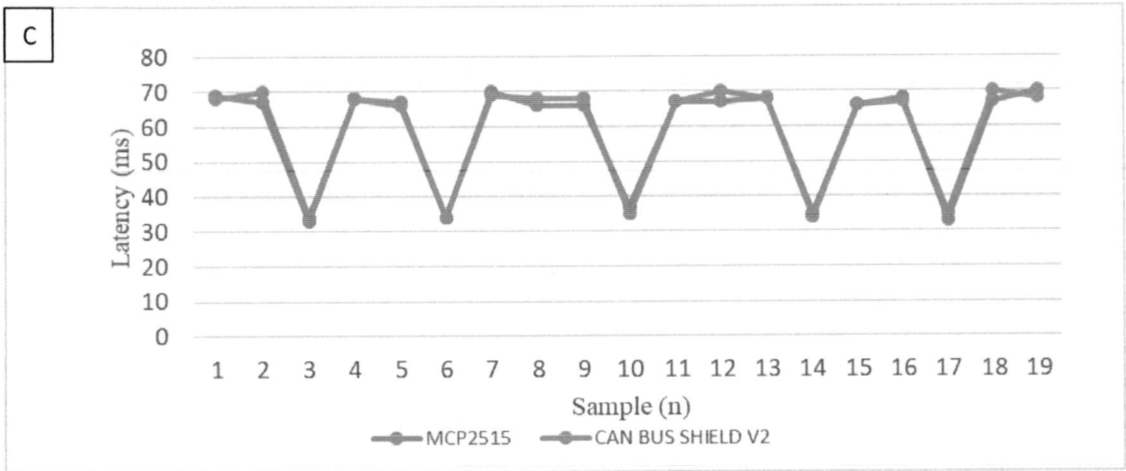

Figure 4. A, B and C are latency data for two different communication converter with different of baud rate at A=250kbps, B=500kbps and C=1000kbps

Based on 20 samples of data, the data loss result from communication converter experiment is separate to three different baud rate 250kbps, 500kbps and 1000kbps as in Figure 5. The result shows decreasing trend of percentage of data loss corresponding to the increasing of baud rate.

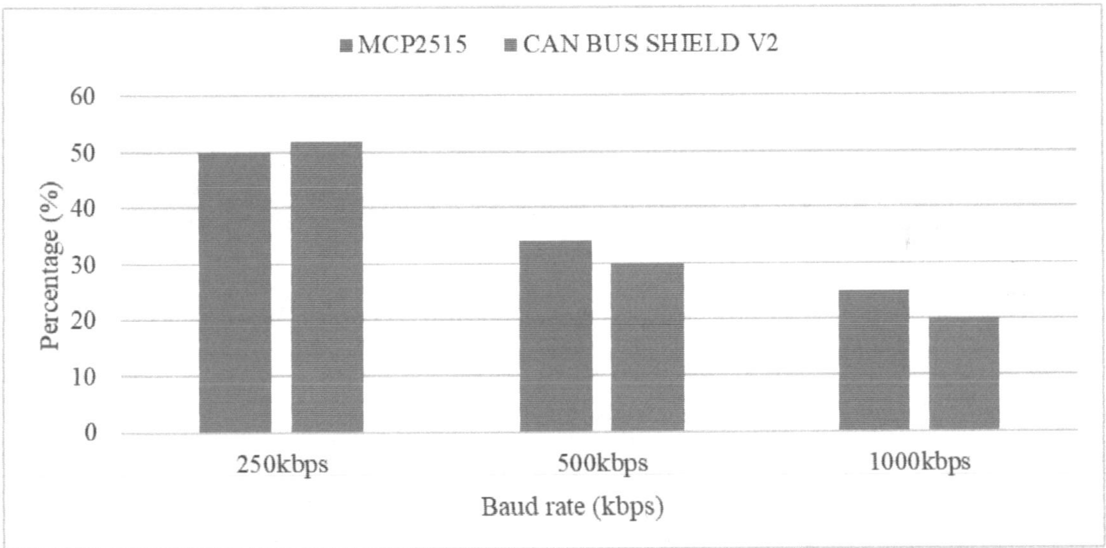

Figure 5. Percentage of data loss between MCP2515 and CAN Bus shield V2.

5. Conclusion

From the experiment result it can be concluded that the best way to collecting data from ECU is by using the CAN-bus shield V2 CAN-bus converter board due to data clocking for every frame of data packet. From the CAN-bus data collection it can be concluded that the use of CAN-bus shield V2 CAN-bus converter will increase system performance due to fast clocking data packet collection. It's easier to segregate the data into sub packet in two byte of data for data conversion method. Then for the CAN-bus data conversion it's more accurate by using string to long integer converter because it will directly convert the hex data without SOF. Besides that, from the result of latency between

MCP2515 and CAN-bus shield V2 showed that CAN-bus shield V2 converter have average low latency than MCP2515. From the study, it has been proven that CAN-bus shield less data loss then MCP2515 due to the previous calculation result.

Acknowledgements

We would like to extend our sincere gratitudes for the funding provided by the Universiti Malaysia Perlis (UniMAP) (Geran Penyelidikan Khas - No.9004-00064) and UniMAP Automotive Racing Team, which enable this research to be carried out.

References

[1] Mi-Jin K, Jong-Wook J 2015 A Study on In-Vehicle Diagnosis System using OBD-II with Navigation *IJCSNS Int. J. Comput. Sci. Netw*

[2] Azarul F A H, Md T A R, S. F. Khan, Abdul H A, Muhajir A R, Norasmadi A R, Muhammad H N I, and Aminurasyid N 2017 Connected car: Engines diagnostic via Internet of Things (IoT*), J. Phys. Conf. Ser.* **908**

[3] Yalian Y, Boalin C, Lin S, and Datong Q 2013 Research and development of hybrid electric vehicles can-bus data monitor and diagnostic system through OBD-II and android-based smartphones, *Adv. Mech. Eng.*

[4] Can P, Jian G, Longfei Z, Jianqi S, Huibiao Z, and Xintun Z, 2014 Modeling and verification of CAN bus with application layer using UPPAAL, *Electron. Notes Theor. Comput. Sci.,* **309**

[5] Li R, Wu J F, Wang H, and Li G 2010 Design method of CAN BUS network communication structure for electric vehicle, *Int. Forum Strateg. Technol.*

[6] Claudia C, Antonio I, Antonella M, Stefeno Y P, and Giuseppe R 2012 An integrated smartphone-based platform to support traffic management applications *First International Workshop on Vehicular Traffic Management for Smart Cities (VTM)*

[7] Johney H 2013 Integrating CAN Bus Communication Interfaces in Various Industries ICPDAS **39**

[8] Wang X J and Guo W C 2009 The Design of RS232 and CAN Protocol Converter Based on PIC MCU *Computer and Information Science* **2**

[9] Kamal I and Trushna S K, Savita S P and Kirti C 2016 Design & Implementation of CAN Bus for Intelligent Vehicle using Sensors System *International Journal for Innovative Research in Science & Technology* **2**

[10] Bonnie C 2013 Ease into the Flexible CANbus Network ADN004 *Microchip Technology Inc*

[11] Vaishya R and Parikshit M 2013 Study of Response Time Analysis *ICS India*

[12] Sheeba J and Vaishnavi T 2018 Review on Range of Controller Area Network (CAN) Protocol In Different Automobile Appliances *International Journal of Pure and Applied Mathematics* **120**

IOP Conf. Series: Materials Science and Engineering **705** (2019) 012012 doi:10.1088/1757-899X/705/1/012012

Analysis on LoRa RSSI in Urban, Suburban, and Rural Area for Handover Signal Strength-Based Algorithm.

Z A Tan, *M T A Rahman, A Rahman, A F A Hamid, N A M Amin, H A Munir and M M M Zabidi

School Mechatronic Engineering, Universiti Malaysia Perlis (UniMAP), Kampus Pauh Putra, 02600 Arau, Perlis, Malaysia

*E-mail: tasyrif@unimap.edu.my

Abstract. This paper focus on analysis on Long Range (LoRa) handover session was conducted in urban, suburban, and rural area in Perlis state, Malaysia. The experiment will consists of observation and analyzed on three area using LoRa communication modules. These three areas were collected on its Received Signal Strength Indicator (RSSI) to know the signal strength over the distance from base station and the differences physical object appear in Line of Sight (LoS) during the transmitting and receiving its signal from LoRa Transmitter to LoRa Receiver. The result at three area of different location have shown various pattern in term of RSSI where the each of the location give the distance that signal of LoRa could reach whilst the motion of vehicle is constant. From this experiment, certain level of RSSI value will be chosen to set as minimum boundary switcher for handover process state indicator and be used handover process algorithm in network switcher.

1. Introduction

Internet of Thing (IoT) comes with several of technique to receive data and send it to the cloud storage [1-7]. LoRa in IoT platform is a new Industrial, Scientific, and Medical (ISM) band wireless technology designed for low power, unlicensed, and Long-Range operation[8]. The frequency of LoRa has given access to the area of the country that could perceive and usable to simultaneously sending and receiving process. The sensor that is installed in modern vehicle currently can be optimized using the latest technology such as On-Board-Diagnostic (OBD-II) system. In the past, people communicate either with each other or with machines. However, with the arrival of the IoT, physical objects have the ability to communicate with each other without human intervention[9]. Thus, the previous non-cellular module being install in all around the smart cities or metropolitan but have their pros and cons in their respective coverage area network.

2. Existing Studies

In past literature, a study regarding the cellular network was executed in the country of India. N. Rakesh and S. K. Srivatsa [10], explain that different radio propagation model was used to determine the suitable method to calculate path loss in three different area. N. Alam, et. al. [11], mentioned radio raging

IOP Conf. Series: Materials Science and Engineering **705** (2019) 012012 doi:10.1088/1757-899X/705/1/012012

technique such as RSSI is very popular due to lower cost and its simplicity compared to other method. Although there are many models viable to analyze the path loss model, not all of them could use in certain unique environments[11].

3. Experimental Setup

To test the reliability and credibility of LoRa network in the communication link layer, communication node hardware was developed using two devices; Arduino Uno, and LoRa module with antenna to communicate between the moving vehicle and a base station as shown in Figure 1.

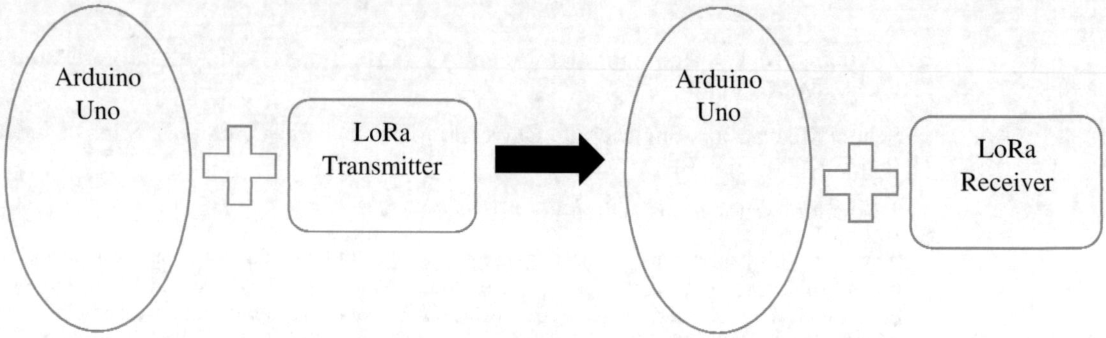

Figure 1. Experimental setup for signal strength

The LoRa device is divided into two communication nodes which is the receiver node and sender node. The placement or the height base for sender node was set to 20m from the ground and the receive node with the height of 1.6m from the ground would take far from the sender node to establish the communication signal. The area that had mention were: (i) rural area, (ii) suburban area and last (iii) urban area. The settings of LoRa was set that to send and receive the RSSI and with that analyze the signal strength with suitable path loss algorithm to ascertain the results. The system build using a microcontroller Arduino Uno, Dorji drf 1276 LoRa and GPS antenna. The RSSI value would be transmited into and kept in the independent database to process the result and set the coverage that need to handover when transition from mobile station away from base station.

(a) Rural area in Perlis Perlis (b) Suburban area in Perlis (c) Urban area in

Figure 2. Location selected for experimental setup

IOP Conf. Series: Materials Science and Engineering **705** (2019) 012012 doi:10.1088/1757-899X/705/1/012012

In Figure 2, the location selected in rural area was Jalan Arau – Changlun which consists trees and uneven hills. For the suburban area taken was from Jalan Pekan Pauh until the city of Arau. The selected location was suitable because half of them were covered in trees, while the other half are building. As for the urban area, the chosen sites were the capital city of Perlis, Kangar due many of buildings and houses.

4. Experimental Result and Discussion

The result obtained from RSSI value from these three significant areas show differently in term of their communicated to other nodes due interference on infrastructure, vehicle noise and etc. From the data RSSI value need to transform into a graphical result which could see in line graph in Figure 3.

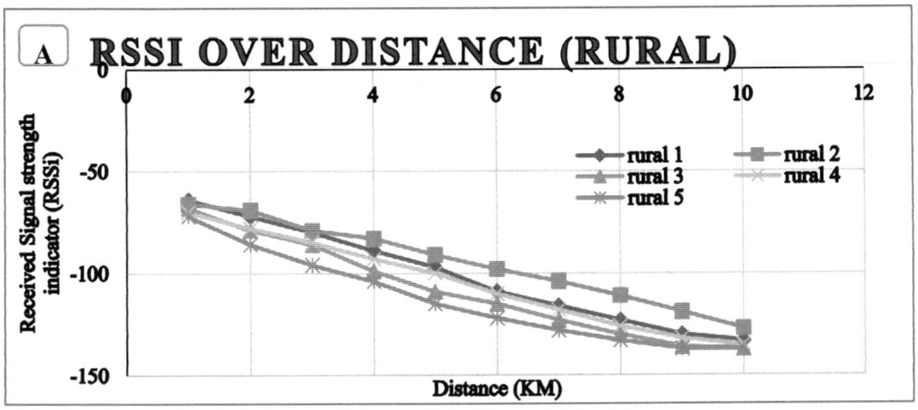

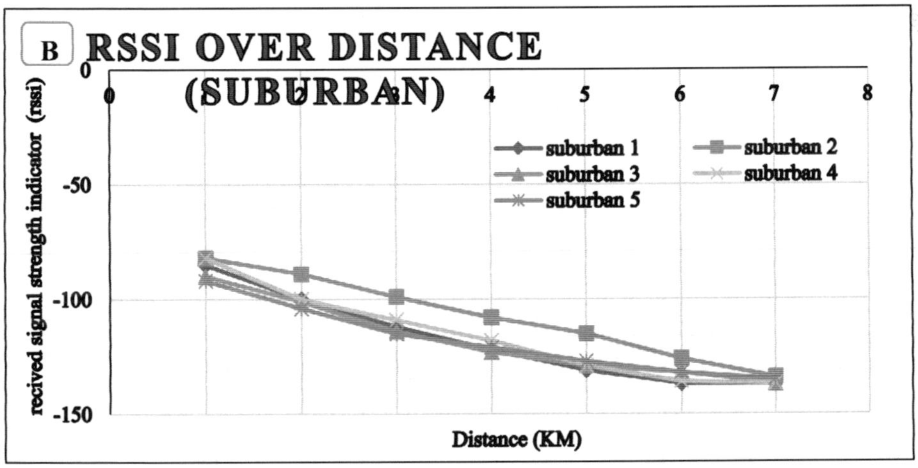

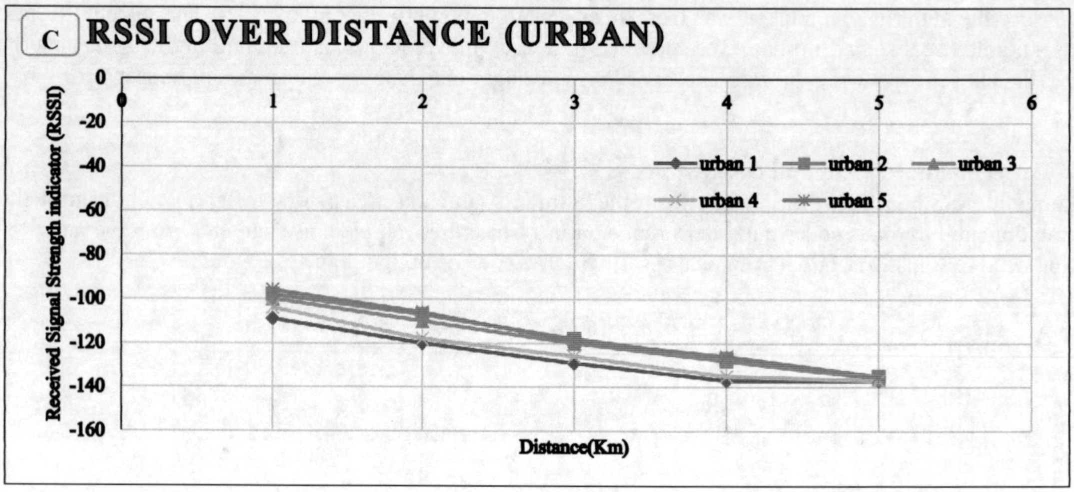

Figure 3. RSSI value from different location in three area

In Figure 3, the RSSI value in rural tested in different location were compare the differences of RSSI which there been involvement of LoS along the communication. The different location of rural area gives the different distance for the maximum RSSI value of LoRa Sender could send and maximum recorded from one of the location is 9.7km. Moreover, the suburban and urban area also used the outcome getting the RSSI value as rural area. The locations involve in suburban area had different distances that the signal strength of LoRa compare in the rural area. The maximum distance that RSSI value from one of the locations that had been recorded was 6.9km where beyond that distance, the signal was not sent any indicator or the strength of signal had no power to thrust the infrastructure between the base station and mobile station. Lastly, for the urban area, five location were tested to compare the maximum distance that the signal strength can received by LoRa receiver. The maximum distance that had recorded in one of the locations in urban area was 4.6km where most of the signals LoRa send to the receiver were diminished when faced the building that past which the height of 38m high. The signal indicator in urban was mostly disrupted by the interference of building height and the motor noise that occurred the recorded RSSI value.

5. Conclusion

In this paper, the analysis on LoRa handover session was determined by the boundary that was set using RSSI value and its connectivity strength during the transition from one base station to another base station. This also include the consideration of the path loss on three area stated and the coverage with the interference in three area while the hand over session perform on multiple base station. From the result show, that condition on using antenna could give a different setting value for the boundary for hand over the communication nodes. However, further analysis is necessary to improve the understanding of the communication link in urban, suburban, rural areas before any conclusions could reached.

Acknowledgements
We would like to extend our sincere gratitude for the funding provided by the Universiti Malaysia Perlis (UniMAP) (Geran Penyelidikan Khas - No.9004-00064) and UniMAP Automotive Racing Team, which enable this research to be carried out.

References

[1] Azarul F A H, Md T A R, Shah F K, Abdul H A, Muhajir A R, Norasmadi A R, Muhammad H N I, and Aminurasyid N 2017 Connected car: Engines diagnostic via Internet of Things (IoT) *J. Phys. Conf. Ser* **908**
[2] Aminurasyid N, Md T A R, Nasrul A M A, Mohd H B, Muhammad H N I, and Azarul F A H 2017 Study of intake manifold for Universiti Malaysia Perlis automotive racing team formula student race car *J. Phys. Conf. Ser* **908**
[3] Mohd L M, Md T A R, Shah F K, Mohd H B, Abdul H A, and M. S. M. Hashim 2017 Design and static structural analysis of a race car chassis for Formula Society of Automotive Engineers (FSAE) event, *J. Phys. Conf. Ser* **908**
[4] Mohd A, Mohd H B, M. F. Hamid, Md T A R, and M. S. M. Hashim 2017 Design optimization of rear uprights for UniMAP Automotive Racing Team Formula SAE racing car, *J. Phys. Conf. Ser* **908**
[5] Muhammad N A Z, M S M Hashim, Md T A R, Mohd H B, Ishak Ibrahim, N S Kamaruddin, Shahriman A B 2017 Analysis of a front suspension system for UniART FSAE car using FEA Analysis of a front suspension system for UniART FSAE car using FEA, *J. Phys. Conf. Ser* **908**
[6] Z. I. Zaman, K. S. Basaruddin, Mohd H B, Md T A R, and R. Daud 2017 Finite element prediction on the chassis design of UniART4 racing car, *AIP Conf. Proc* **1885**
[7] Rahman A, Rahman MTA, Manaf EHA, Rahman ASA 2018 Design and Analysis of Impact Attenuator for a Formula Student Car: A Study between Singular and Bi-tubular Tubes of Varying Geometries. *IOP Conf Ser Mater Sci Eng*
[8] Wixted A J, Kinnaird P, Larijani H, Tait A, Ahmadinia A, and Strachan N 2017 Evaluation of LoRa and LoRaWAN for wireless sensor networks *Proc. IEEE Sensors*
[9] Naoui S, Elhdhili M E, and Saidane L A 2017 Enhancing the security of the IoT LoraWAN architecture *5th IFIP Int. Conf. Perform. Eval. Model. Wired Wirel. Networks, PEMWN*
[10] Rakesh N and Srivatsa S K 2013 A Study on Path Loss Analysis for GSM Mobile Networks for Urban, Rural and Suburban Regions of Karnataka State **4**
[11] Alam N, Balaie A T, and Dempster A G 2010 Dynamic Path Loss Exponent and Distance Estimation in a Vehicular Network using Doppler Effect and Received Signal Strength
[12] Lorca J, Hunukumbure M and Wang Y, On Overcoming the Impact of Doppler Spectrum in Millimeter-Wave V2I Communications
[13] A. Goldsmith 2005 Path Loss and Shadowing *Wirel. Commun.*
[14] Schroder F, Pissolato F and Moreli V L 2016 The adequacy of LoRaWAN on smart grids: A comparison with RF mesh technology *IEEE 2nd Int. Smart Cities Conf. Improv. Citizens Qual. Life, ISC*
[15] Wixted A J, Kinnaird P, Larijani H, Tait A, Ahmadinia A and Strachan N 2017 Evaluation of LoRa and LoRaWAN for wireless sensor networks *Proc. IEEE Sensors*
[16] Noreen U, Bounceur A and Clavier L 2017 A study of LoRa low power and wide area network technology *Proc. - 3rd Int. Conf. Adv. Technol. Signal Image Process. ATSIP 2017*
[17] Bok J and Ryu H Path Loss Model Considering Doppler Shift for High Speed Railroad Communication.

[18] Mikhaylov K and Petäjäjärvi J 2016 Analysis of Capacity and Scalability of the LoRa Low Power Wide Area Network Technology *Univ. Oulu, Cent. Wirel. Commun. Finl.*
[19] Na S, Hwang D, Shin W, and Kim K H 2017 Scenario and countermeasure for replay attack using join request messages in LoRaWAN *Int. Conf. Inf. Network*
[20] Amornthipparat A, Shirai H, Inoue T, and Nakamura Y 2008 Estimation of High Frequency NLOS Path loss in Street-Cell Environment

Classification of different materials for underground object using artificial neural network

H Ali, A Z Ahmad Firdaus, Mohd Shuhanaz Zanar Azalan, S N A M Kanafiah, S H Salman, M R Ahmad[2], T Sarah T Amran[2], M Syafiq M Amin[2]

[1]School of Mechatronic Engineering, Universiti Malaysia Perlis, 02600 Arau, Malaysia
[2]Agensi Nuklear Malaysia, Bangi, 43000 Kajang, Selangor, Malaysia

hasimahali@unimap.edu.my

Abstract. Ground Penetrating Radar (GPR) is widely used for non-destructive investigation of the shallow subsurface exploration especially in locating the buried infrastructure such as pipes, cables and road inspections. However, the interpreting hyperbolic signature of buried object in GPR images remains a challenging task. Therefore, this paper presented the classification of different materials based on GPR images using artificial neural network (ANN). In this research, GPR images so called the B-scan radargram represented by hyperbolic signature are firstly acquired and pre-processed. Then, the hyperbolic signature features are extracted using statistical techniques. The extracted features is then fed up as input to the multilayer perceptron (MLP) neural network classifier. A series of experiments have been conducted based on extracted hyperbolic features of different materials and shapes. Based on the results, the proposed method in classifying different materials based on GPR images using neural network showed promising results.

1. Introduction

GPR is a non-destructive evaluation and imaging tools for shallow subsurface exploration based on the transmitting and receiving of electromagnetic (EM) wave. It is widely used in a variety of applications such as locating buried infrastructure such as pipes [1] and cables, road inspection, land mines [2] and also used in geological/ archeological studies. The GPR system transmits short pulses of high frequency EM from 100 MHz to 1 GHz into the ground. These wave propagate with a velocity that depends on the dielectric property of medium. If the wave hit a buried object with different reactive indices, some of the wave are reflected back and receiver processes them to create a time-series, known as an A-scan [2]. Consecutive patterns in A-scan can be concatenated to create images of the subsurface, called B-scan. Buried objects exhibit characteristic pattern in B-scan so called radargram/map, such as hyperbolas, which can be used for detection [3, 4] as shown in Figure 1.

In the literature, many approaches have been developed in computer vision to analyse hyperbolic signature in detecting and locating the buried object. Some of them utilized histogram of oriented gradient (HOG) [5], scale invariant feature transform [6], finite different time-domain [7], and Hough transform [8, 9]. For instance, Carlotto [8] have used Hough transform in GPR images to find the hyperbolic parameter for buried object. Although Hough transform was commonly used method in detecting hyperbolic signature, however this method is time consuming and low accuracy. Ni et al. [4]

used discrete wavelet transform to detect underground buried pipe based on hyperbolic signature. They used multiresolution analysis of DWT to filter and enhance the GPR data for obtaining more information on the DWT profile. Further, they have used sample of buried pipe made of plastic and metal in their design.

One of the problems in GPR signal is that they easily corrupted with by environmental noise, which might mask the weak reflections from inhomogeneities located deep in the test structure. The scattering amplitudes of wavelengths in the GPR signal are functions of the relative antenna-to-target orientation, target and background properties, scattering angle and target thickness. Moreover deeper objects may be obscured by numerous shallower objects, which appear constructively interfering hyperbolic reflectors. This diffracted energy can mask other reflections of interest and cause misinterpretation of the size and geometry of subsurface object from the GPR raw signature [1]. Therefore, this paper proposed of using multilayer perceptron neural network to classify the materials and the shape based on statistical features extracted from hyperbolic signature. The rationale of using this method is that the neural network has the ability to learn effectively from the data. They have been used successfully in several field such as speech recognition and face recognition.

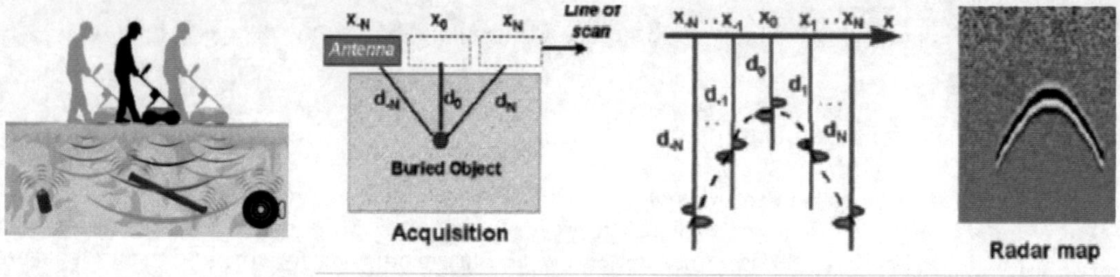

Figure 1. Basic principle of GPR

2. Methodology

The flowchart of the proposed method is depicted in Figure 2. It consists of data acquisition, pre-processing, feature extraction and classification. The detailed of each block is described at the subsequent sections.

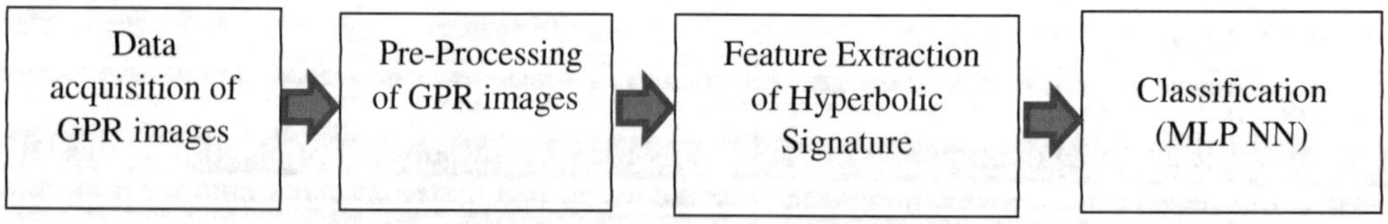

Figure 2: The flowchart of the proposed method.

2.1. Data Acquisition of GPR images

Sample Preparation. Fabrication of different samples having different materials (i.e perspek and alluminium) and different shape (i.e cubic and cylinder) have been conducted. In this work, antenna of 800 MHz transmitter and receiver of Ramac GPR Mala system was used to collect the data. Figure 3 shows the basic GPR components that consist of antenna, battery pack, control unit, and a portable PC.

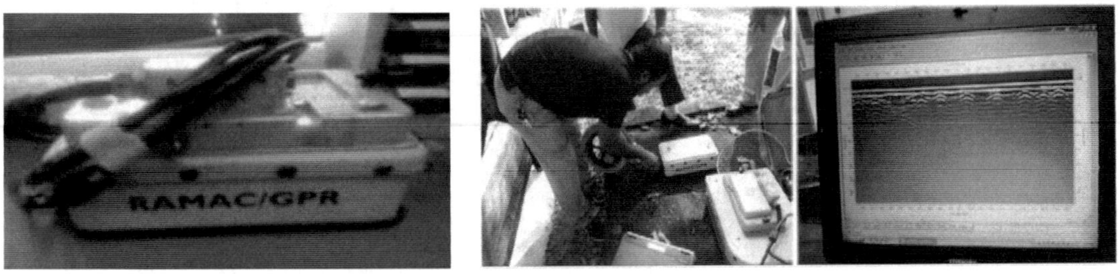

Figure 3: Basic component of GPR system

In this research, two shapes (cylindrical and cube) and two materials (aluminium (metal) and perspex (non-metal)) were used as target samples. These samples are buried in parallel with depth of 200 cm from the ground surface with 100 cm apart between them. Sand was used as soil medium. Figure 4 illustrates the location of the buried sample. The operation GPR system was guided by research officer at Non Destructive Test (NDT), Nuclear Agency Malaysia for collecting the GPR data. The acquisition of data was conducted and repeated with 100 times. The example of hyperbolic signature appear in GPR images with respect to the buried object is depicted in Figure 5.

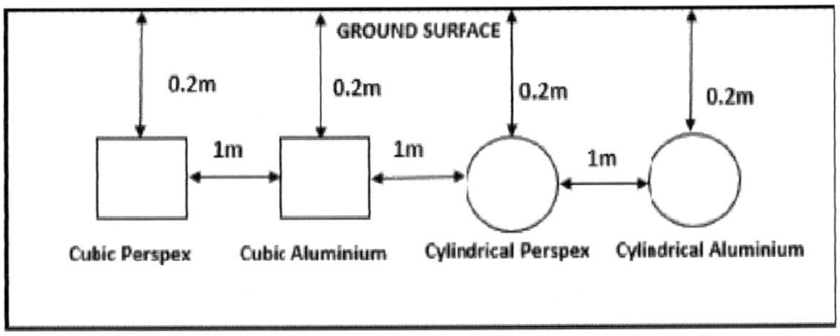

Figure 4: The position of the buried samples

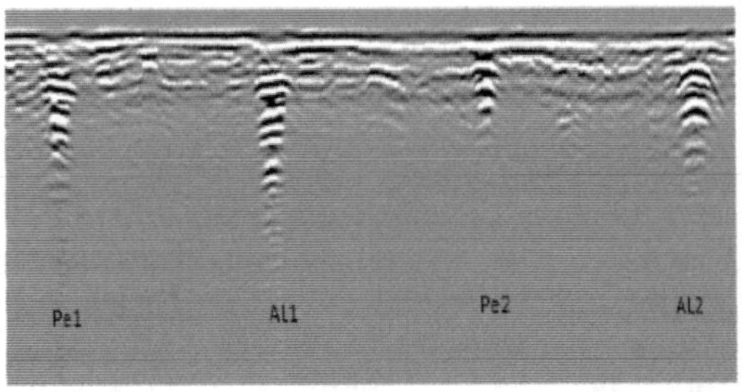

Figure 5: The hyperbolic signature of the buried object

2.2. *Image Pre-Processing*

The acquire image that contains irrelevant parts (noise) should be removed. For the purpose of analysis, the original GPR images needs to be preprocessed. The preprocessed is composed of two step:

IOP Conf. Series: Materials Science and Engineering **705** (2019) 012013 doi:10.1088/1757-899X/705/1/012013

2.2.1. Filtering. Original image firstly converted into grayscale images. Then, in this work, median filter was used to improve the quality of the image by preserving the edge as well as removing unwanted noise. Figure 6 shows the example of applied median filter from grayscale image.

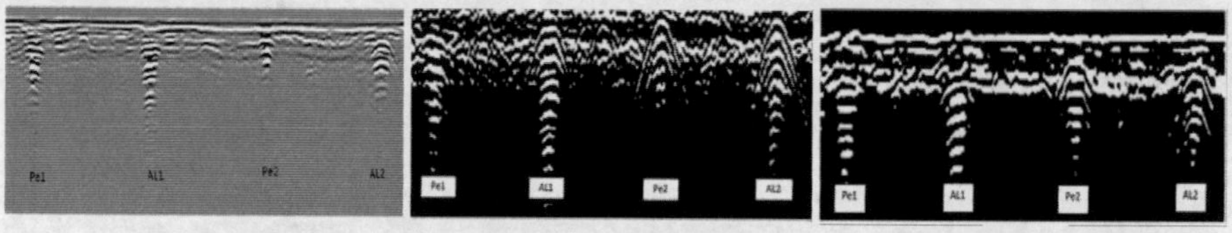

(a) (b) (c)

Figure 6: (a) original GPR images, (b) grayscale image, (c) Applied median filter

2.2.2. Cropping. In this work, the hyperbolic signatures of filtered image of the samples are manually cropped. Only the region of interest of hyperbolic signature will be extracted and used as features for further investigated. The example of cropped hyperbolic signature is depicted in Figure 7.

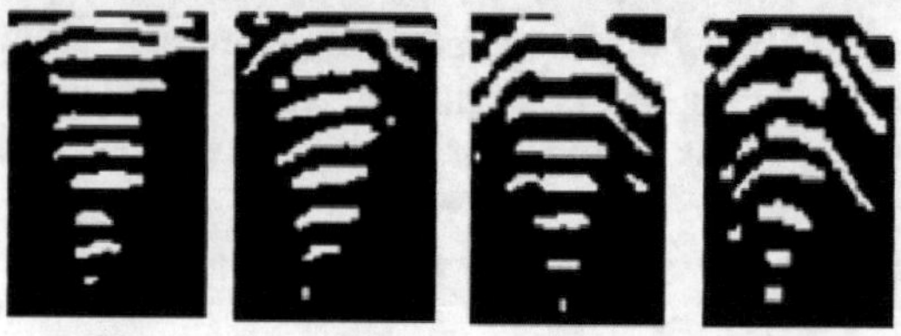

Pe1 (Cubic) Al1 (Cylindrical) Pe2 (Cubic) Al2 (Cylindrical)

Figure 7. Example of cropped hyperbolic signature for different samples

2.3. Feature Extraction

The hyperbolic signature of buried objects exhibit unique pattern with respect to the shape, size, materials and depth. Feature extraction needs to be done to extract important features. In this work, skewness, standard deviation, maximum peak and total number of peak are extracted from hyperbolic signature.

$$Skewness, s = E\left[\frac{(x-\bar{x})_3}{\sigma^3}\right]$$

Where E is expectation operator, $\bar{x}$ is the mean of the data, x in the position of amplitude in the x-axis, s is the skewness and σ is the standard deviation.

2.3.1. Gray Level Co-Occurrence Matrix (GLCM). In statistical texture analysis, texture features are computed from the statistical distribution of observed combinations of intensities at specified positions relative to each other in the image. According to the number of intensity points (pixels) in each combination, statistics are classified into first-order, second- order and higher-order statistics [10]. The Gray Level Co-occurrence Matrix (GLCM) method is a way of extracting second order statistical texture features. In this work, the GLCM features such as contrast, correlation, energy and homogeneity are used to extract the information of the hyperbolic signatures. Contrast is used to measures the local variations in the GLCM, whereas the correlation is used to measures the joint probability occurrence of the specified pixel pairs. Meanwhile, energy provides the sum of squared elements in the GLCM and also known as uniformity or the angular second moment. Then, the homogeneity is used measures the closeness of the distribution of elements in the GLCM to the GLCM diagonal.

2.4. Neural Network Classification

Artificial neural network is one of the techniques in artificial intelligence (AI) which has a functional imitation of the human brain. It is based on simulated nerve cells or neuron, which joined together to form networks. In general, an ANN is characterized by its architecture, learning algorithm and activation functions. Figure 8 shows the example of multilayer perceptron neural network. It consists of an input layer that distributes the input data to the processing elements in next layer. The hidden layers where the nonlinear behavior comes from and the output layer. Both the input and output layers are directly accessible while the hidden layers are not. Each layer contains several processing units which are generally called neurons. In this work multilayer perceptron is used to classify different type of material. The MLP learning algorithm [11] works as follows:

i. Initialize the weights to random values on the interval [-1, +1].
ii. Present the first training pattern, and obtain the output
iii. Compare the network output with the target output.
iv. Propagate the error backwards
   a. Correct the output layer of weights using the following formula: $w_{ho} = w_{ho} + (\eta \delta_o o_h)$
      where w_{ho} is the weight connecting hidden unit h with output unit o, η is the learning rate, o_h is the output at hidden unit h. δ_o is given by the following: $\delta_o = \mathrm{o_o}(1-\mathrm{o_o})(\mathrm{t_o}-\mathrm{o_o})$
      where o_o is the output at node o of the output layer, and $t-o$ is the target output for that node.
   b. Correct the input weights using the following formula: $w_{ih} = w_{ih} + (\eta \delta_h o_i)$
      where w_{ih} is the weight connecting node i of the input layer with node h of the hidden layer, o_i is the input at node i of the input layer, η is the learning rate. δ_h is calculated as follows: $\delta_h = o_h(1-o_h)\sum_o(\delta_o w_{ho})$
      where p is the number of units in the output layer.
v. Calculate the error, by taking the average difference between the target and the output vector. For example the following function could be used:

$$E = \frac{\sqrt{\sum_{n=1}^{p}(t_o - o_o)^2}}{p}$$

vi. Repeat from 2 for each pattern in the training set to complete one *epoch.*
vii. Shuffle the training set randomly. This is important so as to prevent the network being influenced by the order of the data.
viii. Repeat from step 2 for a set number of epochs, or until the error ceases to change.

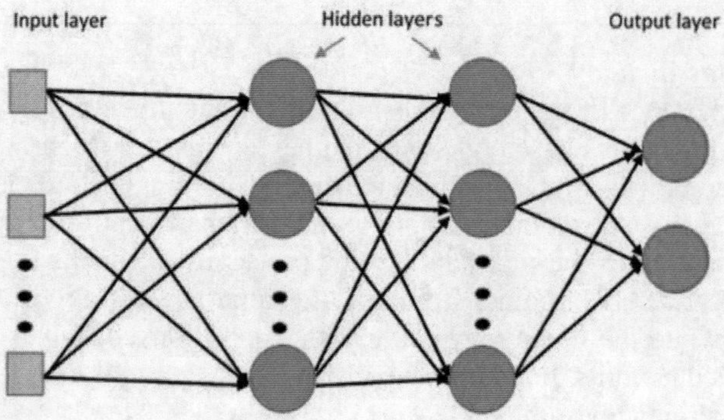

Figure 8: Example of multilayer perceptron neural network structure.

3. Experimental Results & Discussion

In this work, 100 GPR images have been acquired at Malaysia Nuclear Agency using Mala RAMAC GPR with 800 MHz antenna. Firstly, the raw data of GPR image so called radargram is preprocessed by using median filter to improve the quality of image as well as to remove the unwanted noise. Then, the region of hyperbolic signature of GPR images are manually cropped. In this research, eight features of hyperbolic signature of buried object are extracted such as maximum peak, total number of peak, skewness, standard deviation and GLCM features (contrast, correlation, energy and homogeneity). These features serve as input to fed into MLP NN for classification. The effectiveness of the proposed method was evaluated using based on extracted hyperbolic signature features. Further, we also examined four different types of learning algorithm adopted in MLP such as Bayesian Regulation (BR), Scale Conjugate Gradient (SCG), Lavernberg-Marquardt (LM), and Resilient Back-Propagation (RP). Meanwhile, the data was split into training and testing set based on conventional method with (80 %, 20 %), (70 %, 30 %) and (60 %, 40 %) strategy using 10 hidden layer.

Table 1 shows the value of skewness, standard deviation, maximum number of peak and number of peak with respect to the types of material and shape. It can be seen from Table 1 that, the aluminium exhibits higher value of skewness, standard deviation, maximum peak and total number of peak compared to the perspek material. This means that the metal (aluminium) possesses strong reflected value of EM when the EM hits the underground metal object. In addition, in terms of shape, cubic consistently shows higher value as compared to the cylindrical. This is inferred that the shape has unique pattern of the buried object. Based on the results we can summarize that the materials such as aluminium and perspek exhibit different value/pattern based on skewness, standard deviation, peak value and no of peak. In addition, cubic and cylinder are well distinguish in reveal their characteristic.

Table 1: The value of skewness of different sample.

Materials	**Geometry**	**Skewness**	**Standard Deviation**	**Maximum Peak value**	**Total No of Peak**
Aluminium	Cubic	0.64	2289.32	15853	68
	Cylindrical	1.44	598.97	10363	63
Perspek	Cubic	-0.85	632.64	9679	58
	Cylindrical	-0.39	96.06	6242	57

Table 2 presents the recognition rates using MLP NN with various learning algorithm based on aforementioned features. It can be seen from Table 2 that the recognition rates of training set give perfect classification for all strategy. For testing performance, the result of using skewness feature gives the highest recognition rates which is 98.32% using Lavernberg Marquard (LM) learning algorithm with 80% 20% strategy. In contrast, the total number of peak features has the lowest recognition rate which is 74.74% using the same strategy. Meanwhile, standard deviation and maximum no of peak features are comparable which are 94.8% and 90%, respectively. As overall, the computational time of MLP model is relatively short which less than 7s.

Table 3 presents the recognition rates of using GLCM features based on MLP NN with different learning algorithm. It can be seen from Table 3 that, all training performances gives perfect recognition rates. For testing performance, the correlation and homogeneity features achieved the best recognition rates which are 100% using BR, SCG, LM and RBP. Further, energy features contributes the recognition rate of 99.2%. While, the contrast features produced the recognition rate of 96.50 %. This can be inferred that, the GLCM features able to characterize the pattern of hyperbolic signature of buried object having different material and shape distinctively. So, the proposed method shows the promising results in classifying different materials and shape. Further, the average time of MLP NN to train the network only required less than 7 second.

Table 2: The recognition rates using maximum no of peak, no of peak, skewness and standard deviation features MLP NN with different learning algorithm.

L A	Data (%,%)	Max of Peak (%)		AT (s)	Total No. of peak (%)		AT (s)	*Skewness* (%)		*AT* (s)	σ (%)		*AT* (s)
	(Tr, Ts)	*Tr*	*Ts*		*Tr*	*Ts*		*Tr*	*Ts*		*Tr*	*Ts*	
BR	(80, 20)	100	86.50	3.41	100	74.40	4.24	100	95.67	4.60	100	76.80	4.10
	(70, 30)	100	88.80	3.60	100	69.80	3.48	100	96.66	6.68	100	82.80	4.88
	(60, 40)	100	83.70	3.99	100	58.70	3.58	100	90.83	5.23	100	94.20	5.53
SCG	(80, 20)	100	89.00	3.96	100	70.20	3.78	100	93.33	3.92	100	79.20	3.94
	(70, 30)	100	89.30	3.95	100	69.70	3.47	100	94.99	6.46	100	81.30	4.49
	(60, 40)	100	85.20	3.94	100	57.80	3.35	100	91.66	4.96	100	94.80	4.00
LM	(80, 20)	100	85.70	4.24	100	74.00	4.32	100	98.32	3.96	100	78.00	6.77
	(70, 30)	100	87.30	3.68	100	70.20	3.70	100	94.66	6.43	100	81.80	4.99
	(60, 40)	100	87.20	3.42	100	57.50	3.67	100	90.13	4.56	100	94.50	4.07
RBP	(80, 20)	100	90.00	4.55	100	69.00	3.19	100	85.00	5.75	100	79.50	4.29
	(70, 30)	100	88.30	3.51	100	70.50	3.32	100	86.33	3.69	100	81.70	5.44
	(60, 40)	100	83.30	3.62	100	57.20	4.80	100	85.33	4.56	100	93.20	4.81

Note: LA (Learning Algorithm), BR (Bayesian Regulation), SCG (Scale Conjugate Gradient), LM (Lavernberg Marquardt), RBP (Resilient Back-Propagation), Tr (Training), Ts(Testing), AT(Average Time)

Table 3: Recognition rates of GLCM features based on MLP NN using different learning algorithm

L A	Data (%,%)	Contrast (%)		AT (s)	Correlation (%)		AT (s)	*Energy* (%)		*AT* (s)	*Homogeneity* (%)		*AT* (s)
	(Tr, Ts)	*Tr*	*Ts*		*Tr*	*Ts*		*Tr*	*Ts*		*Tr*	*Ts*	
BR	(80, 20)	100	94.0	5.89	100	100.0	5.14	100	93.8	6.73	100	99.3	3.85
	(70, 30)	100	91.7	4.99	100	98.6	4.96	100	96.8	5.53	100	100.0	4.28
	(60, 40)	100	82.2	4.92	100	99.2	4.60	100	97.2	4.95	100	99.5	4.00
SCG	(80, 20)	100	72.5	4.63	100	100.0	4.73	100	97.5	4.08	100	100.0	3.57
	(70, 30)	100	72.0	4.47	100	98.2	5.78	100	97.5	4.28	100	100.0	3.61
	(60, 40)	100	66.3	4.32	100	99.2	4.70	100	97.5	4.12	100	100.0	4.12
LM	(80, 20)	100	96.5	4.96	100	100.0	5.20	100	95.3	6.24	100	99.3	3.61
	(70, 30)	100	93.0	4.61	100	98.3	5.59	100	99.2	4.38	100	99.5	3.80
	(60, 40)	100	80.0	4.55	100	99.2	6.32	100	97.5	4.05	100	99.5	3.52
RBP	(80, 20)	100	74.0	4.62	100	100.0	4.61	100	97.0	5.54	100	100.0	3.55

L A	Data (%,%)	Contrast (%)		AT (s)	Correlation (%)		AT (s)	*Energy* (%)		*AT* (s)	*Homogeneity* (%)		*AT* (s)
	(70, 30)	100	68.0	4.32	100	98.3	5.38	100	97.0	4.68	100	91.0	3.55
	(60, 40)	100	66.0	4.30	100	99.2	5.65	100	97.0	4.53	100	100.0	3.62

Note: LA (Learning Algorithm), BR (Bayesian Regulation), SCG (Scale Conjugate Gradient), LM (Lavernberg Marquardt), RBP (Resilient Back-Propagation), Tr (Training), Ts(Testing), AT(Average Time)

4. Conclusion

In this paper, classification of material using multilayer perceptron neural network based on extracted hyperbolic signature features of GPR images was presented. The extracted hyperbolic features: skewness, standard deviation, maximum peak, total number of peak and GLCM exhibits unique pattern for aluminium and perspex where the value for aluminium was higher. For cubic and cylinder shapes, distinctive features were revealed where cubic exhibited higher value. The use of statistical features was effective way to classify different material (i.e metal and nonmetal) due to unique pattern of hyperbolic signature. Effectiveness of the proposed method was evaluated using MLP NN classifier for various learning algorithm where, the GLCM features (correlation and homogeneity) achieved the best recognition rates which are 100% using BR, SCG, LM and RBP learning algorithm in classifying metal and nonmetal (i.e aluminium and perspek) for cubic and cylinder shapes. Thus, the proposed method shows promising results in crude classification of different material based on GPR images.

References

[1] Ni S H, Huang Y H, Lo K F, Lin D C 2010 *Buried Pipe Detection by Ground Penetrating Radar Using The Discrete Wavelet Transform.* Computers and Geotechnics **37**, pp. 440–448.
[2] Ishitsuka K, Iso S, Onishi K , and Matsuoka T 2018 *Object Detection in Ground-Penetrating Radar Images Using a Deep Convolutional Neural Network and Image Set Preparation by Migration* Hindawi International Journal of Geophysics, Vol. 2018, pp.1-8.
[3] Long Z J, Xing B A, Liu H & Liu Q H 2017 *Hyperbola Recognition from Ground Penetrating Radar Using Deep Convolutional Neural Networks*", Proceeding 2nd International Conference on Artificial Intelligence: Techniques and Applications ISBN: 978-1-60595-491-2.
[4] Syambas N R 2012 *An Approach for Predicting the Shape and Size of a Buried Basic Object on Surface Ground Penetrating Radar System* Hindawi Publishing Corporation International Journal of Antennas and Propagation Vol 2012.
[5] Torrione P A, Morton K D, Sakaguchi Jr R & Collins L M 2014 *Histograms Of Oriented Gradients For Landmine Detection In Ground-Penetrating Radar Data* IEEE Transactions On Geoscience And Remote Sensing, Vol. **52** 3.
[6] Harkat H, Elfakir Y, Bennani S D, Khaissidi G, Mrabti M. 2016 *Ground Penetrating Radar Hyperbola Detection Using Scale-Invariant Feature Transform* Proceeding 2nd International Conference on Electrical and Information Technologies.
[7] Irving J, Knight R 2006 *Numerical Modelling Of Ground-Penetrating Radar In 2-D Using Matlab* Computers & Geosciences **32** 1247–1258.
[8] Carlotto M J 2002 *Detecting Buried Mines in Ground Penetrating Radar Using a Hough Transform Approach* Proceedings Volume 4741, Battlespace Digitization and Network-Centric Warfare II.
[9] Simi A, Bracciali S & Manacorda G 2008 *Hough Transform Based Automatic Pipe Detection for Array GPR: Algorithm Development And On-Site Tests* IEEE Radar Conference.
[10] Gadkari D 2004 *Image Quality Analysis Using GLCM* Msc Thesis, University of Florida.
[11] Noriega L 2005 *Multilayer Perceptron Tutorial* School of Computing, Staffordshire University.

IOP Conf. Series: Materials Science and Engineering **705** (2019) 012014 doi:10.1088/1757-899X/705/1/012014

Cross-correlation analysis of isometric contraction for mechanomyography signals on forearm muscle

M R Mohamad Ismail[1,a], C K Lam[1], K Sundaraj[2] and M H Fazalul Rahiman[1]

[1]School of Mechatronic Engineering, Universiti Malaysia Perlis, Kampus Pauh Putra, 02600 Arau, Perlis, Malaysia

[2]Faculty of Electronic and Computer Engineering, Universiti Teknikal Malaysia Melaka, Hang Tuah Jaya, 76100 Durian Tunggal, Melaka, Malaysia

[a]Corresponding author: razifmohamad144@gmail.com

Abstract. This study examines the cross-correlation coefficient of Mechanomyography (MMG) signals recorded from Palmaris Longus (PL), Flexor Digitorum Superficialis (FDS), Extensor Digitorum (ED) and Extensor Carpi Ulnaris (ECU) of forearm muscles during isometric muscle action. Nine healthy participants using dominant hand (mean±SD: age=24.78±1.79 year) volunteered in this study to perform submaximal to maximal (25%, 50%, 75% and 100%) isometric muscle actions with maximal voluntary contraction force of hand exercise gestures (grip strength supinated, grip strength pronated, finger flexion and pinch grip). During each isometric contraction, four-separated VMG sensor (TSD250A)-sensitive accelerometer type was used to record MMG signals on flexor and extensor side of forearm muscles. Maximum cross-correlation coefficients at zero time lags were analyzed between flexor (PL & FDS) and extensor (ED & ECU) sides for each subject and force level to determine which sides show high level of association during isometric muscle action of forearm with different hand exercise gestures. The results showed maximum cross-correlation coefficients that range from $R_{x,y} = 0.801 - 0.943$ for all hand exercise gestures performed. In addition, from the results obtained between flexor and extensor side indicates that the maximum range of cross-correlation coefficient for extensor side muscle groups were slightly higher than the flexor side muscle groups.

1. Introduction

Recently, the relationship between muscle activities and generated force during muscle contraction is becoming a popular case study among researchers. Various recording techniques have been developed to measure muscle activities from different aspects especially on forearm and leg muscles. Most popular techniques used includes electromyography (EMG) and mechanomyography (MMG) signals. EMG is composed of electrical contribution made by the active motor units during muscle contraction. Meanwhile, MMG is a recording of the low-frequency lateral oscillations of active motor units considered as mechanical contribution during muscle contraction [1].

Several studies are working with both EMG and MMG responses during isometric muscle action on forearm and leg muscles with cross-correlation coefficient to determine the amount of common signal (cross-talk). For example Mogk and Keir [2] used cross-correlation function to determine amount of cross-talk with forearm postures: pronation, neutral and supination. According to the cross-correlation values presented, peak cross-correlation values were not significantly different between target force levels and between grasp and pinch trials. The authors found that the maximum grasp forces were reduced when the forearm was deviated from the mid-prone position, significantly in pronation. In addition, similar effect was not found for pinch forces. As expected, cross-correlation magnitude decreased as the distance between electrode pairs increased.

In addition, T. W. Beck, M. A. Dillon, J. M. DeFreitas, and M. S. Stock [3] examined the cross-correlation of MMG signals detected from rectus femoris muscle in perpendicular and transverse axis during isometric muscle actions. Two MMG sensors used which, one of the sensor was oriented in an axis that was perpendicular to the muscle surface and the other sensor was oriented in an axis that was transverse to the muscle surface. The signal recorded from both axis were correlated to each other obtain the maximum cross-correlation coefficients that are ranged from $R_{x,y} = 0.273 - 0.989$. The present results indicate, a high level of association between the MMG signals detected in both axis based on COHEN interpretation [4].

In other research [5], the authors quantified the level of crosstalk in the MMG signals from the longitudinal, lateral and transverse with different isometric wrist posture: wrist flexion, wrist extension, radial and ulnar deviations. The results indicate, transvers axis of MMG signals between the muscle groups showed significantly less crosstalk for each wrist posture. However, cross-correlation does not always referred to cross-talk. There is evidence that the cross-correlation method used to determine motor unit synchrony as common neural input between two muscles [6]. If two muscles receive common input and response similarly with the muscle action, then it is reasonable to expect that they would demonstrate high correlation. Therefore, cross-correlation can also represent other appliances instead of cross-talk.

Through literature study, more information of isometric contraction are found on leg muscle compared to forearm muscles. Forearm consists of numerous small muscles separated with flexor and extensor side that is in close proximity and contracted to each other when perform muscle actions [7]. Therefore, the purpose of this research was to examine the maximum cross-correlation coefficient of MMG signals recorded from flexor (PL & FDS) and extensor (ED & ECU) side muscles of forearm during isometric muscle action for maximum voluntary isometric contraction of force with multiple hand exercise gestures.

2. Materials and Methods

2.1. Participants

Nine participants, healthy dominant handed male (mean±SD: age=24.78±1.79 y; height= 170.67±5.92 cm; weight= 65.89±6.77) were volunteered with specific criteria for the measurement of the signal. Written consent forms were given to the participants for their participation in this experiment. All volunteered participants were clinically healthy and in good physical condition with no previous or ongoing history of neuromuscular or musculoskeletal disorder specific to the elbow, wrist and/or finger joints.

2.2. Ethical Approval

This study was approved by the local Medical Research & Ethics Committee (MREC), Ministry of Health, Kuala Lumpur, Malaysia via Ref No.: KKM/NIHSEC/P14-1197. The guidelines were followed according to the Declaration of Helsinki due to the involvement of human individuals in experiment.

2.3. Isometric Contraction Protocols

Subject were required to seat comfortably on a chair with adjustable arm support attached to the chair arm. Specifically, participants were required to execute three muscle contraction with three trials for every 4 seconds of interval and each contraction is held for 4 seconds for isometric contraction. The isometric contraction was repeated with four maximal voluntary isometric contractions (25%, 50%,

75%, and 100%). There was 3-4 minutes of rest before executing next hand exercise gestures as shown in Figure 1 to avoid fatigue issue.

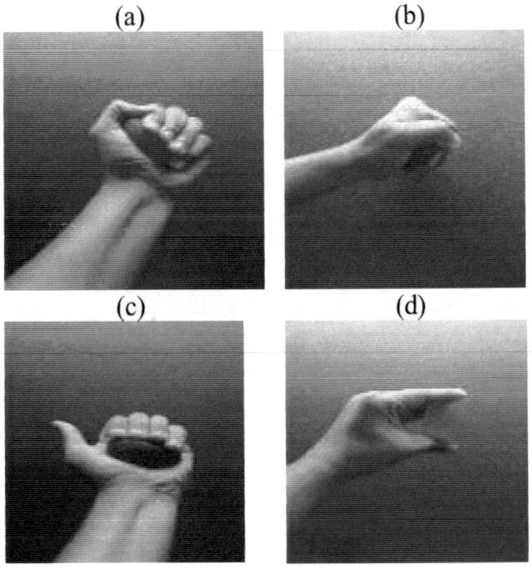

Figure 1. (a) Grip Strength Supinated, (b) Grip Strength Pronated, (c) Flexion Finger (FF) and (d) Flexion Finger (FF)

2.4. *MMG Measurements*

During isometric contraction of forearm muscles, two sides of forearm muscles (extensor and flexor) placed with VMG sensor (TSD250A) to record MMG signals. The TSD250A VMG Transducer is a sensitive accelerometer (32.64 mm diameter) for use with BIOPAC Vibromyography Systems. TSD250A used for measuring absolute muscle force from substantial muscle groups, such as forearm and leg muscles. Four VMG sensors were attached on the skin surface over the muscle bellies of the extensor side (ED and ECU) and flexor side (PL and FDS) with double-sided adhesive tape.

2.5. *Data Acquisition and Signal Processing*

The output of each direction of the four VMG sensors were connected to the data acquisition unit (BIOPAC VMG System) using an MP160 and HLT100C. It was set to record four channel of VMG raw signals as four muscles has been selected. The signals from each muscle were amplified at gain equals to 200, sampled at a rate of 2000 samples/s and stored in a personal computer for subsequent analysis. The filtered signals were selected for a 2-s time period corresponding to the middle 50% of each 4-s muscle contraction as shown in Figure 2. The filtered signals were then cross-correlated between each other (palmaris longus versus flexor digitorum superficialis and extensor digitorum versus extensor carpi ulnaris) by using the cross-correlation with normalization formula as shown below. According to Cohen [4], maximum cross-correlation coefficients between 0.1-0.3 were considered small, those between 0.3-0.5 were considered moderate and those between 0.5-1.0 were considered large.

$$R_{xy}(\tau) = \frac{1}{a \times b \times \omega(\tau)} \sum_{n=0}^{N-1} X_t(n) * Y_t(n+\tau) \, ; 1 - N < \tau < M \quad (1)$$

where $a = \sqrt{\sum_{n=0}^{N-1} X_t^2(n)}$, $b = \sqrt{\sum_{n=0}^{M-1} Y_t^2(n)}$ and ω is weighting factor, M and N are the length of X_t and Y_t, respectively, τ represent time lag between the signals.

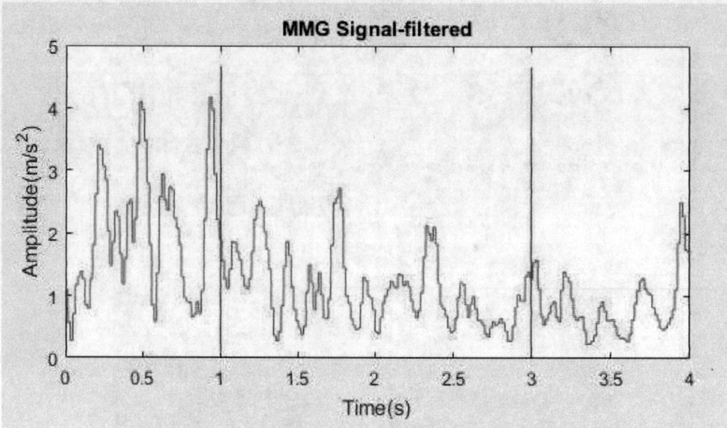

Figure 2. An example of segmented MMG signals from PL muscle during isometric contraction

3. Results and Discussions

Table 1 - 4 show the result of maximum cross-correlation coefficients of isometric contraction for four different hand exercise gestures (GSP, GSS, FF and PG) that range from $R_{x,y} = 0.801 - 0.943$ for all gestures performed. According to the results obtained, for each of muscle comparisons, the maximum cross-correlation coefficients increase with force level for some of the participants, but the trend was not consistent. For each cross-correlation that was performed, the maximum cross-correlation coefficient occurred at a time-shift of 0 s (i.e., $\tau = 0\ s$).

Table 1. Maximum cross-correlation coefficients for grip strength supinated

	PL & FDS				ED & ECU			
Subject	25%	50%	75%	100%	25%	50%	75%	100%
1	0.867	0.890	**0.919**	**0.883**	0.921	0.883	0.867	0.878
2	0.895	0.860	0.873	0.907	0.864	0.889	**0.926**	0.841
3	0.858	0.837	0.851	0.882	0.855	0.838	0.893	0.837
4	0.884	0.876	0.884	0.880	0.877	0.864	0.906	0.900
5	0.871	0.875	0.875	0.883	**0.826**	0.880	0.835	0.856
6	0.865	0.913	0.876	0.857	0.854	0.838	0.875	0.898
7	0.897	0.906	0.869	0.846	0.859	0.884	0.861	0.834
8	0.857	0.833	0.857	0.886	0.891	0.884	0.845	0.882
9	0.866	0.864	0.900	0.842	0.866	0.874	0.879	0.880

Table 2. Maximum cross-correlation coefficients for grip strength pronated

	PL & FDS				ED & ECU			
Subject	25%	50%	75%	100%	25%	50%	75%	100%
1	0.896	0.859	0.832	0.843	0.877	0.877	0.864	0.866
2	0.830	0.866	0.859	0.899	0.837	0.910	0.897	0.882
3	0.843	0.865	0.899	0.838	0.892	0.882	**0.911**	0.894
4	0.868	0.850	0.899	0.882	0.862	0.865	0.897	0.906
5	0.879	0.885	0.870	0.896	**0.834**	0.874	0.872	0.865
6	**0.889**	0.881	0.892	0.886	0.857	0.890	0.891	0.851
7	0.858	0.868	0.858	0.846	0.869	0.849	0.862	0.893
8	0.877	0.853	0.886	0.878	0.860	0.888	0.861	0.880
9	**0.801**	0.839	0.877	0.843	0.876	0.874	0.875	0.861

Table 3. Maximum cross-correlation coefficient for finger flexion

	PL & FDS				ED & ECU			
Subject	25%	50%	75%	100%	25%	50%	75%	100%
1	0.893	0.868	0.870	0.849	0.901	0.874	0.851	0.870
2	0.890	0.901	0.891	0.873	0.861	0.886	0.894	0.901
3	**0.921**	0.819	0.871	0.921	0.849	0.861	0.872	0.869
4	**0.818**	0.889	0.874	0.898	0.877	0.886	0.852	0.836
5	0.837	0.879	0.887	0.826	0.877	0.837	0.886	**0.824**
6	0.827	0.876	0.849	0.826	0.826	0.844	0.865	0.885
7	0.851	0.849	0.866	0.852	0.894	0.871	0.871	0.866
8	0.848	0.842	0.900	0.878	**0.943**	0.831	0.859	0.886
9	0.859	0.852	0.884	0.906	0.849	0.868	0.838	0.876

Table 4. Maximum cross-correlation coefficient for pinch grip

	PL & FDS				ED & ECU			
Subject	25%	50%	75%	100%	25%	50%	75%	100%
1	**0.911**	0.848	0.879	0.871	0.901	0.918	0.889	0.882
2	0.886	0.863	0.890	0.892	0.904	0.860	0.859	**0.921**
3	0.854	0.890	0.865	0.881	0.867	0.879	0.898	0.843
4	0.896	**0.819**	0.854	0.862	0.864	0.885	0.899	0.878
5	0.866	0.907	0.859	0.892	**0.834**	0.858	0.879	0.889
6	0.858	0.851	0.848	0.882	0.893	0.857	0.903	0.872
7	0.868	0.877	0.868	0.904	0.919	0.914	0.888	0.906
8	0.887	0.881	0.877	0.863	0.919	0.858	0.858	0.873
9	0.905	0.827	0.866	0.845	0.857	0.890	0.839	0.872

Based on the summary of maximum cross-correlation function in Table 5, the results indicate that the range level of cross-correlations for extensor side muscle group were higher than flexor side muscle groups for all hand gestures exerted for isometric contraction increment of force (25%, 50% 75% and 100%). This finding is in agreement with the finding of [2] and [5] where extensor muscles group showed high cross-correlation coefficient compared to flexor muscle even for difference of hand gestures or exercise motions specifically on forearm muscle contractions.

In addition, this study found that extensor side muscles were most convenient in determining the cross-talk level of isometric contraction on forearm muscles for MMG signal compared to flexor side muscles based on results obtained. Furthermore, the obtained results show the maximum cross-correlation coefficients did not increase with isometric contraction and force level for both flexor and extensor side muscles for many of the participants. Finally, based on summary on Table 5 it can be concluded that cross-correlations coefficient performed were classified as large based on the recommendations by COHEN since the maximum cross-correlation coefficients for all muscle groups and hand exercise gestures were in the 0.8 to 0.9 range.

Table 5. Summary of maximum cross-correlation coefficients for all gesture

Gestures	Flexor Side (PL & FDS)	Extensor Side (ED & ECU)
GSS	0.883 – 0.919	0.826 – 0.926
GSP	0.801 – 0.899	0.834 – 0.9.11
FF	0.818 – 0.921	0.824 – 0.943
PG	0.819 – 0.911	0.834 – 0.921

4. Conclusion

In this work, maximum cross-correlation coefficient was analyzed for different hand exercise gestures with various force level exerted (25%, 50%, 75% and 100%) to determining which compartmental muscle groups shows high level of association between MMG signals detected in flexor (PL & FDS) and extensor (ED & ECU) side muscles during isometric muscle action of forearm. Presented results of cross-correlation coefficient between flexor and extensor side muscles indicates that the maximum range value of cross-correlation coefficient for extensor side muscle groups were slightly higher than the flexor side muscle groups for all hand exercise gestures exerted. Further studies should examine the cross-talk level quantification of isometric contraction using cross-correlation function as well as isometric contraction during fatiguing muscle actions since cross-talk value can be analysed from the maximum value of cross-correlation coefficient.

Acknowledgments

The authors would like to acknowledge the support from the Fundamental Research Grant Scheme (FRGS) under a grant number of FRGS/1/2015/TK04/UNIMAP/02/5 from the Ministry of Higher Education Malaysia.

References

[1] X. Chen, Y. P. Zheng, J. Y. Guo, Z. Zhu, S. C. Chan, and Z. Zhang 2012 Sonomyographic responses during voluntary isometric ramp contraction of the human rectus femoris muscle *Eur. J. Appl. Physiol.*, vol. **112**, no. 7, pp. 2603–14

[2] J. P. M. Mogk and P. J. Keir 2003 Crosstalk in surface electromyography of the proximal forearm during gripping tasks *J. Electromyogr. Kinesiol.* vol. **13**, no. 1, pp. 63–71

[3] T. W. Beck, M. A. Dillon, J. M. DeFreitas, and M. S. Stock 2009 Cross-correlation analysis of mechanomyographic signals detected in two axes *Physiol. Meas* vol. **30**, no. 12, pp. 1465–71

[4] C. Jacob 1988 Statistical power analysis for the behavioral sciences (New Jersey: Lawrence Erlbaum Associates Inc. Hillsdale)

[5] M. A. Islam, K. Sundaraj, R. B. Ahmad, S. Sundaraj, N. U. Ahamed, and M. A. Ali 2014 Longitudinal, lateral and transverse axec of the forearm muscle influence the crosstalk in the mechanomyographic signals during isomectric wrist postures *PLoS One*, vol. **9**, no. 8, pp. 104-280

[6] M. A. Islam, K. Sundaraj, R. B. Ahmad, S. Sundaraj, N. U. Ahamed, and M. A. Ali 2014 Cross-talk in mechanomyographic signals from the forearm muscles during sub-maximal to maximal isometric grip force *PLoS One*, vol. **9**, no. 5, pp. 1–9

[7] Y. K. Kong, M. S. Hallbeck, and M. C. Jung 2010 Crosstalk effect on the surface electromyogram of the forearm flexors during a static grip task *J. Electromyogr. Kinesiol.*, vol. **20**, no. 6, pp. 1223–29

IOP Conf. Series: Materials Science and Engineering **705** (2019) 012015 doi:10.1088/1757-899X/705/1/012015

Moving Cell Array Prediction for LORAWAN-Handover Based On Received Signal Strength Indicator (RSSI) Algorithm

M M M Zabidi, M T A Rahman*, A Rahman, H A Munir, A H Adom and A F A Hamid

School of Mechatronic Engineering, Universiti Malaysia Perlis (UniMAP), Kampus Pauh Putra, 02600 Arau, Perlis, Malaysia

*Corresponding author: tasyrif@unimap.edu.my

Abstract. This paper discusses the prediction of movement from the vehicles with Long Range Wide Area Network (LoRaWAN) handover communication, based on Received Signal Strength Indicator (RSSI) algorithm. The experiment consists of observation on three different environment area which is rural, sub-urban and urban at road in Kedah and Perlis state in Malaysia. This three-environment area measured based on the past studies. Handover session with a moving vehicle, predicted the movement based on direction of Global Positioning System (GPS) latitude and longitude. The communication session transferring from previous cell area network that pass by the vehicle to a new targeting cell area network with the base station connection. From this experiment and study, it is shown that the predictable or movement from one cell to another cell based on RSSI algorithm with base station connectivity can be achieved.

1. Introduction

Nowadays, Intelligent Transportation System (ITS), rapidly grows and shared any kind of information, communication and technologies that applied to the transport infrastructure in automotive areas. ITS been using widely for real-time information and data observation to improve safety and security to the transportation based on mobile device sensor using the internet of things (IoT) concept and development. A significant research has developing many technologies devices, that can predict or calculate any information based on specific algorithm or method. Long-range-wide-area-network or LoRaWAN is one of the technology devices (non-cellular) using a star network topology to communicate between each node directly with another gateway for wide range of signal [1]. LoRaWAN improves capability of high communication link with low cost and power consumption by using the modulation technique. Chirp spread modulation is the technique used by LoRaWAN to encoded the data.

Besides that, the distance range of communication can be up to (10 kilometer) [2] which can be developed into handover network system for LoRaWAN to predicted the data more accurate and precisely. But the main question is, how the predictable process can be sure whether there is a way of enabling this whole new set of devices with precise location path information by utilizing the fact that these devices are continuously connected to the network. The purposes of this study is to developed handover system LoraWAN using RSSI based algorithm to predicts which base station to transfer communication from previous base station to new base station. The system is achieved by utilizing the mobile speed and direction information from Global Positioning System (GPS) of the latitude and longitude points. Figure 1, show the hexagonal cell area where inside located the base station (BS) of the communication. This base station used for the predictable movement of vehicles from one cell to another cell for the continuous communication of the LoRaWAN.

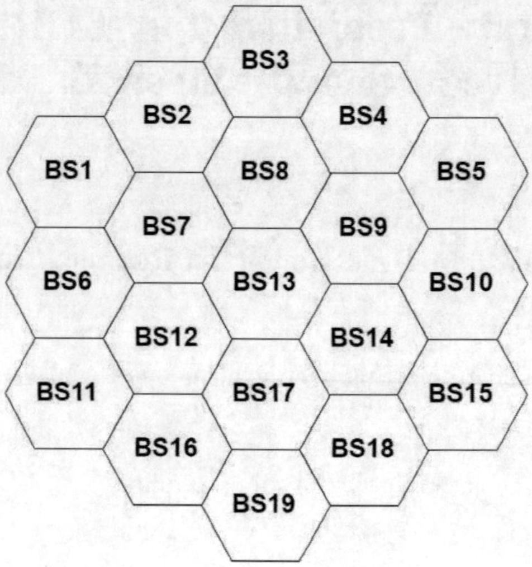

Figure 1. Hexagonal cell system

2. Existing Studies

The study by Radhwan Mohamed Abdullah et al [3] show the improvised vertical handover decision algorithm by integrating multi-criteria within a wireless network. The concept and idea are to propose an enhanced algorithm for vertical handover decision based on multiple criteria, thus enabling the devices make decision for the correct and accurate handover decision. The outcomes from the simulation performed exhibited that the proposed algorithm enhanced the probabilities of handovers, when compared to the conventional algorithm for network decision. Next, study by Wixted et al [4], discusses the performance and reliability of LoRa wireless and LoRaWAN in a typical environment. This was planned as a practical exercise to develop the familiarity of both system at Glasgow.

Besides that, study by Sarra Naoui et al [5] propose solution to enhance the LoRaWAN security protocols, where the established node at the end and the network server is chosen by trustworthy nodes as proxies that can avoid the attack from malicious nodes. This showed that the use of LoRaWAN are in secured connection whether from to server or to the connection to the base station. The others study is using radio raging technique such as RSSI is very popular due to lower cost and its simplicity compared to other method of radio raging technique, although there are many models viable to analyzes the path loss model, not many of them could use in certain unique environments [6].

3. Experiment Setup and Environment Area

The communication node hardware in this study was developed with two devices of Arduino Uno, two modules of LoRa, where one is for transmitting the data and another is for receiving the data. This process connected with the network switcher for the process transferring the communication session from one cell to another. The placement or the height base for sender node is set to 1.5m from ground and the receiving node taken far from the sender node to establish the communication. The environment area for this experiment is tested in three different area in Perlis (Arau-Changlun Road, Kangar Road) and Kedah (Bukit Kayu Hitam-Changlun Road) state in Malaysia as shown in Figure 3. This is to show the signal propagations communication from the LoRa devices with base station from three different area. The settings of LoRa is set that to sending and receiving the RSSI method based on the speed direction of the vehicle and the latitude and longitude from GPS for the prediction movement. Figure 2 show the setup for the experiment.

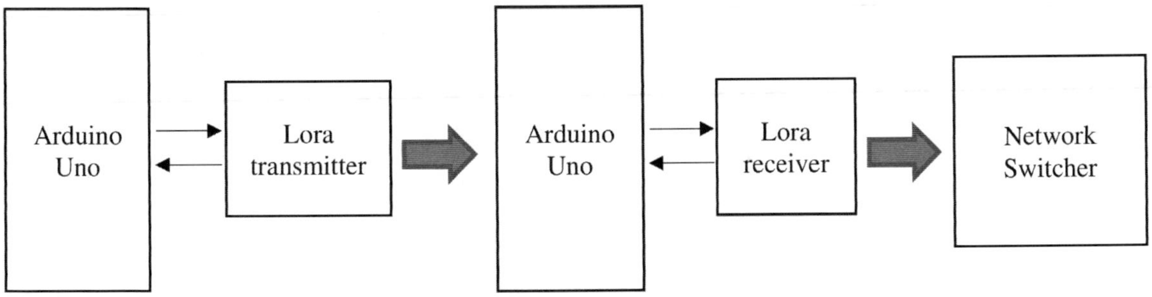

Figure 2. Experiment setup for the LoRa prediction

(a) Rural area (Arau-Changlun Road)

(b) Sub-urban area (Bukit Kayu Hitam-Changlun Road)

(c) Urban area (Kangar Road)

Figure 3. Environment area in Perlis, Malaysia

4. Handover Prediction

The prediction of the vehicle movement from one cell to another cell is based on the previous study of LoRaWAN with RSSI value. The tabulated graph in Figure 4, shown the data measured on RSSI with the distance from 1000 meter to 5000 meters. The RSSI values over the distance for the rural area (black colour) is between -60dBm to -90dBm. When the distance movement of the vehicle increased and the RSSI value is -100dBm to -120dBm, the prediction of the vehicle movement is onto the sub-urban area (blue colour) and from -120dBm to -140dBm is in the urban area (red colour). Next, Figure 5 displays the prediction process flow with hexagon cell, which the (base station 13) green colour is the active base station and the others is the possible vehicle movement to neighbour base station. The optimal RSSI on each end of the wireless link is between -48 dBm and -160 dBm to achieve the highest possible data rates.

For the movement of the vehicle from the active base station are based in the direction methods on the right of the hexagonal cell. The direction is fixed with six possible movement based on direction (in degree value) from the GPS. The process started when the vehicle moves from the active base station (base station 13) to the target base station in red (base station 8). When the GPS detected the direction are 0°, then the prediction movement of the vehicle is towards to the direction 1. The flow process of the prediction movement based on the latitude and longitude from GPS is shown in Figure 6. The prediction is using to know exactly where the movement of the vehicle towards to the base station. For examples if the GPS is detected on 180 degrees movement then the vehicle is move to the direction 4 as shown in figure 5. Therefore, we can predict that the movement from (BS13) of the vehicle are moves to the base station 17.

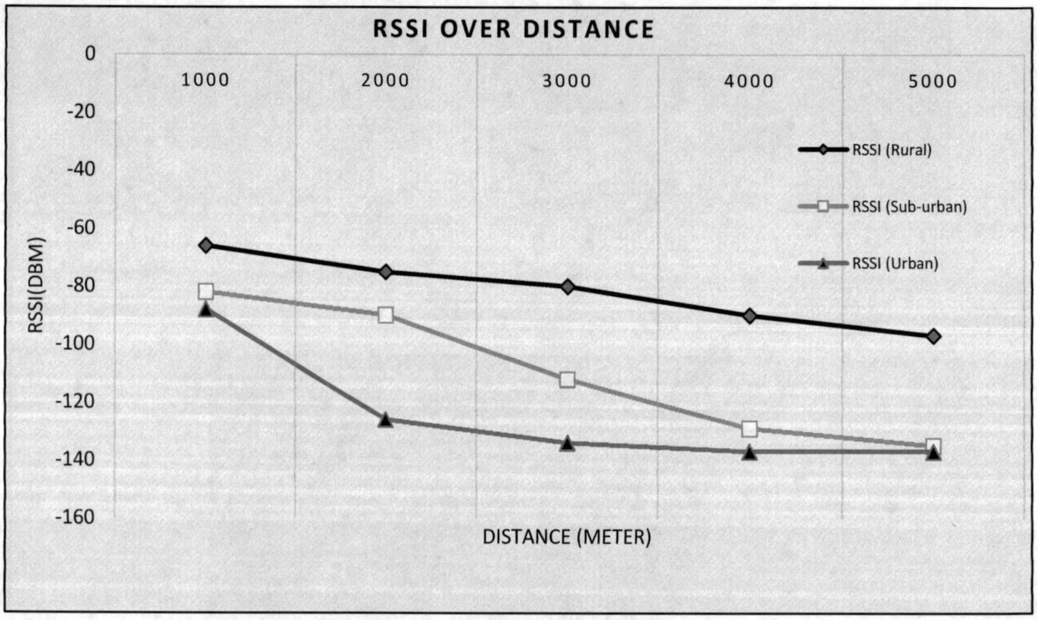

Figure 4. The data of RSSI vs Distance

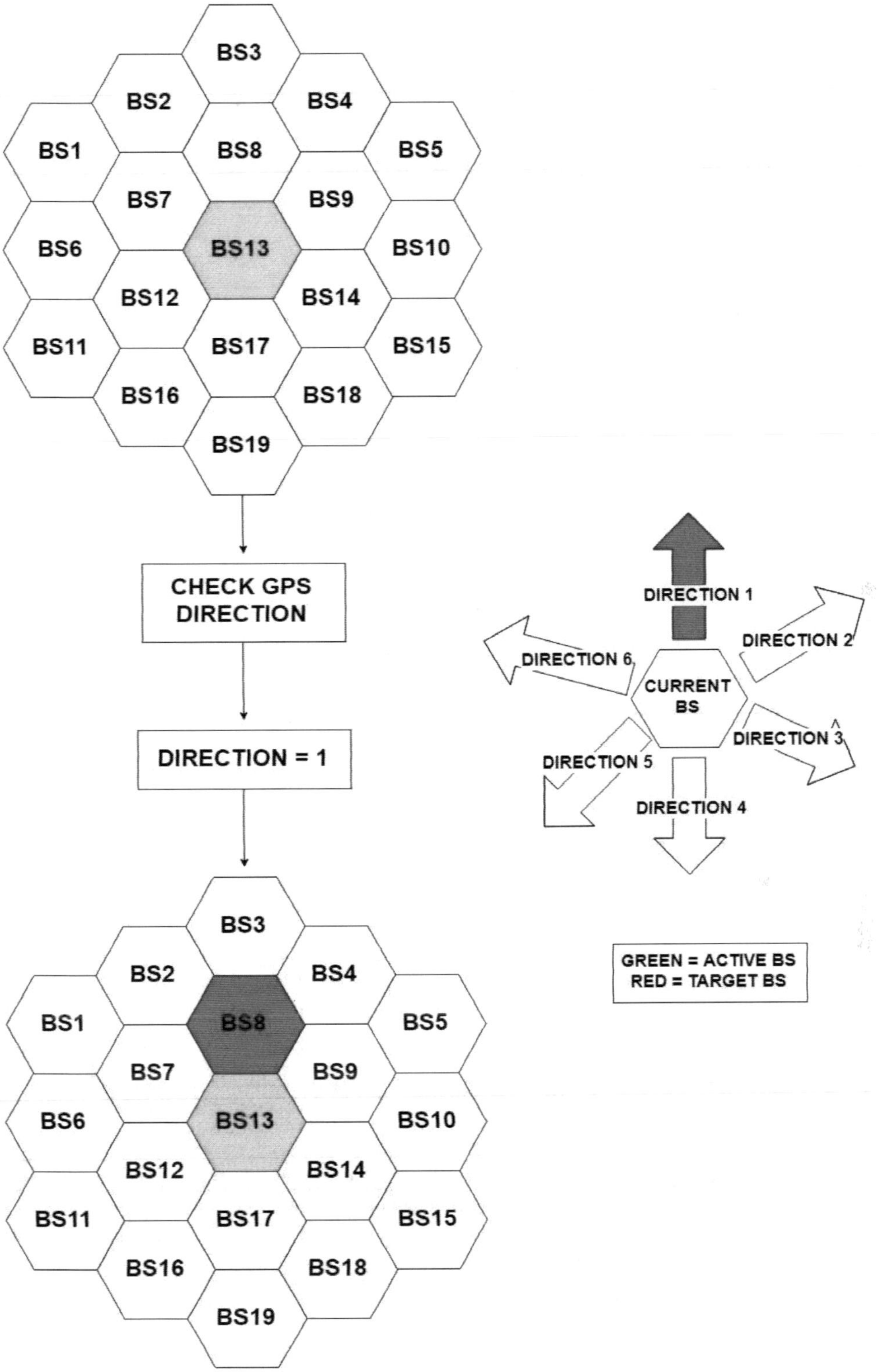

Figure 5. The prediction of vehicle movement from active base station to target base station

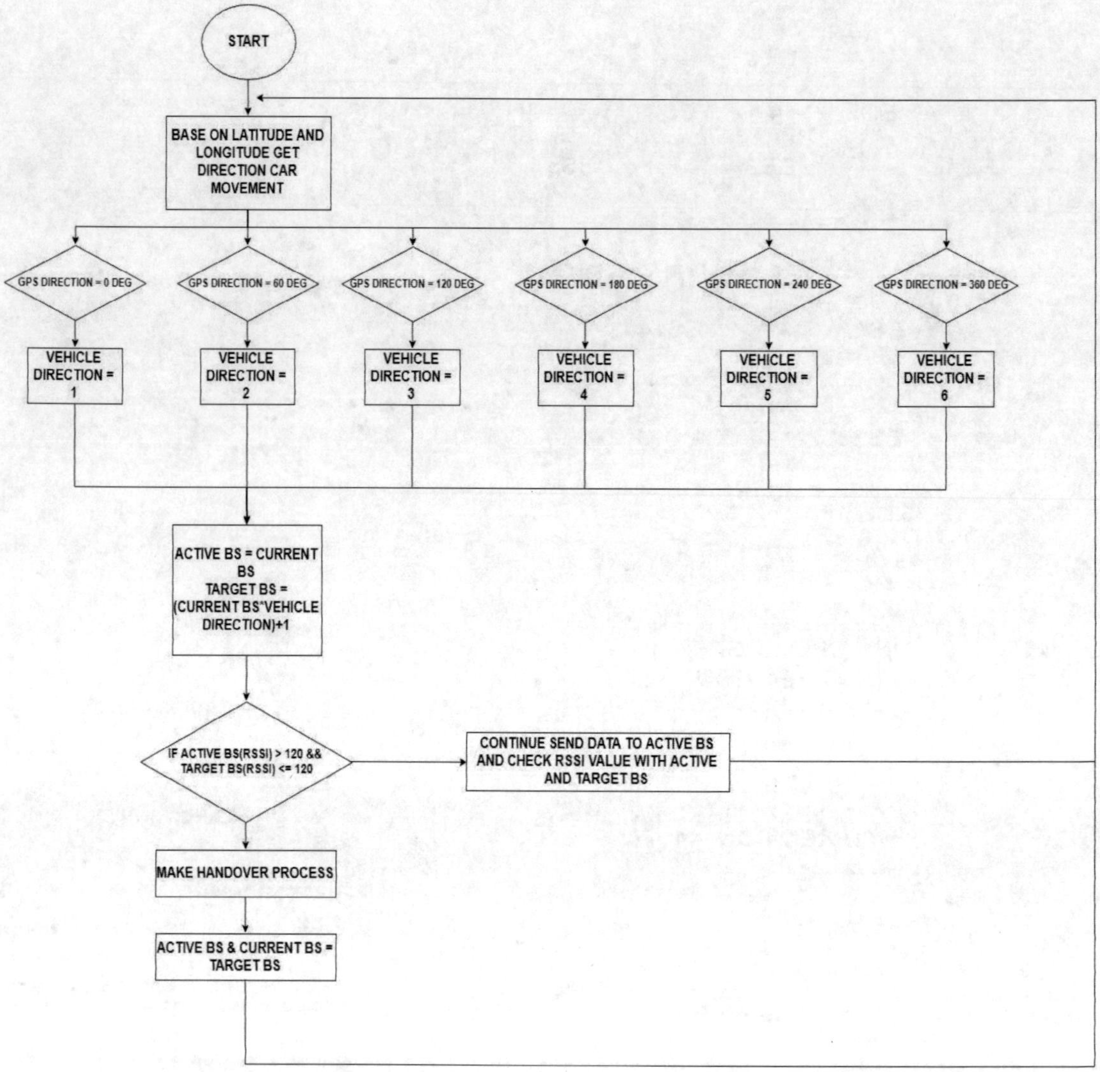

Figure 6. The flowchart process of the prediction based on GPS direction.

5. Result and Discussion

Aside from the process of handover propagation issues in the mobile communication, handover delay will happen when the communication node is moving between two or more cell network that cover by base station. Communication handover is a process when the base station cell network transferring ongoing communication link form current cell network area into a new cell network area that pass by the vehicle like in the Figure 7. During this handover process, the ongoing communication link will also have a delay in data transferring sequence. Figure 8 shown the result of the handover delay occurred when the process of the movement happened by measured the process between two base station is occurs. As from the experimental and measurement based on data from the previous study shown that, the higher RSSI value towards into another cell, the better handover process.

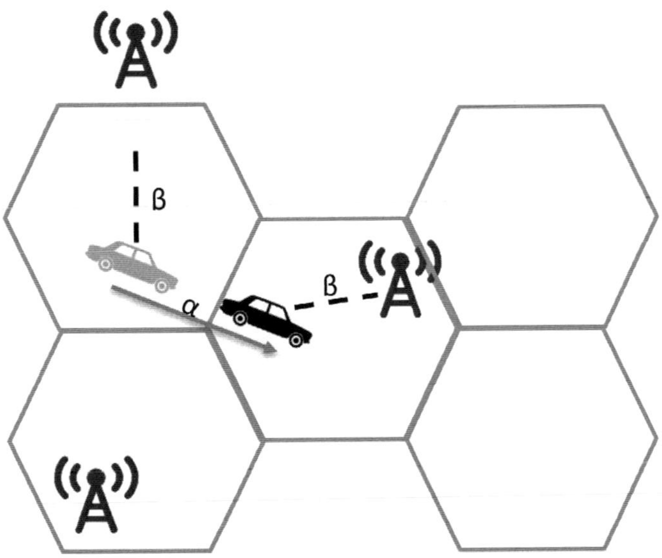

Figure 7. The vehicle that move from one cell network area to another cell network area where the α is direction of vehicle movement and β is communication link between the vehicle and base station.

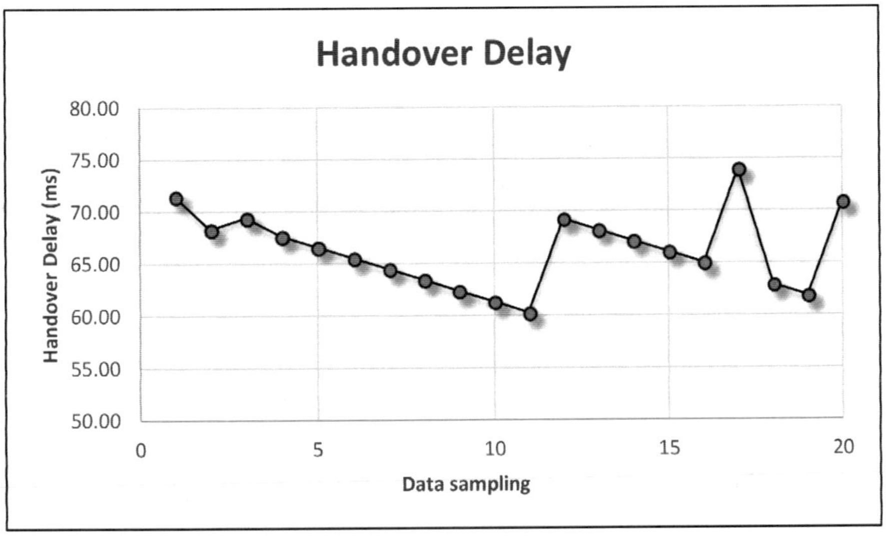

Figure 8. Data sampling for the handover LoRaWAN delay

6. Conclusion

This study explored full potential of the handover LoRaWAN system for the connection and predictable movement of the vehicle by using RSSI value and GPS connection with base station. The studies show that RSSI value is important in this study to exactly predictable the movement of the vehicles from active base station to the targeting base station. In the meanwhile, 5G technology will be introduced in next few years. With this implementation of communication technology, the amount of base station is expected to increase too. With huge amount cell network base station, the performance of handover process will become significantly important. From this study, it can be concluded that the prediction of the vehicle movement from one cell to another targeting cell with handover LoRaWAN device using RSSI algorithm and GPS to get the direction from base station can be predicted and developed.

Acknowledgements
I would like to extend our sincere gratitudes for the funding provided by the Universiti Malaysia Perlis (UniMAP) (Geran Penyelidikan Khas - No.9004-00064) and UniMAP Automotive Racing Team, which enable this research to be carried out.

References

[1] A. Lavric and V. Popa 2017 *A LoRaWAN : Long Range Wide Area Networks Study*
[2] A. Jain and S. Tokekar 2015 *Application Based Vertical Handoff Decision in Heterogeneous Network Procedia Comput. Sci.*vol. 57
[3] R. Mohamed, A. Zaid, and A. A. Alwan 2018 *ScienceDirect ScienceDirect Improved Handover Decision Algorithm Using Multiple Criteria Procedia Comput. Sci.* vol. 141
[4] A. J. Wixted, P. Kinnaird, H. Larijani, A. Tait, A. Ahmadinia, and N. Strachan 2017 *Evaluation of LoRa and LoRaWAN for wireless sensor networks Proc. IEEE Sensors*. vol. 0
[5] S. Naoui, M. E. Elhdhili, and L. A. Saidane 2016 *Enhancing the security of the IoT LoraWAN architecture 5th IFIP Int. Conf. Perform. Eval. Model. Wired Wirel. Networks. PEMWN*
[6] N. Alam, A. T. Balaie, and A. G. Dempster 2010 *Dynamic Path Loss Exponent and Distance Estimation in a Vehicular Network using Received Signal Strength*
[7] M. Ben, H. Frej and C. Bach 2014 *Robust Prediction for Intra and Inter Handoff (RPIIH) Process over CDMA2000*
[8] J. Lian and J. Guo 2009 *Performance Analysis of Soft Handover Based on FFT in CDMA System Int. Conf. Comput. Intell. Softw. Eng*
[9] M. Aref and A. Sikora 2014 *Free Space Range Measurements with Semtech LoRa TM Technology 2nd Int. Symp. Wirel. Syst. within Conf. Intell. Data Acquis. Adv. Comput. Syst.*
[10] O. C. Ozdural and H. Liu 2007 *Mobile Direction Assisted Predictive Base Station Switching for Broadband Wireless Systems*
[11] Y. Chen and J. Chen 2008 *A Fast Subcarrier , Bit , and Power Allocation Algorithm for Multiuser OFDM-Based Systems* vol. 57
[12] A. F. A. Hamid, M. T. A. Rahman, S. F. Khan, A. H. Adom, M. A. Rahim, N. A. Rahim, M.H.N. Ismail, and A. Norizan 2017 *Connected car: Engines diagnostic via Internet of Things (IoT) J. Phys. Conf. Ser.*vol. 908. **12079**
[13] A. Norizan, M. T. A. Rahman, N. A. M. Amin, M. H. Basha, M. H. N. Ismail, and A. F. A. Hamid *2017 Study of intake manifold for Universiti Malaysia Perlis automotive racing team formula student race car J. Phys. Conf. Ser.* vol. 1. **12069**
[14] M. L. Mohamad, M. T. A. Rahman, S. F. Khan, M. H. Basha, A. H. Adom, and M. S. M. Hashim 2017 *Design and static structural analysis of a race car chassis for Formula Society of Automotive Engineers (FSAE) event J. Phys. Conf. Ser.*vol.908. **12042**)
[15] M. Azmeer, M. H. Basha, M. F. Hamid, M. T. A. Rahman, and M. S. M. Hashim 2017 *Design optimization of rear uprights for UniMAP Automotive Racing Team Formula SAE racing car J. Phys. Conf. Ser.* vol. 908. **12051**
[16] Z. I. Zaman, K. S. Basaruddin, M. H. Basha, M. T. A. Rahman, and R. Daud 2017 *Finite element prediction on the chassis design of UniART4 racing car AIP Conf. Proc.* vol. 1885. **1– 10**
[17] Rahman A, Rahman MTA, Manaf EHA, Rahman ASA 2018 *Design and Analysis of Impact Attenuator for a Formula Student Car: A Study between Singular and Bi-tubular Tubes of Varying Geometries IOP Conf Ser Mater Sci Eng*
[18] E. H. A. Manaf, M. T. A. Rahman, A. R. A. Syayuthi, F. Mat., and A. Rahman 2018 *Crashworthiness analysis of empty singular aluminium tubes with varying shapes in LS-Dyna, AIP Conference Proceedings 2030*, **020115**

IOP Conf. Series: Materials Science and Engineering **705** (2019) 012016 doi:10.1088/1757-899X/705/1/012016

Design and Development of Three-Phase Voltage Source Inverter for Variable Frequency Drive

A S Nurhaida[1], A Z Ahmad Firdaus[1], Kamarulzaman Kamarudin[1], S A Azmi[2], M J M Ridzuan[1], S M Hafis[1]

[1]School of Mechatronic Engineering, Universiti Malaysia Perlis

[2]School of Electrical Systems Engineering, Universiti Malaysia Perlis

kamarulzaman@unimap.edu.my

Abstract. Three phase induction motors are the workhorse of industry. Variable speed operation of the motor enables substantial energy savings through implementation of variable frequency drives. This project concerns on the design and implementation of three-phase voltage source inverter (VSI) for variable frequency drive. The focus was to generate variable frequency output suitable to be fed to the induction motor for the purpose of variable speed control. SPWM signal was generated through SIMULINK algorithm. The VSI circuit was designed, fed with SPWM trigger signal and simulated for variable frequency output in terms of line-to-neutral and line-to-line voltage across R-L load. In the experimental setup, the real VSI system was developed, the SPWM trigger signal generated through SIMULINK and interfaced to hardware using Arduino microcontroller. The output is dropped across resistive load and parameters line-to-neutral and line-to-line voltage was measured. Three set of output voltage with frequency of generated at 20Hz, 50Hz and 70Hz. Thus, experimental result validated that variation of frequency of voltage output of the VSI has been successfully achieved.

1. Introduction

Three phase induction motors are the workhorse of industry, constituting to about 85% of all types[1]. This is attributed to their simplicity, ruggedness, low-cost and ease of maintenance[2]. They are mostly utilized to drive compressor, blower fans, pumps and conveyors in most industry ranging from palm oil, sugar and rice mill, to cement and steel plant. Previously, they were only operated as fixed speed due to fixed supply nominal frequency. Compared to fixed-speed drive, variable speed operation of induction motor performed superior with better efficiency, energy savings and significant cost reduction through less frequent downtime for maintenance[1], [2]. In the sugar mills, it is used for centrifugal pump where the speed control have been applied to reduce energy usage whereby, 50% energy savings could be attain by reducing 20% of pump or fan speed[3].

Adjustable speed drive operation could be attained through several ways. Mechanical means however, lacks flexibility of speed change during operation. For wound rotor type of induction motor, external variable resistor connected to slip ring of the motor enabled adjustable speed operation. However, operation through variable resistance was highly inefficient due to high loss[4]. For squirrel cage induction motor though, a more popular method to implement adjustable speed drive operation would be through variable frequency drive.

IOP Conf. Series: Materials Science and Engineering **705** (2019) 012016 doi:10.1088/1757-899X/705/1/012016

The variable frequency drive (VFD) speed control is based upon the concept of varying the frequency of the three-phase alternating current (AC) voltage supply to the stator of induction motor. As the stator frequency is varied, the synchronous speed and thus the rotor speed can be varied[4]. Figure 1 shows a topology of variable frequency drive connected to induction motor. Three phase AC source is rectified to DC using diode, and then fed to three phase voltage source inverter (VSI) where the three-phase output is fed to the stator of induction motor. The frequency of the output can be varied by controlling the switching of the VSI. Previously, Silicon Controlled Rectifiers (SCR), was used as switches. However, SCR lack switching OFF control therefore, extra auxiliary circuitry needed to be integrated to control the VSI. With implementation of MOSFET and IGBT transistors as switches, pulse control becomes manageable and can be generated through computer algorithm. MATLAB/SIMULINK software provides platform of not only simulation of system design but also enables model deployment and hardware integration[5]. In this sense, more complex pulse such as Sinusoidal Pulse Width Modulation (SPWM) can be generated through the program and interfaced as signal output to the VSI hardware circuit.

This project concerns on the design and implementation of three-phase voltage source inverter for variable frequency drive. The focus was to generate variable frequency output suitable to be fed to the induction motor for the purpose of variable speed control. SPWM signal was generated through SIMULINK algorithm. The VSI circuit was designed, fed with SPWM trigger signal and simulated for variable frequency output in terms of line-to-neutral and line-to-line voltage across R-L load. In the experimental setup, the real VSI system was developed, the SPWM trigger signal generated through SIMULINK and interfaced to hardware using Arduino microcontroller. The output is dropped across resistive load and parameters line-to-neutral and line-to-line voltage was measured. Three set of output voltage with frequency of will be generated at 20Hz, 50Hz and 70Hz to demonstrate variable frequency output.

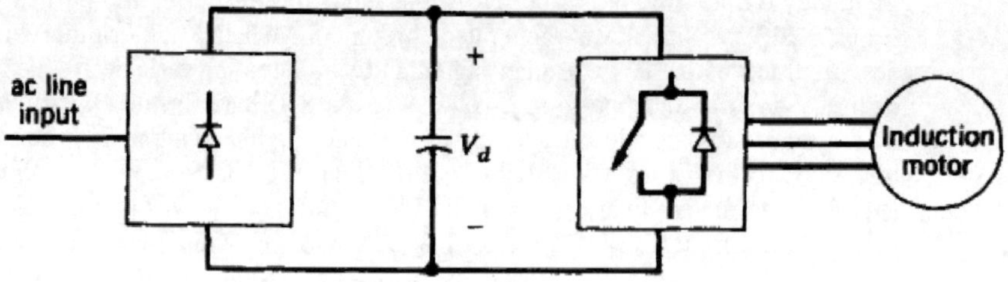

Figure 1: PWM VSI with diode rectifier[6]

2. Switching Control of the Three-Phase Voltage Source Inverter (VSI)

2.1. Three-Phase VSI

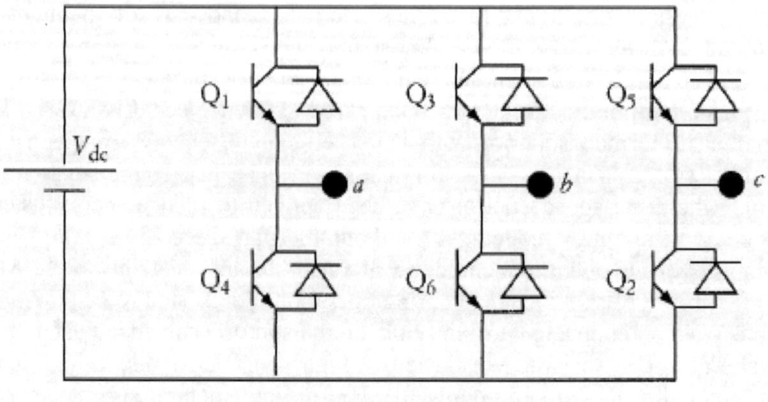

Figure 2: Three-phase VSI circuit[6]

Three-phase voltage source inverter (VSI) topology is shown in Figure 2. The VSI converts DC power to three-phase AC power. The input is DC source (usually rectified DC from nominal three phase AC supply) and the output is through three lines a,b and c. Each of the output line is connected through a leg consisting of top and bottom switches. Apparently six switches are employed where switch Q1 and Q4 for first leg, Q3 and Q5 for second leg and Q5 and Q2 for third leg. For each pair of switches on a leg only alternate switching is allowed to prevent short-circuit[6]. Usually IGBT or MOSFET are employed as switches and can be turned on and off based from gate pulses. The switching can be controlled through either one the following conduction modes; a) 120-degree or b) 180-degree[7], [8]. The switching of three-phase VSI using 120-degree conduction mode and resultant line to neutral voltage is shown in Figure 3. The conduction duration for each of the transistors is 120-degrees for every switching sequence, at every 60-degree duration in one-cycle. Compared to the 180-degree conduction mode, the 120-degree conduction mode requires only two transistors to be turned-on at every 60-degree duration and performs the six mode of operation within one cycle. Thus, major advantage of using 120-degree conduction mode is less susceptible to short circuit[8]. From Figure 3, peak voltage of line to neutral voltage is equal to 2/3 Vdc. In Figure 4, line-to-line voltage is obtained by measuring differential voltage of each of the three line to neutral voltage and the peak voltage is equal to Vdc.

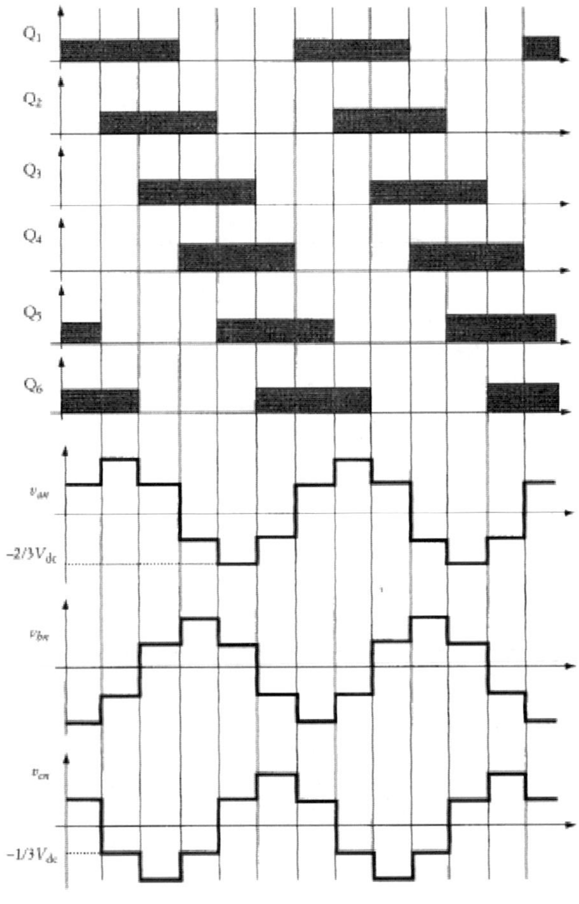

Figure 3: 120-degree conduction mode at top and line to neutral voltage waveform at bottom[6]

2.2. Sinusoidal Pulse Width Modulation (SPWM) Control Signal

If regular square wave pulse were fed through the gates of VSI switches, the line to neutral voltages would be such in Figure 3. However, total harmonic distortion (THD) of these signals would be as high as 30%[9]. Employing pulse width modulation (PWM) signal allows more precise control of voltage through controlled conduction in the pulse-width of the trigger signal[6]. In this case, since desirable output voltage waveform for a variable frequency drive is sinusoidal with low THD, sinusoidal pulse width modulation (SPWM) trigger signal is employed. SPWM is generated through comparison between triangular wave as the carrier signal and the three phase sinusoidal waves as the control signal[4]. Figure 4 illustrates how the comparison is made, in which three reference sinusoidal signal with fundamental frequency was set to be the same with inverter sinusoidal output, denoted as v_{ra}, v_{rb} and v_{rc}, phase-shifted by 120 degree from each other as control element. The carrier element is the triangular wave with much higher frequency. Comparator output denoted by 'g1' and 'g3' shown were two out of three channels SPWM pulse generated. The pulse would be fed individually to the gates of all six switches of the VSI. Figure 5 shows line-to-line voltage of inverter output generated through SPWM trigger.

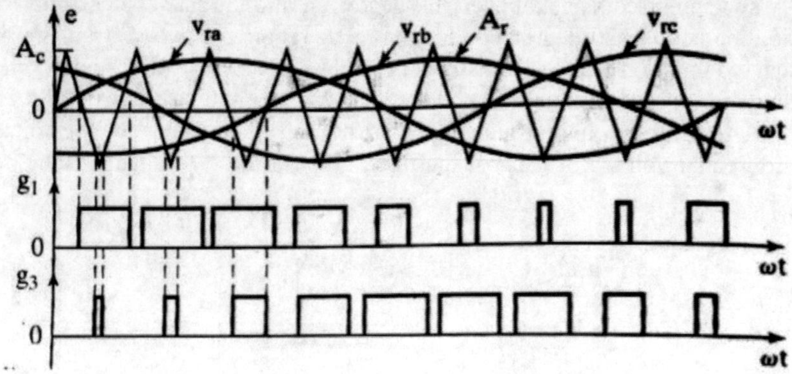

Figure 4: Comparison between carrier signal and three phase-shifted sinusoidal control signals for SPWM trigger pulse generation[4]

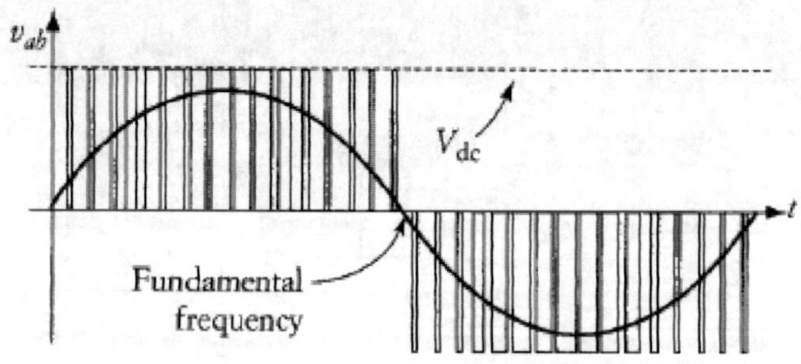

Figure 5: Line-to-line voltage of inverter output triggered by SPWM pulse[6]

2.3. Implementing Switching Algorithm for Three-Phase Voltage Source Inverter (VSI)

Based from method explained in Section 2.1 and 2.2, switching strategy was developed using SIMULINK block diagram as shown in Figure 6. Three control signals (namely phase A, B and C) 120-degree phase-shifted to each other with fundamental frequency set to be equal to inverter output frequency were compared to triangular carrier signal with frequency set to 5kHz. The comparator output consists of three channels containing phase-shifted SPWM signals. Output from phase A was fed to the gate of Q1 while inverted version fed to Q4. Output from phase B fed to Q3 and inverted version to Q6 while output from phase C and its inverted version fed to Q5 and Q2 respectively.

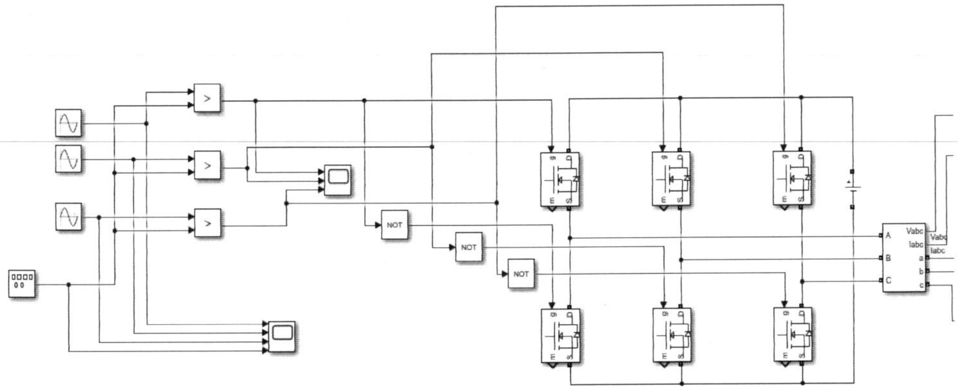

Figure 6: Switching control algorithm implementation for Three Phase VSI using SIMULINK blocks

3. Simulation of Variable Frequency Operation of Three-Phase VSI

In Figure 7, an open loop variable frequency drive system was developed, derived from conceptual diagram in Figure 1. Apparently, VSI output will be connected to drive three phase induction motor. For variable speed operation of induction motor, the VSI must be able to output sinusoidal voltage and current at different frequencies. In the proposed system, variation of frequency will be controlled real-time through SIMULINK input settings. Ultimately, signal generation algorithm will be produced through SIMULINK before being interfaced to hardware as trigger signals through Arduino. VSI was simulated through MATLAB/SIMULINK to generate line-to-neutral and line-to-line inverter output voltage at three different frequencies: 20Hz, 50Hz and 70Hz representing low, medium and high speed of induction motor. Figure 8 shows the simulation block diagram for variable frequency output.

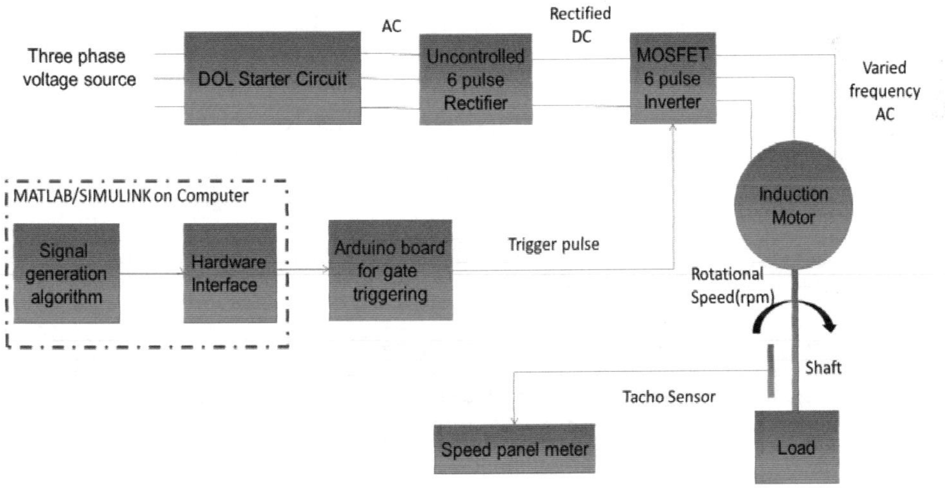

Figure 7: Open-loop architecture for variable frequency drive for three-phase induction motor

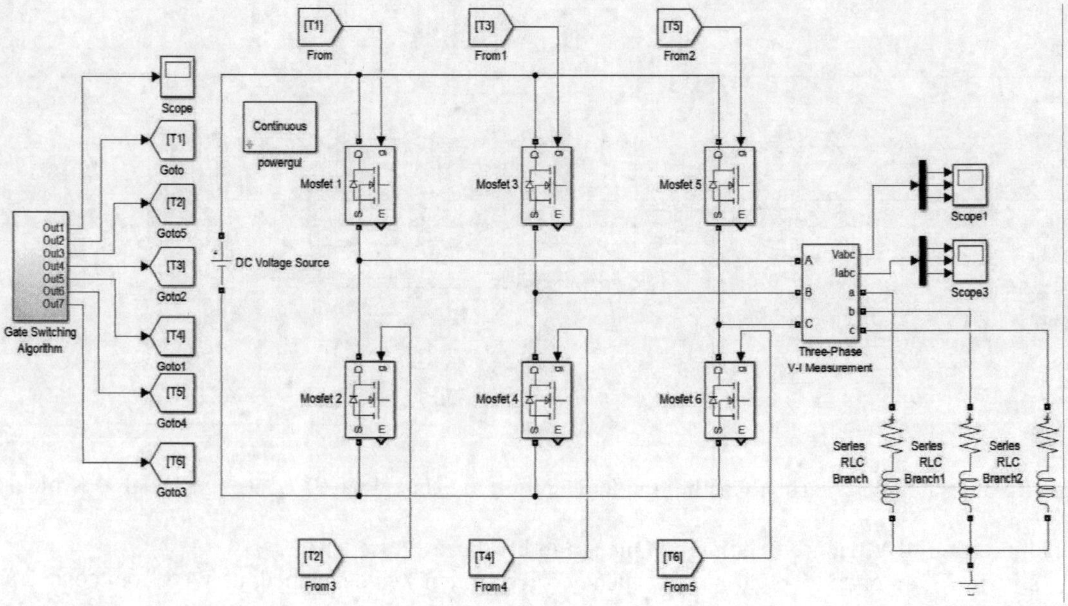

Figure 8: Simulation block diagram of the variable frequency operation of inverter circuit with RL-load

4. Interfacing Switching Algorithm from MATLAB/SIMULINK to hardware through Arduino

The development of the three-phase voltage source inverter (VSI) circuit is tested within SIMULINK for validation of the three-phase switching principle, along with the SPWM switching algorithm. Once the test is successful, the switching algorithm is 'debugged' and 'rebuild' via SIMULINK's code generation and uploaded to the Arduino micro-controller, which consecutively feeds the switching signals to the gates of the MOSFETs of the actual hardware setup. Figure 9 demonstrates the flowchart of how the algorithm was 'build' and uploaded to the micro-controller, while Figure 10 illustrates the switching signals measured from 2 of the six-legged MOSFETS, such that the signals are uploaded from SIMULINK to the Arduino micro-controller and fed to the gates of the three-phase MOSFETs.

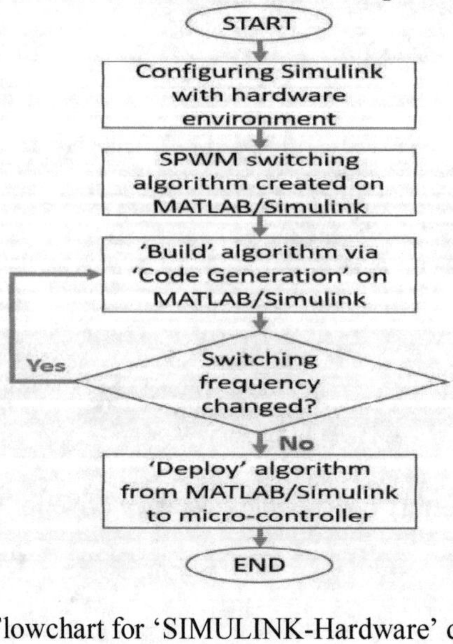

Figure 9: Flowchart for 'SIMULINK-Hardware' deployment

IOP Conf. Series: Materials Science and Engineering **705** (2019) 012016 doi:10.1088/1757-899X/705/1/012016

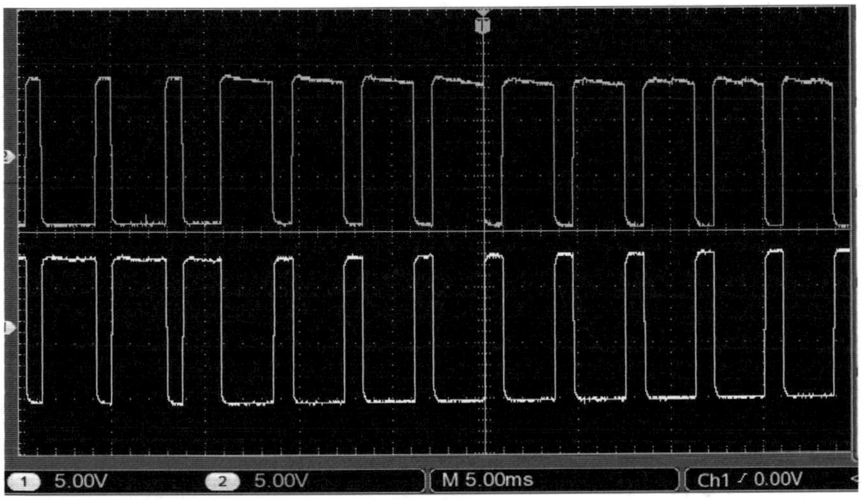

Figure 10: SPWM switching signal measured from three-phase MOSFET gates.

5. Experimental Setup

Figure 11 demonstrates the overall setup of between SIMULINK and the actual hardware of the three-phase voltage source inverter, and the connection between the Arduino micro-controller with the rest of the elements of the three-phase voltage source inverter (VSI). The components used to interface between SIMULINK and three-phase inverter circuit is the Arduino UNO micro-controller. The inverter circuit, which consist of the six-legged MOSFETs and six opto-couplers, in which the opto-couplers are used to isolate the three-phase inverter circuit from the computer section of the interface. Besides that, the opto-couplers were also used as an amplifier to amplify the signal output from the pins of the Arduino UNO microcontroller to an approximate voltage level of 15-volts. Figure 11 also shows how line-to-neutral voltage were measured using digital oscilloscope.

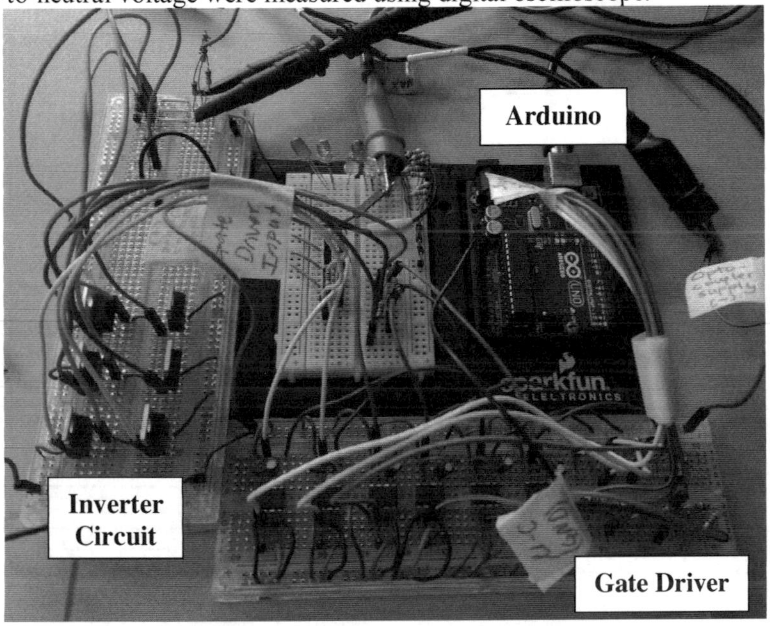

Figure 11: Connection between Arduino with the three-phase voltage source inverter (VSI)

IOP Conf. Series: Materials Science and Engineering **705** (2019) 012016 doi:10.1088/1757-899X/705/1/012016

6. Result and Discussion

6.1. Simulation Result of Three-Phase Inverter Circuit with RL-load

Figure 12 shows the waveform produced by the inverter circuit on SIMULINK, which was measured at the output of the RL-load, is of that replicates the modulation of a sinusoidal waveform. From the same figure, it can be seen that the peak voltage produced is approximately 546 volts (V). The blue, red and green voltage waveforms represent line to line voltage measured between line A-B, line B-C, and line C-A respectively.

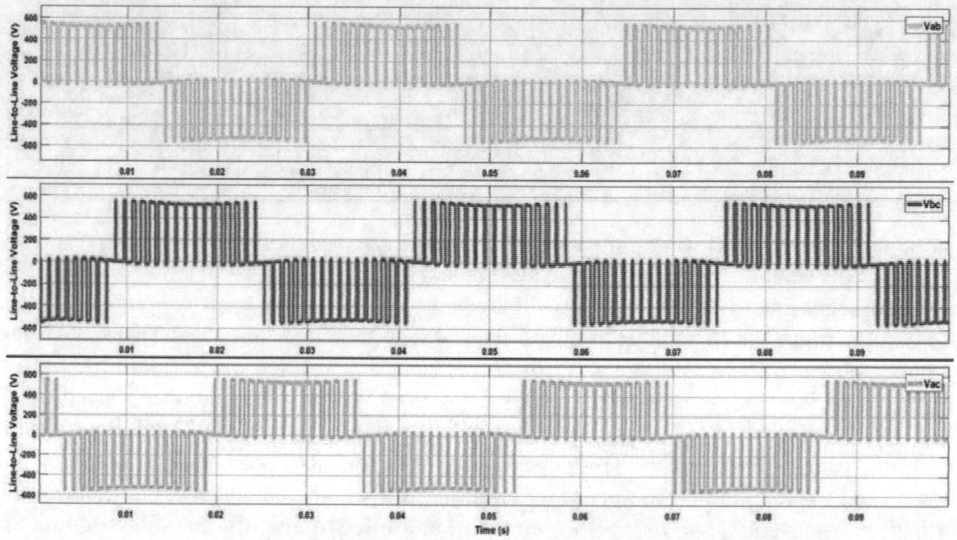

Figure 12: 'Line-to-Line' voltage waveform from three-phase VSI with RL-load

6.2. Comparison Between Simulation and Experimental Result

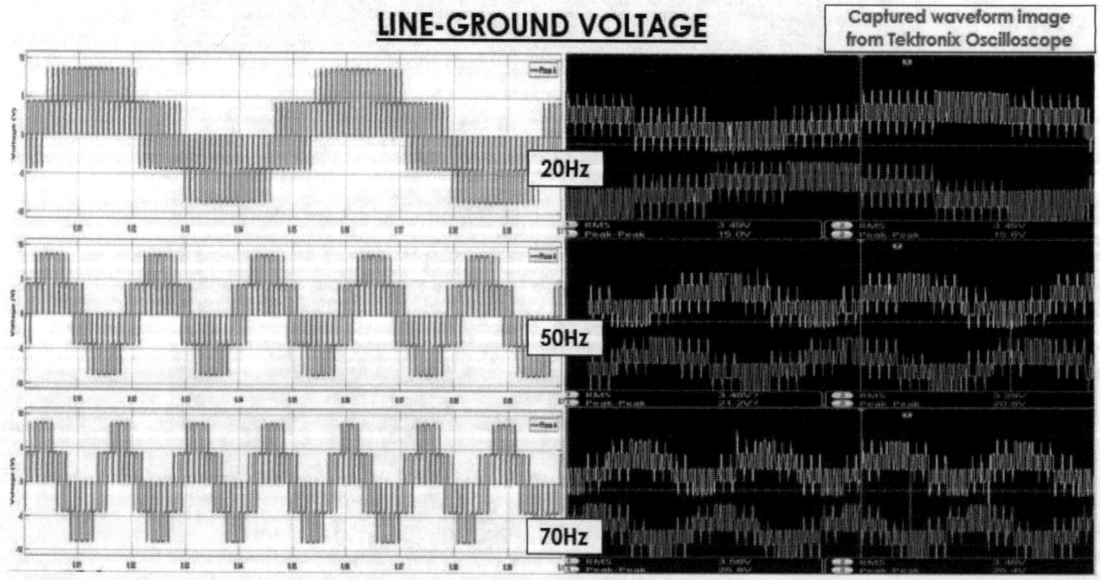

Figure 13: Line to neutral voltage generated through simulation (left) and experimental (right) at20Hz, 50Hz and 70Hz frequency

The simulated results and experimental results, in terms of 'line-to-neutral' voltages are compared and the outcome from the comparison are shown in Figure13.It was observed that the experimental circuit developed managed to generate variable frequency output for line to neutral voltage when tested for 3 different frequency settings: 20Hz, 50Hz and 70Hz. For line to neutral voltage, the experimental output voltage waveform resembles the waveform of simulation output. This validates the reliability of switching control design and hardware interface to generate variable frequency output suitable for variable frequency drive.

7. Conclusion

Experimental circuit developed managed to generate variable frequency output for line to neutral voltage when tested for 3 different frequency settings: 20Hz, 50Hz and 70Hz. For line to neutral voltage, the experimental output voltage waveform resembles the waveform of simulation output. This proves that the ability to generate the SPWM switching signal for a three-phase motor application is plausible through the creation of the switching algorithm via MATLAB/SIMULINK, such that the signals are then 'commenced' via a micro-controller as a 'medium' to begin a digital triggering to the inverter circuit of the comprised element of the overall three-phase variable frequency drive circuit. The limitation of this work is that the inability of the circuit to 'resist' unexpected voltage spikes, which unbeknownst to cause a sudden 'hike' in current flow which leads to component malfunctions, as well as reducing the longevity of the elements of the circuit. Hence, the future work that can be performed for this work is such the improvement of the circuit itself to resist internal electrical 'backlashes', such as the occurrence of the mentioned voltage spikes.

Acknowledgments

The author would like to acknowledge the support from the Fundamental Research Grant Scheme (FRGS) under grant number FRGS/1/2018/TK04/UNIMAP/02/15 from Ministry of Education Malaysia. Credits were also due to laboratory technicians from Automation and Basic Electronics Lab, School of Mechatronics Engineering for technical and equipment support.

References

[1] B. Wu 2006 *High-Power Converters and AC Drives*. Hoboken, NJ, USA: John Wiley & Sons.
[2] N. Mohan, T. M. Undeland, and W. P. Robbins 2003 *Power Electronics : Converters, Applications, and Design* (John Wiley & Sons)
[3] ESKOM 2016 *Variable Speed Drives: Reducing energy costs in the sugar industry*
[4] T. Wildi 2013 *Electrical Machines, Drives, and Power Systems* (Pearson)
[5] Arduino Support from Simulink - Hardware Support - MATLAB & Simulink. [Online]. Available: https://www.mathworks.com/hardware-support/arduino-simulink.html?s_tid=srchtitle. [Accessed: 30-Jul-2019]
[6] M. A. El-Sharkawi 2000 *Fundamentals of Electric Drives* (Brooks/Cole)
[7] D. W. Hart 2011 *Power Electronics* (Tata McGraw-Hill)
[8] M. H. Rashid 2009 *Power Electronics: Circuits, Devices, and Applications* (Pearson)
[9] S. Manias 2016 *Power Electronics and Motor Drive Systems* (Elsevier Science)

Effect of electromagnetic induction spectroscopy of femur bone on electromagnetic signal strength

M. Amran[1*], R. Daud[1], M.K. Ali Hassan[1], Z. Zakaria[1], M. I. Omar[2], F. Mat[1]
[1]Fracture and Damage Mechanics (FDM) SIG, School of Mechatronic Engineering Universiti Malaysia Perlis (UniMAP)
[2]University Health Center, Universiti Malaysia Perlis, Kampus Pauh Putra 02600 Arau, Perlis, Malaysia

*muhammad.bin.amran@gmail.com

Abstract. Bone implant is a promising technique to heal the fracture site where it focusing to stabilize and support the bone. Improvement in bone implant are required in order to ensure better healing process because some failures may happen due to the loss of implant stability. Previous studies showed there are lack of investigation on bone implant. Early detection of screw implant loosening is crucial for faster repairing and a healthy ready-to-operate. The purpose of this study is to investigate the signal strength of electromagnetic field induction on human bone by employing a single channel magnetic induction spectroscopy using flat spiral coil sensor. This study also to evaluate the various type of coil sensitivity of the magnetic induction spectroscopy on femur bone. The electromagnetic field induction will be analyse using the COMSOL Multiphysics simulation software to collect the data of received signal strength. The promising experimental results suggested the feasibility of the proposed electromagnetic induction spectroscopy.

Keywords: electromagnetic field, magnetic induction spectroscopy, bone implant, femur bone.

1. Introduction

Human bone is classified into two types which are cortical (dense and compact) and trabecular bone (spongy, honeycomb-like structure). Femoral (femur) bone is a cortical bone which have to be strong to support the body weight and daily routine activity. Human bone fracture is a common traumatism due to the aging population and the wide incidence of osteoporosis [1]. Proximal femoral Fractures account for a large proportion of hospitalization among trauma cases [2]. Fracture fixation used to stabilize the fractured bone, to enable fast healing of the injured bone, and to return early mobility and full function of the injured extremity [3]. In particular, implant failure remains a problem for unstable fractures, despite improved techniques and various implant modifications [4]. The use of magnetic induction in bone implant is discovered in this paper for further study to analyse bone implant failure in human bone.

The magnetic induction concept recognized as Electromagnetic Induction Spectroscopy (EMIS) was used to detect hidden metals and discriminate among the types of metals using multi-frequency approach based on metal's unique Passive Electrical Properties (PEP) as well its geometry and shape [5]. The conductivities, permeability and permittivity values from the MIS system can be used in medical diagnostic purposes which have been proved by H. Schafetter [6]. Magnetic induction spectroscopy (MIS) is actually a multi-frequency measurement version of the magnetic induction tomography (MIT) [7] where generated magnetic fields and its perturbed signals are gathered by receiver coils in the form of voltages (Vo) and ΔV respectively. Therefore, a voltage difference that is derived from the summation of both aforementioned magnetic signals collected from the receiver coil estimates the ultimate inductive phase shift. MIS technique was then further explored and developed towards detecting pathological conditions, the detection of jaundice in new born baby, a condition where applying magnetics field on baby's finger to detect jaundice level by using transmitter coil and receiver coil which was studied in [8].

2. Methodology

Magnetic Induction Spectroscopy (MIS) and Magnetic Induction Tomography (MIT) has same principle which is the non-invasive measurement [8]. MIT normally applied to construct the change of the conductivity by using alternating magnetic fields in a specified object. The common used of MIT is for non-invasive monitoring for example, monitoring of edema in human brain [9]. Figure 1 below show the ideal concept of MIS system with a transmitter coil and its magnetic field around the coil penetrates into the object near it. To produce the magnetic field, MIS system required the transmitter and receiver coil as the sensor and applied current [10]. The received magnetic field gathered at the receiver was then calculated to measure the magnetic induction phase shift of the MIS applied on the model.

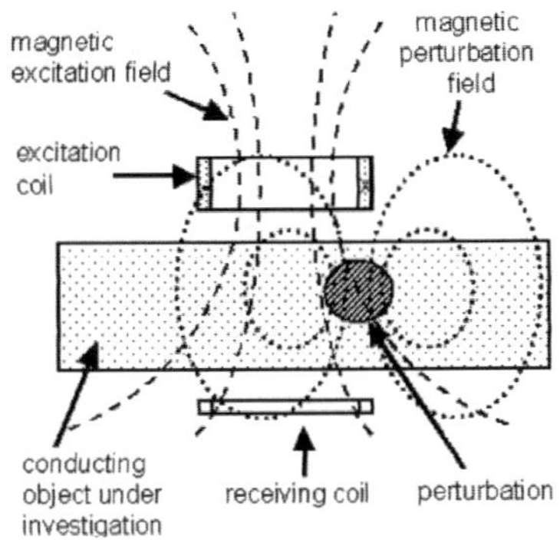

Figure 1: Ideal concept of eddy current in a conductive material [6].

The flowchart in figure 2 shows the step-by-step of the process to detecting the magnetic induction phase shift of the bone by using the COMSOL Multiphysics software. The bone material is set in the software with a different conductivity and permittivity value based on the frequency used which shown in table 1 below. The properties of the coil which is the transmitter and receiver is shown in the table 1. The material used for the transmitter and receiver coil is a copper.

With respect to the MIS principle, when the coil excitation produced the magnetic field which is the primary field, another secondary field which is the eddy current will be induced in the conductive material which is the screw. Eddy currents build their own magnetic field but in the oppositely from the primary field excites from the coil (transmitter)[11]. In this case, currents is an important part as an alternating current which to be fed into the coil to generate magnetic flux around the coil.

Figure 3 below shows the geometry model of the simulation in COMSOL software where the transmitter and receiver is sandwiching the femur bone regardless MIS concept. 1mm gap between sensors (transmitter and receiver) and bone is assign to make it a non-contact measurement.

Table 1: Value of conductivity and permittivity of the femur bone.

Frequency (MHz)	Conductivity	Permittivity
1	0.2122	759.7795
2	0.2162	686.9853
3	0.2169	643.3089
4	0.2203	577.7942
5	0.2264	574.1545
6	0.2267	559.5956
7	0.2279	541.3971
8	0.2243	526.8383
9	0.2273	519.5589
10	0.2304	508.6398

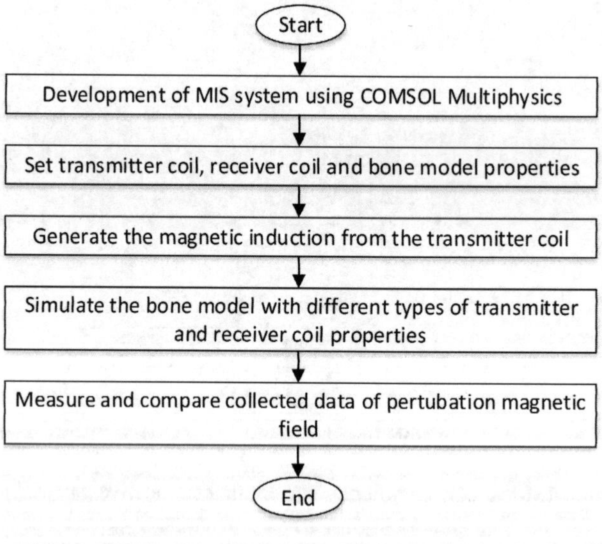

Figure 2: Flowchart of the study.

By using the principle of MIS, spiral coil is used to generate magnetic field at the transmitter and the specification of this coil is very important to generate strong magnetic field to cross the screw which be the interest region and detect the changing of magnetic field strength at the receiver[12]. The range of frequency used in this simulation is 1MHz to 10MHz with 10 step of 1 MHz. Frequency in the range of hundreds kilohertz (kHz) region which occurs at intermediate frequencies is suitable for the MIS where it is common used in many detection in medical field using magnetic induction [10], [12]–[14]. MIS system for this simulation started when the applied current AC pass through transmitter and coil and primary magnetic field is produced[8]. Figure 4 below shows the simulation model with the magnetic field around the sensors.

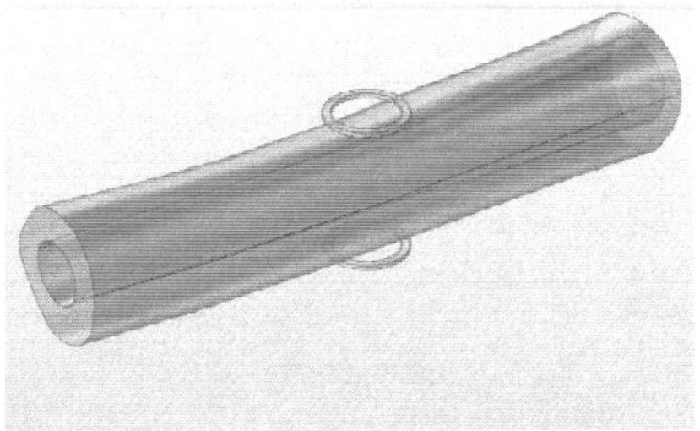

Figure 3: Femur bone with the coils sensor.

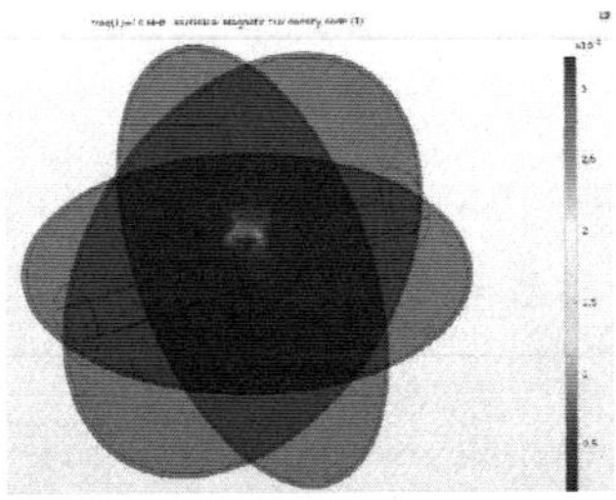

Figure 4: Simulation model at 10MHz frequency.

Table 2: properties of transmitter and receiver coils.

Types	Properties	Transmitter	Receiver
Model 1	Number of turns	5	5
	Current applied	1A	0A
	Coil diameter	10mm	10mm
	Wire diameter	1mm	1mm
Model 2	Number of turns	12	5
	Current applied	1A	0A
	Coil diameter	10mm	10mm
	Wire diameter	1mm	1mm

3. Result and discussion

Simulation process of the MIS system is done in COMSOL Multiphysics including the geometry design of coil sensor and femur bone. This system is perform in single channel magnetic induction and it's useful to measure the total magnetic field strength. Single channel magnetic induction spectroscopy included with one transmitter and one receiver which is opposite to each other [10]. The graph below show the result from the simulation process of different model. Figure 5 below show the result of total magnetic induction phase shift between model 1 and model 2. Figure 6 and Figure 7 shows the effect of multi frequency 1-10MHz on the total magnetic induction phase shift for both model 1 and model 2.

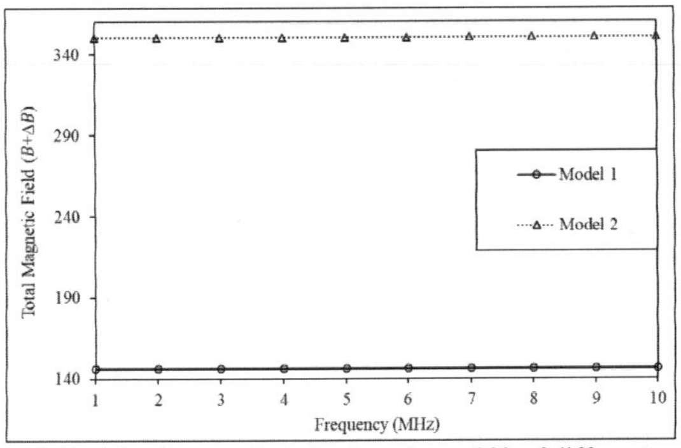

Figure 5: Result of magnetic induction phase shift of different model.

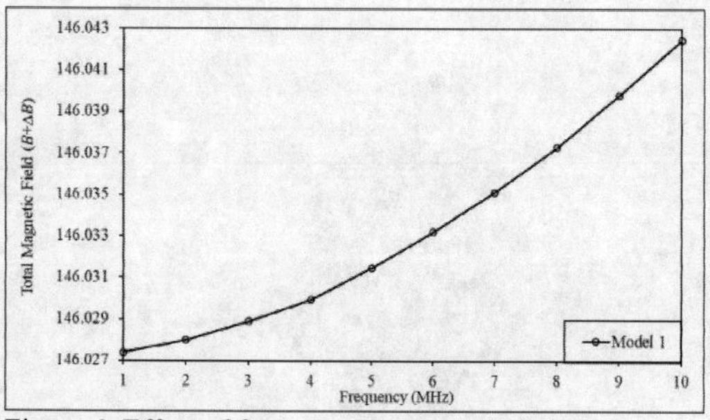

Figure 6: Effect of frequency on magnetic induction phase shift of model 1

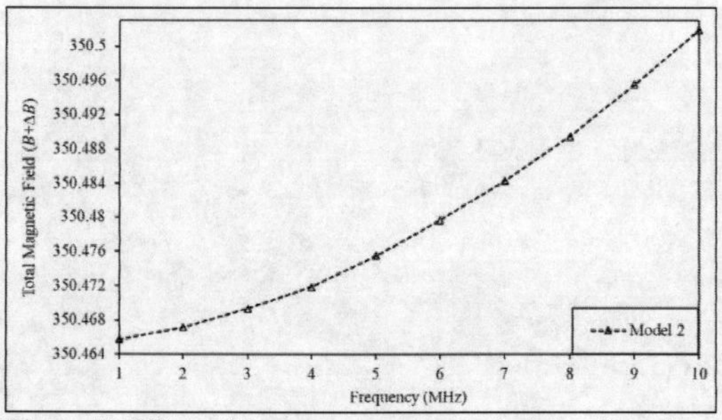

Figure 7: Effect of frequency on magnetic induction phase shift of model 2.

From the graph above, the y-axis represent the total magnetic field (A/m) and the x-axis represent the frequency (MHz) used. The alternating current in the transmitter coil generate the electromagnetic field which is the induction field propagated within the femur bone. The generated eddy current which is the secondary field for magnetic field strength is measured at the receiver. As shown in figure above, the total magnetic field for model 2 shows the higher value compared with the model 1. This is because the higher the number of turns exhibits a higher intensity and more stable signal which cause the changes of total magnetic field strength is higher [15]. By differentiate these two models, it is explained that the higher the number of transmitter coils, the higher the magnetic field generated from the transmitter. When the magnetic field induces in the bone samples, it will generate higher magnetic induction phase shift.

4. Conclusion

Human body is a volumetric with a conductivity material and it made of from several tissues with different conductivity which make the Magnetic Induction Spectroscopy (MIS) easy to be implemented. The application of MIS have been widely used in medical purpose experiment to ensure the healing or health of human body is upgraded to a new level which can save a lot of particular parameter. MIS system can be implemented to human body and also the material of bone implant which bring to the failure of internal fixation can be identified on early stage.

5. Acknowledgement

The author would like to acknowledge the support from the Fundamental Research Grant Scheme (FRGS) under a grant number of FRGS/1/2017/TK03/UNIMAP/02/10 from the Ministry of Higher Education Malaysia.

6. References

[1] M. Marco, E. Giner, J. R. Caeiro-rey, M. H. Miguélez, and R. Larraínzar-garijo 2019 Numerical modelling of hip fracture patterns in human femur *Comput. Methods Programs Biomed.*, vol. 173, pp. 67–75.

[2] K. M. Fox, J. Magaziner, J. R. Hebel, J. E. Kenzora, and T. M. Kashner 1999 Intertrochanteric Versus Femoral Neck Hip Fractures : Differential Characteristics , Treatment , and Sequelae vol. 54, no. 12, pp. 635–640.

[3] M. S. Taljanovic, M. D. Jones, J. T. Ruth, J. B. Benjamin, J. E. Sheppard, and T. B. Hunter 2003 Fracture Fixation *RadioGraphics*, vol. 23, no. 6, pp. 1569–1590.

[4] Y. Gotfried 2004 The Lateral Trochanteric Wall A Key Element in the Reconstruction of Unstable Pertrochanteric Hip Fractures no. 425, pp. 82–86.

[5] B. Gowry, A. B. Shahriman, and Z. Zulkarnay 2016 Inductive phase shift as a future cost effective diagnostic tool in breast cancer localization: A prologue study *AIP Conf. Proc.*, vol. 1774.

[6] H. Scharfetter, R. Casañas, and J. Rosell 2003 Biological Tissue Characterization by Magnetic Induction Spectroscopy (MIS): Requirements and Limitations *IEEE Trans. Biomed. Eng.*, vol. 50, no. 7, pp. 870–880.

[7] Z. Zakaria *et al.* 2012 Advancements in Transmitters and Sensors for Biological Tissue Imaging in Magnetic Induction Tomography pp. 7126–7156.

[8] A. J. Jurimah *et al.* 2016 Development of Magnetic Induction Spectroscopy System in Biomedical Application to Detect Jaundice vol. 050008.

[9] R. Merwa, K. Hollaus, and B. Brandst 2003 Numerical solution of the general 3D eddy current problem for magnetic induction tomography (spectroscopy) vol. 545.

[10] M. Hafiz *et al.* 2016 3D Model Simulation on Magnetic Induction Spectroscopy For Fetal Acidosis Detection Using COMSOL Multiphysics vol. 050003.

[11] I. Pitropakis, H. Pfeiffer, and M. Wevers 2012 Crack detection in aluminium plates for aerospace applications by electromagnetic impedance spectroscopy using flat coil sensors *Sensors Actuators, A Phys.*, vol. 176, pp. 57–63.

[12] Z. Zakaria 2015 Simulation of Single Channel Magnetic Induction Spectroscopy for Fetal Hypoxia Detection Jurnal Teknologi Full paper Simulation of Single Channel Magnetic Induction Spectroscopy for Fetal Hypoxia Detection no. November.

[13] Z. Zakaria *et al.* 2015 Jaundice Assessement of Newborn Baby : A Short Review on Kramel ' s Rule and Magnetic Induction Spectroscopy vol. 6, pp. 103–106.

[14] B. Gowry, A. B. Shahriman, and Z. Zulkarnay 2016 Inductive Phase Shift as a Future Cost Effective Diagnostic Tool in Breast Cancer Localization : A Prologue Study vol. 020001.

[15] H. Tavakkoli, E. Abbaspour-Sani, A. Khalilzadegan, A. M. Abazari, and G. Rezazadeh 2019 Mutual inductance calculation between two coaxial planar spiral coils with an arbitrary number of sides *Microelectronics J.*, vol. 85, no. January, pp. 98–108.

SLAM-GDM Implementation on Robot Operating System for Gas Source Localization

K W Soo[1], K Kamarudin[1,2], V H Bennetts[3], R Visvanathan[2], A S Ali Yeon[2], S M Mamduh[2], A Zakaria[1,2]

[1]School of Mechatronics Engineering, Universiti Malaysia Perlis (UniMAP), Arau, Malaysia,
[2]Centre of Excellence for Advanced Sensor Technology (CEASTech), Universiti Malaysia Perlis, Arau, Malaysia
[3]Applied Autonomous Sensor systems, Orebro University, Orebro, Sweden

kamarulzaman@unimap.edu.my

Abstract. Gas Distribution Mapping (GDM) can be defined as visualization for the spatial distribution of gas concentration in an unknown indoor environment. Recently, incidents of gas leakage have increased gradually worldwide. This will significantly increase the casualty rate at the workplace. This paper focusses on the implementation of Kernel DM based SLAM-GDM on Robot Operating System (ROS), a novel approach that will be useful in the effort of localizing the gas source during gas leakage inspection. This approach has been conducted successfully by using ROS-architecture implemented system, tested in simulation and on a mobile robot. The result shows that this implementation achieved average accuracy of 0.71m and 0.32m and therefore deemed suitable for gas source location prediction.

1. Introduction

Gas related research has been conducted for 25 years. To the authors' knowledge, at least four gas leakage incidents have occurred worldwide in the month of April 2019 [1]–[4]. These have indicated the importance of the gas source localizing technique in this industry. With the advance of technology and mobile robot, gas distribution modelling by integrating gas sensing capability on the mobile robot is believed to provide promising result.

Mobile robot is emerging as a powerful and revolutionary technology for different industries. It is now not only employed in production line [5] but also implemented for education [6], hospitality [7], medical [8] and even household [9]. The adaptation of mobile robot in the industries has improved their performances in different aspect [10]. This is due to mobile robot exhibits certain characteristics better than human to accomplish certain tasks such as environmental monitoring. For an example, a mobile robot can be exposed to dangerous environments. This enables human to access, analyze and carry out the counter measurement for the condition of a lethal environment with lower casualties. Besides that, with the implementation of a statistical approach algorithm such as Kernel DM on the mobile robot, a gas distribution mapping of the environment can be obtained. Its 2D SLAM-GDM representation will significantly assist the localization of the gas source which is very important especially in Urban Search and Rescue (USAR) operation.

This paper proposes an implementation of GDM in ROS environment using gas distribution modelling algorithm, Kernel DM. The algorithm is then integrated with an open source non-odometry dependent SLAM, Hector SLAM for real-time SLAM-GDM solution. The integration of SLAM and GDM is implemented based on the approach proposed by [11]. The expected result would be a representation of gas distribution modelling of mobile robot's environment, capable to assist human to predict and identify gas source location.

2. Related Work

2.1. Simultaneous Localization and Mapping (SLAM)

In a mobile robot related project, it is very important for the robot to be able to localize itself in an unknown environment while building a map of it. Building a map is also known as mapping. It is a task to visualize the geometry features of an unknown environment by using input from sensors such as camera [12], [13], ultrasonic sensor [14], [15], laser scanner [16], [17], and RFID [20].

SLAM algorithm can be classified into various type depends on the sensor used as its input. This is due to different SLAM algorithm will consider different assumption when interface with different kind of sensor to improve the localization and mapping accuracy [18]. For example, Hector SLAM is part of the laser-based SLAM that utilizes input from a laser scanner.

Hector SLAM works by combining both 2D SLAM system based on continuous scan matching and 3D navigation technique with the aid of a inertial sensing system. Besides that, unlike other SLAM approaches such as Gmapping, Hector SLAM is non-odometry dependent. Hector SLAM algorithm is based on occupancy grid maps [17]. This means the algorithms is processing the environment in the form of cells and each cell will carry specific information (i.e unknown, occupied or free) [18].

2.2. Gas Distribution Mapping (GDM)

Gas distribution mapping (GDM) is the task to estimate and visualize the spatial gases distribution with the assistant of a mobile robot. The first study on GDM is conducted by Hayes, Martinoli, and Goodman in the year 2002 where a group of autonomous mobile robot is being used to investigate and localize gas source [19]. However, it is believed that this approach is not suitable for real-world implementation due to its high cost [20]. Besides that, when a group of autonomous mobile robot is moving around, their movement will disturb the air flow in the environment. Hence, worsen the difficulty to localize gas source accurately.

In the year 2003, a method known as Kernel DM has been introduced by A. Lilienthal and Duckett. This proposed method "views" the surrounding as a grid cell and each of the cell consists of the estimated gas distribution mean value. By referring to a single cell, Kernel DM estimates the value of the neighbouring cells based on their distance with the single cell. This means that each of the cells at any time will have the predictive mean for all the measurement. This method has eased the performance of the robot because the robot does not need to move all the place to cover all the region [21]. After six years, the similar authors have proposed an improved version of Kernel DM known as Kernel DM+V which include the computation of predictive variance [22]. With the computation of predictive variance, the accuracy of gas source location prediction improved, compared to solely depends on predictive mean [22].

After Kernel DM+V has been introduced, several researchers have modified and improved the algorithm of Kernel DM+V to suit for their application. For example, [23] has modified the existing Kernel DM+V by replacing the Gaussian kernel function with Epanechnikov kernel. With the implementation of the Epanechnikov kernel, the numbers of cells involved in each gas source prediction will decrease and hence, the scope of prediction will be limited, and accuracy can be increased with less assumption. Besides that, similar author has proposed a new approach, DM+V/T/H that includes both temperature and humidity of the environment with the modified version of DM+V [24]. Despite all the versions proposed [11], [19], [21], [22], there is no implementation made available in ROS and made open source.

2.3. SLAM-GDM Implementation

With the implementation of SLAM-GDM approach, the mobile robot will be capable to perform GDM without prior knowledge about its environment and its position. This approach will enable the mobile robot to perform SLAM and GDM simultaneously hence, output a single grid map based two-dimensional representation consists of both obstacles in the environment and spatial distribution of gas concentration. To the author's knowledge, there are only two works that have implemented SLAM-GDM approach: [11], [25]. A Rao-Blackwellised particle filter approach has been used by [25] to combine both SLAM and GDM while [11] performs the SLAM-GDM approach by replacing the unoccupied cells SLAM's occupancy grid map with the gas distribution.

3. Methodologies

3.1. System Structure

There is a total of two systems required in this approach: the base station system and mobile robot system. An Asus A555L Series laptop with 8GB of RAM and Raspberry Pi 3 Model B (raspi) have been used as the base station and mobile robot, respectively. Both systems are installed with Linux based operating system. Ubuntu 18.04 with ROS Melodic has been installed for base station while Ubuntu Mate with ROS Kinetic for mobile robot.

There are two main components on top of the mobile robot: raspi and Arduino Nano. Raspi is connected to three components:1) camera for teleoperation, 2) RPLIDAR for mobile robot localization and mapping and 3) Arduino Nano. The Arduino Nano is used to control the motors and interface with the gas sensor module. Both motors controlling task and gas sensor module interfacing task cannot be accomplished by raspi. This is due to raspi consists of 3.3V logic signal voltage digital input and output pins only. It cannot interface with the analogue signal as well as sensors and actuator that require 5V logic signal voltage. Additional voltage level shifter and Analogue-to-Digital Converter (ADC) are needed for raspi to be able to accomplish the task. However, to reduce the circuitry complexity in this project, an Arduino Nano has been used to interface with both gas sensor module and motor driver. Both RPLIDAR and gas sensor module data will then be transmitted to the base station for SLAM and Kernel DM computation. Lastly, data fusion will be conducted to integrate SLAM and GDM. Figure 1 shows the system architecture implemented in this project.

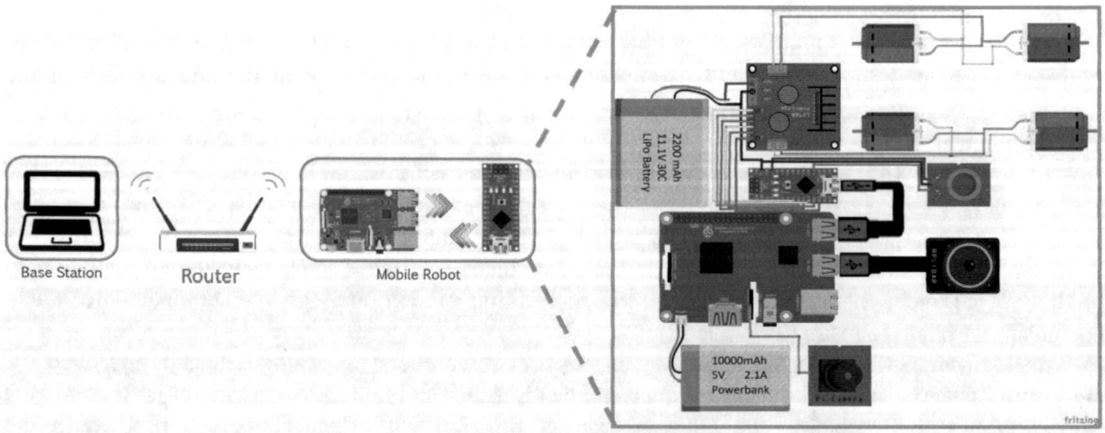

Figure 1: System Architecture

There are three configurations that must be carried out when more than one ROS machine is involved. First, all ROS machines used in this project must share a similar network. Second, only one ROSmaster is allowed in one ROS operation, where the ROSmaster generally responsible for locating and establishing peer-to-peer communication between ROS nodes. Lastly, Network Time Protocol

IOP Conf. Series: Materials Science and Engineering **705** (2019) 012018 doi:10.1088/1757-899X/705/1/012018

(NTP) such as *chrony* and *ntpdate* must be used to synchronize system time of all ROS machines involved in the project.

3.2. Base Station Configuration

Base station serves as the ROSmaster of this project and it is used to perform numeral tasks such as control the movement of the mobile robot as well as processing and visualizing all the data received from the mobile robot. Three main features of base station will be discussed in this section.

3.2.1. Simultaneous Localization and Mapping (SLAM). SLAM is often used in mobile robot application to construct the landscape of an unknown environment and predict the location of the mobile robot on the map. Hector SLAM, an Open-Source non-odometry dependent SLAM approach has been implemented in this project. The advantages of Hector SLAM implementation including SLAM performance will not be affected due to the surface condition the mobile robot moves on and its performance will also not be affected by wheel slip [23]. Hector SLAM module in ROS works by having one of its nodes known as hector_mapping to subscribe to the laser scan topic for calculating and publish two important topics SLAM map topic and robot pose topic. SLAM map topic is a type of OccupancyGrid type topic. It is used to describe the condition of a cell as free, occupied or unknown. Meanwhile, robot pose topic is a PoseStamped type topic that provides an estimated position of the mobile robot. Both topics will then be used by RVIZ to construct a map and locate the location of the mobile robot within the map respectively.

3.2.2. Kernel DM. Kernel DM algorithm has been used in this project to achieve the goal of generating GDM. Kernel DM is a statistical approached algorithm for gas distribution modelling proposed by [22]. Kernel DM algorithm implementation involves four equations:

$$\mathcal{W}^{(k)} = \sum_{i=1}^{n} \mathcal{N}\left(\left|x_i - x^{(k)}\right|, \sigma\right) \quad (1)$$

$$\mathcal{R}^{(k)} = \sum_{i=1}^{n} \mathcal{N}\left(\left|x_i - x^{(k)}\right|, \sigma\right) \cdot r_i \quad (2)$$

$$a^{(k)} = 1 - e^{\frac{-\left(\mathcal{W}^{(k)}\right)^2}{\sigma_{\mathcal{W}}{}^2}} \quad (3)$$

$$r^{(k)} = a^{(k)} \frac{\mathcal{R}^{(k)}}{\mathcal{W}^{(k)}} + \{1 - a^{(k)}\} r_0 \quad (4)$$

Uni-variate Gaussian weighting function $\mathcal{N}$ has been used in Kernel DM to determine how significant location x_i's sensor reading r_i toward cell κ when constructing gas distribution mapping. $\mathcal{W}^{(k)}$ and $\mathcal{R}^{(k)}$ are known as cell confidence summation and cell weighted reading summation, respectively. $x^{(k)}$ stands for cell κ's centre while the kernel width of the algorithm is represented by σ. Cell κ that is located nearer to $x^{(k)}$ tends to have higher value for $\mathcal{W}^{(k)}$ and vice versa. The result of $\mathcal{W}^{(k)}$ is then normalized by using $a^{(k)}$. Hence, all the reading normalized by $a^{(k)}$ will be scaled into the range of 0 to 1. This result of $a^{(k)}$ will then be utilized to calculate the mean concentration estimate $r^{(k)}$ where r_0 can be defined as the overall mean for all the gas concentration reading. Once the mean concentration estimate is obtained for all the cells, a mean map will be obtained.

This has shown that to implement Kernel DM algorithm, several parameters such as robot position and gas sensor reading, must be provided. This can be done by having the Kernel DM node to subscribe to both sensors reading topic and robot pose topics. By computing Kernel DM algorithm, an array of cells position with respective gas concentration level will be obtained.

3.3. SLAM and GDM Integration

A SLAM-GDM map can be defined as a representation for the spatial distribution of gas concentration level by combining both maps of Hector SLAM and Kernel DM. Therefore, it is very crucial to ensure both Hector SLAM and Kernel DM obtain sufficient information for their computation. Figure 2 shows the software architecture of the ROS based environment to generate SLAM-GDM map.

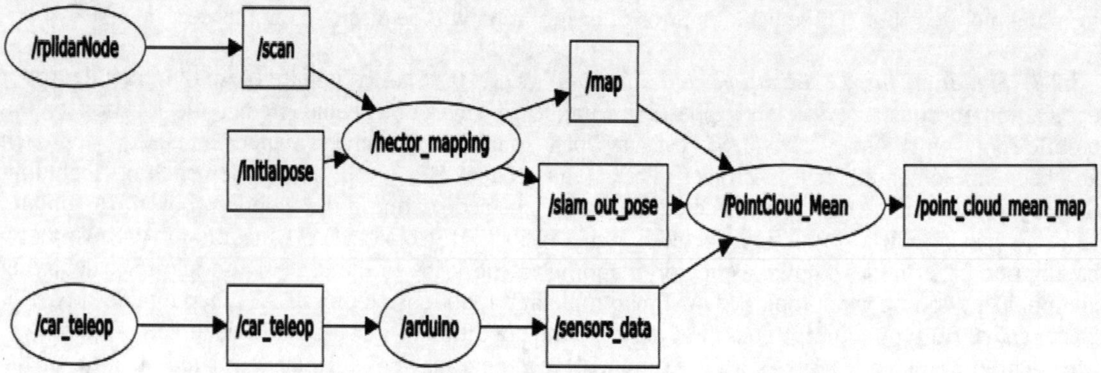

Figure 2: Software Architecture of ROS based System

A ROS node known as rplidarNode has been deployed on the mobile robot's raspi. By having RPLIDAR's data as its input, it will publish all the received data to a topic known as /scan. A ROS node of Hector SLAM, hector_mapping will then subscribe to the /scan topic and /initialpose topic (consists of the initial position of the mobile robot) to compute and publish SLAM map under /map topic. Hector_mapping will also publish a /slam_out_pose topic that provides information about the mobile robot's position and it will be updated at the scan rate of the laser scanner used.

To control the movement of the mobile robot from the base station, a ROS node called car_teleop has been deployed at the base station. It will publish the movement command to /car_teleop topic subscribed by the Arduino (ROS node of Arduino Nano) and hence, the performance of the motors can be controlled from the base station. Meanwhile, Arduino Nano will also publish a custom message under the /sensors_data topic. Since it is a custom message, all the ROS machines involved must have the custom message created to be able to recognize and extract data from it.

By having /map, /slam_out_pose and /sensors_data topics published, PointCloud_Mean ROS node that responsible to generate SLAM-GDM representation will begin its computation. There are two possible approaches can be implemented to achieve the objective of SLAM and GDM integration: (1) overlap the GDM on top of the global map generated by Hector SLAM; (2) reconstruct and merge the global map with GDM to form a single global GDM map. The second approach will be used for this project as it eases the process of data analysis since all the data is store within the same array. To realize this approach, two steps are required: map type conversion and data fusion.

3.3.1. Map Type Conversion. The first step needs to be done to achieve this objective is to convert the global map generated by the Hector SLAM in the type of Occupancy Grid map to Point Cloud. This is due to the Occupancy Grid map consists of three states (free, occupied and unknown) only and it is not suitable to be used to store any value of Kernel DM. Besides that, due to its nature to have three states only, it cannot display other colours besides white (free cell), grey (unknown cell) and black (occupied cell). Therefore, type conversion to Point Cloud type is necessary for the representation able to provide useful information.

3.3.2. Data fusion. With both outputs of Kernel DM algorithm (cells coordinate with its gas concentration level) and Point Cloud SLAM map are ready, both data can now be fused together to generate a SLAM-GDM map based on the approach proposed by [11]. The process begins by storing all the Point Cloud SLAM map data in an array, without publishing it. Next, the Point Cloud data generated from Kernel DM algorithm should be inserted into a similar array to replace the SLAM cells that are free only. This will retain the structure of SLAM map while introducing gas concentration level information onto free cell to achieve the objective of constructing SLAM-GDM representation.

Different colour should be distributed to the Point Cloud geometry to indicate different cell conditions. This will help the user performs a better analysis besides illustrating the relationship between each of the cell. The colour assignation for each cell condition is as follow:

$$[r,g,b] = \begin{cases} [0,0,0], & Occupied, \\ [220,220,220], & Unknown, \\ [255 - H^{(k)}, 255 - L^{(k)}, 0], & Gas\ Concentration\ Level, \end{cases} \quad (5)$$

$$H^{(k)} = \begin{cases} \left((r^{(k)} - r_{ave})(r_{max} - r_{ave})^{-1}\right) * 255, & r^{(k)} > r_{ave}, \\ 0, & otherwise, \end{cases} \quad (6)$$

$$L^{(k)} = \begin{cases} \left((r_{ave} - r^{(k)})(r_{ave} - r_{min})^{-1}\right) * 255, & r^{(k)} \le r_{ave}, \\ 0, & otherwise, \end{cases} \quad (7)$$

For occupied cells and unknown cells, the colour assigned is black and grey respectively. Colour for free cell will depends on its gas concentration level reading. Equation (6) and Equation (7) show the effect of gas concentration reading on the colour of free cell. The lower the gas concentration level reading, the greener the free cell (Equation 7) while the higher the gas concentration level reading, the redness of the free cell increase (Equation 6). The average reading for gas concentration level has been used as the even point for this bipolar scaling approach. When the reading of the free cell is near to the average value of gas concentration level, yellow colour will dominate.

4. Experimental Result

The first part of this section will be focusing on the performance analysis of online SLAM-GDM algorithm by using Rosbag data collected in CEASTech as presented in [11]. Each of the rosbags used has a different gas source location. Therefore, the predicted gas source location is expected not to be similar. Table 1 illustrates the accuracy of the SLAM-GDM by calculating its error in prediction on ethanol gas source location. The purple filled circle and blue filled circle indicate the actual location for ethanol gas source and predicted gas source location respectively. In this work, an average accuracy of below 1m is preferred. Based on the results, the average accuracy of SLAM-GDM representation on predicting gas source location is approximately 0.71m. Hence, the proposed Kernel DM based GDM algorithm is deemed suitable for gas source localization application.

Table 1: Accuracy of SLAM-GDM Representation on Gas Source Location Prediction

	Rosbag 1	Rosbag 2	Rosbag 3
SLAM-GDM Map	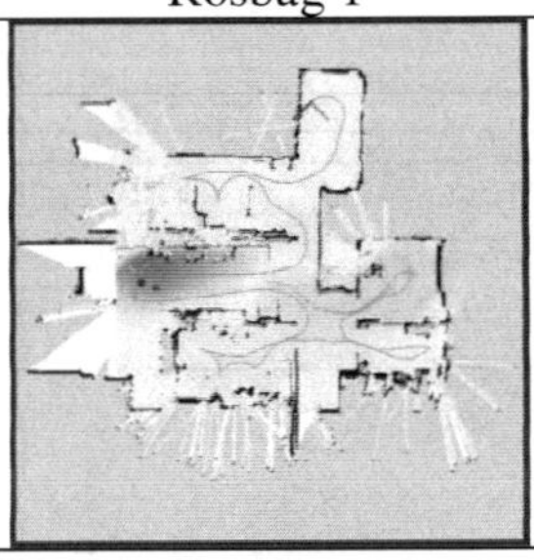	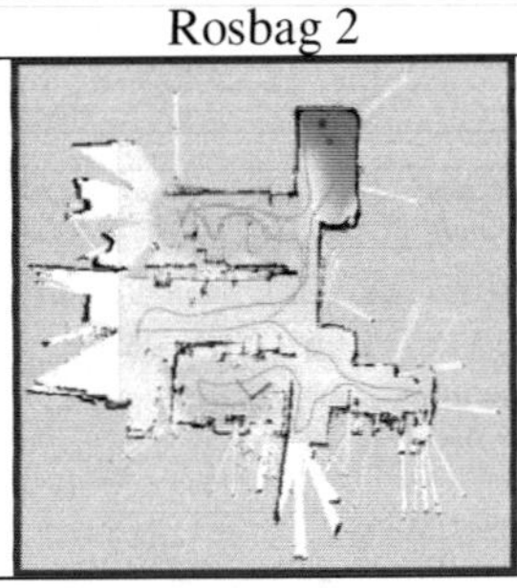	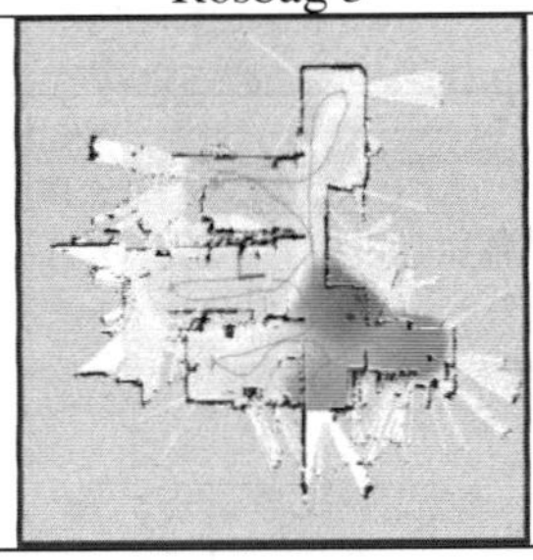
Error (m)	0.61	0.79	0.74

The next experiment conducted is to evaluate the consistency of the SLAM-GDM map in predicting ethanol gas source solution. The experiment is conducted in UniMAP Solution of Universiti

Malaysia Perlis (UniMAP) by utilizing a mobile robot, where the ethanol gas source is fixed at a location throughout the three experimental trials. Table 2 shows the consistency of SLAM-GDM representation to predict a fixed gas source location. From the results of the experiment, the average error in predicting the gas source location is approximately 0.32m. There are two factors detected that have affected the result of this experiment. First, the area of gas dispersing increase with time. This led to significant prediction difference between Trial 1 and Trial 2. Second, a trailing gas concentration level is visualized after the mobile robot pass by the gas source location. This is caused by the memory effect of TGS2600 gas sensor and the high mobile robot movement speed. Memory effect occurs due to the gas sensor used in this project has a long recovery time. It takes several seconds for the gas sensor to readjust its reading back to surrounding gas concentration level. By having a gas sensor module with shorter recovery time and a robot with slower movement, the performance of SLAM-GDM representation can be improved.

Table 2: Consistency of SLAM-GDM Representation on Gas Source Location Prediction

	Trial 1	Trial 2	Trial 3
SLAM-GDM Map	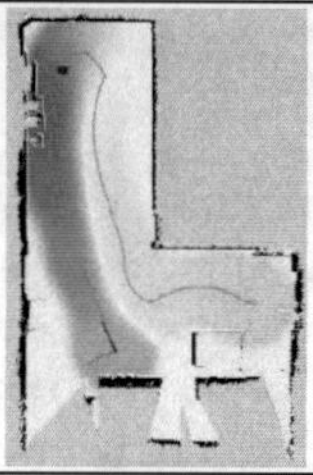	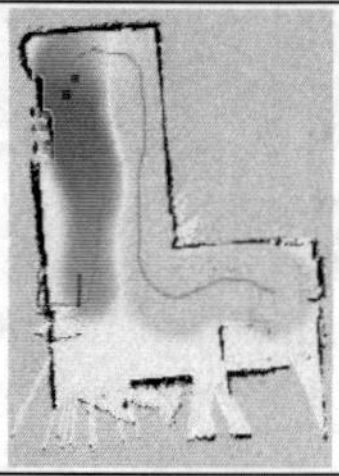	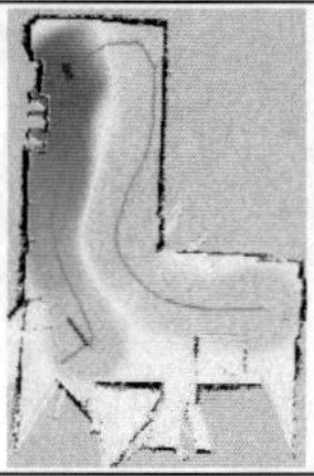
Error (m)	0.10	0.59	0.28

5. Conclusion

Involvement of the mobile robot in the inspection of gas leakage incident surely will become very beneficial for human. This allows the number of humans needed to be exposed to the harmful environment will be reduced and hence, the number of casualties can be kept minimal. With the implementation of SLAM-GDM algorithm on ROS, the mobile robot will also provide better gas source location prediction in an unknown environment.

The conducted experiences have shown that the average accuracy and consistency of SLAM-GDM map performance on predict gas source location are 0.71m and 0.32m respectively which fulfil author's requirement of below 1m. Therefore, the performance of SLAM-GDM representation is reliable. The performance can be further improved by increasing number of sensors implemented such as wind sensor to access the magnitude and direction of air flow surrounding the mobile robot or introducing variance into the Kernel DM algorithm.

6. Acknowledgement

The author would like to acknowledge the support from the Fundamental Research Grant Scheme (FRGS) under a grant number of FRGS/1/2018/TK04/UNIMAP/02/15 from the Ministry of Education Malaysia.

References

[1] Asian Oil & Gas staff 2014 Aasta Hansteen Field Shut after Gas Leak *Asian Oil & Gas* Aasta Hansteen Field.

[2] Bernama 2018 Ammonia leak at ice factory: Two dead, 18 hurt - Nation | The Star Online *The Star Online*. [Online]. Available: https://www.thestar.com.my/news/nation/2018/08/13/ammonia-leak-at-ice-factory-two-dead-18-hurt/. [Accessed: 11-Dec-2018].

[3] R. Mani 2019 Allahabad: Gas leakage in IFFCO plant, four labourers taken ill *Times of India* Allahabad.

[4] United News of India 2019 Ammonia gas leakage: 15 fell ill in Mathura *United News of India* Mathura.
[5] M. Shneier Literature Review of Mobile Robots for Manufacturing Literature Review of Mobile Robots for Manufacturing.
[6] B. S. Barker, G. Nugent, N. Grandgenett, and V. I. Adamchuk 2012 Robots in K-12 Education - A New Technology for Learning *J. Educ. Technol. Soc.*, vol. 19, no. 2, pp. 148–163.
[7] A. Tapus, C. Tapus, and M. Mataric 2006 User-robot personality matching and assistive robot behavior adaptation for *J. Robot. Soc.* ..., pp. 169–183.
[8] R. A. Beasley 2012 Medical Robots: Current Systems and Research Directions *J. Robot.*, vol. 2012, pp. 1–14.
[9] H. Şahin and L. GuÜvenç 2007 Household Robotics Autonomous Devices for Vacuuming and Lawn Mowing *IEEE Control Syst.*, vol. 27, no. 2, pp. 20–96.
[10] Zakri Abdul Hamid 2017 The robot revolution | New Straits Times | Malaysia General Business Sports and Lifestyle News *New Straits Times*. [Online]. Available: https://www.nst.com.my/news/2017/04/228029/robot-revolution. [Accessed: 20-Apr-2019].
[11] K. Kamarudin *et al.* 2018 Integrating SLAM and gas distribution mapping (SLAM-GDM) for real-time gas source localization *Adv. Robot.*, vol. 32, no. 17, pp. 903–917.
[12] Y. Wu, F. Tang, and H. Li 2018 Image Based Camera Localization : an Overview Invited Paper by Visual Computing for Industry , Biomedicine and Art , 2018 *Invit. Rev.*, pp. 1–21.
[13] A. J. Davison, I. D. Reid, N. D. Molton, and O. Stasse 2007 MonoSLAM: Real-time single camera SLAM *IEEE Trans. Pattern Anal. Mach. Intell.*, vol. 29, no. 6, pp. 1052–1067.
[14] P. Yang 2012 Efficient particle filter algorithm for ultrasonic sensor-based 2D range-only simultaneous localisation and mapping application *IET Wirel. Sens. Syst.*, vol. 2, no. 4, pp. 394–401.
[15] L. D'Alfonso, A. Grano, P. Muraca, and P. Pugliese 2013 A polynomial based SLAM algorithm for mobile robots using ultrasonic sensors - Experimental results *2013 16th Int. Conf. Adv. Robot. ICAR 2013*, pp. 1–6.
[16] K. Kamarudin, S. M. Mamduh, A. S. A. Yeon, R. Visvanathan, A. Y. M. Shakaff, and A. Zakaria 2015 Improving Performance of 2D SLAM Methods by Complementing Kinect with Laser Scanner pp. 278–283.
[17] J. M. Santos, D. Portugal, and R. P. Rocha 2013 An evaluation of 2D SLAM techniques available in Robot Operating System *2013 IEEE Int. Symp. Safety, Secur. Rescue Robot. SSRR 2013*.
[18] M. Rojas-fern, M. Dante, M. Matuz-cruz, and L. Diana 2018 PERFORMANCE COMPARISON OF 2D SLAM TECHNIQUES AVAILABLE IN ROS USING A p. 5386.
[19] A. T. Hayes, A. Martinoli, and R. M. Goodman 2002 Distributed odor source localization *IEEE Sens. J.*, vol. 2, no. 3, pp. 260–271.
[20] X. Liu *et al.* 2012 A survey on gas sensing technology *Sensors (Switzerland)*, vol. 12, no. 7, pp. 9635–9665.
[21] A. Lilienthal 2003 Creating Gas Concentration Gridmaps with a Mobile Robot *Nevada Octooer*.
[22] A. J. Lilienthal, M. Reggente, M. Trinca, J. L. Blanco, and J. Gonzalez 2009 A statistical approach to gas distribution modelling with mobile robots - The Kernel DM+V algorithm *2009 IEEE/RSJ Int. Conf. Intell. Robot. Syst. IROS 2009*, pp. 570–576.
[23] K. Bin Kamarudin 2016 An Improved Mobile Robot Based Gas Source Localization with Temperature and Humidity Compensation via SLAM and Gas Distribution Mapping no. 1140610652.
[24] H. W. Liu, S. Ma, W. Li, H. G. Gu, Y. G. Lin, and X. J. Sun 2011 A review on the development of tidal current energy in China *Renewable and Sustainable Energy Reviews*.
[25] A. J. Lilienthal, A. Loutfi, J. Blanco, C. Galindo, and J. Gonzalez-jimenez 2007 A Rao-Blackwellisation Approach to GDM-SLAM – Integrating SLAM and Gas Distribution Mapping (GDM) *Eur. Conf. Mob. Robot.*, no. September, pp. 126–131.

Electrochemical Techniques-Based approaches for Mycobacterium Tuberculosis Detection: Last Decade Review

K Alfarhan[1*], A Zakaria[1], N Yusof[2], S Zakaria[1], J Abdullah[2], L Kamarudin[1,3], U Azmi[2] and N Ariffin[2]

[1]Centre of Excellence for Advanced Sensor Technology (CEASTech), School of Mechatronic Engineering, Universiti Malaysia Perlis (UniMAP), Perlis, Malaysia.

[2]Chemistry Department, Faculty of Science, Univerisiti Putra Malaysia Institute of Advanced Technology (ITMA), UPM.

[3]School of Computer and Communication Engineering, Universiti Malaysia Perlis (UniMAP), Perlis, Malaysia.

[*]Corresponding author: Khudhur.alfarhan@gmail.com

Abstract. Mycobacterium Tuberculosis (MTB) is a common airborne infectious disease that leads to millions of deaths every year worldwide. It is still one of the top ten causes of death and the victims of TB are more than HIV/AIDS in 2017. Traditional approaches for MTB detection are either take a long time, unreliable or high cost. The electrochemical techniques (ECTs) as improved and inexpensive approaches to detect the MTB. Many of ECTs were used in MTB detection such as differential pulse voltammetric (DPV), cyclic voltammetric (CV), square wave voltammetric (SWV), amperometric and impedimetric. Principle of the MTB detection using ECTs depends on DNA hybridization of the MTB on the working electrode of ECTs. The researchers developed biosensors or aptasensors and used them for ECTs analyzing to detect the MTB. They developed various biosensors from various composite and DNA probes but all the developed composite of the biosensors were used to coat the electrodes that used in ECTs. Many types of electrodes and electrolytes were used in MTB detection. The most used ECTs in MTB detection is DPV and CV while the least used is amperometric. The ECTs for MTB detection achieved high sensitivity, reliability, low detection time and very low detection limits.

1. Introduction

Mycobacterium Tuberculosis (MTB) is responsible for Tuberculosis (TB) disease. TB is a common airborne infectious disease that leads to millions of deaths every year worldwide [1]. In 2014, about one million children got TB out of which 14% died [2]. TB is still one of the top ten causes of death and the victims of TB are more than HIV/AIDS in 2017 [3]. Usually, MTB attacks and affects lungs but some times it affects bones, abdomen, lymph nodes and genitourinary tract [3]. Traditional approaches for MTB detection are either take a long time or high cost. Approaches such as the chest X-ray [4], sputum smear microscopy and skin test is lack specificity [5]. Also, there are other approaches for MTB detection but these approaches are considered as expensive such as antibody-based Enzyme-Linked Immunosorbent Assay (Antibody ELISA) [6] and polymerase chain reactions (PCR) [7].

Recently, many studies directed to use electrochemical techniques (ECTs) as improved and inexpensive approaches to detect the MTB. A bunch of ECTs were used in MTB detection such as differential pulse voltammetric (DPV), cyclic voltammetric (CV), square wave voltammetric (SWV), amperometric and impedimetric. The ECTs are electrical measurements through chemical solutions. These measurements are classified into three types:

- Potential difference (voltammetric) like DPV, CV, SWV and so on.
- Changes in current (amperometric).
- Changes in impedance (impedimetric).

Generally, the basic principle of ECTs measurements as shown in Figure 1 is based on measuring the current passed through the chemical solution (electrolyte). This current measured between two electrodes; namely, the working electrode (WE) and the counter electrode (CE). There is a third electrode also is dipped in the same chemical solution and it is called the reference electrode (RE) except the amperometric does not have this electrode or in sometimes it is shortened with WE. RE and WE are used to provide the excitation electrical signal to the chemical solution while the current is measured between WE and CE. In the amperometric, the measured current between WE and CE is used directly for the analyzing. Voltammetric is the study of the current as a function of applied potential that means the measured current through WE and CE electrodes is converted to a voltage. Impedimetric is the ratio of an incremental change in voltage to the resulting change in current that means the measured current through the WE and CE is used to find the impedance in the chemical solution [8, 9].

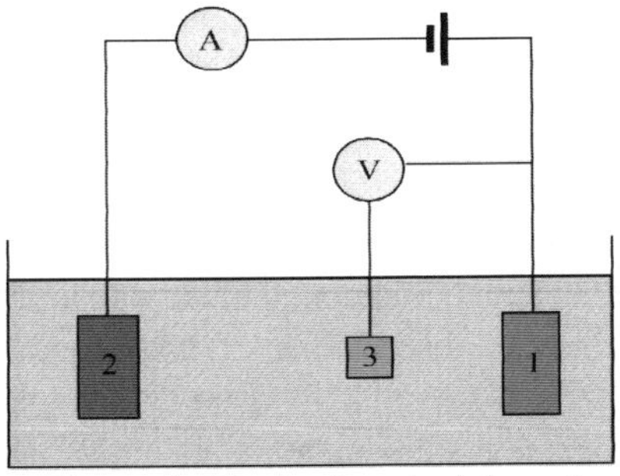

Figure 1. The basic principle and setup of ECTs (1- WE, 2- CE and 3- RE) [10]

In this paper, the ECTs that were used for MTB detection using DNA will be reviewed. We will focus on only three ECTs types were used in MTB detection; namely, voltammetric, amperometric and impedimetric. Some techniques of voltammetric will be reviewed in details and these techniques are DPV, CV and SWV. Also, the types and materials of the electrodes will be reviewed.

2. Detection of MTB Using Electrochemical Techniques

ECTs are rapid, easy and inexpensive approaches of MTB detection. For these reasons, many researchers worked on developing new and advanced approaches for MTB detection using ECTs. All studies in our literature focused on developing a DNA hybridization and immobilization (using DNA probes) for MTB detection using ECTs. The principle behind MTB detection using ECTs depends on DNA hybridization of the MTB on the working electrode of ECTs [11]. Hybridization of the targeted DNA on the electrode changes the impedance value of the electrode so that leads to change in the measured electrical current [11]. In MTB detection using ECTs, the researchers developed biosensors or aptasensors and used them for ECTs analyzing to detect the MTB. The researchers developed various biosensors from various composites and DNA probes but all the developed composites of the biosensors were used to coat the electrodes that used in ECTs [12]. There are four common type of working electrodes used in MTB detection screen-printed carbon electrode (SPCE) [12, 13, 14, 15, 16, 17], graphite [18] and a mechanical pencil as a graphite electrode (PGE), graphene or graphene oxide electrode (GOE) [19, 16] and bare gold disk electrode (BGE) [1]. To increase the surface area of the electrodes, the gold in nano-forms was used like gold nanoparticles (AuNPs) [13, 14, 19, 15, 11] and gold nanotubes (AuNTs) [20]. As mentioned before, the ECTs used for chemical analyzing by measure the electrical current through the electrolyte, in MTB detection also the electrolyte or indicator is an important part of ECTs and play a vital rule in MTB detection. Methylene blue (MB) widely used as an indicator/electrolyte in MTB detection [12, 4, 18, 20, 21]. By monitoring the changes in MB oxidation current can indicate the MTB, because MB has a strong affinity for the free guanine bases [22]. Also, MB has high stability and extraordinary ability to transfer electrons[4]. There are other studies used other types of electrolytes like phosphate buffered saline (PBS) [23, 24] and potassium chloride (KCl) [17]. ECTs like DPV, CV, SWV, Amperometric and impedimetric will be reviewed and explained in brief below:

2.1. Differential Pulse Voltammetric (DPV)

DPV is a method of voltammetric with a small influence of capacitive current. It has better signal-to-noise ratio by attenuating the background currents. In DPV, the excitation electrical signal has constant magnitude and linear ramped electrical pulses [25]. The current sample is measured twice in each pulse period (once at the beginning of the pulse, and at the pulse's end), and the difference between these two current values is used in the measurements [25, 26]. The electrical wave of the DPV is shown in Figure 2. DPV is widely used in MTB detection [11, 7, 19, 27] with various setups like voltage range and scan rate.

2.2. Cyclic Voltammetric (CV)

The CV is a potential-controlled reverse electrochemical process [28] also, it is widely used ECT in MTB detection [29, 11, 19]. Normally, the experiment of CV starts by scanning the potential from a user-defined initial potential valve to final user-defined potential value then reverse the scanning to the initial value [28], as shown in Figure 3. There are many researchers used the CV in MTB detection with various setups like initial and final potential values, scanning rate and so on [11, 7, 19, 30].

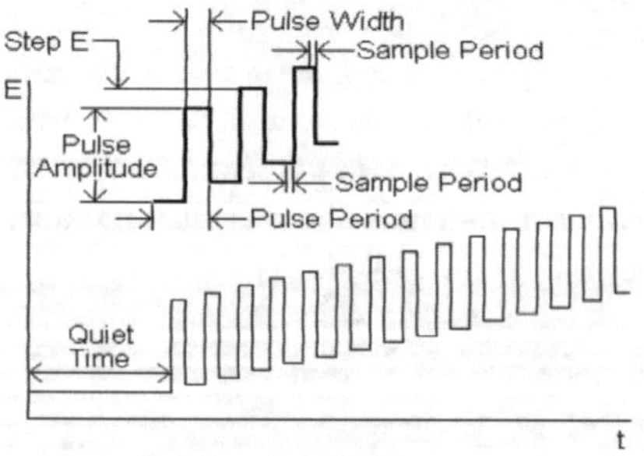

Figure 2. The electrical wave of the DPV [26]

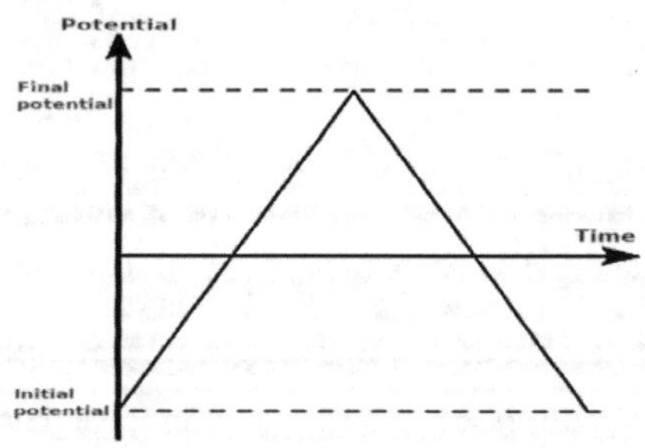

Figure 3. The wave form of the CV [31]

2.3. Square Wave Voltammetric (SWV)

The SWV is a linear potential sweep voltammetric technique that uses a constant amplitude square wave formed in a staircase manner. In the SWV the current is measured at the end of each peak, and the current measured on the button peak (I_b) is subtracted from the current measured on the top peak (I_t). This difference current (I_t - I_b) is displayed as a function of the applied potential [32, 26]. The waveform of the SWV is shown in Figure 4. Since SWV and DPV have similar voltage sequences and exhibit similar peak shapes [33] it is a commonly used method in MTB detection [34, 35, 36].

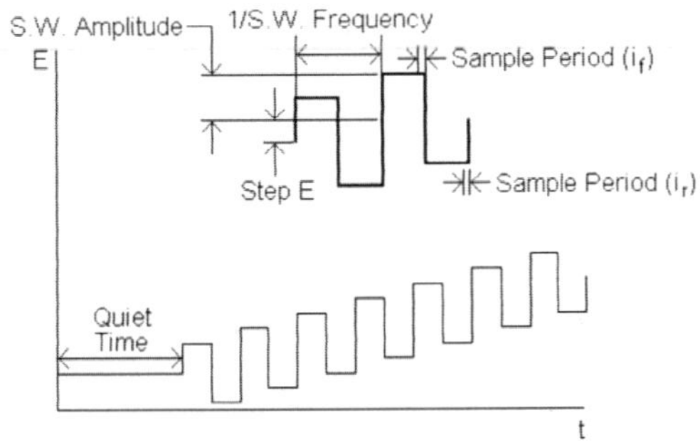

Figure 4. The waveform of the SWV [26]

2.4. Impedimetric

Impedimetric or electrochemical impedance spectroscopy (EIS) is an ECT used to measure the electrical impedance in the electrolyte. The impedimetric uses a range of frequencies to scan the electrolyte and this range of frequencies can be set by the user [37]. In our review, there are many studies that used the impedimetric in MTB detection [38, 1].

2.5. Amperometric

It is simple and basic ECT that used in chemical analyzing and it is depending on the changing in the direct electrical current between two electrodes dipped in the electrolyte (WE and CE) [39]. Amperometric is used in MTB detection for its simplicity, and it gives good results for MTB detection by using various configuration [7, 40, 30].

3. Summary

In this section, we will summarize (in Table 1) all studies that we reviewed about MTB detection using ECTs and some details about these studies like types of ECTs, potential range in volt (V), scan rate in millivolt.second^{-1} (mV.sec^{-1}), electrodes' materials, electrolytes, DNA detection limits and other parameters like the range of frequencies of EIS. Only, working electrodes will be mentioned in the table because most studies use Ag/AgCl as a reference electrode and platinum as a counter (auxiliary) electrode.

Table 1. The summary of the reviewed studies about MTB detection using ECTs

Ref.	ECTs	Potential range (V)	Scan rate (mV.sec^{-1})	WE material	Electrolyte and/or indicator	Detection limit	Other parameters
[1]	EIS CV	0.01 -0.2 to +0.7	50	Bare gold disk electrode (BGE)	Ferricyanide and KCl	6 ng.µL^{-1}	EIS frequency =0.1 to 100^5 Hz
[12]	DPV	0.01	20	SPCE	MB	15 µg.mL^{-1}	-
[18]	DPV	0.01	20	PGE	MB	15 µg.mL^{-1}	-
[38]	CV DPV EIS	-0.5 to 0.5 -0.5 to 0.5 -	100 50 -	BGE	Ferricyanide and KCl	20 fg.mL^{-1}	-
[11]	CV DPV	−0.2 to 0.6 −0.4 to 0.3	100 50	BGE	Phosphate buffered saline (PBS) and	50 fM	-

IOP Conf. Series: Materials Science and Engineering **705** (2019) 012019 doi:10.1088/1757-899X/705/1/012019

					KCl		
[7]	DPV CV EIS	− 0.4 to 0.3 -	50 -	Glass carbon electrode (GCE)	PBS	0.33 fM	-
[22]	CV DPV	-1.5 to +0.7	20	BGE	MB and PBS	0.065 ng.µL^{-1}	-
[17]	DPV CV	−0.2 to +0.8 −1 to 0.8	- 100	SPCE	MB	7.853×10^{-7} M	-
[24]	DPV	-1 to +1	50	BGE	PBS	2.7046 µg.mL^{-1}	accuracy=99.22%
[41]	DPV EIS	0.1 to −0.3 0.01	50 -	BGE	PBS Ferricyanide and KCl	10 fg·mL^{-1}	EIS freq.=0.01 to 10^5 Hz
[19]	DPV CV	-0.3 to +0.4 -0.2 to 0.6	- 100	GCE	PBS	1×10^{-15} M	-
[16]	DPV CV EIS	−0.6 to −0.1 +0.1 to +0.7 -	- 50 -	SPCE	MB and PBS	8.9×10^{-13} M	EIS freq.=10^{-4} to 0.1 Hz
[42]	DPV	0.15 to 0.45	20	SPCE	PBS	1 CFU	-
[15]	DPV	− 1 to +1	50	BGE	PBS	3.47 µg.mL^{-1}	-
[23]	DPV EIS	0.2 to 0.8 -	60 -	GOE	PBS	0.9 fg·mL^{-1}	EIS freq.=0.1 to 10^5 Hz
[2]	DPV EIS	0.2 to 0.6 0.01	50 -	Carbon nanotubes (CNT)	PBS	0.5 ± 0.2 fg.mL^{-1}	EIS freq.=0.1 to 10^5 Hz
[13]	DPV CV EIS	0 to 0.4 0 to 0.4 0.24	50 50 -	Fabricated	PBS	1.25 ng.mL^{-1}	EIS freq.=0.01 to 10^5 Hz
[14]	DPV	0 to 1.25	33.5	SPCE	-	0.01 ng.mL^{-1}	-
[43]	DPV	-100 to 400	100	Silk screen printing ITO	PBS	0.5 pg.mL^{-1}	-
[44]	Amperometric	-8 to -1	-	cobalt-titanium nanotube array	Methyl	0.018 mg.mL^{-1}	-
[45]	CV	0.7 to -1.1	50	BGE	Sodium chloride	-	-
[27]	DPV CV EIS	-0.6 to 0 -0.6 to 0.8 -	- - -	indium–tin–oxide (ITO)	PBS	1×10^{-12} M	EIS freq.=0.01 to 10^5 Hz
[46]	DPV CV	-0.4 to 0.1 -0.6 to 0.6	50 50 to 250	ITO	PBS	0.00078 µM	Sensitivity=6.38×10^{-6} AµM^{-1}
[47]	DPV CV	- -0.4 to 0.8	- -	SPCE	MB and PBS	10 pg	Sensitivity=95%
[40]	Amperometric	-2.5 to 2.5	-	BGE	PBS	10 pg.mL^{-1}	-
[17]	DPV CV	-0.2 to 0.8 -1 to 0.8	- 100	SPCE	MB and PBS	7.853×10^{-7} M	-
[48]	CV EIS	-0.4 to 0.6 -	- -	BGE	PBS	10^4 cfu/mL	EIS freq.=0.1 to 10^5 Hz
[49]	CV EIS	-0.70 to 0.30 -0.20 to 0.60	100 100	GCE	PBS	3.3×10^{-5} ng·mL^{-1}	-
[50]	SWV	-0.5 to 0	-	ITO	MB	28 copies µL^{-1}	-

[51]	SWV CV	-1 to 1 -1.5 to 1.5	30 100	SPCE	PBS and methyl	238 μM.L^{-1}	SWV freq.=20 Hz
[36]	SWV CV	0.02 −0.4 to 0.9	- 100	GCE	PBS	0.3 fM	SWV freq.=50 Hz
[52]	CV SWV EIS	-0.6 to 1 -0.5 to -0.1 -	10 to 100 60 -	ITO	PBS	0.1 fM	SWV freq.=150 Hz
[30]	Amperometric	-0.5 to -1	-	Titania nanotubes arrays (TNA)	Methyl	10 mM	-
[53]	CV SWV	-0.55 to 0.35 50	100 50	SPCE	Tris-acetate buffer	0.46 ng.μL^{-1}	SWV freq.=15 Hz
[54]	CV	-0.2 to 0.7	50	ITO	PBS	0.1fM	-
[35]	CV SWV	-0.8 to 2 -0.5 to 0.4	100 10	PGE	PBS	1.3 nM	SWV freq.=15 Hz
[55]	EIS	0.01	-	GOE	PBS	3.3×10^{-16} M	EIS freq.=0.1 to 10^4 Hz
[56]	SWV	-0.65 to -0.2	200	self-assembled mono-layers	PBS	300 pM	SWV freq.=12.5 Hz
[57]	SWV EIS	-0.9 to -0.3 -	25 -	GCE	PBS	8.7×10^{-15} M	EIS freq.=0.1 to 10^5 Hz SWV freq.=15 Hz

4. Conclusion

In this paper, we reviewed the studies about MTB detection using ECTs. Five types of ECTs were reviewed; namely, DPV, CV, SWV, EIS and amperometric. Also, parameters of the ECTs, electrodes composites, electrolytes and DNA detection limits were reviewed. The most used ECTs in MTB detection are DPV and CV then EIS followed by SWV and finally amperometric. The ECTs for MTB detection achieved high sensitivity, reliability, low detection time and very low detection limits. Also, ECTs for MTB detection become widely used in point-of-care as a commercial MTB detection approaches.

5. Acknowledgment

This work financially supported under grant (UPM/800-4/11/MRUN/2018/5539230).

References

[1] Costa M, Andrade C, Montenegro R, Melo F and Oliveira M 2014 Self-assembled monolayers of mercaptobenzoic acid and magnetite nanoparticles as an efficient support for development of tuberculosis genosensor *J Colloid Interface Sci* **433** 141

[2] Thakur H, Kaur N, Sareen D and Prabhakar N 2017 Electrochemical determination of M. tuberculosis antigen based on Poly (3,4-ethylenedioxythiophene) and functionalized carbon nanotubes hybrid platform *Talanta* **171** 115-123

[3] World Health Organization (WHO) 2018 Global Tuberculosis Report

[4] Lavania S, Das R, Dhiman A, Myneedu V, Verma A, Singh N, Sharma T and Tyagi J 2018 Aptamer-Based TB Antigen Tests for the Rapid Diagnosis of Pulmonary Tuberculosis: Potential Utility in Screening for Tuberculosis *ACS Infect Dis* **4** 1718

[5] Tufa L, Oh S, Tran V, Kim J, Jeong K, Park T, Kim H and Lee J 2018 Electrochemical immunosensor using nanotriplex of graphene quantum dots, Fe3O4, and Ag nanoparticles for tuberculosis *Electrochimica Acta* **290** 369

[6] Zhu C, Liu J, Ling Y, Yang H, Liu Z, Zheng R, Qin L and Hu Z 2012 Evaluation of the clinical value of ELISA based on MPT64 antibody aptamer for serological diagnosis of pulmonary tuberculosis *BMC Infect Dis* **12** 96

[7] Chen Y, Guo S, Zhao M, Zhang P, Xin Z, Tao J and Bai L 2018 Amperometric DNA biosensor for Mycobacterium tuberculosis detection using flower-like carbon nanotubes-polyaniline nanohybrid and enzyme-assisted signal amplification strategy *Biosens Bioelectron* **119** 215-20

[8] Zoski C 2007 Handbook of electrochemistry 1st ed. (Amsterdam : Boston: Elsevier) p 892

[9] Wang Y, Ye Z and Ying Y 2012 New Trends in impedimetric biosensors for the detection of foodborne pathogenic bacteria *Sensors* **12** 3449-3471

[10] Voltammetry 2019 Wikipedia Available from: https://en.wikipedia.org/w/index.php?title=Voltammetry&oldid=892933254

[11] Chen Y, Li Y, Yang Y, Wu F, Cao J and Bai L 2017 A polyaniline-reduced graphene oxide nanocomposite as a redox nanoprobe in a voltammetric DNA biosensor for Mycobacterium tuberculosis *Microchim Acta* **84**1801-1808

[12] Issa R, Hamdan N and Noh M 2019 Differential pulse voltammetric determination of dna hybridization using methylene blue on screen printed carbon electrode for the detection of mycobacterium tuberculosis *Biotechnology* **9** 304-311

[13] Thiruppathiraja C, Kamatchiammal S, Adaikkappan P, Santhosh D and Alagar M 2011 Specific detection of Mycobacterium sp. genomic DNA using dual labeled gold nanoparticle based electrochemical biosensor *Anal Biochem* **417** 73-79

[14] Torres E and Alocilja E 2011 Nanoparticle based DNA biosensor for tuberculosis detection using thermophilic helicase-dependent isothermal amplification *Biosens Bioelectron* **26** 4614-4618

[15] Ng B, Xiao W, West N, Wee E, Wang Y and Trau M 2015 Rapid, Single-cell electrochemical detection of mycobacterium tuberculosis using colloidal gold nanoparticles *Anal Chem* **87** 10613–1618

[16] Mat M, Abdullah J, Yusof N, Sulaiman Y, Wasoh H, Noh M and Issa R 2017 PNA biosensor based on reduced graphene oxide/water soluble quantum dots for the detection of Mycobacterium tuberculosis Sens *Actuators B Chem* **241** 1024–1034

[17] Mohamad F, Zaid M, Abdullah J, Zawawi R, Lim H, Sulaiman Y and Rahman N 2017 Synthesis and characterization of polyaniline/graphene composite nanofiber and its application as an electrochemical dna biosensor for the detection of mycobacterium tuberculosis *Sensors* **17** 2789

[18] Hamdan N, Issa R, Noh M and Zin N 2012 Electrochemical technique using methylene blue with pencil graphite electrode for optimum detection of mycobacterium tuberculosis DNA *Curr Res Tuberc* **4** 1-12

[19] Liu C, Jiang D, Xiang G, Liu L, Liu F and Pu X 2014 An electrochemical DNA biosensor for the detection of Mycobacterium tuberculosis, based on signal amplification of graphene and a gold nanoparticle–polyaniline nanocomposite *The Analyst* **139** 5460–5465

[20] Torati S, Reddy V, Yoon S and Kim C 2016 Electrochemical biosensor for Mycobacterium tuberculosis DNA detection based on gold nanotubes array electrode platform *Biosens Bioelectron* **78** 483-488

[21] Tahir M, Bajwa S, Mansoor S, Briddon R, Khan W, Scheffler B and Amin I 2018 Evaluation of carbon nanotube based copper nanoparticle composite for the efficient detection of agroviruses *J Hazard Mater* **346** 27-35

[22] Das M, Sumana G, Nagarajan R and Malhotra B 2010 Zirconia based nucleic acid sensor for mycobacterium tuberculosis detection *Appl Phys Lett* **96** 133703
[23] Thakur H, Kaur N, Sabherwal P, Sareen D and Prabhakar N 2017 Aptamer based voltammetric biosensor for the detection of Mycobacterium tuberculosis antigen MPT64 *Microchim Acta* **184** 1915-1922
[24] Gaffar S, Nurmalasari R and Hartati Y 2017 Voltammetric DNA Biosensor using gold electrode modified by self assembled monolayer of thiol for detection of mycobacterium tuberculosis *Procedia Technol* **27** 74-80
[25] Amin M, Isa M, Sidek R and Yusof N 2015 An embedded processing of differential pulse voltammetry (DPV) data using ARM processor (LPC1768) *IEEE International Circuits and Systems Symposium (ICSyS)* (Langkawi: Malaysia IEEE) p 80-84
[26] Bioanalytical Systems Inc Pulse Voltammetric Techniques Available from: https://www.basinc.com/manuals/EC_epsilon/Techniques/Pulse/pulse
[27] Das M, Sumana G, Nagarajan R and Malhotra B 2010 Application of nanostructured ZnO films for electrochemical DNA biosensor *Thin Solid Films* **519** 1196-1201
[28] Vivian J 2013 Towards HIV Sensing: The Development of electrochemical DNA/RNA aptamer biosensors on dendrimer-gold platforms *University of Johannesburg*
[29] Rueda D, Furukawa R, Fuentes P, Comina G, Castro N, Requena D, Gilman R, Sheen P and Zimic M 2018 A novel inexpensive electrochemical sensor for pyrazinoic acid as a potential tool for the identification of pyrazinamide-resistant Mycobacterium tuberculosis *Int J Mycobacteriology* **7** 275
[30] Smith Y, Bhattacharyya D, Mohanty S and Misra M 2016 Anodic functionalization of titania nanotube arrays for the electrochemical detection of tuberculosis biomarker vapors *J Electrochem Soc* **163** B83-B89
[31] Denis A 2008 Cyclic Voltammetry (New York: John Wiely & Sons publication) p 3-12
[32] Mirceski V, Skrzypek S and Stojanov L 2018 Square-wave voltammetry *ChemTexts* **4** 1-14
[33] Ainla A, Mousavi M, Tsaloglou M, Redston J, Bell J, Teresa M and Whitesides G 2018 Open-source potentiostat for wireless electrochemical detection with smartphones *Anal Chem* **90** 6240-6246
[34] Yan Z, Gan N, Zhang H, Wang D, Qiao L, Cao Y, Li T and Hu F 2015 A sandwich-hybridization assay for simultaneous determination of HIV and tuberculosis DNA targets based on signal amplification by quantum dots-PowerVision polymer coding nanotracers *Biosens Bioelectron* **71** 20713
[35] Yesil M, Donmez S and Arslan F 2016 Development of an electrochemical DNA biosensor for detection of specific Mycobacterium tuberculosis sequence based on poly (L-glutamic acid) modified electrode *J Chem Sci* **128** 1823-1829
[36] Miodek A, Mejri N, Gomgnimbou M, Sola C and Korri H 2016 E-DNA Sensor of mycobacterium tuberculosis based on electrochemical assembly of nanomaterials (MWCNTs/PPy/PAMAM) *Anal Chem* **128** 1823-1829
[37] Liju Y and Anthony G 2008 Impedimetric biosensors for nano and microfluidics, encyclopedia of microfluidics and nanofluidics (Boston: Springer)
[38] Bai L, Chen Y, Bai Y, Chen Y, Zhou J and Huang A 2017 Fullerene-doped polyaniline as new redox nanoprobe and catalyst in electrochemical aptasensor for ultrasensitive detection of mycobacterium tuberculosis MPT64 antigen in human serum *Biomaterials* **133** 11-19
[39] Lakshminarayanaiah N 1976 Membrane Electrodes (Academic Press) p 378
[40] Evans D, Papadimitriou K, Greathead L, Vasilakis N, Pantelidis P, Kelleher P, Morgan H and Prodromakis T 2017 An Assay system for point-of-care diagnosis of tuberculosis using commercially manufactured PCB technology *Sci Rep* **7**

[41] Li N, Huang X, Sun D, Yu W, Tan W, Luo Z and Chen Z 2018 Dual-aptamer-based voltammetric biosensor for the Mycobacterium tuberculosis antigen MPT64 by using a gold electrode modified with a peroxidase loaded composite consisting of gold nanoparticles and a Zr(IV)/terephthalate metal-organic framework *Microchim Acta* **185**

[42] Mohd U, Yusof N, Kusnin N, Abdullah J, Suraiya S, Ong P, Raston N, Rahman V and Fathil M 2018 Sandwich electrochemical immunosensor for early detection of tuberculosis based on graphene/polyaniline-modified screen-printed gold electrode *Sensors* **18** 3926

[43] Zhang Y, Yan Y, Zhang B, Zhu W, He Y, Huang H, Li J, Jiang Z and Tan S 2015 Fabrication of an interferon-gamma-based ITO detector for latent tuberculosis diagnosis with high stability and lower cost *J Solid State Electrochem* **19** 3111-3119

[44] Bhattacharyya D, Smith Y, Mohanty S and Misra M 2016 Titania nanotube array sensor for electrochemical detection of four predominate tuberculosis volatile biomarkers *J Electrochem Soc* **163** B206-B214

[45] Chiu N, Huang T, Kuo C, Lee W, Hsieh M and Lai H 2012 Single-layer graphene based SPR biochips for tuberculosis bacillus detection *Biophotonics: Photonic Solutions for Better Health Care III* p 84273M

[46] Das M, Dhand C, Sumana G, Srivastava A, Nagarajan R, Nain L, Iwamoto M, Manaka T and Malhotra B 2011 Electrophoretic fabrication of chitosan−zirconium-oxide nanobiocomposite platform for nucleic acid detection *Biomacromolecules* **12** 540-547

[47] Das R, Dhiman A, Mishra S, Haldar S, Sharma N, Bansal A, Ahmad Y, Kumar A, Tyagi J and Sharma T 2019 Structural switching electrochemical DNA aptasensor for the rapid diagnosis of tuberculous meningitis *Int J Nanomedicine* **14** 2103-2113

[48] He F, Xiong Y, Liu J, Tong F and Yan D 2016 Construction of Au-IDE/CFP10-ESAT6 aptamer/DNA-AuNPs MSPQC for rapid detection of mycobacterium tuberculosis *Biosens Bioelectron* **77** 799-804

[49] Li L, Yuan Y, Chen Y, Zhang P, Bai Y and Bai L 2018 Aptamer based voltammetric biosensor for Mycobacterium tuberculosis antigen ESAT-6 using a nanohybrid material composed of reduced graphene oxide and a metal-organic framework *Microchim Acta* **185** 379

[50] Luo J, Fang X, Ye D, Li H, Chen H, Zhang S and Kong J 2014 A real-time microfluidic multiplex electrochemical loop-mediated isothermal amplification chip for differentiating bacteria *Biosens Bioelectron* **60** 84-91

[51] Metters J, Kampouris D and Banks C 2014 Fingerprinting breath: electrochemical monitoring of markers indicative of bacteria mycobacterium tuberculosis infection *J Braz Chem Soc* **25** 1667-1672

[52] Prabhakar N, Solanki P, Kaushik A, Pandey M and Malhotra B 2010 Peptide nucleic acid immobilized biocompatible silane nanocomposite platform for mycobacterium tuberculosis detection *Electroanalysis* **22** 2672-2682

[53] Tsaloglou M, Nemiroski A, Camci G, Christodouleas D, Murray L, Connelly J and Whitesides G 2018 Handheld isothermal amplification and electrochemical detection of DNA in resource-limited settings *Anal Biochem* **543** 116-121

[54] Navin M, Vikrant S, Raj S and Dhanraj T 2016 Reduced graphene oxide nanoribbon immobilized gold nanoparticles based electrochemical DNA biosensor for detection of mycobacterium tuberculosis *J Mater Chem B* **6** 5181-5187

[55] Yu X, Chai Y, Jiang J and Cui H 2012 Sensitive ECL sensor for sequence-specific DNA from mycobacterium tuberculosis based on N-(aminobutyl)-N-ethylisoluminol functionalized gold nanoparticles labeling *J Photochem Photobiol Chem* **241** 45-51

[56] Zhang Q, March G, Noel V, Piro B, Reisberg S, Tran L, Hai L, Abadia E, Sola C, et al. 2012 Label-free and reagentless electrochemical detection of PCR fragments using self-assembled quinone derivative monolayer: application to mycobacterium tuberculosis *Biosens Bioelectron* **32** 163-168

[57] Zhang C, Lou J, Tu W, Bao J and Dai Z 2015 Ultrasensitive electrochemical biosensing for DNA using quantum dots combined with restriction endonuclease *The Analyst* **140** 506-511

IOP Conf. Series: Materials Science and Engineering **705** (2019) 012020 doi:10.1088/1757-899X/705/1/012020

National Instruments LabVIEW Biomedical Toolkit for Measuring Heart Beat Rate and ECG LEAD II Features

W L Khong*, M Mariappan, N S V Kameswara Rao

Faculty of Engineering, Universiti Malaysia Sabah

*Corresponding author: weileongkhong@gmail.com

Abstract. Electrocardiogram (ECG) is undoubtedly the gold method for assessing heart activities. However, the ECG machine for measuring ECG is mostly only available in health care center especially in the third-world countries. Therefore, this research proposed to utilize the product of National Instruments i.e. LabVIEW Biomedical Toolkit and data acquisition device as an alternative approach to acquire ECG features and heart beat rate (HBR). As for performances verification, the NI-LabVIEW instruments are tested simultaneously with two types of ECG machine which are Philips PageWriter TC 30 and KenZ ECG 108. The results indicated that the ECG features and HBR obtained from NI-LABVIEW instruments are agreed well with the measurements attained from both of the ECG machines. Hence, this study proved that NI-LabVIEW instruments are highly potent for measuring ECG features and HBR. Moreover, the computerized NI-LabVIEW instruments with the self-interpretation function can always outperform most of the conventional ECG machines that are still required cardiologists for manually interpreting the results printed on the ECG paper.

1. Introduction

Human cardiovascular system is the foundation for various physiological parameters where it consists of five major components i.e. heart, arteries, capillaries, veins and blood. Cardiovascular system can be further divided into two types of circulatory system, which are systematic circulation and pulmonary circulation [1]. These circulatory systems are behaved like a closed loop system, which is controlled by the heart. In every cardiac cycle, the cardiac cells in the heart generate bio-potential. Thus, various heart activities which also included the measurement of heart beat rate (HBR) can be determined via analyzing the generated bio-potential.

Electrocardiogram (ECG) is a graphical method established for measuring the bio-potential generated by the heart. It is considered a contact based non-invasive method because the electrodes are required to attach on the correct position of the body parts for acquiring the bio-potential signals. According to [3], three types of electrodes i.e. electrode with adhesive gel (disposable), clamp electrodes (reusable) and suction electrode (reusable) are commonly used for ECG measurement. Although ECG method has been developed many decades ago, the twelve-leads ECG system is still the gold standard for measuring heart activities in clinical units [2, 3]. Among the twelve-leads ECG system, Lead II is the most preferable lead utilized for determining HBR and continuously monitoring the patient's condition because it can closely resemble the route of the sequence bio-potential generated in the heart [4, 5, 6].

The equipment used to acquire ECG is known as electrocardiograph or more simply ECG machine [4]. Unfortunately, ECG machine mostly only available in health care center and required well-trained health professionals to operate. Furthermore, some of the ECG machines are unable to provide continuous HBR and ECG measurements. Thence, in this research, National Instruments (NI) LabVIEW Biomedical Toolkit, which is associated with the NI data acquisition (DAQ) device are proposed as an alternative method in continuously acquiring HBR and ECG features.

2. NI-LabVIEW Hardware Setup

In this research, the model of NI USB-6281 DAQ device is selected for acquiring the bio-potential generated by the heart through electrodes as the sensors. According to [7], NI USB-6281 DAQ device is under categories of NI M-series, which is a multifunctional USB DAQ device that is optimized to achieve high accuracy measurements at fast sampling rates. Since NI USB-6281 DAQ device is a high accuracy DAQ device, it is able to log the voltage faster and accurately, which is essential for acquiring ECG signals. Figure 1 displays the NI USB-6281 DAQ device with its screw terminal pinouts for connecting the sensors.

The 16 analog inputs of the NI USB-6281 DAQ device either can be used as differential channels or single ended channel to acquire signals. Since the placements of the electrodes are based on the Lead II configuration to find the bio-potential difference between the two electrodes, hence in this research only two analog inputs are used to acquire the ECG signals.

This NI USB-6281 DAQ device supports three types of ground reference settings for the analog input. In this research, the ground reference setting of differential mode is selected for measuring the bio-potential through the Lead II configuration. In order to use differential mode, the input signal is required to be low level which is less than 1 V, therefore it is very suitable for measuring the bio-potential generated by heart, which only consists of millivolts.

As mentioned earlier, the differential mode is used to measure the potential between two analog inputs, hence it is necessary to pair the analog inputs for differential measurements. In this study, the pair of the analog input AI 4 and AI 12 is used to record the Lead II ECG signals. AI 4 acts as the positive lead is connected to the electrode placed at left leg (LL), while AI 12 acts as the negative lead is connected to the electrode placed at right arm (RA) as shown in Figure 2.

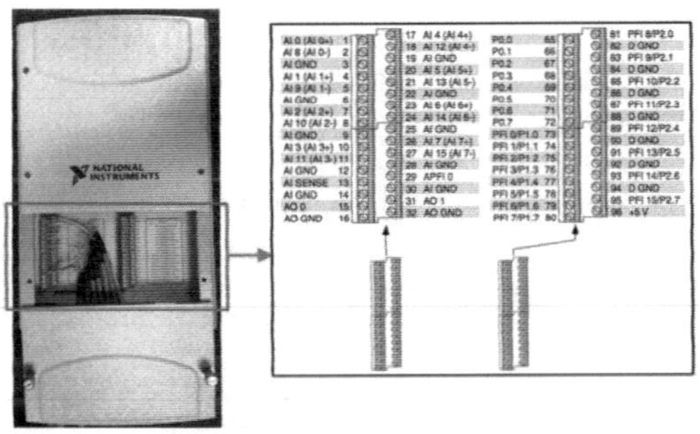

Figure 1. NI USB-6281 DAQ device with its screw terminal pinouts [5].

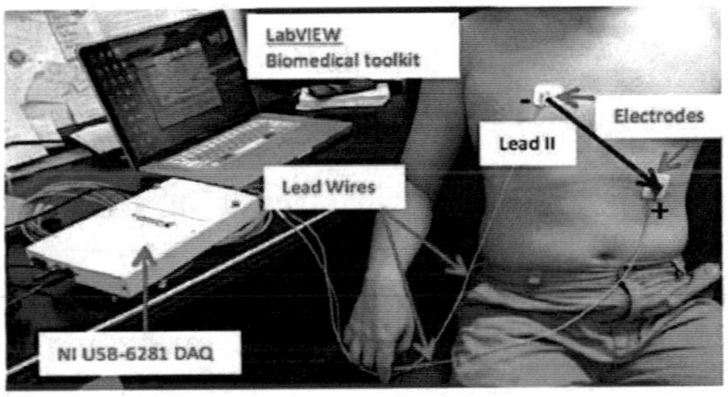

Figure 2. Hardware setup for acquiring Lead II ECG signals.

2.1. *Logger and Player Workbench of Ni-LabVIEW Biomedical Toolkit Configuration*

After set up the hardware and connected the electrodes to the NI USB-6281 DAQ device, the NI-LabVIEW software needs to be configured for acquiring the ECG signals. In this study, the Logger and Player workbench in LabVIEW Biomedical Toolkit is used to log the Lead II ECG signals. In the logger setting, the user is requested to select the corresponding DAQ device used for logging the signals. According to [8], the sampling rates that are typically used for NI DAQ device to acquire raw ECG signals are either 125 Hz or 250 Hz. In this study, 250 Hz is chosen as the sampling rate because the modern ECG machines usually operate at this rate to record the ECG signals. Subsequently, the terminal configuration is set to differential, so that the DAQ device can record the potential between the two analog inputs.

The Logger and Player workbench provides various customized digital filters in the virtual channel settings for selection. If the user prefers to use analog filter to filter the high frequency of the raw ECG signals, this step can be discarded. Otherwise, virtual channel can be used to attenuate the high frequency noises. In order to set up the virtual channel for filtering, the types of noise need to be identified. Power line interference, electrode contacting noise and electromyographic noise (caused by muscle movement) are the high frequency noises that could influence the ECG signals during signals acquisition [9]. However, among these various types of noise, only power line interference is the most significant noise that could contaminate the ECG signals during the recording process.

Power line interference is mainly caused by improper grounding, electrode cables, or nearby devices, which normally centered at high frequency of 50 Hz or 60 Hz with the narrow bandwidth of fewer than 1 Hz. Hence, in this study a virtual channel is set to filter the power line interference using notch filter to preprocess the Lead II ECG signals acquired from the NI USB-6281 DAQ device. Notch filter is selected because it is able to filter the frequencies with the narrow bandwidth of the interfering noise [10, 11].

2.2. ECG Feature Extractor Workbench of Ni-LabVIEW Biomedical Toolkit Configuration

ECG Feature Extractor is one of the workbenches in NI-LabVIEW Biomedical Toolkit that is responsible to interpret the ECG waveform for ECG features extraction. There are two ways ECG Feature Extractor imports the ECG signals. The first way is to acquire the real time ECG signals by switching the toggle to the direction towards DAQ settings. Unfortunately, it is unable to save the ECG signals for future clarification. The second way is to import the files that are saved in the computer for the ECG features extraction by switching the toggle to file settings. However, it does not support the real time measurement.

Various ECG features such as R wave amplitude; QRS, PR, QT and RR intervals can be extracted using ECG Feature Extractor. To begin with, it detects the R waves in the ECG waveform and followed by other features because R waves with the highest amplitude are the easiest features to be identified. Therefore, accuracy in detecting R waves becomes very essential. In order to enhance the quality of the logged ECG signals, this workbench also allows the user sets the feature settings to filter and rectify the entire logged ECG signals. After enhancing the ECG signals, the exact locations of the R waves in the entire ECG waveform will be more obvious and distinguishable. According to [12, 13], R wave normally has a frequency between 10 – 25 Hz. Hence, the user can apply the digital band pass filter that is provided by ECG Feature Extractor to filter the entire ECG signals for R waves detection. Subsequently, the threshold adjustment factor is the threshold used by the workbench to detect the peak of the R wave. The default value for the threshold adjustment factor is 0.1, if this factor is set to a larger value, the R waves can be detected will reduce due to the ignoring of the R waves that with smaller amplitude. Furthermore, the rough highest heart rate is the value for initializing the internal parameter in the first several cycles for obtaining more accurate results. The default value given by NI is 60 bpm. In addition, number of beats for initialization is used to skip the R waves at the beginning of the ECG signals. The default value is 8, which means the workbench will skip 8 beats because the ECG signals at the beginning are normally not yet steady. Besides, user is required to tick the box of "remove noise from raw signal" for more accurate measurements. Lastly, the isoelectric limit is set to 0.1 mV to refine the P wave and T wave of ECG signals [14]. If the difference between the maximum and minimum amplitude within the time intervals of PR segment and ST segment is less than the isoelectric limit, it is considered as isoelectric. Figure 3 indicates the feature settings to filter and rectify the entire logged ECG signals and Figure 4 shows the filtered and rectified ECG signals.

After the R peaks are being detected, other ECG features will be extracted using the R waves as the reference. Figure 5 displays the ECG features being extracted using LabVIEW ECG Feature Extractor workbench. Besides, the HBR can also be calculated by using the average of the RR intervals [15].

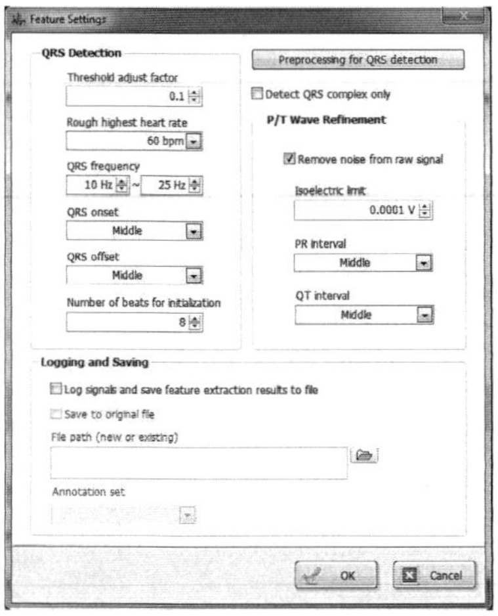

Figure 3. Feature settings.

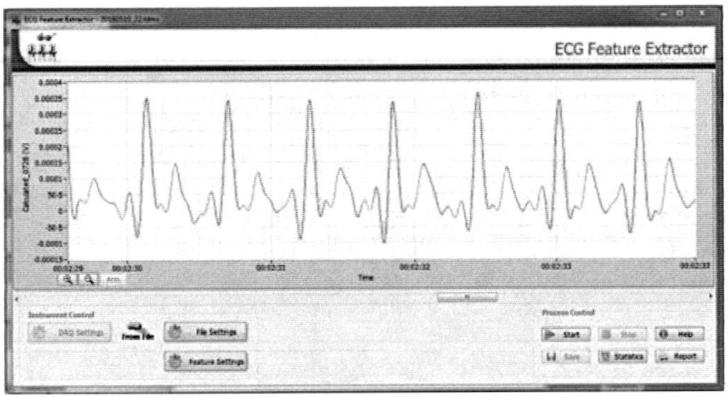

Figure 4. Filtered and rectified ECG signals.

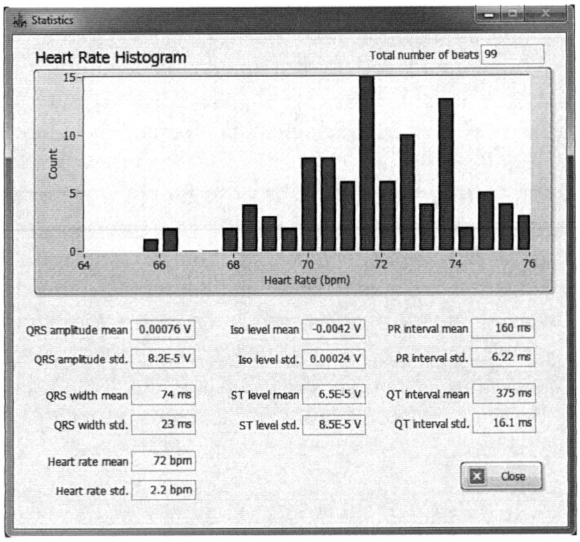

Figure 5. Extracted ECG Features

3. Results and Discussions

The ECG features obtained from the NI-LabVIEW Biomedical Toolkit are compared with the commercial ECG machines, which are used in the health care center. In this study, two types of ECG machines (Philips PageWriter TC30 ECG machine and Kenz ECG 108 ECG machine) are used for performance evaluation. Both of these ECG machines also used twelve-leads ECG system to acquire ECG signals, while Lead II is used for ECG monitoring and HBR calculation. In order to produce ECG results that are comparable between the mentioned devices, the ECG signals are recorded simultaneously using the NI-LabVIEW instruments and ECG machines as shown in Figure 6. The evaluation process of NI LabVIEW instruments in measuring ECG involved two persons. For Case 1, the ECG test is performed simultaneously using Philips PageWriter TC30 ECG machine and NI LabVIEW instruments. For Case 2, the ECG test is done using Kenz ECG 108 ECG machine and NI LabVIEW instruments.

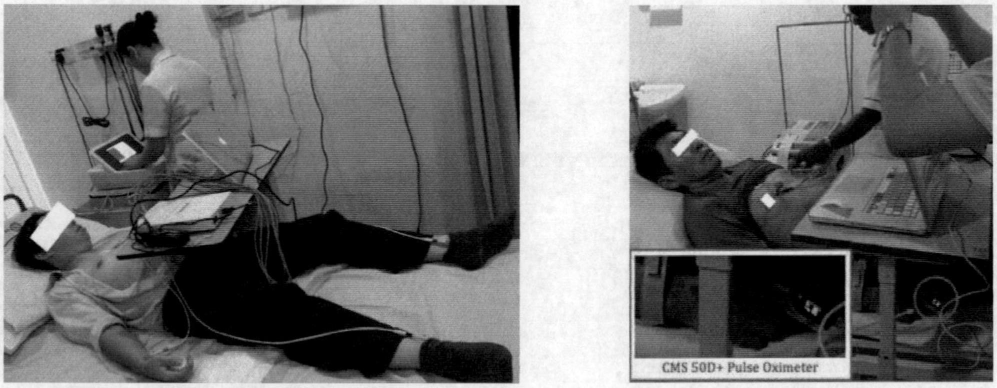

Case 1 Case 2

Figure 6. NI USB-6281 DAQ device with LabVIEW Biomedical Toolkit acquires the ECG signals simultaneously with ECG machines.

For Case 1, Table 1 shows the results comparison between NI-LabVIEW instruments and Philips PageWriter TC30 ECG machine. From the results shown in Table 1, it can be seen that NI-LabVIEW instruments show a good agreement with the results from health center using Philips PageWriter TC30 ECG machine by achieving only 1 bpm difference for the HBR measurement. Although other ECG features like PR interval, QT interval and QRS interval obtained are slightly different between NI-LabVIEW instruments and Philips PageWriter TC30 ECG machine but they are still in close range with just a maximum of 14 ms difference. This is because Philips PageWriter TC30 not only filtered the power line interference but it also has the extra function to filter the baseline wander of the ECG signals, in which the ready-to-execute NI-LabVIEW Biomedical Toolkit does not occupy this function. Baseline wander is the low frequency noise that below the frequency of 0.5 Hz. The sources for creating the baseline wander are respiration, the loose contact of the electrodes with the skin surface or movement of the subject during the recording process. Hence, there is a slight difference in the time interval of the ECG features between these two instruments. Although the ready-to-execute NI-LabVIEW Biomedical Toolkit is unable to set the baseline wander filter, but the results obtained are able to fulfill the allowable error (±20 ms), which is in ANSI/AAMI/EC13:2002 for ECG features intervals [16].

For Case 2, Kenz ECG 108 ECG machine is used to assess the NI-LabVIEW instruments in measuring ECG features. This ECG machine acquires the ECG signals using suction electrodes, hence it is not suitable for continuous ECG monitoring because the suction electrode will cause pain to the subject after a few minutes. This Kenz ECG 108 ECG machine is unable to interpret the ECG waveforms for ECG features calculation, but it does provide the HBR during the entire ECG signals acquisition process. Hence, manual calculation is required in this case to extract the ECG features from the ECG waveforms printed on the ECG graph paper. In order to assess the performance of the NI-LabVIEW instruments, the ECG features are calculated from the Lead II ECG waveform and the results are listed in Table 2. In this case, the CMS 50D+ pulse oximeter also measures the subject's pulse rate simultaneously as shown in Case 2 of Figure 6.

From the results shown in Table 2, the HBR for Kenz ECG 108 ECG machine and NI-LabVIEW instruments also 72 bpm, while other ECG features time interval are able to satisfy the allowable error (±20 miliseconds) stated in ANSI/AAMI/EC13:2002. Again, it shows that the results obtained using NI-LabVIEW instruments show a good agreement with the commercial ECG machine that are used in health care centers. As a conclusion, NI-LabVIEW Biomedical Toolkit associated with the NI USB-6281 DAQ device can be used as an alternative method to acquire, process and analyze the ECG signals for various ECG features and HBR measurements. Furthermore, the pulse rate obtained using CMS50D+ pulse oximeter is also agreed well with the HBR obtained from Kenz ECG 108 ECG machine and NI-LabVIEW instruments. The difference is only 0.7 bpm for this case.

Table 1. ECG features comparison between Philips PageWriter TC30 and NI-LabVIEW instruments.

ECG Features	Philips PageWriter TC30 ECG machine	NI-LabVIEW instruments	Difference	% Accuracy
Heart Beat Rate	98 bpm	97 bpm	1 bpm	98.98%
RR interval	612 ms	619 ms	7 ms	98.86%
PR interval	135 ms	121 ms	14 ms	89.63%
QT interval	326 ms	314 ms	12 ms	96.32%
QRS interval	96 ms	110 ms	14 ms	85.42%

Table 2. ECG features comparison between Kenz ECG 108 and NI-LabVIEW instruments.

ECG Features	Kenz ECG 108 ECG machine	NI-LabVIEW instruments	Difference	% Accuracy	CMS 50D+ pulse oximeter
Heart Beat Rate	72 bpm	72 bpm	-	100.00%	72.7 bpm
RR interval	829 ms	833 ms	4 ms	99.52%	-
PR interval	152 ms	160 ms	8 ms	94.74%	-
QT interval	373 ms	375 ms	2 ms	99.46%	-
QRS interval	80 ms	74 ms	6 ms	92.50%	-

4. Conclusion

In this study, NI-LabVIEW Biomedical Toolkit associated with DAQ device is presented for measuring ECG features and HBR. The extracted ECG features and HBR from NI-LabVIEW instruments have been compared to the features extracted from two types of ECG machine i.e. Philips PageWriter TC 30 and KenZ ECG 108. The results show that the performance of NI-LabVIEW instruments is very convincing by satisfying the allowable error stated in ANSI/AAMI/EC13:2002 for both cases in this study. Thus, NI-LabVIEW instruments are highly potent as an alternative method to acquire ECG features and HBR for diagnosing cardiovascular diseases in future. In addition, the computerized NI-LabVIEW instruments with the self-interpretation function can always outperform most of the conventional ECG machines that are still required cardiologists for manually interpreting the results displayed on the ECG paper.

Acknowledgements

The authors would like to thank Polyclinic UMS and Kinta Medical Centre provide the services for ECG test. The authors also would like to acknowledge the support of Faculty of Engineering and Research & Innovation Management Centre, Universiti Malaysia Sabah under the SLB research grant, grant no. SLB0188-2019.

References

[1] Marieb E N and Hoehn K N 2014 *Human Anatomy & Physiology 10th Edition* (Pearson)
[2] Kisilevsky B S and Brown C A 2016 Comparison of Fetal and Maternal Heart Rate Measures Using Electrocariographic and Cardiotocographic Methods *Infant Behavior & Development* **42** 142-151
[3] Scalise L 2012 *Advances in Electrocardiograms – Methods and Analysis* (InTech)
[4] Beasley B M 2003 *Understanding EKGs: A Practical Approach Second Edition* (Prentice Hall)
[5] Director of Nursing 2013 *Continuous Electrocardiography (ECG) Monitoring Practice Guideline – Sydney Children's Hospital (SCH)*
[6] Clinical Nurse Educator 2014 *Electrocardiographic (ECG) Monitoring Procedure – Children's Hospital Westmead (CHW)*
[7] National Instruments Corporation 2014 *High-Accuracy M Series Multifunction DAQ for USB – 18-Bit, up to 625 kS/s, up to 32 Analog Inputs*
[8] Deshmukh A and Gandole Y 2015 ECG Feature Extraction using NI Lab-VIEW Biomedical

IOP Conf. Series: Materials Science and Engineering **705** (2019) 012020 doi:10.1088/1757-899X/705/1/012020

Workbench *International Journal of Recent Scientific Research* **6** 5603-5607

[9] Velasco M B, Weng B and Barner K E 2008 ECG Signal Denoising and Baseline Wander Correction Based On The Empirical Mode Decomposition *Computers in Biology and Medicine* **38** 1-13

[10] Wan Mahmud W M H and Malarvili M B 2011 Comparative Analysis of Preprocessing Techniques for Quantification of Heart Rate Variability *Proceedings of International Federation for Medical and Biological Engineering* 415-419

[11] Najarian K and Splinter R 2006 *Biomedical Signal and Image Processing* (Taylor & Francis Group)

[12] Srinagesh M, Sarala P and Aparna K D 2013 ECG Wireless Telemetry *International Journal of Engineering and Innovative Technology* **2** 75-78

[13] Elmansouri K, Latif R, Nassiri, B and Elouaham S 2013 New Electrocardiogram Signal Analysis in a Research Laboratory Using LabVIEW *International Journal of Interdisciplinary Research and Innovations* **1** 15-21

[14] Ansari S, Farzaneh N, Duda M, Horan K, Andersson H B, Goldberger Z D, Nallamothu B K and Najarian K 2017 A Review of Automated Methods for Detection of Myocardial Ischemia and Infarction Using Electrocardiogram and Electronic Health Records *IEEE Reviews in Biomedical Engineering* **10** 264-298

[15] Khong W L, Rao N S V K and Mariappan M 2017 National Instruments LabVIEW and Video Imaging Technique for Health Status Monitoring *Journal of Fundamental and Applied Sciences* **9** 858-886

[16] Association for the Advancement of Medical Instrumentation 2002 *American National Standard ANSI/AAMI/EC13:2002: Cardiac Monitors, Heart Rate Meters, and Alarms*

IOP Conf. Series: Materials Science and Engineering **705** (2019) 012021 doi:10.1088/1757-899X/705/1/012021

pH Prediction of Perlis Sunshine Mango Using NIR Spectrometer

P S Woo[1] and M F Ibrahim[*1,2]

[1]School of Bioprocess Engineering, Universiti Malaysia Perlis, Perlis, Malaysia
[2]Centre of Excellence for Advance Sensor Technology, Universiti Malaysia Perlis, Perlis, Malaysia

[*]Corresponding author: firdausibrahim@unimap.edu.my

Abstract. To evaluate the various internal quality attributes of fruits, NIR spectroscopy techniques are undoubtedly quick and non-destructive tools. Acidity (pH) is one of the main quality attributes in mango fruits. Generally, it is being predicted in destructive way. The aim of this research was to develop calibration model and prediction of pH in Perlis Sunshine mangoes using NIR spectrometer. The transmission spectra of Sunshine mangoes were acquired in the wavelength range from 300 to 1000 nm. The effects of different types of pre-processing methods and spectra treatments, such as baseline correction, multiplicative scatter correction (MSC), Savitzky-Golay (SG) smoothing, second order derivative (SG) and normalisation were analyzed. The prediction models were developed by partial least squares (PLS) regression. The coefficient of determination (R^2) of pH was 0.928 and the standard error of cross-validation (SECV) was 0.153. The results indicated that by using the NIR measurement system, in the suitable spectral range, it is possible to predict the pH of mango fruits by non-destructively.

1. Introduction

Mango (*Mangifera indica L.*) is abundantly available in subtropical and tropical countries. It carries a good price in the world market due to its delicious flavor, high nutritional value, low calories and excellent aroma [1]. Perlis Sunshine mango also known as 'Sala' mango is a type of mango which grows in Perlis, Malaysia. Normally, the length of Sunshine mango is around 12 – 15cm while the average weight is between 300 – 450 g. Sunshine mango is green in colour when it is unripe. Its colour turns to yellowish green when it is ripe. Fascinatingly, the taste for both ripe and unripe Sunshine mango is sour.

Nowadays, the agro-food industry is growing rapidly around the world. It is crucial to select fruit with the suitable degree of maturation in order to ensure the production of high-quality fruits. By measuring size, colour and firmness, maturity of mango could be predicted [2]. Various researchers have considered maturity from different point of views [3]. In mango fruits, acidity (pH) is one of the internal quality attributes used in determining the maturity. However, destructive methods are generally carried out to measure this internal quality. Hence, it is necessary to develop effective non-destructive methods so that the internal quality of fruits such as acidity (pH) can be measured.

Near-infrared (NIR) spectrometry is a rapid and non-destructive tool which has been widely used to evaluate various internal quality attributes of fruits [4]. The radiation of NIR spectrometry covers a spectra wavelength ranging from 780 to 2500 nm [5]. Compared to other conventional methods, NIR spectrometry offers decisive benefits. It involves simple sample preparation, fast analysis, chemical-free, pollution-free as well as can be carried out on-line [6, 7].

IOP Conf. Series: Materials Science and Engineering **705** (2019) 012021 doi:10.1088/1757-899X/705/1/012021

Partial least square (PLS) regression is the computation of the optimal least-squares fit to part of a correlation or covariance matrix. It is normally carried out to relate spectral data to quality attributes. PLS is quite identical to Principal Component Regression (PCR). Despite, it defines the latent variables (principal components) based on the covariance between the independent and dependent variables, rather than on the variance in the independent variables alone [8]. Hence, the aim of this research was to develop calibration model and prediction of pH in Perlis Sunshine mangoes using NIR spectrometer. The prediction models were developed by partial least squares (PLS) regression.

2. Experimental methods

2.1. Design of storage box

A simple storage box was designed to store Perlis Sunshine mango fruits under ambient condition. Temperature and humidity sensor (DHT22) were used to measure temperature and air humidity inside the storage box. The main purpose of designing the storage box was to control temperature and air humidity inside the storage box. This was because the temperature and humidity might affect the spectral data collection of NIR spectrometry [9]. Arduino is a microcontroller that was used to control the temperature and humidity inside the storage box. The bulb was fixed on top of the storage box. Its function was to act as the heat source. Two 12V DC fans were fixed at both sides of the storage box. The fans were used to control the temperature inside the storage box by taking cold air in and hot air out. The storage box was designed with the dimensions of 550 mm x 400 mm x 320 mm. Figure 1 shows the simple circuit diagram of mangoes storage box. Figure 2 shows the 3D drawing of mangoes storage box and Figure 3 shows the top view drawing of mangoes storage box.

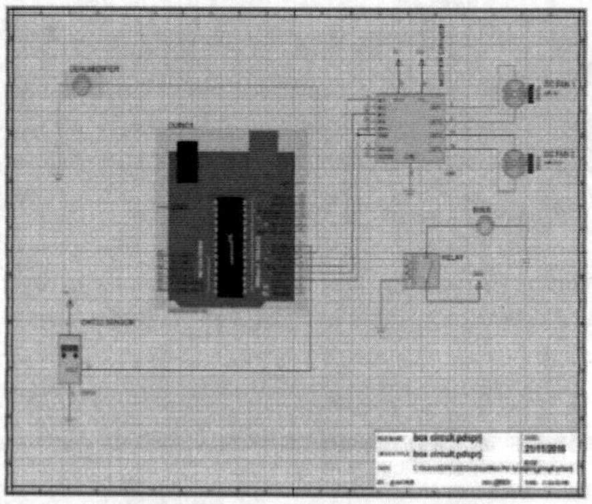

Figure 1. Circuit diagram of mangoes storage box.

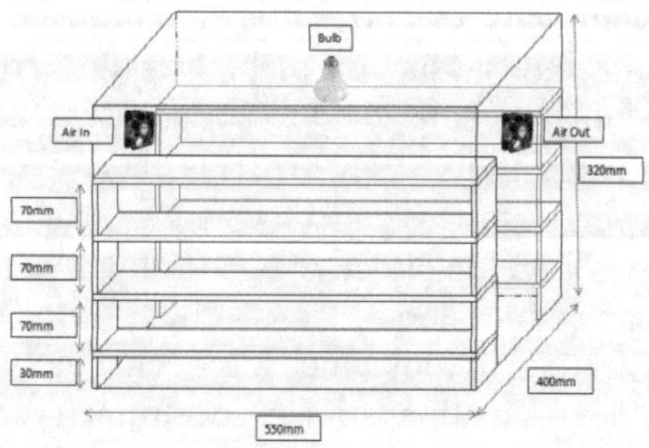

Figure 2. 3D drawing of mangoes storage box.

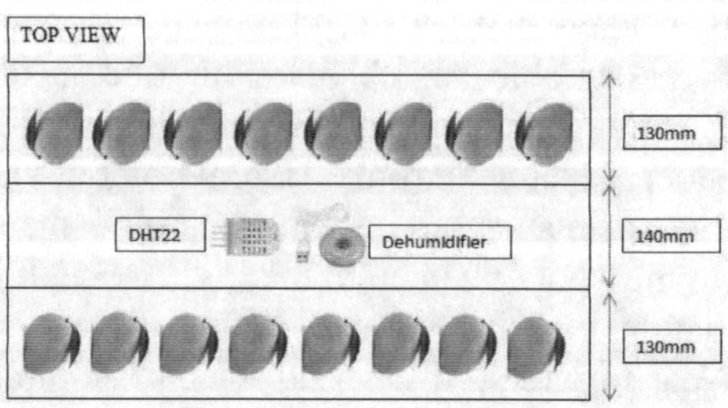

Figure 3. Top view drawing of mangoes storage box.

2.2. Sample preparation
After the designing the storage box, a total of 120 unripe Sunshine mangoes were used. At first, 40 unripe mango samples were stored inside the storage box with temperature of 27 ± 3 °C and relative humidity of 65 ± 5 %. A total of three blocks replication for the sample preparation were carried out. 40 unripe mango samples were used for each replication.

2.3. NIR spectra acquisition
The acquisition of spectra was performed using a NIR spectrometer (Ocean Optics USB4000-XR1-ES), equipped with a 400 μm reflection probe fiber bundle (6 illumination fibers around 1 read fiber). The fiber optic variable attenuator was used to control the amount of light transmitted between two fibers. Tungsten halogen light beam (360 – 2000 nm) with power rating of 5-V/12-W was applied to the selected sampling area and then a stray-light protection cover was closed. It was used as lighting for this instrument. After that, a flexible rubber pad was fixed to the end of the probe as to prevent any light noise and light radiation leakage from outside.

The transmission spectra of mango samples were acquired in the range of 300 – 1000 nm with an optical resolution of 2 nm. Each day, a set of NIR reflectance spectra was acquired for every 5 mango samples. Three scans were taken at three spots in a single side of the mango in order to remove the spatial variability [10]. The readings were averaged to obtain one spectrum signal for each sample. The white reference measurement which has reflectivity above 98% for 250 – 1500 nm was obtained using Polytetrafluoroethylene (PTFE) Diffuse Reflectance Standard. This PTFE standard was specifically designed for reflection studies. Before carrying out the measurement process, the reference was taken only once.

2.4. Determination of actual pH value
After acquiring the spectra, the actual pH value of mango samples was determined using standard destructive measurement for reference values. Three portions of 15 mm width and 15 mm deep mango pulp were taken from three different spots which have been illuminated by the NIR radiation. The mango pulp was homogenized for 3 minutes using a small blender. After that, a piece of muslin cloth was used for filtration as to avoid any suspended solid on the juice. Then, a digital pH meter (Hanna HI model 2213, Padova, Italy), calibrated with pH 4.0 and 7.0 buffers was used to measure the actual pH value of mango juice and the average values were noted. After acquired the NIR spectra and actual pH value, the mango samples used were discarded. All mango samples were not used and examined repeatedly.

2.5. Spectra data analysis
Spectral data and actual pH value obtained were imported to Unscrambler (Version 10.3, Camo, Oslo, Norway) software to carry out multivariate analysis. Partial least squares regression (PLS) was used to develop the models for predicting the pH of mango samples. The wavelength ranges between 300 – 1000 nm were analyzed to select the best wavelength ranges that would obtain the best correlation between the spectral data and pH of mango samples.

Before carried out the modelling, it was necessary to pre-process spectral data in order to obtain stable and accurate calibration models [11]. Irrelevant information such as uncertainties, noise and unrecognized features were needed to be discarded. Several pre-processing techniques such as such as baseline correction, multiplicative scatter correction (MSC), Savitzky-Golay (SG) smoothing, second order derivative (SG) and normalisation were used to improve the predictability of PLS model. With the help of software, a few outlier samples were identified and eliminated for further improvement in the model.

3. Results and discussion

3.1. Spectral characteristics of Perlis Sunshine mango

Figure 4 shows the original NIR diffuse reflectance spectra ranged 300 – 1000 nm for 120 Perlis Sunshine mangoes. Among these mango samples, there were many cross-over points and significant number of overlapping with each other. Through visual inspection, the original spectra showed a homogenous trend without any outliers. The general outline of the absorption spectra for Sunshine mangoes was very much alike to that of other kinds of fruit materials such as lemon [12] and kiwi [13]. The selection of wavelength range was important in developing a strong calibration model. The spectral data which produced noise and little predictive ability was discarded. With original spectra, it was very hard to determine the relationship between NIR spectra characteristics and pH of the mango samples. Baseline drift and noise were existed in the NIR absorbance spectra in the wavelength range of 300 – 500 nm and 950 – 1000 nm. These features were quite common to be found in NIR spectra which collected by diffuse reflectance method. Thus, several pre-processing techniques were used in the analysis of NIR absorbance spectra.

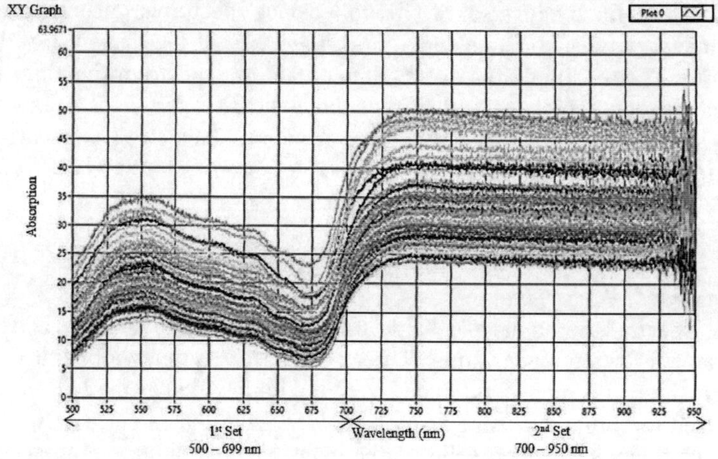

Figure 4. Original NIR diffuse reflectance spectra ranged 300 - 1000 nm for 120 Perlis Sunshine mangoes.

Figure 5 shows the original NIR diffuse reflectance spectra ranged 300 – 1000 nm for 3 randomly selected Perlis Sunshine mangoes with low, medium and high pH value. Based on Figure 5, the NIR reflectance spectra taken from the mango sample with high pH value showed the most significant fluctuation compared to the mangoes with medium and low pH value. This showed that the relationship between NIR spectra characteristics and pH of Perlis Sunshine mangoes might have existed between each other. However, noises could be observed at the beginning (300 – 500 nm) and at the end of the spectral data (950 – 1000 nm). Thus, the wavelength data within these two spectral ranges were eliminated in the selection of wavelength group with the best performance so that the accuracy of the results can be enhanced.

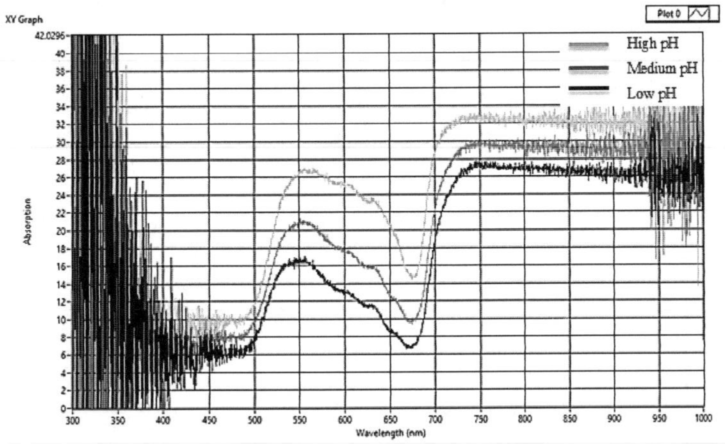

Figure 5. Original NIR diffuse reflectance spectra ranged 300 - 1000 nm for 3 randomly selected Perlis Sunshine mangoes with low, medium and high pH values.

3.2. Selection of wavelength group with the best performance

The spectral range of (500 – 950 nm) was calibrated using partial least squares (PLS) regression. The models were developed in order to predict the pH of mango samples non-destructively. By using PLS regression, coefficient of determination (R^2) of the calibration and validation set were 0.880 and 0.663 respectively. R^2 value obtained for the calibration set was acceptable, however, for the validation set was quite low. Table 1 shows the statistical parameters and results of PLS model for pH prediction using 500 – 950 nm wavelength range.

Table 1. Statistical parameters and results of PLS model for pH prediction using 500 - 950 nm wavelength range.

Wavelength range, nm	Calibration		Validation	
	SEC	R^2	SECV	R^2
PLS model				
500 - 950	0.096	0.880	0.163	0.663

Due to the low R^2 value obtained for the validation set, the models were then tested using a smaller wavelength range. The number of wavelengths can be minimized by selecting the wavelength group with the best performance in order to increase the model's accuracy and reduce errors of the results [14]. The wavelength range of 500 – 950 nm was divided into two sets. The first set wavelength range was started from 500 to 699 nm while the second set was started from 700 to 950 nm. These wavelength range was separated based on the visible spectrum (500 – 699 nm) and NIR spectrum (700 – 950 nm). Table 2 shows the statistical parameters and results of PLS model for pH prediction using different NIR wavelength range. Figure 6 and Figure 7 show the scatter plot of pH using PLS model in the wavelength range of 500 – 699 nm and 700 – 950 nm.

Table 2. Statistical parameters and results of PLS model for pH prediction using different NIR wavelength range.

Wavelength range, nm	Calibration		Validation	
	SEC	R^2	SECV	R^2

PLS model				
500 - 699	0.089	0.897	0.184	0.571
700 - 950	0.075	0.928	0.153	0.705

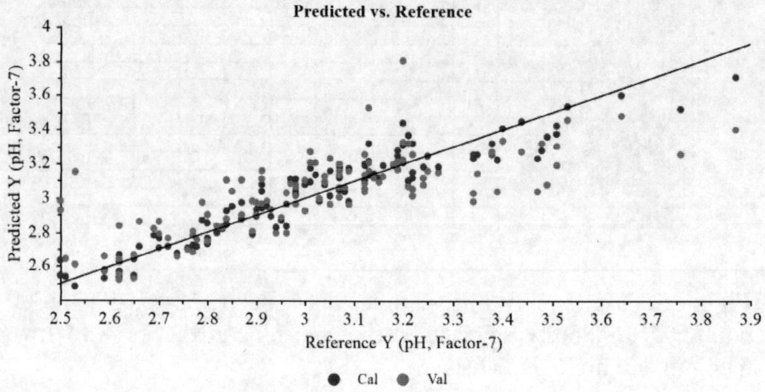

Figure 6. Scatter plot of pH using PLS model in the wavelength range of 500 - 699 nm.

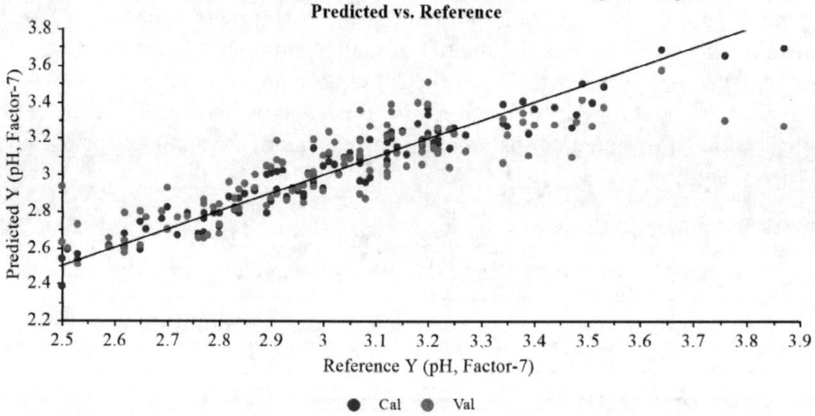

Figure 7. Scatter plot of pH using PLS model in the wavelength range of 700 - 950 nm.

For wavelength range of 700 – 950 nm, the R^2 value for the calibration and validation set using PLS model were 0.928 and 0.705 respectively. Based on Table 2, the result indicated that the PLS model tested using the wavelength range of 700 – 950 nm was more suitable to predict the pH of mango samples compared to the wavelength range of 500 – 699 nm. This was because the wavelength range of 700 – 950 nm produced better statistical results. The results showed that the values of SEC and SECV decreased with an increasing of R^2 values for both calibration and validation set.

Based on Figure 6 and 7, the scatter plot of pH using PLS model in the wavelength range of 700 – 950 nm showed stronger correlation between the predicted and measured pH compared to the wavelength range of 500 – 699 nm. This was due to fewer outliers were detected in the scatter plot of pH using the wavelength range of 700 – 950 nm. The truncation of wavelength tested from 700 – 950 nm was found to improve the results of PLS model. Thus, the results proved that the performance of PLS model can be enhanced by selecting the wavelength group with the best performance. Same improvement was also found in the pH prediction of mangoes from seven different mango cultivars [14].

The removal of visible wavelength region of 500 – 699 nm proved that there was no significant correlation existed between the skin colour and pH of mango samples. Based on Figure 4, the peak in the NIR diffuse reflectance spectra indicated that there were strong reflectance characteristics existed in the wavelength range of 700 – 950 nm. The peak was used to create relationship between the NIR spectra and pH of mango samples. Thus, the NIR wavelength range of 700 – 950 nm was used to develop the next PLS model by using different pre-processing methods. Similar wavelength range (700 – 1100 nm) was also being used by [15] to determine the acidity (pH) of mango 'Nam Dokmai'.

3.3. Prediction of pH using different pre-processing methods

There were several pre-processing methods that can be processed together with PLS model such as baseline correction, multiplicative scatter correction (MSC), Savitzky-Golay (SG) smoothing, second order derivative (SG) and normalisation [14]. These pre-processing methods were usually used to remove the irrelevant information from spectra which derived from unknown sources such as irregular surface morphology as well as varied distance between sample and detector [16]. Table 3 shows the results of PLS model with different pre-processing methods for pH prediction using the wavelength range of 700 – 950 nm.

Table 3. Results of PLS model with different pre-processing methods for pH prediction using the wavelength range of 700 - 950 nm.

Pre-processing method	**No. of PLS factor**	**Calibration**		**Validation**	
		SEC	**R^2**	**SECV**	**R^2**
PLS model (700 - 950 nm)					
No data treatment	5	0.114	0.833	0.169	0.640
	6	0.096	0.880	0.160	0.677
	7	0.075	0.928	0.153	0.705
Baseline correction	5	0.116	0.827	0.175	0.617
	6	0.101	0.867	0.167	0.652
	7	0.074	0.929	0.161	0.677
Multiplicative scatter correction (MSC)	5	0.081	0.916	0.177	0.597
	6	0.065	0.945	0.175	0.606
	7	0.053	0.964	0.175	0.608
Smoothing (SG)	5	0.135	0.763	0.182	0.577
	6	0.104	0.861	0.161	0.668
	7	0.090	0.894	0.158	0.680
2nd order derivative (SG)	5	0.078	0.921	0.187	0.560
	6	0.067	0.939	0.190	0.546
	7	0.055	0.961	0.198	0.507
Normalisation	5	0.102	0.866	0.172	0.614
	6	0.081	0.916	0.172	0.616
	7	0.068	0.941	0.171	0.620

According to [17], the suitable number of regression factor to be used for the PLS model was selected based on the minimum standard error of cross-validation (SECV). Based on Table 3, the PLS model with no data treatment showed the minimum SECV which was 0.153. Thus, the performance of PLS model processed with other types of pre-processing methods were evaluated based on the regression factor of 7.

For the calibration set, PLS model with regression factor of 7 treated with MSC pre-processing method showed the best correlation between the predicted and measured pH. The R^2 value obtained was 0.964 and SEC was 0.053. The results showed that SEC decreased with an increasing of R^2. For baseline correction, MSC, second order derivative (SG) and normalisation, the R^2 values obtained for the calibration set were slightly higher than the PLS model with no data treatment. However, the R^2 values obtained for the validation set by using these four types of pre-processing methods were all lower than the PLS model with no data treatment. Low R^2 values indicated that a weak correlation was existed between the predicted and measured pH of mango samples.

For the validation set, the PLS model with no data treatment showed the lowest SECV and highest R^2 value compared to the other types of pre-processing methods. The SECV obtained was 0.153 and R^2 value was 0.705. Besides, the value of SECV and SEC obtained for the PLS model with no data treatment was closed to each other. This result indicated that the loss in the accuracy was small. A low standard error indicated that the PLS model with no data treatment showed reasonably accurate results in predicting the pH of mango samples. [18] discovered that PLS model was suitable to be used for pH prediction of mangoes due to good calibration results were found in their research.

The application of different types of pre-processing methods such as baseline correction, MSC, smoothing (SG), second order derivative (SG) and normalisation on the PLS model were found to be less or no effect in improving the predictability of the calibration and validation set of mango samples. Similar results were also reported by [14] and [12] for the pH prediction of mango and lemon samples using NIR spectroscopy.

Figure 7 shows the scatter plot of pH using PLS model with no data treatment in the wavelength range of 700 – 950 nm. There was a strong positive correlation existed between the predicted and measured pH of mango samples. From the scatter plot, most of the data points were closed to the target line and the slope of the line fitted to the data points was closed to 45°. This indicated that the predicted and measured pH were highly correlated with each other.

4. Conclusion

PLS regression was used to determine the correlation between the measured and predicted pH of mango samples. The results showed that the PLS model with no data treatment in the wavelength range of 700 – 950 nm produced the best regression coefficient values compared to the other types of pre-processing methods. The application of several types of pre-processing methods such as baseline correction, MSC, smoothing (SG), second order derivative (SG) and normalisation on PLS model were found to be less or no effect in improving the predictability of the calibration and validation set of mango samples. The results indicated that PLS model with no data treatment in the wavelength range of 700 – 950 nm yielded good calibration and validation results. The R^2 value obtained for calibration set was 0.928 and validation set was 0.705. There was a strong correlation existed between the predicted pH and measured pH of mango samples. Therefore, it is concluded that by using the NIR measurement system, in the suitable spectral range, it is possible to predict the pH of mango fruits by non-destructively.

Acknowledgment

The authors acknowledge financial support from Universiti Malaysia Perlis (UniMAP) through 9001-00582 Internal Social Project Via Research (INSPIRE) 2019.

References

[1] Schmilovitch Z, Mizrach A, Hoffman A, Egozi H and Fuchs Y 2000 Determination of mango physiological indices by near-infrared spectrometry *Postharvest Biology and Technology* **19** 245-52

[2] Jha S, Kingsly A and Chopra S 2006 Physical and mechanical properties of mango during growth and storage for determination of maturity *Journal of Food engineering* **72** 73-6

[3] Peacock B 1986 *Influence of Harvest Maturity of Mangoes on Storage Potencial and Ripe Fruit Quality*

[4] Costa RC and de Lima KM 2013 Prediction of parameters (soluble solid and pH) in intact plum using NIR spectroscopy and wavelength selection *Journal of the Brazilian Chemical Society* **24** 1351-6
[5] Sheppard N, Willis H and Rigg J 1985 Names, symbols, definitions and units of quantities in optical spectroscopy (Recommendations 1984) *Pure and Applied Chemistry* **57** 105-20
[6] Saleh B 2012 Biochemical and genetic variation of some Syrian wheat varieties using NIR, RAPD and AFLPs techniques *Journal of Plant Biology Research* **1** 1-11
[7] Pissard A, Baeten V, Romnée J-M, Dupont P, Mouteau A and Lateur M 2012 Classical and NIR measurements of the quality and nutritional parameters of apples: a methodological study of intra-fruit variability *BASE*
[8] Wold S, Sjöström M, Eriksson L 2001 PLS-regression: a basic tool of chemometrics. Chemometrics and intelligent laboratory systems **58** 109-30
[9] Burns DA and Ciurczak EW 2001 *Handbook of Near-Infrared Analysis* (CRC press)
[10] Blahovec J and Kutílek M 2003 *Physical Methods in Agriculture: Approach to Precision and Quality* (Springer Science & Business Media)
[11] Cen H and He Y 2007 Theory and application of near infrared reflectance spectroscopy in determination of food quality *Trends in Food Science & Technology* **18** 72-83
[12] Reddy NS, Nivetha D and Yadav B 2016 Non-destructive quality assessment of citrus fruits using ft-near-infrared spectrocopy *International Journal of Science, Environment and Technology* **5** 1850-60
[13] Lee JS, Kim S-C, Seong KC, Kim C-H, Um YC and Lee S-K 2012 Quality prediction of kiwifruit based on near infrared spectroscopy *Korean Journal of Horticultural Science and Technology* **30** 709-17
[14] Jha SN, Jaiswal P, Narsaiah K, Gupta M, Bhardwaj R and Singh AK 2012 Non-destructive prediction of sweetness of intact mango using near infrared spectroscopy *Scientia Horticulturae* **138** 171-5
[15] Rungpichayapichet P, Mahayothee B, Nagle M, Khuwijitjaru P and Müller J 2016 Robust NIRS models for non-destructive prediction of postharvest fruit ripeness and quality in mango *Postharvest Biology and Technology* **111** 31-40
[16] Lu R 2001 Predicting firmness and sugar content of sweet cherries using near–infrared diffuse reflectance spectroscopy *Transactions of the ASAE* **44** 1265
[17] Moghimi A, Aghkhani MH, Sazgarnia A and Sarmad M 2010 Vis/NIR spectroscopy and chemometrics for the prediction of soluble solids content and acidity (pH) of kiwifruit. *Biosystems Engineering* **106** 295-302
[18] Purwanto YA, Zainal PW, Ahmad U, Sutrisno M, Makino Y, Oshita S, Kawagoe Y, Kuroki S 2013 Non destructive prediction of pH in mango fruit cv. Gedong Gincu using NIR spectroscopy *International Journal of Engineering and Technology IJET-IJENS* **13** 70-3

IOP Conf. Series: Materials Science and Engineering **705** (2019) 012022 doi:10.1088/1757-899X/705/1/012022

Study on the Mobility Performance of Wheelchair Using Smartphone Accelerometer

L K H'ng, M A Rahim*, N A Rahim
School of Mechatronic Engineering, Universiti Malaysia Perlis (UniMAP), Pauh Putra Campus, 02600 Arau, Perlis, Malaysia

*Corresponding author: muhajir@unimap.edu.my

Abstract. Wheelchair is very important as a means of transport by a person who is unable to walk as a result of illness or injury. On the other hand, smartphones have become common device that allow users to remotely control many electrical and electronic appliances. This paper presents a study on the mobility performance of wheelchair using smartphone accelerometer. The powered wheelchair is presented by a two-wheeled robot car which emphasized on the control layout platform for the smartphone's tilting directions with suitable tolerance values. A basic motion's algorithms for the movement of robot car have been developed, and its mobility performances on the cornering and straight-line direction are analysed.

1. Introduction

A wheelchair is a crucial modality for rehabilitation of people with mobility impairments. Two common wheelchairs implemented in industries are manual self-propelled and powered wheelchair. Jorge L. Candiotti et al. [1] proposed in designing their powered wheelchair namely Mobility Enhancement roBotic (MEBot) wheelchairs to enhance the navigation skills of the users in common terrains. V. Santhanam et al [2] had designed an automated wheelchair controlled by tilting smartphone with 3 axis accelerometer inbuilt and Bluetooth connection.

Some studies involved experiment with such applications to test mobility performance of wheelchair. N. Viswanath et al [3] proposed experiment with operation of four-wheel drive wheelchair that done by tilting the smartphone. It considered with maneuverability for the performances moving in different directions and also the optimization of ergonomics in designing a better wheelchair. R.M.A. van der Slikke et al [4] proposed to measure wheelchair sports kinematics based on inertial sensors with wheel skid correction. Furthermore, there are some studies using human manually test on wheelchair such as C. Usma-alvarez et al and David S. Haydon et al [5,6] who proposed an Illinois agility test and sprinting test. They mentioned that tangential velocity and turning radius were the factors that affect performance of rugby athletes where it only involved with skilful sportsmen that trained for competition manually control the wheelchairs.

However, to the best of our knowledge the study on mobility performance has not been reported in any literature to date. Therefore, the main objective of this project is to create a two-wheeled robot car chassis that represents wheelchair using smartphone accelerometer controlled to study on its mobility performance.

The structure of this paper starts in section 2 which provides brief description prototype's implementations. Section 3 presents the methods in conducting experiments for cornering and straight line performances. Section 4 discusses the results obtained and analysis for both experiments conducted. Finally, the conclusion of this project is given at the end of this paper.

2. Prototype Implementations

2.1. Developing Hardware

For the hardware section, the Arduino microcontroller is connected with a motor driver, L298N which allows speed and direction control for the two DC motors and a Bluetooth module, HC-05 for Bluetooth communication with smartphone. The electric voltage supply is from the lithium polymer battery that has capacity of 11.1V and 1100mAh.

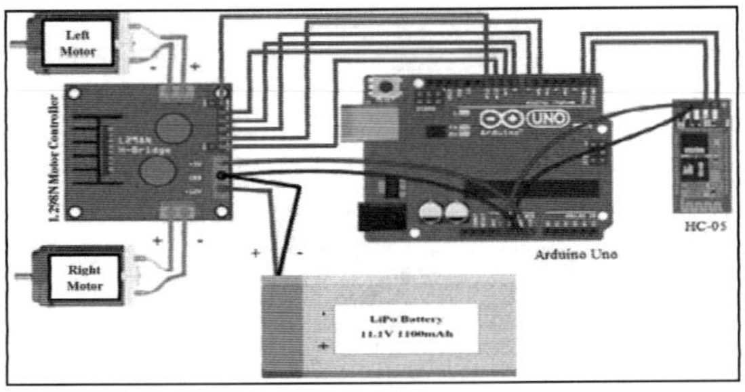

Figure 1. Hardware Connection on Robot Car.

2.2. Developing Software

This control layout design implemented using MIT App Inventor for the smartphone tilting that sends command to the robot car. A series of building block designed for the tilting algorithms in this section. Furthermore, Bluetooth connection is involved for the communication between smartphone and Arduino microcontroller implanted in robot car.

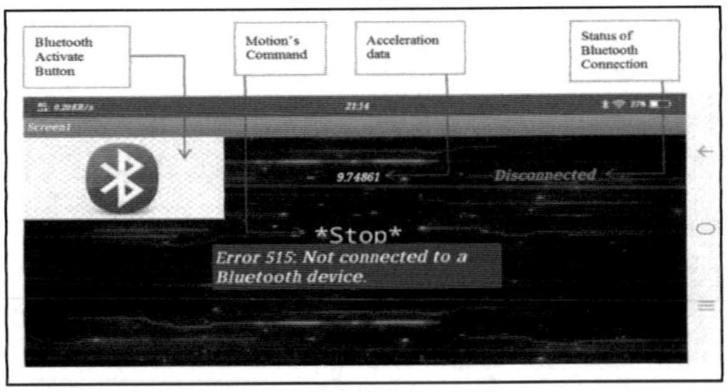

Figure 2. Control Layout Designed for Smartphone Tilting.

3. Experiment on Mobility Performance

3.1. Cornering Performance

Different PWM signals input among both wheels for different configurations. A pencil attached behind the robot car will draw out different sizes of circles due to different configurations when the robot car moves. Since, the pencil and unsteady small front wheel can affect the size of circles drawn, each measurement for the circle's radius will be repeated for 3 times to obtain the average value.

IOP Conf. Series: Materials Science and Engineering **705** (2019) 012022 doi:10.1088/1757-899X/705/1/012022

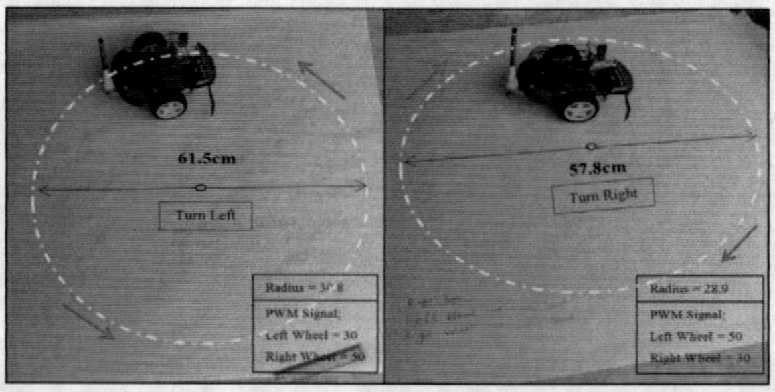

Figure 3. Experiment on Left and Right Turning of Robot Car.

3.2. Straight Line Performance

A test with different recorded time for robot car to reach a distance of 150cm has been conducted. Both wheels are set with same PWM signal to give forward motion and the PWM signal has been increased constantly to obtain the recorded time taken.

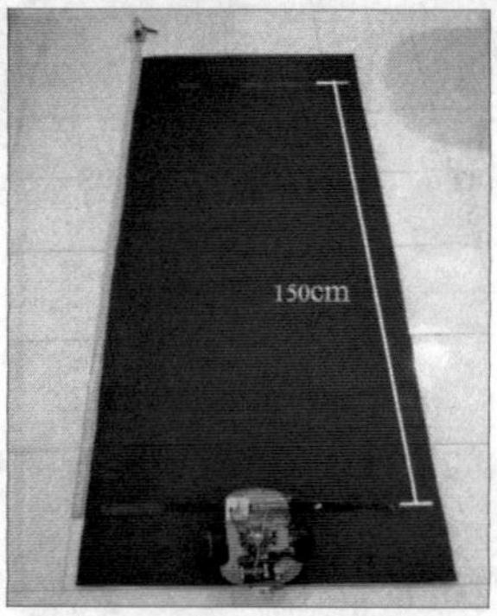

Figure 4. Experiment on Straight Line Performance.

4. Results and Discussion

4.1. Results on Cornering Performance

The configurations with increment in PWM signals of 10 (30, 40, 50 and 60) implemented on left wheels, radii measured become shorter in left turning. Regarding to similar configurations, radii measured become longer for right turning.

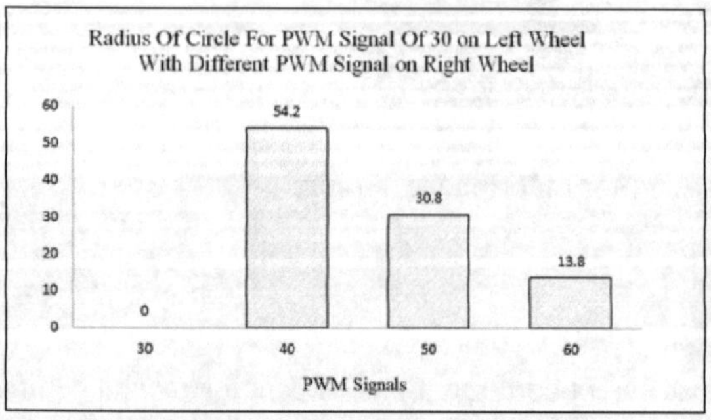

Figure 5. Radius Measured for PWM Signal of 30 on Left Wheel.

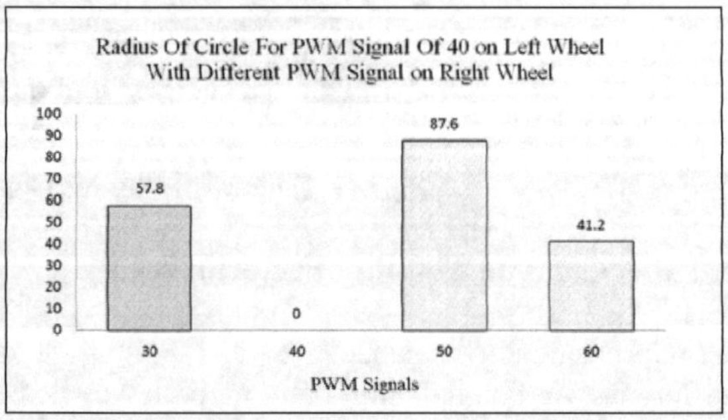

Figure 6. Radius Measured for PWM Signal of 40 on Left Wheel.

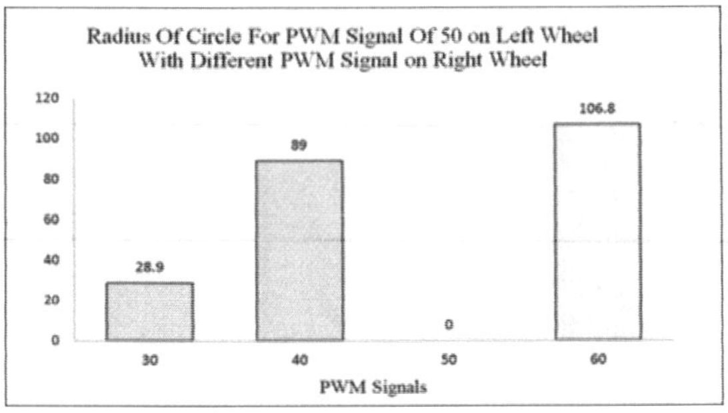

Figure 7. Radius Measured for PWM Signal of 50 on Left Wheel.

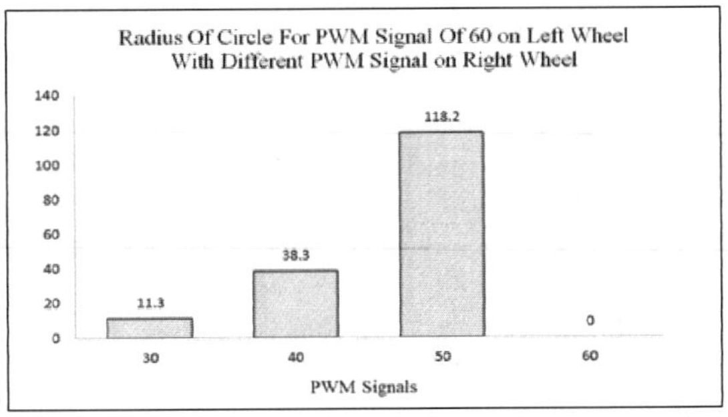

Figure 8. Radius Measured for PWM Signal of 60 on Left Wheel.

(***Remarks**: Blue bar is right turning and yellow bar is left turning of robot car.)

After all, it clearly shows that huge difference in PWM signal values between left and right wheels provide the robot car to draw shorter radius of circle which indicates better cornering performance. In order to perform left turning of robot car, the PWM signal on right wheel must always higher than that on the left wheel or vice versa. The zero measurement in each configuration will be carried out for the following straight line experiment.

4.2. Results on Straight Line Performance

Robot car with similar PWM signal inputs in both wheels but has the highest inputs values compared to other configurations can reach the distance of 150cm in the shortest time. The line graph inclined downwards and almost inversely proportional. This has proved that robot car with higher PWM signal inputs can move in a faster pace to reach its destination.

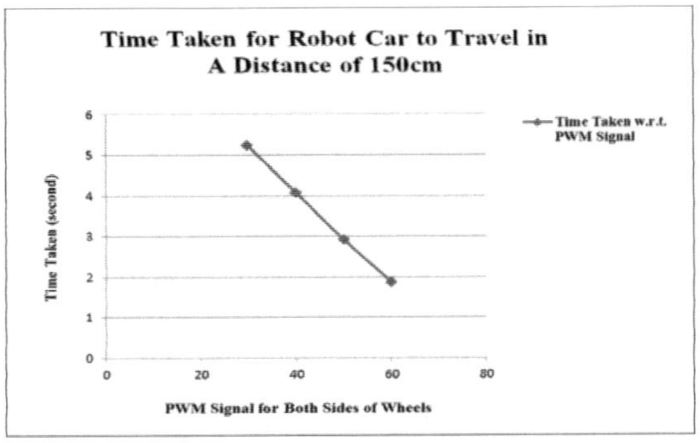

Figure 9. Time Taken for Robot Car to Travel in 150cm.

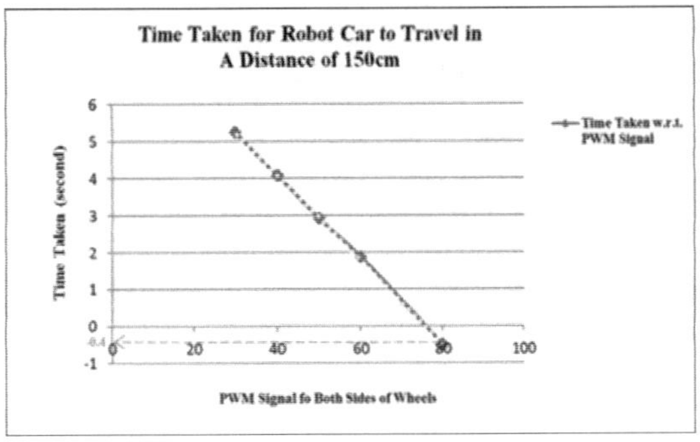

Figure 10. Graph of Estimated Equation for Time Taken.

IOP Conf. Series: Materials Science and Engineering **705** (2019) 012022 doi:10.1088/1757-899X/705/1/012022

Due to light weight of robot car that will make it turn over, an estimated equation has been derived with y=mx+c as following:
To obtain gradient using points $(X_1, Y_1) = (30, 5.24)$ and $(X_2, Y_2) = (50, 1.87)$:
Gradient, m = (5.24-1.87) / (30-60)
m = -0.1123
Hence,

$$y = -0.1123x + c \tag{1}$$

Using the point estimated points (80, -0.4), substitute into equation (1):

$$-0.4 = -0.1123(80) + c$$

Then, get c = 8.584
Substitute again c = 8.584 into equation (1) to get:

$$y = -0.1123x + 8.584 \tag{2}$$

After inserting zero time taken (y=0) into equation (2), the PWM signal value is approximately 76 which means the implemented robot car in this project will immediately turn over or collapsed at this PWM signal inputs on both wheels.

5. Conclusions

In cornering experiment, it implied that huge difference in PWM signal values between both wheels provide a sharp cornering for robot car. Different PWM signals input on both wheels with different speed rotations determine the direction of turning in robot car. Meanwhile, the straight line experiment shows robot car with same higher PWM signals input on both wheels allows the robot car to move faster in forward motion. A limitation due to light weight of robot car is determined with estimated equation derived to indicate the maximum PWM signal it can afford with.

6. References

[1] Candiotti J L, Kamaraj D C, Daveler B, Chung C S, Grindle G G and Cooper R 2009 *Usability Evaluation of a Novel Robotic Power Wheelchair for Indoor and Outdoor Navigation* **100** 627-637

[2] Santhanam V and Viswanathan V 2013 Smartphone Accelerometer Controlled Automated Wheelchair 3rd *International Conference on Electronics, Biomedical Engineering and its Applications (ICEBEA'2013)* 57–61

[3] Viswanath N, Anbarasan M and Jaisiva S 2017 Android Based Automated Wheel Chair Control for Physically Challenged Person *Asian J. Appl. Sci. Technol.* **1** 26–29

[4] van der Slikke R M A, Berger M A M, Bregman D J J and Veeger H E J 2015 Wheel Skid Correction is a Prerequisite to Realiably Measure Wheelchair Sports Kinematics Based on Inertial Sensors *Procedia Engineering* **112** 207-212

[5] Usma-alvarez C C, Jang J, Chua C and Fuss F K 2011 Advanced performance analysis of the Illinois agility test based on the tangential velocity and turning radius in wheelchair rugby athletes *Sports Technology* 201–214

[6] Haydon A D S, Pinder R A, Paul N and Robertson W S P 2018 Test Design and Individual Analysis in Wheelchair Rugby *Journal Science and Medicine in Sports* 1–18

IOP Conf. Series: Materials Science and Engineering **705** (2019) 012023 doi:10.1088/1757-899X/705/1/012023

Triangular Separation Distance Effects on Wave Electrical Energy Harvester Performance

M N F M Kazim[1,2], M R M Rasani [2], A A Abd Rahman[1], M Z Nuawi [2], Z Harun[2], N A M Amin[1]

[1] School of Mechatronic Engineering, Universiti Malaysia Perlis, Pauh Putra Campus, 02600 Arau, Perlis, Malaysia.
[2] Centre for Integrated Design for Advanced Mechanical Systems (PRISMA), Faculty of Engineering and Built Environment, Universiti Kebangsaan Malaysia, 43600 Bangi, Selangor, Malaysia

E-mail: fakhzan@unimap.edu.my

Abstract. Energy scavenging has become a topic of interest toward green and environmentally friendly. Vortex generated in the wake region behind a bluff body is potential to be harvested and transform the mechanical vortices into electrical energy. Therefore, the study of the fluid flow and piezoelectric material is highly potent and a promising research area. Thus, the vortex-induced vibration (VIV) energy harvesting of two bluff bodies in a tandem arrangement is investigated using the 3D simulation. There are four sets of bluff bodies with various distance ratio, x/D from 3 to 6, were analyzed in this simulation. The coupling analysis was analyzed using ANSYS Transient Structural and computational fluid dynamic Fluent module. The triangular width, to 100 mm and the fluid average flow velocity is 1.46 ms^{-1}, which corresponds to Reynold number (Re) of 10k. Four types of analysis have, out; velocity vector, turbulence kinetic energy, piezoelectric tip deformation, and generated output voltage. The results show the bluff shape with spacing ratio 6 has better vortex shedding formation and turbulence kinetic energy formed on both upstream and downstream triangular bluff body. The tips mesh displacement for piezoelectric is highest at space ratio 6 with 7.0 x 10^{-5} essential, and after 10s the beam deflection is continuing at around 1 x 10^{-5} m. While the voltage generated is, at spacing ratio 6 with 0.2 v produced. The obtained results are essential to VIV analysis of the dual triangular cylinder with piezoelectric film, which is vital in energy scavenging knowledge.

1. Introduction

Many researchers are interested in studying about energy harvesting in a fluid using Vortex-induced Vibration (VIV). VIV is a non-linear resonance that occurs when the body oscillates in resonance system struck by the fluid flow with a certain velocity and harvest a vortex shedding frequency that is close to the natural frequency of the body [1–6]. In other words, VIV are motions induced on bodies interacting with an external fluid flow, produced by periodical irregularities or the motion producing periodical irregularities on this flow. Its occur in many fields of engineering, as mention by Williamson *et al.* [1] and Xu Xu *et al.* [5]. Xu Xu *et al.* [5] also mention that the effect caused by VIV is considered as an undesirable effect in the certain engineering field. This effect may lessen the structure lifetime by reduction its performance, its integrity, or fatigue loading. But, the effect of the VIV is useful in some field. It has led to many fundamental studies, many of which are reviewed by Williamson and Govardhan [1], Sarpkaya [3], and Bearman [4]. Bernitsas *et al.* [7]

proposed to obtain the electricity by converting the hydrokinetic energy from the ocean or river streams using VIV. Further studies in this field have been reported by Sanchez-Sanz and Velazquez [8], Barrero-Gil *et al.* [6], Nishi [9], Mackowski and Williamson [10], Meliga *et al.*[11], Stappelbelt *et al.* [12], Song *et al.* [13], Mehmood *et al.* [14] and Dai *et al.* [15, 16].

On the other hand, the study of piezoelectric in the VIV in the airflow have been carried out by many researchers. Mehmood [14] has carried out scientific research on VIV of a circular cylinder with piezoelectric energy harvester on Re number between 96 and 118. Gao [17] has experimented laminar and turbulence flow for piezoelectric energy harvester system. He also proposed to examine the blunt-body with different cross-section area in the VIV. Dai [16] focus his study on the circular cross-section cylinder based on the base excitation and the VIV. The VIV enhance study in the wind have been testified by Zhao [18], Sirohi [19,20], and Abdelkefi [21]. The idea is to convert the kinetic energy of the incoming fluid flow into the oscillatory mechanical energy via VIV.

The flow of the fluid can affect the solid structure; even the effect is minor. The interaction of the fluid and structure is called fluid-structure interaction (FSI). FSI is the interaction of a structure that moved or deformed with fluid flow, either internally or surrounding [22]. It is a multiphysics coupling between the fluid region and the structural region. FSI can be stable or oscillatory. In oscillatory interactions, the strain-induced to the solid structure will cause it to move and at the same time causing the strain to reduce, and the structure reappearance to its former state for the repetition process [23]. The deformations of the structure can be quite large or very small. It depends on the flow pressure and velocity and the structure material properties. To hypothetically treat the fluid-structure interaction associated with VIV, this study adopts wake oscillator model [2]. Tamura and Matsui [24] and Nishi *et al.* [25] has created a model that represents the real phenomenon of viscous flow around a moving body. They describe the change in hydrodynamic force using the equation derived by introducing the oscillatory motion of a rigid bar. Di Silvio [26] shown that the wake area of a body exposed to a steady stream displays swing – like motion and periodically lengthens and shortens alongside growth and shedding of vortices. In the models of Tamura and Matsui [24] and Nishi *et al.* [25], they introduce the wake oscillator is a virtual rigid bar with the mechanical motion of a bar to replace the periodic phenomenon of viscous flow. This simplification allows the model to be applied for computing hydrodynamic forces acting on a long elastic structure. Previous studies of this case have been done by Violette *et al.* [27] and Xu *et al.* [28] without computational expenses.

Achenbach and Heinecke [29] have identified six regimes for the flow around a circular cylinder which are the laminar flow, steady separated region, periodic laminar wake, subcritical regime, critical regime, supercritical regime, and transcritical regime. The classification of the flow was made based on visual observation and changes in the Strouhal number. Further studied by Zdravkovich [30], he classified the fluid flow according to the boundary layer characteristics over the cylinder surface and the separated shear layer. Fifteen types of flow regime have been classified, and Most of the previous studies of VIV have focused on relatively low Reynolds numbers, less than 3×10^4 [31]. Previously, most of the experiments conducted lie in the transition of the shear layer 2 (TrSL2) regime ($1 \times 10^3 \leq Re \leq 4 \times 10^4$). This regime is classified under the subcritical regime and according to Raghavan and Bernitsas [32], the completed transition in the shear layer of TrSL2 regime from laminar to turbulent. Therefore, in this analysis, the Reynolds number used is 1×10^4 which fall in the TrSL2.

Many works have been done to enhance the production of vortex shedding in VIV, including the efficiency of the VIV. Based on previous studies, to improve the efficiency of VIV, the mass – damping ratio should be reduced so that the amplitude and the frequency are increases [33]. The effectiveness of the VIV can affect energy extraction in the energy harvesting system, and it is one of the factors to enhance the energy harvesting system based on VIV.

Another factor that can affect the energy extraction in energy harvesting is the cross-section used in the system. The bluff body uses to create the vortex shedding. Many types of cylinder cross-section have been used to optimize the vortex shedding. According to Ordia *et al.* [34], the characteristics of a good vortex shedder is the stability, and the strength of the vortex generated minimum power losses and the least dependence of Strouhal's number. All these characteristics depend on the vortex shedder geometry. Ding *et al.* [35] proved that a cylindrical and trapezoidal-shaped was performed better in the movement induced excitation vortex-induced vibration (MIEVIV) harvester system out of all shapes designed. But triangular and trapezoidal are the most commonly used shapes as those are easy for manufacturing. Ordia *et al.* [34] observed that the wake width increases with the streamwise length of the bluff body. They also studied different bluff body shapes to find the factor that influences the performance of the blunt-body. Ding *et al.* [37] have studied a single bluff body VIV with different cross-section via numerical simulations. The amplitude and frequency of the VIV vary with the cross-section of the bluff body. Thus, we can improve the energy conversion efficiency of the VIVACE converter by choosing the suitable cross-section of the bluff body. Nevertheless, the triangular cross-sectional bluff body is preferable in this research.

Vortex shedding also can be enhanced by looking at the number of the bluff body in the system. In the past few years, many researchers give a great effort to improve energy harvesting by using VIV. Thus, they come up with one method of developing this system which by using two bluff bodies in tandem formation. A narrow gap separates the bluff bodies. Based on the above approach, Song *et al.* [37] show that there is an improvement in power production when two cylinders are combined with piezoelectric harvesters. The Strouhal number of the system also changes. Bachal and Shrivastava [38] show that dual bluff body has higher Strouhal number than the single bluff body. The increasing of the Strouhal number is due to the increase of the oscillating flow mechanisms. In another case, Zhang *et al.* [39] investigated the VIV response will be higher if the intensity of vortex around downstream is greater than the intensity of vortex around the upstream. In term of pressure fluctuation, Tam Nguyen *et al.* [40] found that the dual bluff body in tandem arrangement gives a higher pressure fluctuation behind the bluff bodies than that of the single bluff body. This finding to be valuable in the design of energy harvesting. From their work, this condition can be achieved by using a dual bluff body.

In the VIV energy harvesting model, vortex excitation can be achieved only with the deformation of the plate of the energy harvester device. The flow shedding that excites VIV on the plate can be generated by installing the bluff body upstream of the system [41], and the position of the piezoelectric film is at the downstream. Ordia *et al.* [34] that studied the vortex shedding using the visualization method. They show that visualization of the vortex flow can help us to understand that the piezoelectric film should be located at the vortex formation length for a particular body. Still, this research does not state the specific location of the piezoelectric film. But, Bachal and Shrivastava [38] have stated the location of the piezoelectric film in their experiment. They specify the position of the piezoelectric film is 3 mm behind the bluff body.

However, this analysis will focus on dual triangulate bluff body that arranges in tandem formation with the piezoelectric film placed downstream. The analysis will perform in a wind tunnel environment and air as a test medium. The simulation run in ANSYS 17.0 using two modules; Transient Structural and Fluid Flow (Fluent). Both modules coupled by using fluid-structure interaction (FSI) technique. The purpose of this analysis is to look into the effect of the energy converter performance by applying the different distance between the bluff bodies.

2. Methodology

The simulation run using the fluid-structure interaction (FSI) simulation technique in ANSYS software to investigate the vibrational performance of the coupled bluff body and the piezoelectric film.

2.1. Theoretical & Navier-Stoke Equation

A reliable coupling method of FSI numerical operational is governed by the incompressible Navier-Stokes equation coupled with a transient structural model through a partitioned approach. The flow motion is governed by the incompressible Navier Stokes equation. An accelerated frame of reference attached to the cylinder is used to solve the incompressible Navier-Stokes equation. The momentum equation is added with a new non-inertial force, which is known as the acceleration's reference frame. The governing equations of the fluid flow are given by

$$\frac{\partial u_j}{\partial x_j} = 0 \qquad (1)$$

$$\frac{\partial u_i}{\partial t}+\frac{\partial}{\partial x_j}(u_j u_i) = -\frac{\partial p}{\rho \partial x_i}+v\frac{\partial^2 u_i}{\partial x_j \partial x_j}-\ddot{x}_i \quad (2)$$

where i, j = 1,2,3, u_i represent the Cartesian velocity components (u, v, w), ρ, v, and p are the density, kinematic viscosity of the fluid, and pressure, respectively. Here, $\ddot{x}_i$ = a is used to denote the cylinder acceleration in the cross-flow direction.

2.1.1. Basic Element

In this project, two ANSYS module was used, which are Transient Structural and Fluid Flow (Fluent). In the Transient Structural module, the piezoelectric film is analyzed in terms of its deformation, total velocity, energy probe, and voltage. These characteristics generated to analyze the performance of the energy converter device. The total deformation shows the deflection of the film when the flow strikes its total velocity shows the speed of the deflection. Lastly, the energy probe and the voltage show the power generated during the analysis. For Fluid Flow (Fluent), this module is used to analyze the flow of the air. Specifically, this analysis is to analyse the vortex shedding according to the different distance of the bluff bodies.

2.1.2. Geometry

The model of the energy harvesting system was developed in a 3D model. The bluff body in this analysis is limited to the triangular cylinder shape. The dual bluff bodies arranged in the tandem formation. The piezoelectric as an energy converter device placed behind the second bluff body, as shown in Figure 1.

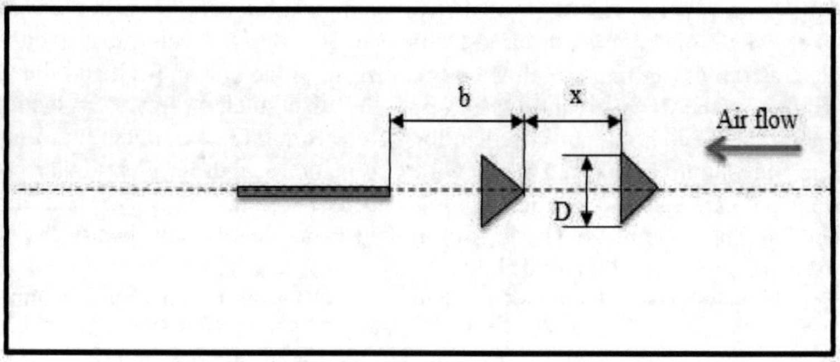

Figure 1. Vortex-induced vibration energy converter system

The characteristic length, D fix at 0.1 m. The separation length, x is various, as shown in Table 1, to optimum ratio of the separation length to the characteristic length, x/D. While distance b is fixed at 0.1 m [38]. The details of the length of b, D, and x are in Table 1. In the analysis, a uniform velocity profile was applied at the inlet. Its flow along the direction of the inlet flow. The flow velocity determined using a derivation of Reynolds number formula.

$$v=\frac{ReV}{\rho D} \quad (3)$$

The velocity, v is 1.46 ms^{-1} equivalent to Re number 10k, V is 1.789×10^{-5} $kgm^{-1}s^{-1}$, ρ is 1.223 kgm^{-3} and D is 0.1 m, is the Reynolds number in turbulent condition, dynamic viscosity and density of the air and the characteristic length of the bluff body, respectively. This analysis is considering the non-slip conditions (v = 0 ms^{-1}) alongside the wind tunnel's walls, and the incompressible flow.

Table 1. Parameter that used in the analysis

Characteristic length, D (m)	Separation length, x (m)	The ratio between separation length and the bluff body width, x/D	b (m)
0.1	0.3	3	0.1
	0.4	4	
	0.5	5	
	0.6	6	

2.1.3. Modeling with ANSYS

The geometry of the model was design in ANSYS Design Modeller. The model designed on z – plane. Then, it was extruded according to the dimension shown in Figure 3. The piezoelectric film assigned as a solid domain. While wind tunnel and the bluff bodies assigned as a fluid domain. The simulation is carried out in the fluid domain with 3 m length x 1.5 m width x 0.5 m height wind tunnel and air as the test medium. The dimension of the bluff body is given by 0.1 m × 0.1 m × 0.5 m height. The height of the bluff body is set to be the same as the wall height to avoid the vortex shedding from the top and bottom of the bluff body. The analysis just considered the vortex shedding that directly hit the piezoelectric film horizontally.

2.1.4. Mesh Generation

The mesh for this analysis is generated using ANSYS build – in a meshing module. The mesh generates in both Structural and Fluent modules by focusing on the different part to run the analysis. The Structural module will be a focus on the piezoelectric film while the Fluent module will be a focus on the fluid-solid interface.

In the ANSYS Structural module, the analysis will be focused on piezoelectric film. The meshing for the model runs automatically by setting up the mesh sizing. In the sizing section, the relevance center and the span angle center were set to fine with high smoothing. The size of the element was set to default.

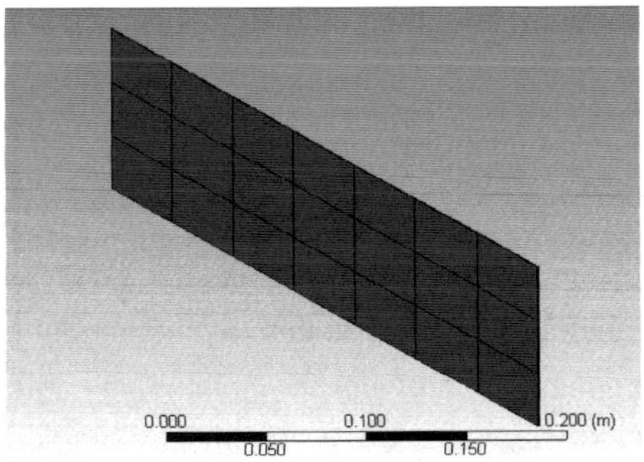

Figure 2. Meshing model in the ANSYS Structural module

As a result, the structural part shown in Figure 2 possesses 200 number of nodes and 21 elements. The mesh skewness for the structure is near to zero which the average value is 1.3061×10^{-10}.

2.1.5. The setting of Dynamic Meshing

A fluid domain grid was applied with 0.005 m to 0.06 m of unstructured tetrahedral mesh, as shown in Figure 3. Another meshing method also applied in this domain, which is edge sizing method. This method was applied on the fluid-solid interface, the bluff body surface, and the fluid wall domain as shown respectively.

On the fluid-solid interface, all eight edges; vertical and horizontal, are selected to be mesh. The mesh type was specified to be element size. It was set to be 0.003 m of element size at all edges.

Figure 3. Unstructured fluid domain mesh

For the bluff body surface, the all vertical edges were selected, and the size of the element was specified to be 0.03 m. Finally, the fluid domain wall was also set to apply the edge sizing mesh method. All the edges of the air domain wall were selected and have been specified to be 0.05 m of the element size which generates 45,348 nodes and 248,513 elements in fluid domain mesh with the maximum skewness achieving 0.79876.

2.2. Boundary Conditions

The wall of the wind tunnel, the surface of the bluff bodies and the structural model, were established as a wall boundary. The wall boundary that set at the wind tunnel is a non-slip wall boundary, while at the structural model is a fluid-solid interface boundary. The inlet velocity is 1.46 ms^{-1} at the inlet boundary. Overall, the fluid domain has established to be a stationary domain except for the fluid and the fluid-solid interface. Both zones were established as a deforming domain and system coupling domain, respectively. Therefore, it can allow the movement of the grid when it interacts with the displacement of the structural model.

2.3. Fluid-Structural Interaction

The fluid-structure interaction (FSI) technique was used in this analysis to analyze the relation of the structural body and the fluid flow. FSI technique applied by combining two modules; Transient Structural and Fluid Flow (Fluent) module. The turbulence model applied in this simulation was the two-equation k-ε model with scalable wall function. The ANSYS System Coupling module was used to carry out FSI analysis. The FSI coupling was a steady two-way coupling approach. This approach allows the data exchange between the structural and CFD module occurred in a single coupling iteration. The converged solution was attained before moving to the next time step. The time step size was set at 1×10^{-2} s to run for 2×10^{1} s with 1 to 10 coupling iteration for each time step. A maximum of 100 repetitions in CFD solver and ten iterations in coupling solver were set for each time step.

3. Result and Discussion

3.1. Flow Structure

The flow structure of the velocity vector around the bluff bodies are presented in Figure 4. The VIV responses of upstream bluff bodies strongly influenced the responses of the downstream bluff bodies. The spacing between the bluff bodies gives effect on the vortex shedding development.

a) x/D=3

b) x/D=4

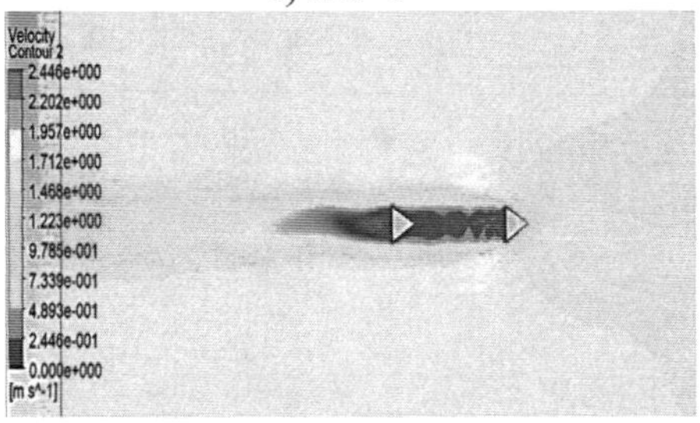

c) x/D=5

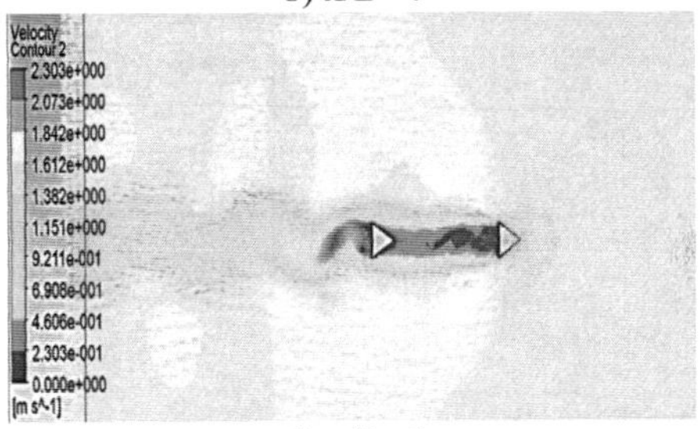

d) x/D=6

Figure 4. The velocity vector

From the observation, the different spacing ratio results in the various concentration velocity vector at the vortex shedding. Figure 4 a) and b) with spacing ratio x/D of 3 and 4 shows no shedding pattern after the second bluff body. The length of the tail of the non-shedding area is 4D. The vortex shedding starts to develop when the spacing ratio x/D is greater than 5. Spacing ratio x/D of 6 in Figure 4 d) gives the active vortex shedding, and the velocity circulation is seen at the downstream of the piezoelectric film.

The velocity contour shows the fluid flow in the system. Only Figure 4 d) shows the deflection of a fluid flow at the downstream of the second bluff body, which at the piezoelectric film position. From the observation, the piezoelectric film deflected by this flow and generate the electricity. The other figure only shows the same trend of the fluid flow. From this case, it can be concluded that the spacing ratio of the bluff body from 3 to 5 is too small.

Another factor is that the flow structure is the turbulence kinetic energy figure, as shown in Figure 5. From the observation, Figure 5 (a) and Figure 5 (b) shows that vortices are close to each other. The intensity of vortex at the downstream of the second bluff body is lower than the downstream of the first bluff body. Thus, it results in lower VIV response for the spacing ratio of 3 and 4.

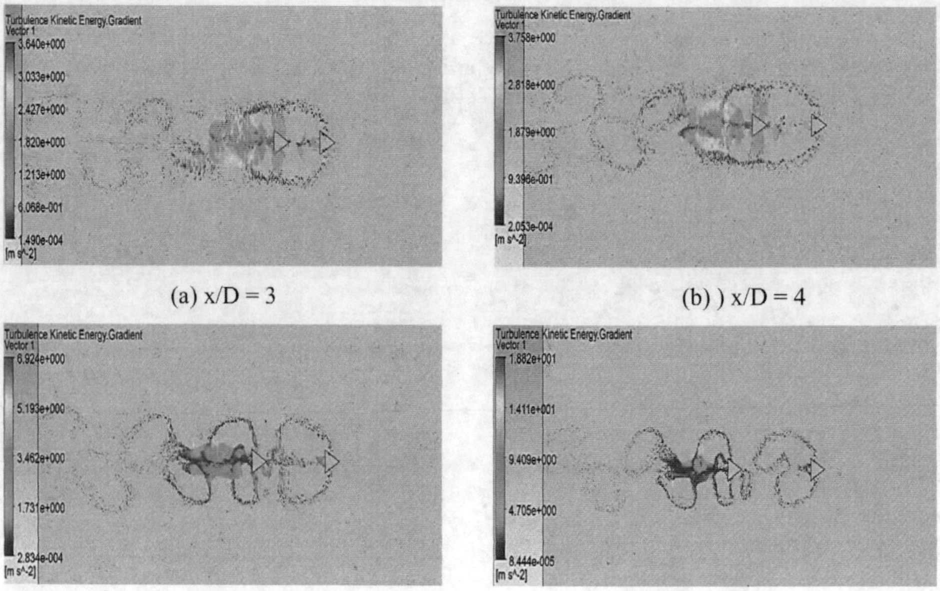

(a) x/D = 3 (b)) x/D = 4

(c) x/D = 5 (d) x/D = 6

Figure 5. Turbulence kinetic energy

Figure 5 c) shows that the vortex amplitude at the downstream of the second bluff body is the same as the first bluff body. However, the vortices are shorter than the vortices at the downstream of the first bluff body. For Figure 5 d) cases, the vortex shedding is almost similar. The intensity of the vortex around the downstream bluff body is higher than the upstream bluff body. Therefore, it results in higher VIV response.

3.2. Deformation of the Piezoelectric Film

The deformation of the piezoelectric film shows the performance of the power produced where the piezoelectric film is deflected by the fluid flow to generate electricity. Figure 6 shows the trend of deflection.

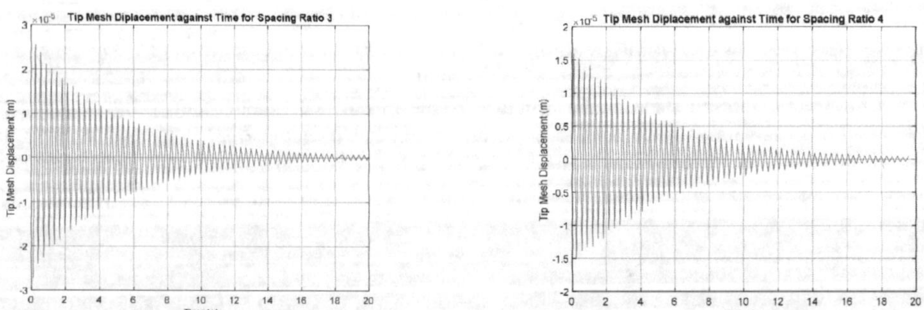

a) x/D = 3 b) x/D = 4

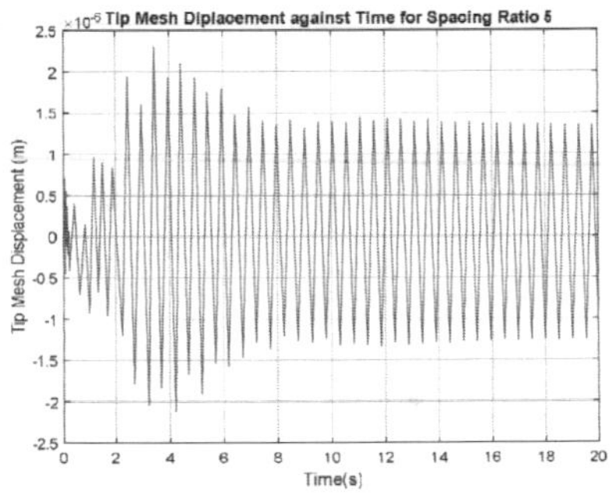

c) x/D = 5

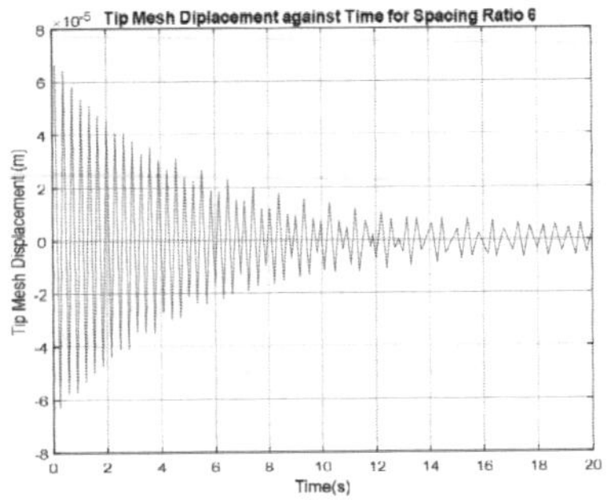

d) x/D = 6

Figure 6. The deformation of the piezoelectric film

Figure 6 (a), (b) and (d) shows the deformation of the piezoelectric structure decreases. An attenuation trend is observed for deflection of the piezoelectric film due to the energy exchange in the fluid-structure model. It is noticeable that the velocity circulation weakens during the whole period. The vorticity decrement is due to the rotational kinetic energy carried by the vortex disperses all the time from the beginning to the end. Figure 6 (c) shows the trend to be consistent. Figure 6 (d) shows the decreases in trend, but consistently toward the end of the simulation. The graph at the initial simulation is high because of the development of the vortex shedding. It decreases throughout the time because of the stability of the fluid flow at the downstream. However, for the spacing ratio 3 and 4, the deformation is decreased to zero because the spacing of the bluff body is too close. Thus, small vortex shedding developed after a few seconds.

3.3. Voltage Analysis

Theoretically, the distance of the bluff bodies can affect the performance of the energy converter due to the vortex shedding development. Therefore, Figure 7 is presented to show the differences in the voltage with the difference spacing ratio of the bluff bodies.

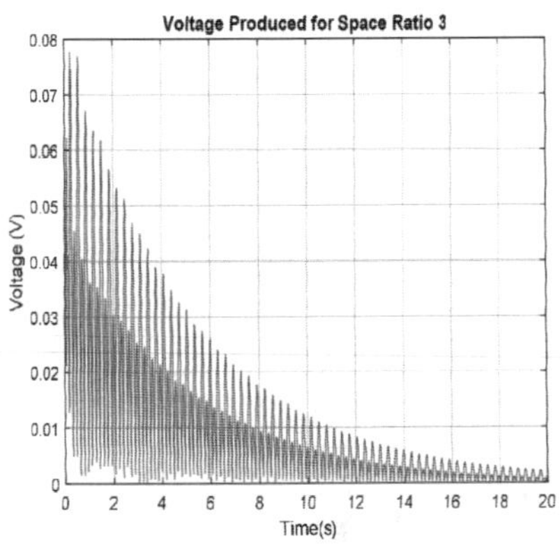

(a) x/D = 3

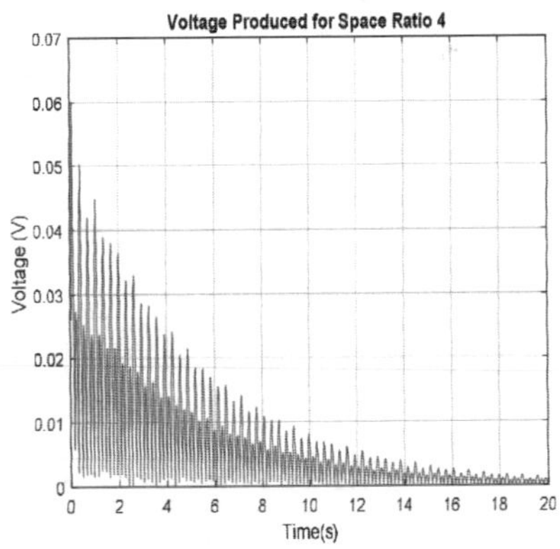

(b) x/D = 4

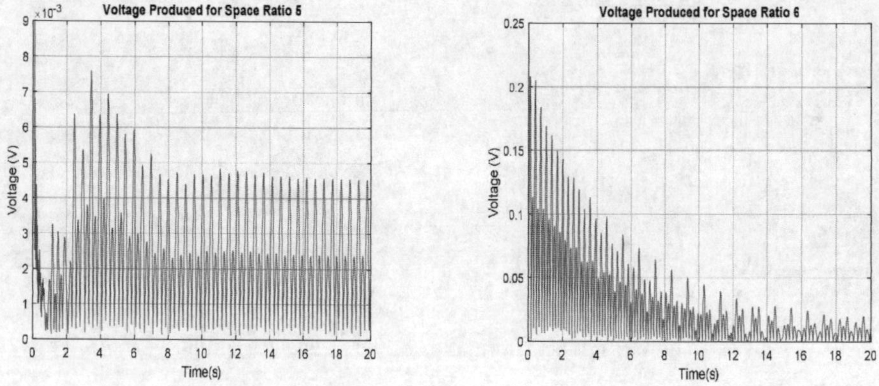

(c) x/D = 5 (d) x/D = 6

Figure 7. Voltage produced

From Figure 7, the voltage produces by Figure 7 (c) is consistent. For the spacing ratio of 3 and 4, Figure 7 (a) and (b) produced the same trend of the voltage which high voltage produced during the initial simulation but decreases throughout the time and its end close to zero-voltage produced. While spacing ratio 6 gives a better result. Although it decreases according to time, the trend shows the voltage produce consistently at the final simulation.

4. Conclusion and Recommendation

The result shows that spacing ratio 6 gives a better performance compared to the other spacing ratio. Spacing ratio 6 gives high VIV response compared to another spacing ratio, and this is because of the vortex shedding development at the downstream of the bluff bodies. The conclusion is spacing ratio 3 and 4 are too small. Thus it cannot develop a perfect vortex at the downstream of the bluff body and gives lower VIV response. Therefore, for this simulation, the dual bluff body with the spacing ratio of 6 gives better performance in developing the vortex and producing voltage.

In this research, a few distances of the dual bluff bodies are analyzed to find the optimum distance to maximize the energy converter. The simulation had not correctly done due to some unexpected issues.

References

[1] Williamson C H K and Govardhan R 2008 A brief review of recent results in vortex-induced vibrations *Journal of Wind Engineering and Industrial Aerodynamics* **96** pp 713–35
[2] Nishi Y 2013 Power extraction from vortex-induced vibration of dual mass system *Journal of Sound and Vibration* **332** pp 199–212
[3] Sarpkaya T 2004 A critical review of the intrinsic nature of vortex-induced vibrations *Journal of Fluids and Structures* **19** pp 389–447
[4] Bearman P W 1984 Vortex Shedding from Oscillating Bluff Bodies *Annu. Rev. Fluid Mech.* **16** pp 195–222
[5] Xu-Xu J, Barrero-Gil A and Velazquez A 2016 Dual mass system for enhancing energy extraction from Vortex-Induced Vibrations of a circular cylinder *International Journal of Marine Energy* **16** pp 250–61
[6] Barrero-Gil A, Pindado S and Avila S 2012 Extracting energy from Vortex-Induced Vibrations: A parametric study *Applied Mathematical Modelling* **36** pp 3153–60
[7] Bernitsas M M, Raghavan K, Ben-Simon Y and Garcia E M 2008 VIVACE (Vortex Induced Vibration Aquatic Clean Energy): A New Concept in Generation of Clean and Renewable

Energy from Fluid Flow *Journal of Offshore Mechanics and Arctic Engineering* **130** 041101-041101–15

[8] M. Sanchez-Sanz, B. Fernandez and A. Velazquez 2009 Energy-Harvesting Microresonator Based on the Forces Generated by the Karman Street Around a Rectangular Prism *Journal of Microelectromechanical Systems* **18** pp 449–57

[9] Nishi Y, Ueno Y, Nishio M, Quadrante L A R and Kokubun K 2014 Power extraction using flow-induced vibration of a circular cylinder placed near another fixed cylinder *Journal of Sound and Vibration* **333** pp 2863–80

[10] Mackowski A W and Williamson C H K 2013 An experimental investigation of vortex-induced vibration with nonlinear restoring forces *Physics of Fluids* **25** 087101

[11] Meliga P, Chomaz J-M and Gallaire F 2011 Extracting energy from a flow: An asymptotic approach using vortex-induced vibrations and feedback control *Journal of Fluids and Structures* **27** pp 861–74

[12] Stappenbelt B, Johnstone A D and Anger J D L 2016 Vortex-Induced Vibration Marine Current Energy Harvesting *Fluid-Structure-Sound Interactions and Control* ed Y Zhou, A D Lucey, Y Liu and L Huang (Springer Berlin Heidelberg) pp 401–6

[13] Song R, Shan X, Lv F and Xie T 2015 A study of vortex-induced energy harvesting from water using PZT piezoelectric cantilever with cylindrical extension *Ceramics International* **41** S768–73

[14] Mehmood A, Abdelkefi A, Hajj M R, Nayfeh A H, Akhtar I and Nuhait A O 2013 Piezoelectric energy harvesting from vortex-induced vibrations of circular cylinder *Journal of Sound and Vibration* **332** 4656–67

[15] Dai H L, Abdelkefi A, Yang Y and Wang L 2016 Orientation of bluff body for designing efficient energy harvesters from vortex-induced vibrations *Applied Physics Letters* **108** 053902

[16] Dai H L, Abdelkefi A and Wang L 2014 Piezoelectric energy harvesting from concurrent vortex-induced vibrations and base excitations *Nonlinear Dynamics* **77** 967–81

[17] Gao X, Shih W-H and Shih W Y 2013 Flow Energy Harvesting Using Piezoelectric Cantilevers With Cylindrical Extension *IEEE Trans. Ind. Electron.* **60** 1116–8

[18] Zhao L, Tang L and Yang Y 2013 Comparison of modeling methods and parametric study for a piezoelectric wind energy harvester *Smart Mater. Struct.* **22** 125003

[19] Sirohi J and Mahadik R 2012 Harvesting Wind Energy Using a Galloping Piezoelectric Beam *J. Vib. Acoust.* **134** 011009

[20] Sirohi J and Mahadik R 2011 Piezoelectric wind energy harvester for low-power sensors *Journalof Intelligent Material Systems and Structures* **22** 2215–28

[21] Abdelkefi A, Yan Z and Hajj M R 2014 Performance analysis of galloping-based piezoaeroelastic energy harvesters with different cross-section geometries *Journal of Intelligent Material Systems and Structures* **25** 246–56

[22] Dowell E H and Hall K C 2001 Modeling Of Fluid-Structure Interaction *Annu. Rev. Fluid Mech.* **33** 445–90

[23] Lavooij C S W and Tusseling A S 1991 Fluid-structure interaction in liquid-filled piping systems *Journal of Fluids and Structures* **5** 573–95

[24] Tamura Y and Matsui G 1980 Wake-Oscillator Model Of Vortex-Induced Oscillation Of Circular Cylinder *Wind Engineering* ed J E CERMAK (Pergamon) pp 1085–94

[25] Nishi Y, Kokubun K, Hoshino K and Uto S 2009 *Quasisteady theory for the hydrodynamic forces on a circular cylinder undergoing vortex-induced vibration* vol 14

[26] DiSilvio G 1969 Self-Controlled Vibration of Cylinder in Fluid Stream *Journal of the Engineering Mechanics Division* **95** 347–62

[27] Violette R, de Langre E and Szydlowski J 2007 Computation of vortex-induced vibrations of long structures using a wake oscillator model: Comparison with DNS and experiments *Computers & Structures* **85** 1134–41

[28] Xu W-H, Zeng X-H and Wu Y-X 2008 High aspect ratio (L/D) riser VIV prediction using wake oscillator model *Ocean Engineering* **35** 1769–74

[29] Achenbach E and Heinecke E 1981 On vortex shedding from smooth and rough cylinders in the range of Reynolds numbers 6×10^3 to 5×10^6 *Journal of Fluid Mechanics* **109** 239–51

[30] Gerrard J H 1997 Flow around Circular Cylinders; Volume 1. Fundamentals. By M. M. Zdravkovich. Oxford Science Publications, 1997. 672 pp. £120. - *Journal of Fluid Mechanics* **350** 375–8

[31] Williamson C H K and Govardhan R 2004 Vortex-Induced Vibrations *Annu. Rev. Fluid Mech.* **36** 413–55

[32] Raghavan K and Bernitsas M M 2011 Experimental investigation of Reynolds number effect on vortex induced vibration of rigid circular cylinder on elastic supports *Ocean Engineering* **38** 719–31

[33] Rostami A B and Armandei M 2017 Renewable energy harvesting by vortex-induced motions: Review and benchmarking of technologies *Renewable and Sustainable Energy Reviews* **70** pp 193–214

[34] Ordia L, Venugopal A, Agrawal A and Prabhu S V 2013 Influence of after body shape on the performance of blunt shaped bodies as vortex shedders **7** p 6

[35] Ding L, Zhang L, Wu C, Mao X and Jiang D 2015 Flow induced motion and energy harvesting of bluff bodies with different cross sections *Energy Conversion and Management* **91** 416–26

[36] Ding L, Zhang L, M. Bernitsas M and Chang C-C 2016 Numerical simulation and experimental validation for energy harvesting of single-cylinder VIVACE converter with passive turbulence control *Renewable Energy* **85** 1246–59

[37] Song R, Shan X, Lv F, Li J and Xie T 2015 *A Novel Piezoelectric Energy Harvester Using the Macro Fiber Composite Cantilever with a Bicylinder in Water* vol 5

[38] Bachal A S and Shrivastava D R 2018 Study of Strouhal's number for dual trapezoidal bluff body **13** p 9

[39] Zhang B, Song B, Mao Z, Tian W and Li B 2017 Numerical investigation on VIV energy harvesting of bluff bodies with different cross sections in tandem arrangement *Energy* **133** 723–36

[40] Tam Nguyen H-D, Pham H-T and Wang D-A 2013 A miniature pneumatic energy generator using Kármán vortex street *Journal of Wind Engineering and Industrial Aerodynamics* **116** 40–8

[41] Chin W, Ong Z, Kong K, Khoo S, Huang Y-H and Chong W 2017 Enhancement of Energy Harvesting Performance by a Coupled Bluff Splitter Body and PVEH Plate through Vortex Induced Vibration near Resonance *Applied Sciences* **7** p 921

IOP Conf. Series: Materials Science and Engineering **705** (2019) 012024 doi:10.1088/1757-899X/705/1/012024

A Review on Remote Sensing-based Method for Tree Detection and Delineation

S N H Syed Hanapi[1,a], S A A Shukor[1,b], J Johari[2]

[1]School of Mechatronic Engineering, Kampus Pauh Putra, Universiti Malaysia Perlis, 02600 Arau, Perlis, Malaysia
[2]Jalal Johari Consultant Sdn Bhd, No. 6-3, Jalan 9/23A, Off Jalan Usahawan, 53200 Setapak,Kuala Lumpur,Malaysia

Email: [a]nurulhusna@unimap.edu.my, [b]shazmin@unimap.edu.my

Abstract. Tree detection and delineation has become one of the important factors that contribute to precision agriculture specifically in plantation industry and efficient forestry management. However, this requires tools and technology that would give reliable information and high accuracy data processing. Recent researches aimed at providing this goal by utilizing the advancement of available remote sensing technology and integrate various algorithms. This article reviews these researches with a focus on algorithms applied to remote-sensing imagery for the purpose of tree detection and delineation. It categorizes and evaluates those methods with the respective types of remote sensing imagery used in the area to evaluate the influence of these factors on the method. Finally, it summarizes the finding of the current algorithms, and suggests on the new development that can be expected in the future.

1. Introduction

Studies in tree detection and delineation for forestry and precision agriculture such as for oil palm plantations have been growing over the last few decades particularly adapting remote sensing. In forestry, remote sensing is generally used for mapping, tree crown identification and delineation, assessment of tree species composition and also for forest monitoring [1]–[11].While for precision agriculture, the application of remote sensing is mostly for land cover detection, tree detection and counting, age estimation, pest or disease detection, land monitoring and mapping [1], [12]–[20].

Throughout the decades, various methods were applied for tree detection and delineation from the traditional image processing in the early years to the latest deep learning-based method in the last 5 years. The rapid development of new algorithm has recently been parallel to the technological advances in the field of remote sensing. However, this assumption is still unclear in showing the relationship between certain algorithms to certain remote sensing data since each study has its own considerations. Thus, this paper has the following objectives: (1) to provide an overview of remote sensing used in forestry and plantation, and (2) to provide an overview of the methodology used for various type of dataset of remote sensing in forestry and plantation. The remaining article is organized as follows: Section 2 starts with the type of remote sensing data acquisition in the field of tree detection, followed by the methodology used covers in Section 3, some discussion in Section 4 and Section 5 concludes on the finding of this review paper.

2. Remote Sensing-based Data Acquisition

As the name implies, remote sensing is how to sense an object from a distance to obtain some information or data which is usually done by using aircraft or satellites [21]. The process of sensing or acquiring data can be divided into two main categories which are by using passive sensor or active sensor. Sensors that collect information from the already available reflected light and radiation are called passive sensors while sensors that initiate its own signal to the object to get the reflection information are called active sensors.

Various types of data using remote sensing were collected to detect and delineate trees particularly for plantations and forestry. Since sensor and computation technologies have been growing rapidly, remote sensing applications in tree detection and delineation also have been evolved from conventional aerial photography or manned aircraft to satellite imagery to the latest technology of unmanned aerial vehicle (UAV). Figure 1 summarizes the data acquisition used for tree detection as presented in the literature from 1996 to 2019. The data used in the studies (total number of included papers N = 50 excluding 10 review paper) are classified into three categories: airborne, satellite and UAV imagery. For the first decade (1996-2005), satellite and UAV were not so popular and the studies about tree detection and delineation were also very limited. The use of UAV and satellite imagery dramatically increased starting from 2011 until 2019 and this trend is expected to increase as the technology is improving.

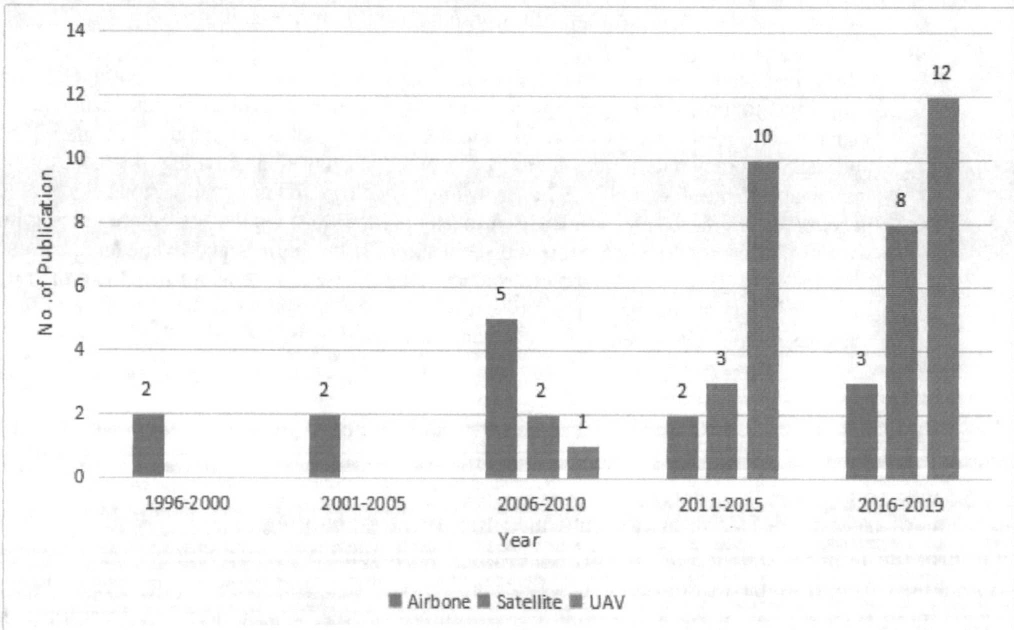

Figure 1. Number of publications studied (1996-2019) on various types of remote sensing data acquisition.

2.1 Airborne / Aerial Imagery

The airborne or aerial imagery usually referred to image captured by using manned aircraft, plane or helicopter. This airborne system is equipped with sensors such as electro-optical sensor [22], infrared sensor [2], [7], spectrometer [18], aerial laser scanner (ALS) [8], Light Detection and Ranging (LiDAR) [23]–[25], multispectral sensor [9] and hyperspectral sensor [4]. In those studies, the altitude of the flight was set from as low as 50m to as high as 3000m above ground. The different flight altitude and speed can give a significant effect to the quality of image acquisition process other than the spatial resolution of the camera used. While for sensor selection, the sensors being used would depend on the purpose of the study. For instance, LiDAR and ALS sensors are useful for estimating height of the trees which give more details information for tree species classification.

2.1. Satellite Imagery
The most common satellite imagery used for tree detection and delineation in the reviewed studies were WorldView-2 [17], [26], Worldview-3 [25][26], Quickbird [12][15][27]–[29], PALSAR [31] and Google Earth [16], [32] imagery. Most of this satellite imagery is categorized as multispectral (MS) imaging which has typically three to fifteen spectral bands compared to hyperspectral imaging which contain hundreds of contiguous spectral bands. For WorldView-2, there are eight multispectral bands and one panchromatic band with a nominal resolution of 1.84m and 0.46m respectively. While for WorldView-3, the eight MS bands has a nominal resolution of 1.24m and 0.31m which is better that the previous one [26]. However, only five of the eight Worldview-2 bands were found relevant for the analysis of tree vegetation which are Red, Green, Blue, Red edge and Near infrared band.

2.2. Unmanned Aerial Vehicle (UAV)
In recent years, Unmanned Aerial Vehicle (UAV) or mostly known as 'drone' has been widely used in many applications including precision agriculture and forestry. With the availability of wide range of sensors, users can define high spatial resolution of imagery and generate 3D data which lead to the advancement of UAV data acquisition. UAV can be programmed to fly autonomously or remotely control from the ground and the performance of fixed wing and rotary wing UAV would be different and specific for certain kind of application. For instance, fixed wing is preferable for large land coverage while rotary wing may be more suitable for high spatial resolution of data acquisition [33].
There are variety of payload sensing instruments that can be carried by UAV including Radar, LiDAR, thermal infrared (TIR), near infrared (NIR), shortwave infrared (SWIR) and visible light sensors [33].The development of lightweight camera sensor for UAV has enabled a very high spectral and spatial resolution for aerial measurement with ground sampling distance (GSD) less than 10cm. The GSD is the distance between the center points of image taken on the ground, where the bigger the GSD value, the lower spatial resolution of the image would be.

3. Methodology
Over the past few decades, existing methods for detecting tree from remote sensing can be classified into four categories: (1) the image processing-based methods; (2) the machine learning-based methods (3) the dense point cloud method and (4) the deep learning-based methods. Based on the literature, the type of data acquisition would somehow affect the kind of methods used for tree detection. Thus, this paper will give some review on the different methods applied for different data acquisition method.
The image processing-based method includes local maximum filter, edge detection and segmentation, morphological operation, template matching and the vegetation indices for feature extraction method. Many studies in the early years (1996-2005) that used digital aerial imagery such as [22],[2] and [7] applied this traditional image processing-based method.[22] use template matching (TM) method which involve geometric and radiometric aspects by developing a synthetic model for different crown size and shape to be matched with the aerial images. [7] use a multiple-scale algorithm for edge contours and curves in high spatial resolution infrared images for the delineation of individual tree crowns. [2]designed local transects extending outward algorithm from tree apex resulting 91% accuracy of tree detection. The need for very high-resolution imagery in delineated tree crown diameter has been proved by testing the variation of image resolution.

Using airborne high spatial resolution (1m) image,[18] proposed an oil palm tree extraction method based on texture and spectral analysis, segmentation of edge and morphological operation and blob analysis for tree counting obtaining to achieve 95% of counting accuracy. A quite similar approach was also done for processing high resolution multispectral satellite imagery including edge detector using Sobel, texture analysis, high-pass filter, opening filter and morphological operation of dilation and erosion [15].

IOP Conf. Series: Materials Science and Engineering **705** (2019) 012024 doi:10.1088/1757-899X/705/1/012024

The image processing-based method is also applied in many studies [12], [17], [34]–[36] that use various vegetation index values for feature selection. Different vegetation index has been defined for specific task and some indices cannot be computed depending on the spectral data available. [12] applied several vegetation indices which are derived from the multi-spectral satellite imagery for local peak detection of oil palm. The indices such as Normalized Difference Vegetation Index (NDVI), Normalized Difference Index (NDI), Green Normalized Difference Vegetation Index (GNDVI) and Transformed Vegetation Index (TVI) are used. The best index is chosen by comparing the histograms generated from all the index values which maximize the dissimilarity between the oil palm tree and the background.[35] successfully determines that the low-cost UAV platform and RGB digital compact camera with modified filter (capture NIR) able to be used by smallholders in monitoring the palm tree growth condition, where the higher the value of reflectance, the higher the greenness (healthier) of the oil palm tree. This result can be derived from the combination of both Soil Adjusted Vegetation Index (SAVI) and Normalized Difference Vegetation Index (NDVI).

The machine learning-based (ML) tree detection methods usually require feature extraction, image segmentation, classifier training, and prediction [17][20][30][31][36][37]. The classifier which is the most important part in ML can be categorized as supervised or non-supervised. Random Forest (RF), Support Vector Machine (SVM), Decision Tree (DT) and Neural Network are some examples of supervised classifier while K-Nearest Neighbour (k-NN) and K-means clustering are non-supervised classifier. Supervised ML requires images database labelled as positive and negative with corresponding categories (i.e. tree and background) which is used for training samples. From the ALOS-PALSAR Orthorectified Mosaic images,[31] proposed three classification algorithms which are Decision Tree, SVM, and K-Means clustering to map oil palm plantations in Cameroon. Results show that SVM has good overall performance with accuracy of 86% to 92% depending on training samples sizes while Decision Tree algorithm outperforms the others in terms of speed for large scale mapping. The use of Google Earth Engine (GEE) is demonstrated by [32] using Classification and Regression Trees (CART), Random Forests (RF) and Minimum Distance (MD) classifier for detection of industrial oil palm. [20] proposed a novel method using Scale Invariant Feature Transform (SIFT) to extract key points from UAV image. These key points were then analysed using Extreme Learning Machine (ELM) classifier to distinguish between palm trees and non-palm trees.

The airborne imagery that fused with LiDAR data provide a three dimensional (3D) high spatial resolution image in a form of point cloud data which can generate many features that are difficult to obtain from spectral imagery such as tree height, quartile heights and crown base height [39]. This point cloud features were useful in obtaining a Digital Surface Model (DSM) and Digital Terrain Models (DTM) which were further use for the derivation of Canopy Height Model (CHM). In a few studies, this attributes was combined with hyperspectral [4] and multispectral data [40] for tree species classification. The fusion of active sensor (LiDAR) and passive sensor (multispectral, hyperspectral, RGB camera) data is also done for the satellite and UAV imagery. [27] proposed a method from Worldview 3 satellite imagery and LiDAR data using an integrated OBIA height model and SVM analysis for oil palm counting and age estimation. [41] studies the comparison of ALS and structure from motion (SfM) point cloud for assessing individual tree information including the terrain height and the horizontal and vertical distribution of forest canopy. Both studies show that the LiDAR and ALS point cloud data can provide tree height information accurately which complements the geometrical and spectral information from multispectral and RGB imagery.

Many studies also show that the dense point cloud data can be derived from RGB digital images by using SfM which mostly applied for UAV imagery [1][3][41]–[45]. For instance, [44] use a data fusion of UAV based point cloud data and hyperspectral image in boreal forest for individual tree detection and classification. The results show 3D point cloud derived from UAV photogrammetric data produced good result for the individual tree detection. However, it did not improve the result of tree species classification compared to the spectral features derived. As an alternative to LiDAR data, [43] proposed feature extraction algorithm using SIFT to extract features point from UAV imagery but the dense point cloud generated is not sufficient to extract tree parameters accurately. The usefulness of modern software such as Agisoft Photoscan Professional v1.0.0 software to produce 3D point cloud data resulting higher accuracy of more than 85% in the study of individual tree detection in open canopy forest [42]. The CHM was derived from the SfM algorithms on RGB photographs based on the location of the images with respect to each other as well as the objects viewed within them.

In recent years, several studies applied the deep learning-based method including fruit counting [47], land cover mapping [48] and tree detection [25][27][29][49]. The deep learning-based methods for tree detection studies are mostly applied to high resolution satellite images based on the sliding window technique combined with the CNN. [28] manually selected 5000 palm tree samples and 4000 background sample and randomly selected 7200 of this as training samples. The sliding window of 17x17 pixels was tested to predict labels in image dataset which result in accuracy of more than 96%. While in the other study which involve a large scale study area (around 5km^2), [30] proposed a two-stage convolutional neural network-based method for oil palm tree detection. Four classes of 20,000 samples were trained including oil palm, background, cloud and other vegetation for object classification and land cover classification. In their study, [30] compare their deep learning-based method with earlier methods (the template matching, the ANN, and the local maximum filter method) in all selected regions in their previous studies [28]. The results show that the CNN-based method outperforms other oil palm detection methods by 3 to 10 percentage points in terms of the F1-Score.

4. Discussion

The study of tree detection and delineation either in forestry or plantation such as oil palm plantation is an important issue for researchers especially in remote sensing community. This paper has reviewed 60 papers which consist of 50 research papers and 10 review papers from the year 1996 till 2019. The contribution of researchers in terms of objectives, remote sensing data acquisition, methodology and accuracy assessment for tree detection in forestry and plantation were considered. The effect of remote sensing data acquisition to various method used for tree detection has been studied. Table 1 summarizes the advantages and disadvantages of specific algorithm used in each method group.

Table 1. The advantages and disadvantages of specific algorithm used in each method group.

Method	Example of Algorithm Used	Advantages	Disadvantages
Image processing-based method	Watershed segmentation	Easier for crown detection.	Can be over segmented for non-similar crown size trees.
	Template matching	Suitable for well separated tree.	Not suitable for dense tree canopy.
	Vegetation Index	Well suited for crop health and tree growth monitoring.	Requires more spectral data such as NIR band.
Machine learning-based method	Support vector machine	High accuracy for small and good training dataset.	Need longer time to train dataset.
	K-means clustering	Need no training dataset and runs faster.	Less accurate.
Point cloud-based method	LiDAR /ALS	Accurate estimation for vertical tree structure	High cost.
	Structure from motion (SfM)	Low cost, provide greater spatial coverage.	Cannot measure terrain altitude in dense canopy area.
Deep learning-based method	Convolution Neural Network	High accuracy in any difficult condition; can cover large scale area.	Requires thousands of training dataset and expensive GPU.

The image processing-based method can be concluded as the most versatile method since it can be applied for any remote sensing data acquisition and can be fused with many other advance methods. This method not only involve traditional image processing such as image filtering, segmentation, binarization and feature extraction, but a template matching, and vegetation index-based feature extraction are also included as part of the category. The advantages of image processing-based method are it is relatively not complicated, easy to understand, and does not require many tools. It is also suitable for most spatial and spectral resolution of image data. However, the major limitation of this method is the performance of tree crown or tree top detection deteriorates for denser tree area and overlapping tree canopies [14].

Some of the studies using remote sensing not only focus on tree crown detection but also for tree delineation and identifying tree species. The machine learning-based methods can be applied to assess various classifiers for this purpose. Using a supervised or non-supervised machine learning method would depend on the dataset available. However, for best accuracy, this method requires high resolution image, multispectral, hyperspectral or thermal imagery image which can provide wide range of spectral data [5].

In recent years, deep learning has increased attention of many researchers as it shows great success in various studies. This is because deep learning can learn the features from images automatically and can be said as a fully data driven scheme [50]. Studies also prove that the deep learning-based method has outperformed most of the other traditional tree detection method. However, deep learning requires thousands of sampling data set and to train it requires complex data models which can be extremely expensive. Moreover, expensive GPUs and hundreds of machines would also increase the cost.

In term of data acquisition, all remote sensing platform discussed in this paper has their own advantages and disadvantages. Satellite imagery allows much greater areas of coverage and can provide wide range of spectral data such as panchromatic, multispectral, and hyperspectral data. While for aerial and UAV imagery, the spatial and spectral data acquired would depend on the type of sensor and camera used and the area coverage is limited. However, with the development of lightweight multispectral and hyperspectral camera available, with lower cost and more up-to- date, UAV can still be a better choice for small stakeholders or personal commercial uses than satellite imagery. The methodologies of processing the image data for UAV and satellite imagery in most of the studies are almost the same. The major difference is for airborne imagery equipped with LiDAR sensor, which can provide deeper information of forest canopies penetration. Thus, it can be said that for vertical stratification layer information, LiDAR point cloud data is more efficient compared to satellite and UAV imagery which are more reliable in capturing spectral information of the surface or top layer. Some studies show that the algorithm for generating point cloud LiDAR data processing can also be used for processing the derived UAV-based point cloud data. However, the derived photogrammetric point cloud cannot produce accurate detection results as compared to LiDAR point cloud and need further development in the future.

5. Conclusion

Tree detection and delineation studies greatly contribute to both forestry and plantation inventory needs. However, there is still room for improvement and development especially in terms of the methods used. With the increasing availability of remote sensing data, studies would be focusing more on the practicability of the method used with lower cost and higher quality result. The accuracy of the tree detection would not rely on the algorithm only, but also depends on the quality of data acquisition. The fusion of passive and active sensors data might improve the tree detection accuracy and could provide more advance method in the future.

References

[1] Torres-Sánchez J, López-Granados F, Serrano N, Arquero O and Peña J M 2015 High-throughput 3-D monitoring of agricultural-tree plantations with Unmanned Aerial Vehicle (UAV) technology *PLoS One* **10** 6

[2] Pouliot D A, King D J, Bell F W and Pitt D G 2002 Automated tree crown detection and delineation in high-resolution digital camera imagery of coniferous forest regeneration *Remote Sens. Environ.* **82** 322–334
[3] Zahawi R A, Dandois J P, Holl K D, Nadwodny D, Reid J L and Ellis E C 2015 Using lightweight unmanned aerial vehicles to monitor tropical forest recovery *Biol. Conserv.* **186** 287–295
[4] Maschler J, Atzberger C and Immitzer M 2018 Individual Tree Crown Segmentation and Classification of 13 Tree Species Using Airborne Hyperspectral Data *Remote Sens.* **10** 1218
[5] Tuominen S, Näsi R, Honkavaara E, Balazs A, Hakala T, Viljanen N, Pölönen I, Saari H and Ojanen H 2018 Assessment of classifiers and remote sensing features of hyperspectral imagery and stereo-photogrammetric point clouds for recognition of tree species in a forest area of high species diversity *Remote Sens.* **10** 714
[6] Hassaan O, Nasir A K, Roth H and Khan M F 2016 Precision Forestry: Trees Counting in Urban Areas Using Visible Imagery based on an Unmanned Aerial Vehicle *IFAC-PapersOnLine* **49** 16–21
[7] Brandtberg T and Walter F 1998 Automated delineation of individual tree crowns in high spatial resolution aerial images by multiple-scale analysis *Machine Vision and Applications* **11** 64–73
[8] Sääjoki R, Packalén P, Maltamo M, Vehmas M and Eerikäinen K 2008 Detection of aspens using high resolution aerial laser scanning data and digital aerial images *Sensors* **8** 5037–5054
[9] Katoh M and Gougeon F A 2012 Improving the precision of tree counting by combining tree detection with crown delineation and classification on homogeneity guided smoothed high resolution (50 cm) multispectral airborne digital data *Remote Sens.* **4** 1411–1424
[10] Guldogan O, Rotola-Pukkila J, Balasundaram U, Le T H, Mannar K, Chrisna T M and Gabbouj M 2016 Automated tree detection and density calculation using unmanned aerial vehicles *Visual Communications and Image Processing (VCIP)* 1–4
[11] Bulatov D, Wayand I and Schilling H 2016 Automatic tree-crown detection in challenging scenarios *International Archives of the Photogrammetry, Remote Sensing and Spatial Information Sciences - ISPRS Archives* **41** 575–582
[12] Srestasathiern P and Rakwatin P 2014 Oil palm tree detection with high resolution multi-spectral satellite imagery *Remote Sens.* **6** 9749–9774
[13] Kalantar B, Oludare Idrees M, Mansor S and Abdul Halin A 2017 Smart Counting-Oil Palm tree inventory with UAV *Coordinates* **13** 17–22
[14] Aliero M M, Mansur M A and Al-Doksi J 2014 The Usefulness Of Unmanned Airborne Vehicle (UAV) Imagery For Automated Palm Oil Tree Counting *Journal of Forestry. Researchjournali* **1** 1–12
[15] Santoso H, Tani H and Wang X 2016 A simple method for detection and counting of oil palm trees using high-resolution multispectral satellite imagery *Int. J. Remote Sens.* **37** 5122–5134
[16] Maillard P and Gomes M F 2016 Detection and Counting of Orchard Trees from VHR Images Using a Geometrical-Optiacal Model and Marked Template Matching *ISPRS Annals of the Photogrammetry, Remote Sensing and Spatial Information Sciences* **3** 75–82
[17] Chemura A, van Duren I and van Leeuwen L M 2015 Determination of the age of oil palm from crown projection area detected from WorldView-2 multispectral remote sensing data: The case of Ejisu-Juaben district, Ghana *ISPRS J. Photogramm. Remote Sens.* **100** 118–127
[18] Shafri H Z M, Hamdan N and Saripan M I 2011 Semi-automatic detection and counting of oil palm trees from high spatial resolution airborne imagery *Int. J. Remote Sens.* **32** 2095–2115
[19] Al-Ruzouq R, Shanableh A, Gibril M B A and Al-Mansoori S 2018 Image segmentation parameter selection and ant colony optimization for date palm tree detection and mapping from very-high-spatial-resolution aerial imagery *Remote Sens.* **10** 1413
[20] Malek S, Bazi Y, Alajlan N, AlHichri H and Melgani F 2014 Efficient framework for palm tree detection in UAV images *IEEE J. Sel. Top. Appl. Earth Obs. Remote Sens.* **7** 4692–4703

[21] US Department of Commerce *What is remote sensing*
[22] Pollock R 1996 The Automatic Recognition of individual Trees in Aerial Images of Forests based on a Synthetic Tree Crown Image Model *Diss. University of British Columbia*
[23] Strîmbu V F and Strîmbu B M 2015 A graph-based segmentation algorithm for tree crown extraction using airborne LiDAR data *ISPRS J. Photogramm. Remote Sens.* **104** 30–43
[24] Garrido M, Perez-Ruiz M, Valero C, Gliever C J, Hanson B D and Slaughter D C 2014 Active optical sensors for tree stem detection and classification in nurseries *Sensors (Switzerland)* **14** 10783–10803
[25] Secord J and Zakhor A 2007 Tree detection in urban regions using aerial lidar and image data *IEEE Geosci. Remote Sens. Lett.* **4** 196–200
[26] Kleinn C, Freudenberg M, Nölke N, Wörgötter F, Agostini A and Urban K 2019 Large Scale Palm Tree Detection In High Resolution Satellite Images Using U-Net *Remote Sens.* **11** 312
[27] Rizeei H M, Shafri H Z M, Mohamoud M A, Pradhan B and Kalantar B 2018 Oil Palm Counting and Age Estimation from WorldView-3 Imagery and LiDAR Data Using an Integrated OBIA Height Model and Regression Analysis *J. Sensors* 1–13
[28] Li W, Fu H, Yu L and Cracknell A 2017 Deep learning based oil palm tree detection and counting for high-resolution remote sensing images *Remote Sens.* **9** 22
[29] Vibha L, Shenoy P D, Venugopal K R and Patnaik L M 2009 Robust technique for segmentation and counting of trees from remotely sensed data *IEEE International Advance Computing Conference, IACC* 1437–1442
[30] Li W, Dong R, Fu H and Yu L 2019 Large-scale oil palm tree detection from high-resolution satellite images using two-stage convolutional neural networks *Remote Sens.* **11** 11
[31] Li L, Dong J, Tenku S N and Xiao X 2015 Mapping oil palm plantations in cameroon using PALSAR 50-m orthorectified mosaic images *Remote Sens.* **7** 2
[32] Lee J S H, Wich S, Widayati A and Koh L P 2016 Detecting industrial oil palm plantations on Landsat images with Google Earth Engine *Remote Sens. Appl. Soc. Environ.* **4** 219–224
[33] Tang L and Shao G 2015 Drone remote sensing for forestry research and practices *Journal of Forestry Research* **26** 791–797
[34] Mckinnon T and Hoff P 2017 Comparing RGB-Based Vegetation Indices With NDVI For Drone Based Agricultural Sensing 1–8
[35] Estimation B, Tropical I N, Using R, Of I and Lidar A 2016 Oil palm tree growth monitoring for smallholders by using unmanned aerial *Jurnal Teknologi* **77** 17–27
[36] Berni J A J, Zarco-Tejada P J, Suárez L, González-Dugo V and Fereres E 2009 Remote Sensing of Vegetation from UAV Platforms using Lightweight Multispectral and Thermal Imaging Sensors *Int. Arch. Photogramm. Remote Sens. Spatial Inform. Sci*. **38** 6
[37] Miserque Castillo J Z, Laverde Diaz R and Rueda Guzmán C L 2016 Development of an aerial counting system in oil palm plantations *IOP Conference Series: Materials Science and Engineering* **138** 012007
[38] Yuan Y and Hu X 2016 Random forest and objected-based classification for forest pest extraction from uav aerial imagery *International Archives of the Photogrammetry, Remote Sensing and Spatial Information Sciences - ISPRS Archives* 1093–1098
[39] Zhen Z, Quackenbush L J and Zhang L 2016 Trends in automatic individual tree crown detection and delineation-evolution of LiDAR data *Remote Sensing* **8** 333
[40] Heinzel J N, Weinacker H and Koch B 2008 Full automatic detection of tree species based on delineated single tree crowns–a data fusion approach for airborne laser scanning data and aerial photographs *Proceedings of SilviLaser* 76–85
[41] Wallace L, Lucieer A, Malenovskỳ Z, Turner D and Vopěnka P 2016 Assessment of forest structure using two UAV techniques: A comparison of airborne laser scanning and structure from motion (SfM) point clouds *Forests* **7** 62
[42] Mohan M, Silva C, Klauberg C, Jat P, Catts G, Cardil A, Hudak A and Dia M 2017 Individual tree detection from unmanned aerial vehicle (UAV) derived canopy height model in an open

canopy mixed conifer forest *Forests* **8** 340

[43] Tao W, Lei Y and Mooney P 2011 Dense point cloud extraction from UAV captured images in forest area *ICSDM 2011 - Proceedings 2011 IEEE International Conference on Spatial Data Mining and Geographical Knowledge Services* 389–392

[44] Nevalainen O, Honkavaara E, Tuominen S, Viljanen N, Hakala T, Yu X, Hyyppä J, Saari H, Pölönen I, Imai N and Tommaselli A 2017 Individual Tree Detection and Classification with UAV-Based Photogrammetric Point Clouds and Hyperspectral Imaging *Remote Sens.* **9** 185

[45] Rokhmana C A 2015 The Potential of UAV-based Remote Sensing for Supporting Precision Agriculture in Indonesia *Procedia Environ. Sci.* **24** 245–253

[46] Zhang J, Hu J, Lian J, Fan Z, Ouyang X and Ye W 2016 Seeing the forest from drones: Testing the potential of lightweight drones as a tool for long-term forest monitoring *Biol. Conserv.* **198** 60–69

[47] Liu X, Chen S W, Aditya S, Sivakumar N, Dcunha S, Qu C, Taylor C J, Das J and Kumar V 2018 Robust Fruit Counting: Combining Deep Learning, Tracking, and Structure from Motion *International Conference on Intelligent Robots and Systems (IROS)* 1045–1052

[48] Hu Y, Zhang Q, Zhang Y and Yan H 2018 A Deep Convolution Neural Network Method for Land Cover Mapping: A Case Study of Qinhuangdao, China *Remote Sens.* **10** 2053

[49] Cheang E K, Cheang T K and Tay Y H 2017 Using Convolutional Neural Networks to Count Palm Trees in Satellite Images *arXiv preprint arXiv:1701.06462*

[50] Wang S, Quan D, Liang X, Ning M, Guo Y and Jiao L 2018 A deep learning framework for remote sensing image registration *ISPRS J. Photogramm. Remote Sens.* **145** 148–164

IOP Conf. Series: Materials Science and Engineering **705** (2019) 012025 doi:10.1088/1757-899X/705/1/012025

Image reconstruction for liquid-gas regime identification based on multiple excitation sources in electrical resistance tomography system

Y A Wahab*, M F Fuad, N M A Ghani, Z M Zain, M M Noh, M A Zawawi and M S Najib

Faculty of Electrical & Electronics Engineering, Universiti Malaysia Pahang, 26600 Pekan, Pahang

*Corresponding author: yasmin@ump.edu.my

Abstract. Electrical resistance tomography (ERT) reconstructs internal resistance images of the field from electrical measurements on the surface. The objective of this paper is to obtain the tomogram of the liquid-gas regime of ERT system by using multiple excitation sources. A linear back projection algorithm was implemented as a basic algorithm. Then, it was modified to suit the implementation of multiple excitation sources. A liquid-gas regime in the steel pipe was used as a tested medium. The analysis and performance of the image reconstructed which applies multiple excitation sources were compared with the single excitation technique. As a common strategy applied a single excitation source and produce problem especially at the center of the pipe, this paper mainly focuses on image reconstruction of multiple excitation sources. As a conclusion, 50% of the excitation source cannot produce the image as expected. Tomograms that produced by using the single excitation source were much better than the images obtained by using 50% of the excitation source.

1. Introduction

ERT is considered a promising process tomography technology due to its advantages such as high speed, low cost, and non-intrusive sensors [1]. The ERT system consists of sensing electrodes, a data collection system, and image reconstruction and visualization unit [2]. The operation mode of ERT applies a current or voltage on an excitation electrode and measure a potential difference or a current at the detection electrodes [3]. A current is excited at a pair of electrodes in the measurement section and the sensing field. In the case of conduction pipes, the electrodes need to be insulated from the conducting wall [4].

ERT system can be categorized as one of the soft-field tomography [5]. The nature of soft-field behavior is that it produces a curve line for each of the projection in the forward problem. It means that, when the image is reconstructed, the possibility to get a quality tomogram of the medium of interest especially at the center of the region of interest also reduced. One of the methods that commonly used to solve the problem is by increasing the number of electrodes applied in the system. However, by increasing the number of electrodes it also increases the cost of the system because the cost of the sensor fabrication become more expensive. Simultaneously, that approach applies a single source excitation. Therefore, multiple excitation sources technique was introduced in [6]. The technique was introduced to avoid the increment number of electrodes in the ERT system and hence can reduce the sensor fabrication cost. The results showed that by increasing the number of excitation sources up to half of the total electrodes in time will increases the sensor reading performance. Based on this paper, the authors believed that the increment of the sensor reading performance will also increase the quality tomogram obtained in the image reconstruction part.

Thus, based on [6] this paper is focused on the linearization approach of the image reconstruction part for the ERT system. The main objective is to prove either it is true or not that the multiple excitation technique can increase the quality of tomogram obtained especially at the center of the pipe.

2. Image reconstruction of ERT system

2.1. Forward problem solving

The forward problem is identified as the system's sensitivity map [7]. The linearization solution is used by the ERT system when the sensor output is assumed to be a linear approximation [8]. The general mathematical equation for the forward problem of ERT system can be applied by using equation (1) [9].

$$M_{i,j}(x,y) = \int_{A(x,y)} \frac{E_i(x,y)}{I_i} \cdot \frac{E_j(x,y)}{I_j} dxdy \tag{1}$$

Where $M_{i,j}(x,y)$ is the sensitivity map between pair electrode i and j at position (x,y) within area $A(x,y)$, E_i and E_j are the electric fields due to the current or voltage driven owing to electrode i and j. The dot multiplication between Ei and Ej produce a pair for each of i-j electrode.

COMSOL Multiphysics was applied as the software platform to solve the forward problem. The 2-dimensional electric current, ec physic in the COMSOL software was chosen for this purpose. Then, the geometry that mimics a real pipe as well as the material for each domain, were set. Sixteen rectangular electrodes (E1-E16) had been installed and attached along the inner circumference of the steel pipe (see Figure 1). The electrodes were positioned equidistantly to ensure that the system can withdraw as much information as possible. The applied parameters in the simulation are shown in Table 1.

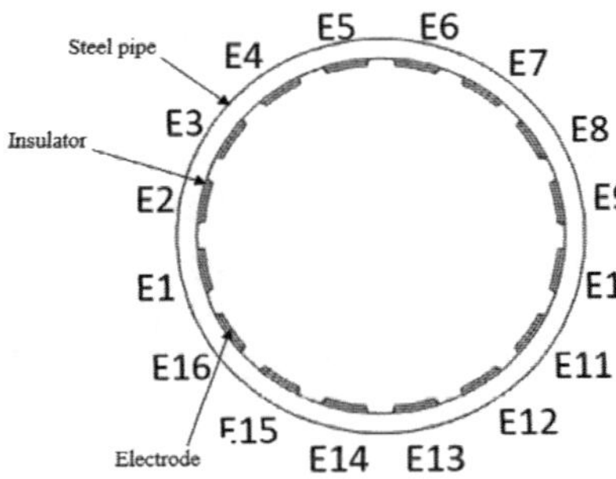

Figure 1. Geometry of ERT system in COMSOL Multiphysics

Table 1. Properties and specific dimension [10]

Item	Value
Inner diameter of steel pipe	54mm
Thickness of wall	3mm
Material of electrode	Silver
Number of electrodes	16
Excitation current	20mA
Electrode's width	6.4mm

The respective channels out of sixteen electrodes based on the multiple excitation sources or single excitation source were chosen for getting an individual map. Here, the electric current source was set at the boundary condition of the respective channels. Then, the model was mesh using normal mesh, and the data was obtained. Before exporting the data, the electric field expression and the desired numbers of pixels for the data were set. The electric field expression used in COMSOL is shown in equation (2). Besides, the number of pixels was applied by using 64 x 64 pixels values of the regular grid data of x and y points.

$$e_{(x,y)} = \sqrt{{E_x}^2 + {E_y}^2} \tag{2}$$

where $e_{(x,y)}$ is defined as the velocity vector of electric current between the electric field of x-axis and y-axis. E_x is the electric component in the x-axis of the electric field while E_y is the electric component in the y-axis of the electric field. Finally, the output data from COMSOL Multiphysics was exported into MATLAB. The selection data that only consists of the electric field values concerning the number of the pixels was plotted by using 'pcolor plotting' options in the MATLAB. After plotting the respective channels of the sensitivity map of the ERT system, the process was repeated until all the channels have been plotted. Finally, the pre-processing of the data of the sensitivity map was done in the MATLAB.

2.1.1. Normalized map for single excitation source

For the single excitation source technique, each of the pair projection maps was normalized to get the standardized map. This process of normalization sensitivity distribution, $\bar{M}_{i,j(single)}(x,y)$ was done by using equation (3). Figure 2 shows the example of the normalized sensitivity map when channel 1 was set as the source in the single excitation source technique.

Figure 2. Example of the normalized map for single excitation source technique

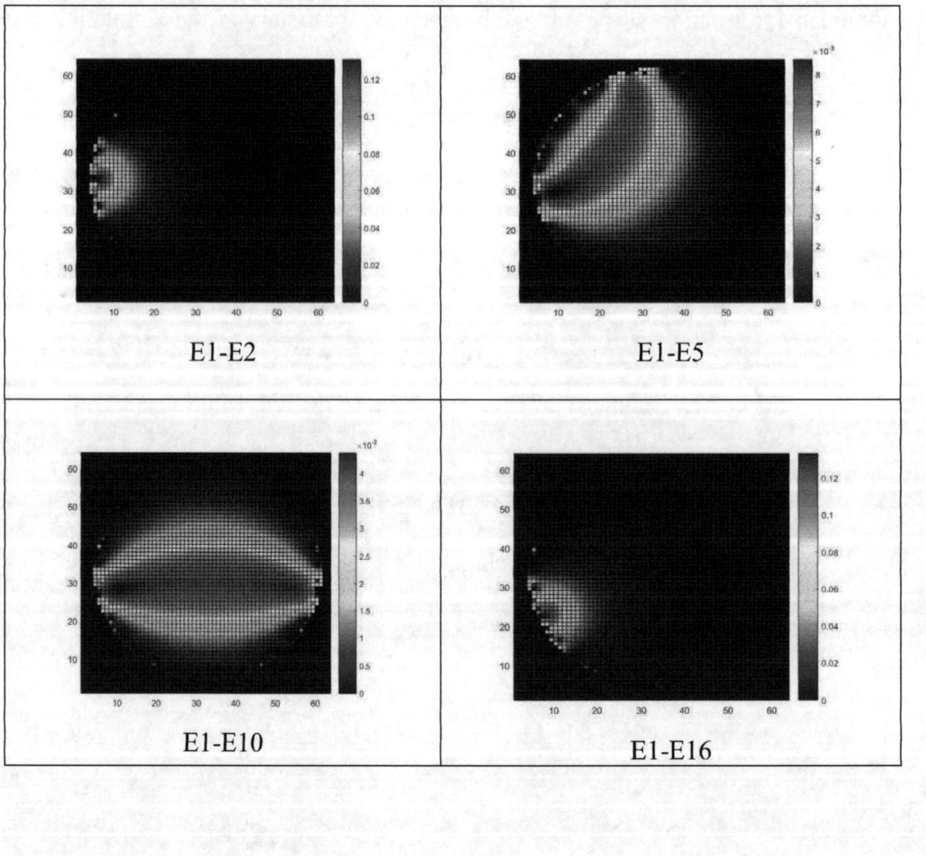

$$\bar{M}_{i,j(single)}(x,y) = \frac{M_{i,j}(x,y)}{\sum_{i=1}^{16}\sum_{j=1}^{16} M_{i,j}(x,y)} \tag{3}$$

2.1.2. Normalized map for multiple excitation sources

For multiple excitation sources technique, the same steps were similar to sub-topic 2.1.1, but the difference is the number of the electrode for exciting and detecting were 8 electrodes, respectively. For the pair projection mapping, the multiple excitation sources would be paired up with the multiple receivers to produce a pair projection map. After the pair projection map, $M_{a,b(multiple)}(x,y)$ is obtained, the next step was to produces the normalized map, $\bar{M}_{a,b(multiple)}(x,y)$ as in equation (4). The a and b are the pair electrodes for each of the pair projection map. The difference between equation (4) and equation (3) is in term of the number of transmitter and receiver used in the ERT system. Figure 3 illustrates the normalized map for multiple excitation sources technique.

$$\bar{M}_{a,b(multiple)}(x,y) = \frac{M_{a,b(multiple)}(x,y)}{\sum_{a=1-8}^{16-7}\sum_{b=1-8}^{16-7} M_{a,b(multiple)}(x,y)} \tag{4}$$

Figure 3. Example of the normalized map for multiple excitation sources technique

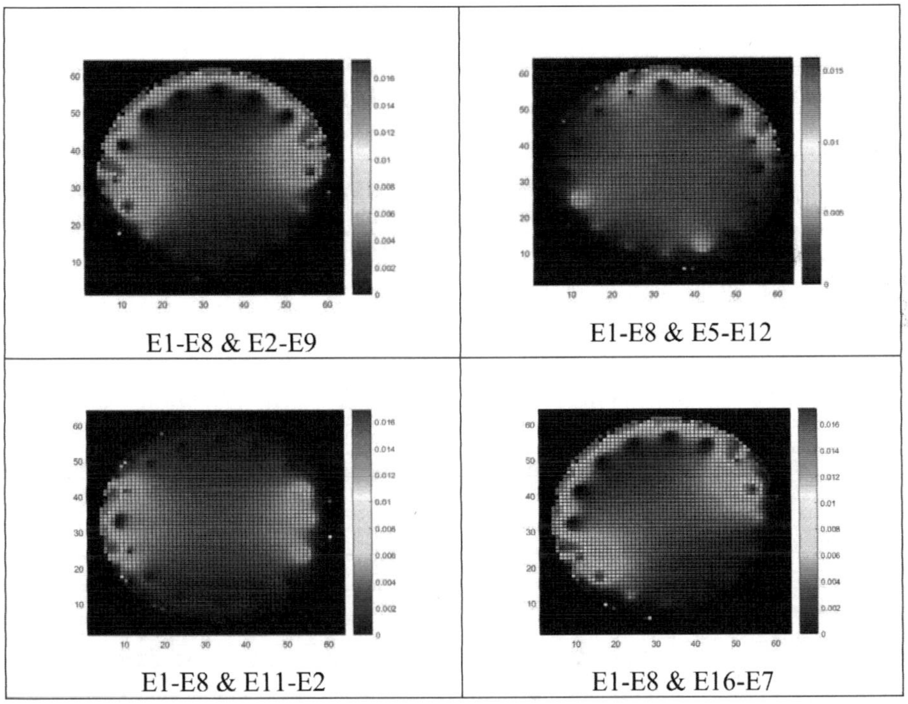

Based on Figure 2 and Figure 3, it can be seen that the single excitation method produced a better-normalized map compared to the multiple excitation technique. It can be observed that the normalized map produced by using the single excitation method had a clear curved line. The curved line presented owing to the soft-field behavior of the electrical resistance tomography. For multiple excitation method, the sensitivity distribution produced was not uniform and even. It is believed that the increasing number of excitation source affect the nature of soft field behavior. As for example between the group of source channels, E1-E8 and the group of receiver channels, E16-E7; due to the very high values of electric field at the group of source channels and a very small electric field values at the group of receiver channels, the pairing results caused the highest value of electric field at only at the source part and consequently it did not produce a clear curve line between that pair projection. Hence, it is predicted that the tomogram that can be obtained in the image reconstruction part would also not clear for multiple excitation source method. To proof, this hypothesis, the normalized map for both methods were implemented in the inverse problem solving to reconstruct the image of the ERT system.

2.2. Inverse problem solving

In this project, the linear back projection (LBP) algorithm was applied as a basis for the image reconstruction part. LBP algorithm is widely used in many tomography processes [9]. The advantages of the LBP algorithm are its supply of low computational complexity and can produce an image at a high speed [11]. In reconstructing the image using the LBP algorithm, each of the corresponding sensor reading is being multiplied with each of the sensitivity matrices. Thus, the disadvantage of the LBP algorithm is it generates a blurred effect or also known as smearing effect. The tomogram can be produced based on equation (5) [11].

$$V_{LBP(single)} = \sum_{T_x=1}^{m} \sum_{R_x=1}^{m} \bar{S}_{T_x,R_x} \times \bar{M}_{T_x,R_x} \tag{5}$$

where

$$\bar{S}_{T_x,R_x} = \frac{S_{FullLiquid_{T_x,R_x}} - S_{Object_{T_x,R_x}}}{S_{FullLiquid_{T_x,R_x}}} \tag{6}$$

$V_{LBP(single)}$ is defined as a voltage distribution for a single excitation technique, m is the number of electrodes, while $\bar{S}_{T_x,R_x}$ is the sensor loss values between the transmitter, T_x and receiver, R_x ,and $\bar{M}_{T_x,R_x}$ is the normalized sensitivity maps of the transmitter and receiver. The equation (5) is the common LBP algorithm that mainly used in many tomography processes especially in the common method strategies or also known as a single excitation source. As for equation 3, $S_{FullLiquid_{T_x,R_x}}$ is the sensor values of full liquid applied into the steel pipe while $S_{Object_{T_x,R_x}}$ is the sensor values when the steel pipe consist of the object. Equation (6) was used to obtain the standardized value for the sensor loss performance of the steel pipe application.

The tomography image for multiple excitation source method is produced based on equation (7). The equation (7) was a modified equation based on the equation (5).

$$V_{LBP(multiple)} = \sum_{T_x=1-8}^{n} \sum_{R_x=1-8}^{n} \bar{S}_{T_x,R_x} \times \bar{M}_{T_x,R_x} \tag{7}$$

where $V_{LBP(multiple)}$ is defined as voltage distribution using multiple excitation source method, n is 50% number of electrodes out of the total electrodes for each process, while $\bar{S}_{T_x,R_x}$ is the sensor loss value of transmitter and receiver, and $\bar{M}_{T_x,R_x}$ is the normalized sensitivity maps of transmitter and receiver. The distinguish from its common method is that the channel 1 till channel 8 will group as an excitation source and the measurement is obtained at the group of the detection channels (channel 9 till 16). Here, the detection channels that redundant with the source channel were set as a zero value. Next, the process will be repeated until all channels have become as transmitter and receiver.

3. Results & discussion

The image reconstruction for liquid-gas regimes at three different positions and sizes using common technique and multiple excitation technique were done to identify the regime. The sensor values for each of the tested medium were done by using dummy data. It means that, if there is an obstacle, the receiver will receive 0 value and vice versa. Later, the multi-scale structural similarity (MSSIM) [12] was applied in the analysis part. The indexes output of MSSIM results is between zero (0) to one (1) [12]. It means that, the larger the value of the MSSIM index, the closer the reconstructed image to the reference image.

Figure 4 shows the tomogram for 10 mm bubble at 3 different positions. As provided in Figure 4, between both methods, it can be seen that the reconstructed images produced by the single excitation source method were better than the multiple excitation sources method. Besides, the bubble at three different positions can be recognized easily using the single excitation source technique. The MSSIM index also indicates that the single excitation source method gave a higher value rather than the multiple excitation method. Also, the images produced by the multiple excitation method did not give any closer images to the reference images.

Figure 4. Bubble (10mm) at 3 different positions for single versus multiple excitation technique

Reference image	Single excitation	Multiple excitation	MSSIM
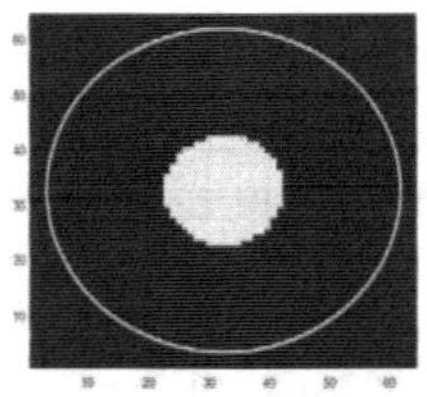	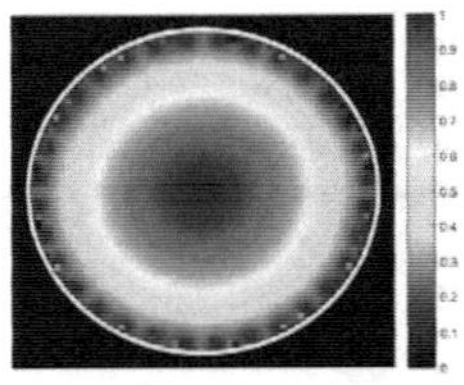	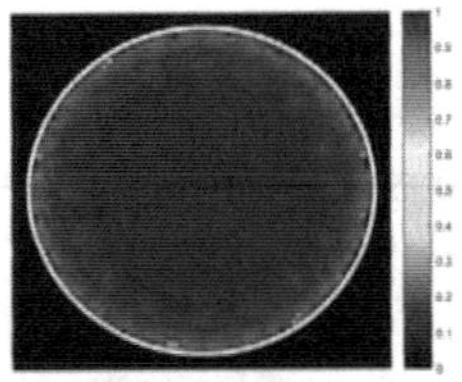	Single=0.1086 Multiple=0.0682
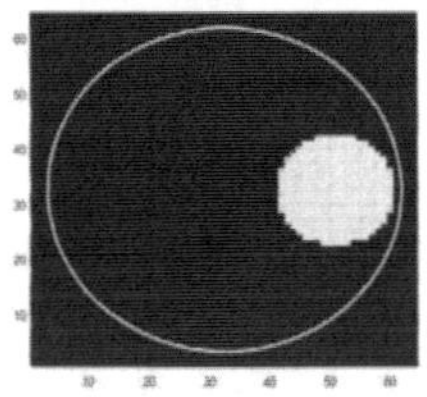	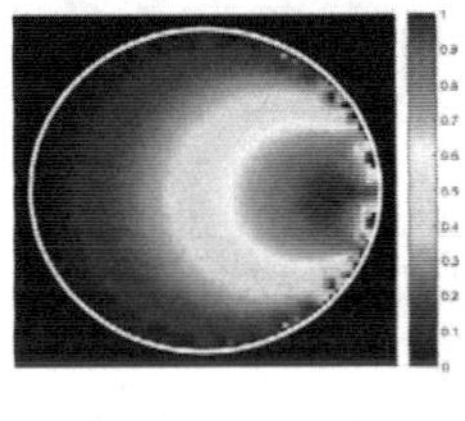	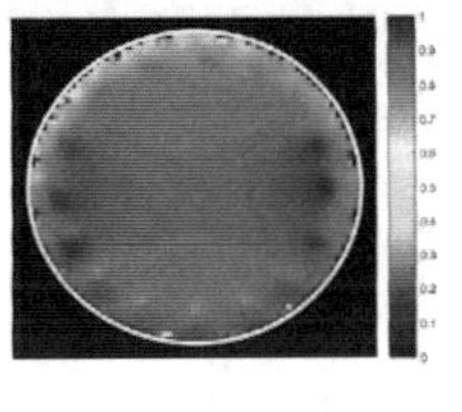	Single=0.1203 Multiple=0.0726
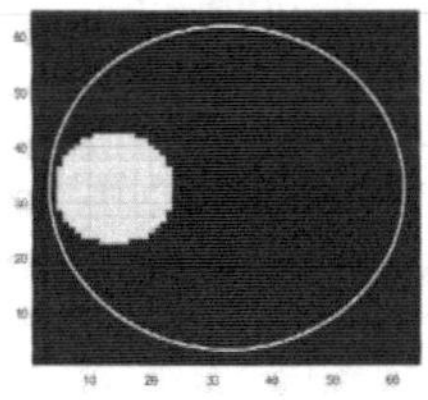	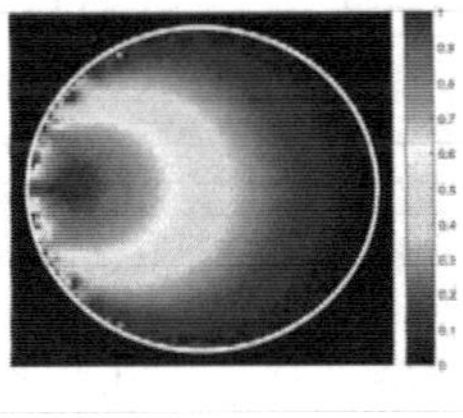	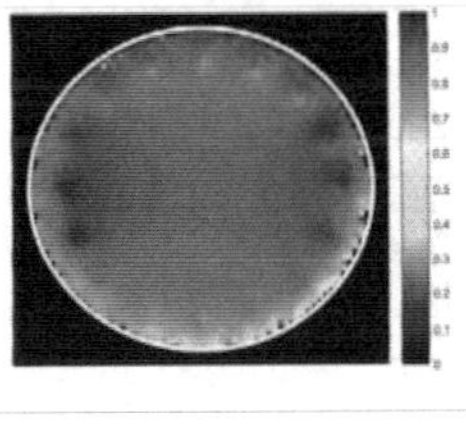	Single=0.1228 Multiple=0.0729

Moreover, tomogram at three different sizes which were 5mm, 10mm, and 20mm in diameter were also compared for both methods. Based on the results shown in Figure 5, again it was proved that the single excitation method produced a better tomogram than the other technique. Images reconstructed by the single excitation source method had a lot of similarity with the reference images. Apart from that, MSSIM index for a single excitation source method also gave a higher value if compared with the

IOP Conf. Series: Materials Science and Engineering **705** (2019) 012025 doi:10.1088/1757-899X/705/1/012025

multiple excitation source approach. Multiple excitation source method sat at the lowest bottom of the chart when it comes to the tomogram and as well as the MSSIM index values.

Figure 5. Bubble at the center with different sizes for single versus multiple excitation technique

Reference image	Single excitation	Multiple excitation	MSSIM
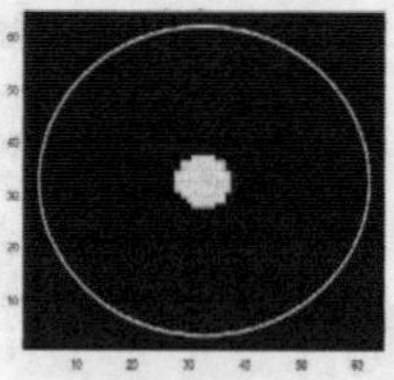 5 mm	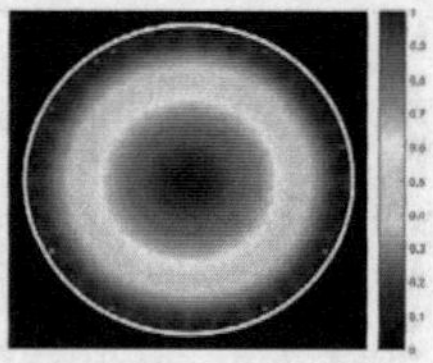	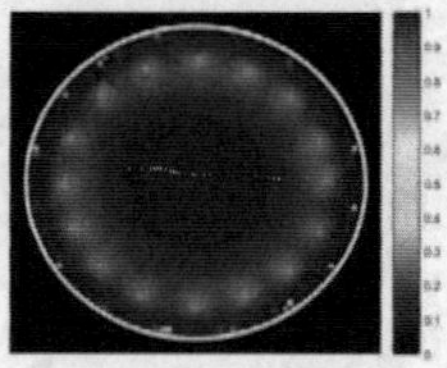	Single=0.0723 Multiple=0.0352
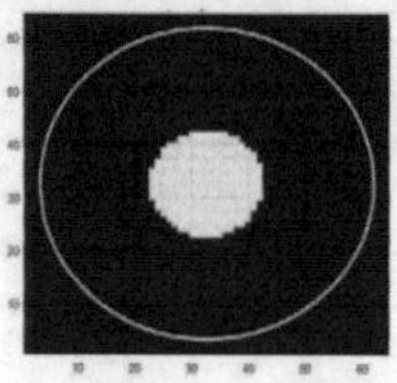 10 mm	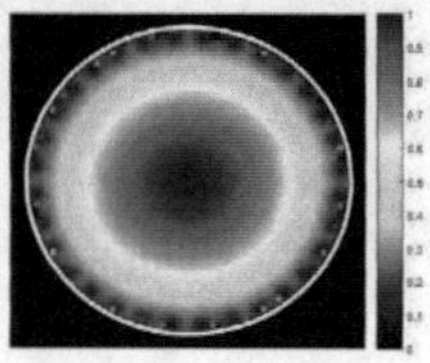	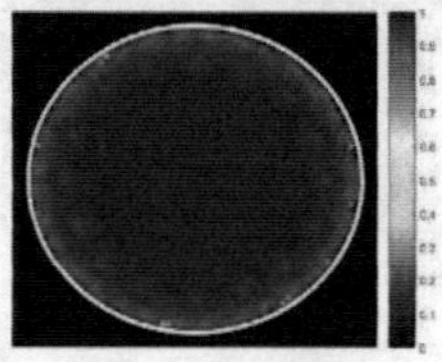	Single=0.1086 Multiple=0.0682
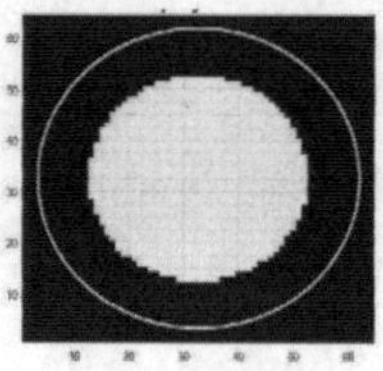 20 mm	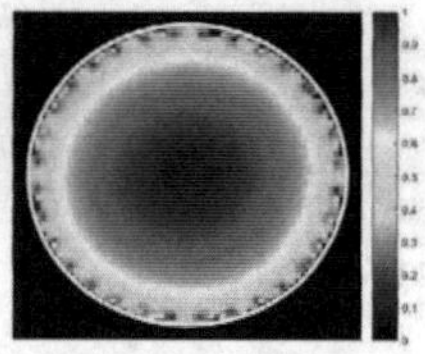	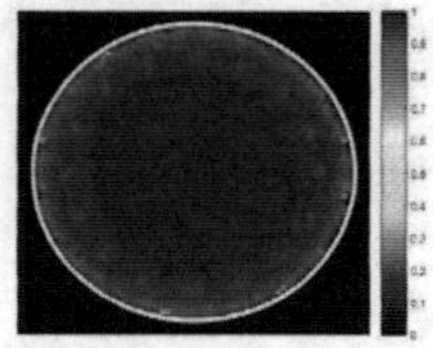	Single=0.2778 Multiple=0.2986

4. Conclusion

Throughout this research, the main objective of this research is to investigate whether the multiple excitation sources can produce a good and high quality of the reconstructed image or not. Research had been done, sensitivity map and comparison of the reconstructed images between the single excitation source, and multiple excitation sources also had been done, and various types of bubbles also had been implemented throughout this research. Many things had been considered throughout this paper, and all those things lead to one conclusion, which was the multiple excitation sources cannot improve the tomogram of the ERT system. It was recommended to use a single excitation source to reconstruct the image of the medium of interest. Even though the sensor reading performance was increased as stated in [6] by using this idea, but the image reconstructed showed them otherwise. The reconstructed image by using this method was not similar at all with the reference images. However, the outcomes of this research are the preliminary result and a further investigation is needed in this research. A different approach of the algorithm can be considered for the future work to get a better tomogram of the medium of interest.

Acknowledgement
This research was supported by the Universiti Malaysia Pahang internal grant (RDU170368).

IOP Conf. Series: Materials Science and Engineering **705** (2019) 012025 doi:10.1088/1757-899X/705/1/012025

References

[1] Abdul Wahab Y, Abdul Rahim R, Fazalul Rahiman M H, Abdul Rahim H, Ridzuan Aw S, Mohd Fadzil N S and Jamaludin J 2014 A Review of Process Tomography Application in Inspection System *J. Teknol.* **70** 35–9

[2] Ridzuan Aw S, Abdul Rahim R, Fazalul Rahiman M H, Mohd Yunus F R and Goh C L G 2014 Electrical Resistance Tomography: A Review of The Application of Conducting Vessel Walls *Powder Technol.* **254** 256–64

[3] Deng X, Dong Z, Chen L and Liu H 2014 Research on Excitation & Measurement Strategy of Tuber Electrical Resistance Tomography System *Appl. Mech. Mater.* **577** 701–4

[4] Ridzuan Aw S, Abdul Rahim R, Fazalul Rahiman M H, Mohamad E J, Mohd Yunus F R, Abdul Wahab Y, Fadzli N S and Jamaludin J 2015 Simulation Study on Electrical Resistance Tomography using Metal Wall for Bubble Detection *J. Teknol.* **73** 31–5

[5] Wahab Y A, Rahim R A, Fazalul M H, Aw S R, Rahman F, Yunus M, Muzakkir N, Ayob N, Ling L P, Abdul H, Ahmad I L, Jonet A, Seng C K and Sek T K 2017 Inverse Problem : Comparison between Linear Back- Projection Algorithm and Filtered Back-Projection Algorithm in Soft-Field Tomography *Int. J. Integr. Eng.* **9** 32–6

[6] Abdul Wahab Y, Shakyean S A, Md Zain Z, Abd Ghani N and Mat Noh M 2018 Initial Study of Multiple Excitation Source for Electrical Resistance Tomography in Steel Pipe Application *Proc. of the 10th National Technical Seminar on Underwater System Technology* **538** 353–60

[7] Saad F O, Sdayria A, Madiouli J, Sghaier J and Fudym O 2017 Quadrupole method: A New Approach for Solving The Direct Problem of Electrical Resistance Tomography *J. King Saud Univ. - Sci.* **29** 360–70

[8] Shaw G R, Goussard Y and Guardo R 1993 Linearization of The Forward Problem in Electrical Impedance Tomography *Proc. of the IEEE Int. Conf. on Engineering in Medicine and Biology Society* 82–3

[9] Abdul Wahab Y 2017 *2 MHz Electrical Resistance Tomography for Static Liquid- Solid Profile Measurement* (Universiti Teknologi Malaysia)

[10] Shakyean S A 2018 *Initial Study on Multiple Excitation Source of Electrical Resistance Tomography for Liquid Gas Regime Identification in Steel Pipe* (Universiti Malaysia Pahang)

[11] Wahab Y A, Rahim R A, Hafiz M and Rahiman F 2017 Image Reconstruction for Solid Profile Measurement in ERT Using Non-invasive Approach *Indones. J. Electr. Eng. Comput. Sci.* **15** 1554–64

[12] Wang Z, Bovik A C, Sheikh H R and Simmoncelli E P 2004 Image Quality Assessment: From Error Visibility to Structural Similarity *IEEE Trans. Image Process.* **13** 600–12

IOP Conf. Series: Materials Science and Engineering **705** (2019) 012026 doi:10.1088/1757-899X/705/1/012026

Arabic Fonts Representation in Cubic Bézier Curve using Different Soft Computing Algorithm

N F I C Amat*, Z R Yahya, N A Rusdi, N A Helmee and W Z A W Muhamad

*Corresponding author: fatiahizzati@gmail.com

Institute of Engineering Mathematics, University Malaysia Perlis, Kampus Pauh Putra, 02600 Arau, Perlis, Malaysia

Abstract. This study presents the comparison between two meta-heuristic algorithms for curve reconstruction of Arabic fonts. The two algorithms namely Sine Cosine Algorithm (SCA) and Whale Optimization Algorithm (WOA) are used for comparison purposes. These algorithms have been applied to optimize the solution of curve fitting problem by using cubic Bezier curve. There are several pre-processing steps involve before curve fitting process has been done. Firstly, extract the boundary. Then, detect the corner of outline fonts and the third step is chord length parameterization. The aim of this study is to find the best curve that fit the data. Sum Square Errors (SSE) has been used to calculate the error given by original image and parametric curve. The finding of this study shows that, WOA gives better curve approximation based on the errors result than SCA. This algorithm significantly improves the shape of the corner of the outline Arabic fonts.

1. Introduction

Curve reconstruction is a crucial field in computer graphic especially for the curves which are not easily computed. Generally, it is a process of reconstructing a curve or a mathematical function, which has the best fit to a set of data points [1]. The main challenging parts of curve reconstruction are to get the smoothness and accurateness of the curve [2]. It has been grabbed much attention by researchers in the last few years [3–6]. There are numerous aspects of curve reconstruction such as font designing[7–10], data visualization[11], capturing hand-drawn images on computer screens [6] and medical image [12]. The important phase of curve reconstruction is curve fitting. The goal of curve fitting is to spot the set of data points that can approximately represent the given curve. Before curve fitting has been done, a few pre-processing phase (boundary extraction, corner point detection and chord length parameterization) needed in order to interpolate the data.

There are some frequently curves had been chosen by the researchers such as B Spline[3], cubic Spline[13, 14], cubic Bézier curve [4–6, 12, 15], and quartic Bézier curve [9] for solving curve fitting problems. Cubic Bézier curve had been chosen as an appropriate curve because of its advantages and interesting properties which make it differ from other curves [13]. The control points of the curve can be adjusted to effectively produce a good result.

Soft Computing is basically an optimization technique to find the best solution in solving various types of problem especially in complex engineering problems [1]. There are several example of soft computing technique such as Genetic Algorithm (GA) [2, 16], Differential Evolution (DE) [1, 10, 17], Artificial Bee Colony algorithm (ABC) [5, 12, 18], Whale Optimization Algorithm (WOA) [6, 19] and Sine Cosine Algorithm (SCA) [20]. According to [5], Artificial Bee Colony Algorithm has been used to reconstruct the Arabic font and presents great result since the fitted cubic Bézier curve approximates the original image. Based on [6], WOA proposed to explore curve fitting for handwriting image and the error obtained quite large but most of the fitted cubic Bézier curve is on its original boundary. SCA is a simple effective optimization algorithm based on sine and cosine mathematical functions developed by Mirjalili to solve optimization problems. Numbers of researches shown SCA has been successfully employed in many optimization problems such as feature selection [21], manuscript image binarization [22], modeling optimization [23] and data clustering [24]. In this study, the comparison of two different soft computing algorithms which are SCA and WOA with cubic Bézier curve has been discovered in order to find the best curve in solving curve fitting problem.

This paper can be divided into six sections. Section 2 discussed the Bézier curve. Then it was followed by explaining the process of SCA and WOA. Next, in section four the pre-processing process will be discussed in details. The elaboration on curve fitting process for both algorithms had been done in Section 5. Lastly, the paper has been concluded in Conclusion section.

2. Cubic Bézier Curve

Bézier curve was introduced by the French engineer Pierre Bézier are widely used in computer graphics to model smooth curves [25]. Using the Bernstein polynomials as a basis function, a Bézier curve can be defined of degree n (in terms of Bernstein polynomials) to be a curve in parametric form $S(t)$. The general form of Bézier curve of degree n is represented by

$$S(t)=\sum_{i=0}^{n} B_i^n(t)P_i \qquad 0\le t\le 1 \tag{1}$$

where P_i is a control point and B_i^n are blending function, known as Bernstein polynomial basis function and t is parameter. Basis function of Bernstein is then defined as

$$B_i^n(t)={}^nC_i(1-t)^{n-i}t^i \tag{2}$$

In this study, Bezier curve of degree three was used. Notes that, cubic Bézier curve is determined by four control points P_0, P_1, P_2 and P_3. From (1), cubic Bézier curve can be illustrated as $S(t)=\sum_{i=0}^{3} B_i^3(t)P_i$.

3. Sine Cosine Algorithm (SCA)

This chapter discussed the Sine Cosine Algorithm (SCA), which is a recently meta-heuristic that applies a model based on trigonometric functions to estimate the global optima of optimization problems [20]. The optimization process is generally split into two phases: exploration and exploitation. In each iteration, the position update of X_i^t formula can be obtained by the following equation:

$$\overrightarrow{X_i^{t+1}} = \begin{cases} \overrightarrow{X_i^t} + r_1\sin(r_2)\left|r_3\overrightarrow{P_i^t} - \overrightarrow{X_i^t}\right| & r_4 < 0.5 \\ \overrightarrow{X_i^t} + r_1\cos(r_2)\left|r_3\overrightarrow{P_i^t} - \overrightarrow{X_i^t}\right| & r_4 \ge 0.5 \end{cases} \tag{3}$$

where $\overrightarrow{X_i^t}$ represents the position of current solution in i th dimension at t th iteration, $\overrightarrow{P_i^t}$ is position of the destination point in the i th dimension at the t th iteration, while || specifies the absolute cost. SCA used 4 main random parameters namely r_1, r_2, r_3 and r_4. These parameters which effectively balances the exploration and exploitation ability of the algorithm in solution space. Parameter r_1 decides that search agent is going to do whether exploration or exploitation. Then, parameter r_2 decides how far the solution's movement should be towards or outwards the destination. In order to stochastically emphasize ($r_3 > 1$) or deemphasize ($r_3 < 1$) the effect of destination in finding the distance, parameter r_3 assign a random weight for the destination. Finally, r_4 ranging from 0 to 1 decided whether sine or cosine formula is going to be used.

The parameter r_1 is the main stochastic variable defining the movement direction of the solution towards or outwards the destination. In population-based algorithms to converge a point at the end of optimization, the exploration phase should be decreased. The parameter linearly r_1 decreases in SCA as follows:

$$r_1 = a - t\frac{a}{T} \tag{4}$$

where t is the current number of iterations, T is the maximum number of iterations, a is a constant.

4. Whale Optimization Algorithm (WOA)

Whale Optimization Algorithm (WOA) is a bio-inspired meta-heuristic algorithm inspired by hunting mechanism of humpback whales [19]. This special hunting mechanism called bubble net feeding method. This algorithm includes three typical behavior of humpback whale and models them mathematically. The mathematical model of encircling prey, spiral bubble-net feeding maneuver and search for prey is described in the following section:

4.1. *Encircling Prey*

Humpback whales at first will spot the position of prey and then they encircle the prey. Target prey or close to the optimum is assumed as current best candidate solution. The other search agents will hence try to update their position towards the best search agent after the best search agent is defined. This action is demonstrated by the following equations:

$$\vec{D} = |\vec{C} \cdot \overrightarrow{X^*}(t) - \vec{X}(t)| \tag{5}$$

$$\vec{X}(t+1) = \overrightarrow{X^*}(t) - \vec{A} \cdot \vec{D} \tag{6}$$

where t indicates the current iteration, $\vec{A}$ and $\vec{C}$ are coefficient vectors, $\overrightarrow{X^*}$ is the position vector of the best solution, $\vec{X}(t)$ is position vector, $\vec{D}$ is distance between i^{th} whale and the prey, $|\ |$ is absolute cost, and ⊔ is an element wise multiplication. $\overrightarrow{X^*}$ should be updated in each iteration if there is a better solution. The vectors $\vec{A}$ and $\vec{C}$ are calculated as follows:

$$\vec{A} = 2\vec{a} \cdot \vec{r} - \vec{a} \tag{7}$$

$$\vec{C} = 2 \cdot \vec{r} \tag{8}$$

where $\vec{a}$ is linearly decreased from 2 to 0 over the course of iterations (in both exploration and exploitation phases) and r is a random vector in $[0,1]$.

4.2. *Bubble-net Attacking Method (Exploitation Phase)*

There are two approaches in this hunting method that has been mathematically model.

1. Shrinking encircling mechanism: This technique employed by decreasing linearly the value of $\vec{a}$ from 2 to 0. Random value for a vector $\vec{A}$ in the range between $[-1,1]$.
2. Spiral updating position: Mathematical spiral equation for position update between humpback whale and prey that was helix-shaped movement as follows:

$$\vec{X}(t+1) = \overrightarrow{D'} \cdot e^{bl} \cdot \cos(2\pi l) + \overrightarrow{X^*}(t) \tag{9}$$

where l is a random number $[-1,1]$, b is constant defines the logarithmic shape, $\vec{D} = \overrightarrow{X^*}(t) - \vec{X}(t)$ express the distance between i^{th} whale to the prey (best solution so far).
Note: Assume that there is a probability of 50 % that whale either follow the shrinking encircling or logarithmic path during optimization. The mathematical model as follows:

$$\vec{X}(t+1) = \begin{cases} \overrightarrow{X^*}(t) - \vec{A} \cdot \vec{D} & if\ p < 0.5 \\ \overrightarrow{D'} \cdot e^{bl} \cdot \cos(2\pi l) + \overrightarrow{X^*}(t) & if\ p \geq 0.5 \end{cases} \tag{10}$$

where p is a random number in $[0,1]$.

4.3. Search for Prey (Exploration Phase)

The vector $\vec{A}$ can be used for exploration to search for prey. Vector $\vec{A}$ also used random values greater than 1 or less than -1. At this phase, the update position of search agent is according to a randomly chosen search agent. If $|\vec{A}| > 1$ emphasize exploration where a random search agent is chosen and it allow the WOA algorithm to perform a global search. If $|\vec{A}| < 1$ emphasize best solution is selected for updating the position of the search agent. The mathematical model is as follows:

$$\vec{D} = |\vec{C} \cdot \overrightarrow{X_{rand}} - \vec{X}| \tag{11}$$

$$\vec{X}(t+1) = \overrightarrow{X_{rand}} - \vec{A} \cdot \vec{D} \tag{12}$$

where $\overrightarrow{X_{rand}}$ is a random position vector (a random whale) chosen from current population.

5. Pre-Processing Steps

There are several pre-processing phases involved before the curve fitting process, which include boundary extraction, corner point detection and chord length parameterization. This section will explain in detail the role of each phase. The pre-processing phases and curve fitting process can be summarized in a flowchart as shown in Figure 1.

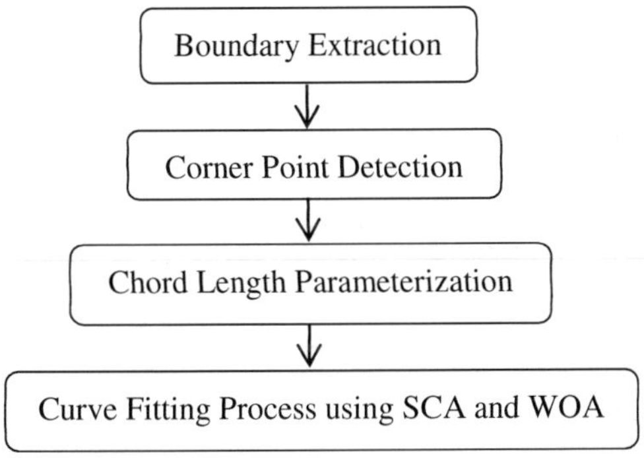

Figure 1. Flow chart pre-processing phases and curve fitting process

5.1. Boundary Extraction and Corner Point Detection

The first process to reconstruct the curve starts with the boundary extraction of the original image. The boundary of the image is extracted to represent an object's shape in graphical representation (number of boundary points) [17]. There are numerous algorithms has been used to extract the boundary of the generic shape such as [26, 27]. In this work, boundary of digitized image is extracted by using MATLAB function that called Boundaries in order to extract the boundary of the font images. The

IOP Conf. Series: Materials Science and Engineering **705** (2019) 012026 doi:10.1088/1757-899X/705/1/012026

function returns number of pieces and for each piece number of boundary points and values of these boundary points $P_i = (x_i, y_i), i = 1,...,n$.

The corner points of the image are detected, after finding out the boundary points. Corner points are defined as the point that separates the outline into various segments before curve fitting. Corner points play an important role in image representation [7] and preprocessing phase of outline capturing systems [29]. Various corner detection algorithms have been discovered such as Davis Algorithm, Harris Corner Detection Algorithm, Rosenfeld and Johnston (RJ73) Algorithm and SAM06 Algorithm [30]. In this paper, SAM06 algorithm was applied to detect the corner points. The boundary extraction and corner detection of the Arabic font are represented in Table 1.

Table 1. Boundary extracted and detection of corner points.

Name	Lam Alif	Kha	Ka
Original Images			
Boundary Extraction and Corner Detections	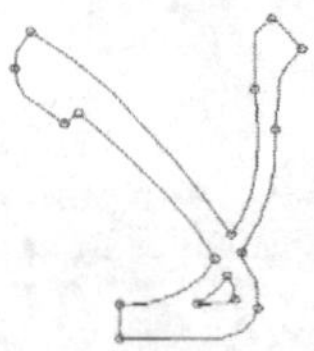	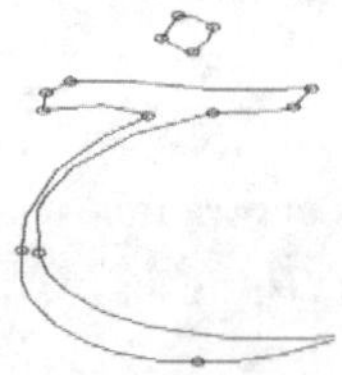	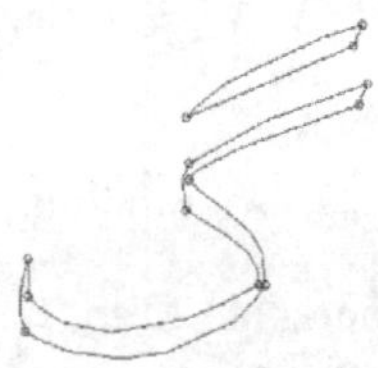

5.2. Chord Length Parameterization

In section 2, cubic Bézier curve is defined by the beginning point and the ending points, which are interpolated and have two interior points that control the shape of the curve. This cubic Bezier curve consists of four control points: P_0, P_1, P_2 and P_3. The initial and final points of the curve are P_0 and P_3 respectively. The interior control points P_1 and P_2 are used to control the shape of the curve. The cubic Bezier curve can be written in a matrix form by expanding the analytic definition of the curve into its Bernstein polynomial coefficients, and then writing these coefficients in a matrix form using the polynomial power basis [25]. That is,

$$S(t) = \sum_{i=0}^{3} B_i(t)P_i$$

$$S(t) = B_0^3(t)P_0 + B_1^3(t)P_1 + B_2^3(t)P_2 + B_3^3(t)P_3$$

$$= (1-t)^3 P_0 + 3t(1-t)^2 P_1 + 3t^2(1-t)P_2 + t^3 P_3$$

$$= [(1-t)^3 \quad 3t(1-t)^2 \quad 3t^2(1-t) \quad t^3\,] \begin{bmatrix} P_0 \\ P_1 \\ P_2 \\ P_3 \end{bmatrix} \tag{13}$$

$$= \begin{bmatrix} 1 & t & t^2 & t^3 \end{bmatrix} \begin{bmatrix} 1 & 0 & 0 & 0 \\ -3 & 3 & 0 & 0 \\ 3 & -6 & 3 & 0 \\ -1 & 3 & -3 & 1 \end{bmatrix} \begin{bmatrix} P_0 \\ P_1 \\ P_2 \\ P_3 \end{bmatrix}.$$

Chord-length parameterization which been receiving attention in CAGD is commonly adopted for simulating curve arc-length parameterization in discrete point data interpolation and approximation [5]. Chord-length parameterization is used to estimate the parametric value t associated with each point p_i as follows:

$$t_i = \begin{cases} 0 & \text{if } i = 1 \\ \dfrac{|p_1p_2| + |p_2p_3| + ... + |p_{i-1}p_i|}{|p_1p_2| + |p_2p_3| + ... + |p_{n-1}p_n|}, & \text{if } i \le n\text{-}1 \\ 1 & \text{if } i = n \end{cases} \tag{14}$$

The goal is to approximate the boundary of the original image by a parametric curve in an optimal way.

6. Fitting the Curve using Sine Cosine Algorithm (SCA) and Whale Optimization Algorithm (WOA)

The main part of curve reconstruction is called curve fitting. The three selected Arabic fonts Lam Alif (لا), Kha (خ) and Kaf (ک) were employed in this study. The aim of this research is to compare the results obtained by using SCA and WOA in curve fitting phase. The same pre-processing steps were adopted in order to interpolate the data. At this phase, the both soft computing algorithms were applied to approximate the two middle points P_1 and P_2. Then, these points will be used to find the parametric curve that gives the best optimal results for given values of t.

There are some parameters included in this phases which are population size, number of iterations, limit, and dimension. In this study, population size is set to be 50 with 100 numbers of iterations for both algorithms. The middle points P_1 and P_2 consist of (x_1, y_1) and (x_2, y_2), respectively. So, the dimension is set to be four since it expressed x_1, x_2, y_1 and y_2.

The fitted cubic Bézier curve is obtained, when the best value of points P_1 and P_2 are found. The goal of this study is to minimize the distance between boundaries of the original images and fitted cubic Bézier curve obtained by using SCA and WOA. In order to achieve the different result, Sum Square

IOP Conf. Series: Materials Science and Engineering **705** (2019) 012026 doi:10.1088/1757-899X/705/1/012026

Error (SSE) has been used to calculate errors between the parametric points of generated curve and original image.
The result of SCA will be compared with WOA. The final results presented visually and numerically are illustrated in Figure 2 and Table 2 respectively. Figure 2 shows the demonstration of fitted cubic curve (blue line) over the original curve (dotted line). The fit of the cubic Bézier curve is on boundary of bitmap image. The reconstruction have similar image visually while the Table 2 represents the errors obtained between Sine Cosine Algorithm (SCA) and Whale Optimization Algorithm (WOA). The result shows that the performance of WOA is better than SCA since it produced a small error.

Figure 2. The visualization result of fitted cubic Bézier curve (blue line) over the original curve (dotted line)

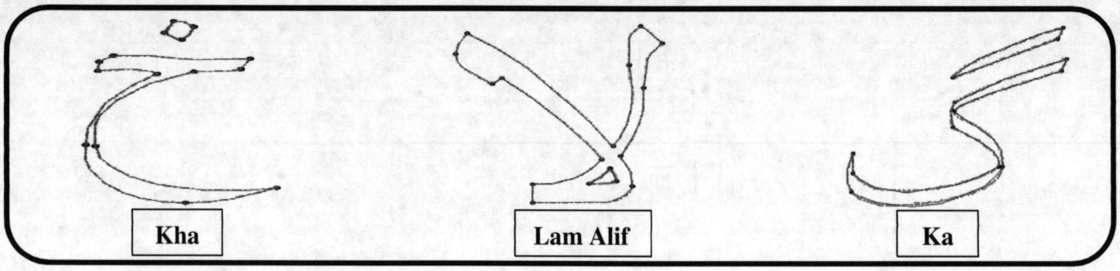

Table 2. The error obtained by each image.

Image	Sine Cosine Algorithm (SCA)	Whale Optimization Algorithm(WOA)
Lam Alif (لا)	10.3980	10.1805
Kha (خ)	5.5936	2.6588
Ka (گ)	16.8602	16.6891

7. Conclusion

This paper applied two different soft computing algorithms to solve curve fitting problem. The two algorithms are SCA and WOA. Sum Square Error (SSE) was used in order to estimate the errors between the boundary of bitmap images and the parametric curve. This study shows that WOA is more effective method than SCA in solving this problem since it produced an accurate representation with small error. The result illustrated, the reconstruction have similar image visually, while the error shown that the WOA lead all from the three selected image. In conclusion, both algorithms remain as the interesting algorithm that can be explored in more details and can be applied in numerous problems.

Acknowledgement

The authors highly acknowledge the financial sponsorship provided by Fundamental Research Grant Scheme FRGS/1/2017/STG06/UNIMAP/03/2.

References

[1] Hasan Z A, Yahya Z R, Rusdi N A and Roslan N 2018 Curve Reconstruction In Different Cubic Functions Using Differential Evolution *MATEC Web of Conference* **150** 06030

[2] Majeed A and Piah A R M 2014 Image reconstruction using rational ball interpolant and genetic algorithm *Applied Mathematical Sciences* **8**(74) 3683-3692

[3] Haron H, Rehman A, Adi D I S and Saba T 2012 Parameterization Method on B-Spline Curve *Mathematical Problems in Engineering* 640472

[4] Rusdi N A and Yahya Z R 2015 Reconstruction of generic shape with cubic Bézier using least

square method *AIP Conference Proceedings, AIP Publishing* **1660** 050004

[5] Rusdi N A and Yahya Z R 2016 Reconstruction of Arabic Font using Artificial Bee Colony Algorithm *ARPN J Eng Appl Sci 2016* **11** 10761–10767

[6] Helmee N A, Yahya Z R and Hasan Z A 2018 Implementation of whale optimization algorithm for handwriting representation *AIP Conference Proceeding AIP Publishing* **2013** 020007

[7] Yahya F, Md Ali J, Majid A A and Ibrahim A 2006 An automatic generation of G1 curve fitting of arabic characters *International Computer on Graphics, Imaging and Visualisation: Techniques and Applications, CGIV'06.* IEEE, 542-547

[8] Poon S H and Majid B A 2016 Interpolation between Outline Font Characters *2016 First International Conference on Multimedia and Image Processing (ICMIP)* IEEE 61–66

[9] Rusdi N A and Yahya Z R 2015 Reconstruction of Arabic Font with Quartic Bézier Curve (Pembinaan Semula Fon Arab menggunakan Lengkung Bézier Kuartik) *Sains Malaysiana* **44** 1209–1216

[10] Roslan N and Yahya Z R 2015 Reconstruction of Font with Cubic Bezier Using Differential Evolution *Sains Malaysiana* **44** 1203–1208

[11] Yahya Z R, Piah A R and Majid A 2012 Curve Fitting Using Conic by Evolutionary Computing *Journal of Mathematics and Statitics* **8** 107-110

[12] Rusdi N A, Yahya Z R, Roslan N, Muhamad W and Azman W Z 2018 Reconstruction of Medical Images Using Artificial Bee Colony Algorithm *Mathematical Problems in Engineering* 8024762

[13] Sarfraz M and Al-Dabbous N A 2013 Curve representation for outlines of planar images using multilevel coordinate search *WSEAS Transactions on Computers* **12** 62–73

[14] Irshad M, Sarfraz M and Hussain M Z 2015 Reverse Engineering of Planar Objects Using Imperialist Competitive Algorithm *19th International Conference on Information Visualisation.* IEEE, 430–435

[15] Roslan N and Yahya Z R 2016 Parameterization method on cubic Bézier curve fitting using differential evolution *AIP Conference Proceeding, AIP Publishing* **1775** 030075

[16] Sarfraz M, Irshad M and Hussain M Z 2013 Reverse Engineering of Planar Objects using Genetic Algorithm *Sains Malaysiana* **42** 1167-1179

[17] Hasan Z A, Yahya Z R and Piah A R M 2015 Differential Evolution for Reverse Engineering Using Cubic Ball Curve *Educ J Sci Math Technol* 2015 **2** 58–68

[18] Rusdi N A and Yahya Z R 2016 Artificial Bee Colony algorithm for curve reconstruction *AIP Conference Proceeding, AIP Publishing* **1775** 030077

[19] Mirjalili S and Lewis A 2016 The Whale Optimization Algorithm *Advances in engineering software* **95** 51-67

[20] Mirjalili S 2016 SCA: A Sine Cosine Algorithm for solving optimization problems *Knowledge-Based Systems* **96** 120–133

[21] Hafez A I, Zawbaa H M, Emary E and Hassanien A E 2016 Sine cosine optimization algorithm for feature selection *International Symposium on Innovations in Intelligent Systems and Application INISTA,* IEEE. 1–5

[22] Elfattah M A, Abuelenin S, Hassanien A E and Pan J S 2016 Handwritten Arabic manuscript image binarization using sine cosine optimization algorithm *Advances in Intelligent Systems and Computing, Springer, Cham* 273–280

[23] Kaur S and Prashar S A 2016 A novel sine cosine algorithm for the solution of unit commitment problem *International Journal of Science, Engineering and Technology Research* **5** 12

[24] Kumar V and Kumar D 2017 Data Clustering Using Sine Cosine Algorithm I*n Handbook of research on machine learning innovations and trends, IGI Global* 715–726

[25] Pastva T A 1998 Bezier Curve Fitting *Naval Postgraduate School Monterey CA*

[26] Itoh K and Ohno Y 1993 A curve fitting algorithm for character fonts. *Electronic Publication* **6**

[27] Sarfraz M and Khan M A 2002 Automatic outline capture of Arabic fonts. *Information Sciences* **140** 269–281

[28] Jähne B 2005 Digital Image Processing *New York: Springer-Verlag*, Chapter 1

[29] Sarfraz M, Masood A and Asim M R 2004 A Web Based System for Capturing Outlines of 2D Objects *Proc. of International Conference on Information & Computer Science, Dhahran, Saudi Arabia*
[30] Sarfraz M 2008 Corner Detection for Curve Segmentation In: *Interactive Curve Modeling*

IOP Conf. Series: Materials Science and Engineering **705** (2019) 012027 doi:10.1088/1757-899X/705/1/012027

Intelligent control of a modular power assisted stretcher for indoor application

T Dechathanat[1], C Seksan[1], S Pramot[2], N Chote[3] and A H Ismail[4]

[1]Facultyof Engineering, Pathumwan Institute of Technology, Bangkok Thailand.
[2]Faculty of Engineering, Kasem Bundit University, Bangkok Thailand.
[3]Faculty of Medicine, Chiangmai University, Chiangmai Thailand.
[4]School of Mechatronic Engineering, Universiti Malaysia Perlis, Malaysia.

Email: thewan.thong@gmail.com; seksan@pit.ac.th; p_srinoi@hotmail.com; oic2535@gmail.com; ihalim@unimap.edu.my

Abstract. This paper presents the development of an intelligent control of a modular power assisted stretcher movement on flat and sloped floors for indoor environment. Conceptual design techniques are used to select the best concept for prototyping power assist systems. The operator's movement intention and related voltage data is sensed by an Eight-axis hall effect sensor. It proposes the use of PID controller to optimize and maintain the velocity of the stretcher adapted to different weight and floor conditions. A stepping motor is applied to drive each wheel to respond appropriately to different floor and steering conditions (flat or sloped, straight section or turns). Without assistance it would be impractical for an operator to move a loaded stretcher (with patient and accessories) weighing up to approximately 100 kg on 15 degrees straight slope-up / slope-down floor sections. The power assist application turns this into an easy operation.

1. Introduction

Over the years, many countries, including Thailand have seen an increase of elderly people and a reduction in number of working age citizens [1][2]. Medical Health Service providers such as nurses, nursing assistants, that are tasked with the moving of patients have a heavy load to handle [3][4]. Moving a patient-loaded stretcher needs the assistance of at least two nurses (especially on sloped floors) [5]. Therefore, the auxiliary power system is an effective solution to reduce the number of stretcher operators to only one person per bed. In addition, in this study, the auxiliary power system is applied to the modular drive system, cancelling the need for drive units on other patient stretchers. As such, implementing this system results in a significant cost reduction. Therefore, the goal of this study is to provide a system to reduce the workload and the number of workers in hospitals and health care facilities.

2. Design Concept

There are two main concepts applied in this research. One is the prototype construction of a complete stretcher with power-assist system assembly. Another is a prototype of a modular stand-alone power assist system without bed. Table 1 shows the design concepts according to the type of function. A selection process based on these design concepts will lead to further prototyping.

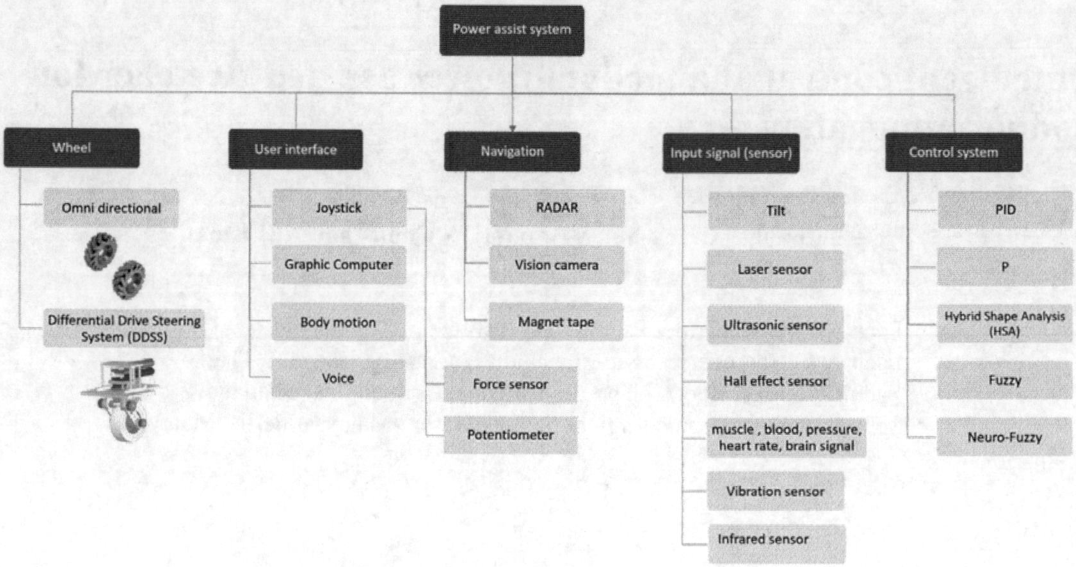

Figure 1. Architecture of power assist system today.

Different components make up power assist systems, including wheels (such as Omni directional or differential drive steering wheels), user interfaces (such as joystick [9], body motion, voice), navigation systems (such as radar, vision, magnet tape), different sensor types depending on applications, requirements and control systems as shown in Figure 1.

2.1 Functional of conceptual design

Table 1. Functional details of the design concepts.

System	Design 1	Design 2
User interface	Potentiometer joystick	Hall effect sensor joystick
Localization	none	WPS
Controller	Electric DC motor Controller	Arduino controller board
Power Motor	DC brushless motor	Stepping motor

The selection process on the design concepts depends on important principles, including hierarchical framework, priority analysis and by determining the parameters/criteria that are used to make decisions at the top level by showing the overall objective or goal [6]. A pairwise comparison is conducted based on a relative evaluation score level as shown in Table 2. Finally, the most suitable option is selected [7][8]. The parameters evaluated in this comparison are functionality-, safety-, price- and flexibility-targets. The results of the pairwise comparison of the criteria shown in Table 3 based on the score level outlined in Table 2.

Table 2. Relative score level of pairwise comparison.

Scale	Definition
1	Equal value
3	Slightly more value
5	Stronger value
7	Very strong value

Table 3. Pairwise comparison of requirements.

Requirement	Functionality	Safety	Price	Flexibility
Functionality	1	1/7 = 0.14	5	1
Safety	7/1 = 7	1	5	3
Price	1/5=0.2	0.2	1	0.2
Flexibility	1	1/3 = 0.33	5	1
Total	9.2	1.64	16	5.2

The relative importance of two selected criteria using pair-wise comparison is shown in Table 3. It shows the relative score level for pairwise comparison as shown in Table 2. For instance, if safety is rated to be a very strong value over functionality, then we got 7/1 = 7. The data for relative importance is extracted via questionnaires and knowledge about power assist systems.

2.2 Evaluation of the priority of functionality

The data from Table 3 is then synthesized to find the priority of the target requirements. This is calculated from the following equation (1):

$$W_i = \frac{1}{n}\sum_{j=1}^{n}\frac{a_{ij}}{\sum_i^n a_{ij}} \quad when\ i,j = 1,2,\ldots,n \qquad (1)$$

Table 4. Synthesized matrix related to functionality.

Requirement	Functional	Safety	Price	Flexibility	SUM	Priority
Functional	0.11	0.085	0.313	0.192	0.70	0.175
Safety	0.761	0.61	0.313	0.577	2.261	0.560
Price	0.022	0.122	0.063	0.038	0.245	0.061
Flexibility	0.11	0.20	0.313	0.192	0.815	0.204
					$\sum$	1.000

The results of the synthesis in Table 4 show the importance and weight priority that can be sorted as follows: safety (0.560), flexibility (0.204), functional (0.175) and price (0.061).

2.3 Pairwise comparison of sub-criteria

The next step is comparison of the level of selection of sub-devices such as user interface, control system and motor drive system. Score 1 or 2 was assigned to evaluate or compare which design is evaluated worst (score=1) or better (score=2). Table 5, Table 6 and Table 7 present the pairwise comparison of sub-criteria.

Table 5. Pairwise comparison of the user interface unit of *design 1* and *design 2* (from Table 1).

Requirement (Design1)	(Design2) Functionality	Safety	Price	Flexibility
Functionality	1/2=0.5			
Safety		1/2=0.5		
Price			1/2=0.5	
Flexibility				1/2=0.5

Table 6. Pairwise comparison of the control system unit of *design 1* and *design 2* (from Table 1).

Requirement (Design1)	(Design2) Functionality	Safety	Price	Flexibility
Functionality	1/2=0.5			
Safety		1/2=0.5		
Price			2/1=2	
Flexibility				1/2=0.5

Table 7. Pairwise comparison of the motor drive system of *design 1* and *design 2* (from Table 1).

Requirement (Design1)	(Design2) Functional	Safety	Price	Flexibility
Functional	1/2=0.5			
Safety		1/2=0.5		
Price			2/1=2	
Flexibility				1/1=1

The importance or weight priority is used for calculation and selection of the best design from this equation.

$$Y_i = W_i * \sum_{j=1}^{m} b_{ij} \quad when\ i = 1,2,\ldots,n\ ;\ j = 1,2,\ldots,m \tag{2}$$

$$Z = (\sum_{i=1}^{n} Y_i)/m \quad when\ i,j = 1,2,\ldots,n \tag{3}$$

Where

Y_i is total score requirement-i of *design 1* to *design 2*

b_{ij} is summation of pairwise score of *design 1* to *design 2* from sub criteria-j

Z is the final comparison score of *design 1* to *design 2*

From equation (2) then we have,

- Functionality: 0.175 x [0.5+0.5+0.5] = 0.2625
- Safety: 0.56 x [0.5+0.5+0.5] = 0.840
- Cost: 0.061 x [0.5+2.2+2.2] = 0.2745
- Flexibility: 0.204 x [0.5+0.5+1.0] = 0.408 respectively.

And from equation (3) Z is the summation score for *design 1* versus *design 2.*

- Z = (0.2625 + 0.840 + 0.2745 + 0.408)/3 = (1.785/3) = 0.595

Based on the synthesis of the criteria comparison results, it was found that the concept design 1 per concept design 2 was 0.595, which is not more than half (2 is half), therefore it can be concluded that the concept design 2 is appropriate to be used as a prototype concept for creating an intelligent control of an indoor modular power assist stretcher.

IOP Conf. Series: Materials Science and Engineering **705** (2019) 012027 doi:10.1088/1757-899X/705/1/012027

3. Prototype Architecture

The components to be provided for prototype assembly are presented here below:

- A Stepping motor is applied to drive each wheel (Front R-L and Rear R-L)
- Three - axis accelerometer is used for feeding back the angle of sloped floors.
- A vibration sensor 801S is used to feedback the signal for improving the smoothness of the stepping motor's movement.
- An Eight-axis hall effect sensor joystick is used for controlling the direction and velocity of stepping motor.
- An Arduino 2560 controller unit to apply PID[10] controlling algorithm.
- Transformer Power source which transforms the 220 VAC power to 24 VDC.

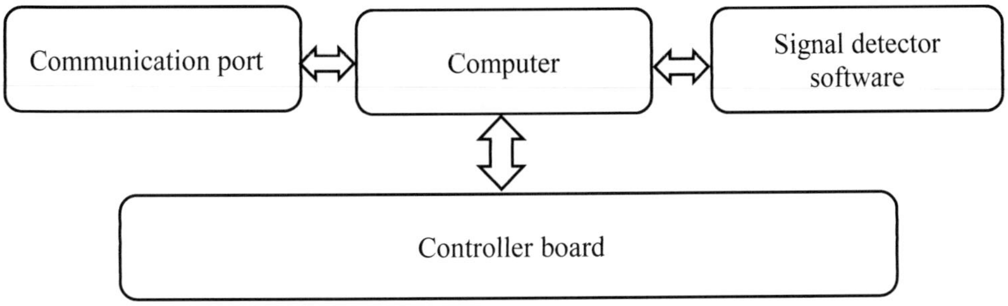

Figure 2. Communication diagram.

The software for detecting speed, acceleration and vibration signals has been created specifically for this work. The program analyzes the behavior of signals generated by the PID control system.

4. Experiment

The experiment is done on a fully loaded system (adding a person's weight) of approximately total 100 kilograms. Hereby one operator controls the stretcher with a joystick, to move on flat and sloping hospital floors.

A mobile computer used to observe the operation signal of joystick, sensor and the behavior of motors. Figure 4 and figure 5 shows a signal graph with movement zoning A,C,D and F resulting from moving forward on a flat floor. The results show that the stretcher can move in the specified direction with relatively stable speed and acceleration. The speed is quite uniform in every direction of movement. The acceleration is found to have a constant ripple caused by the automatic adjustment with every 2% per degree of inclination. The uneven experimental conditions cause the sensor to send acceleration compensation signals to the control circuit resulting in the signal as shown in the graph.

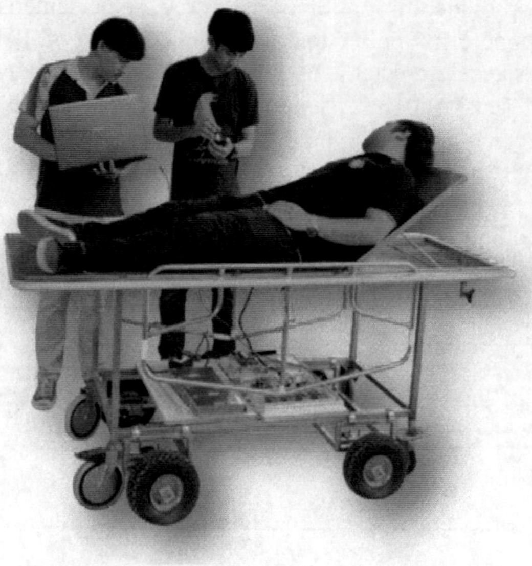

Figure 3. Testing of the control system and motion behaviors.

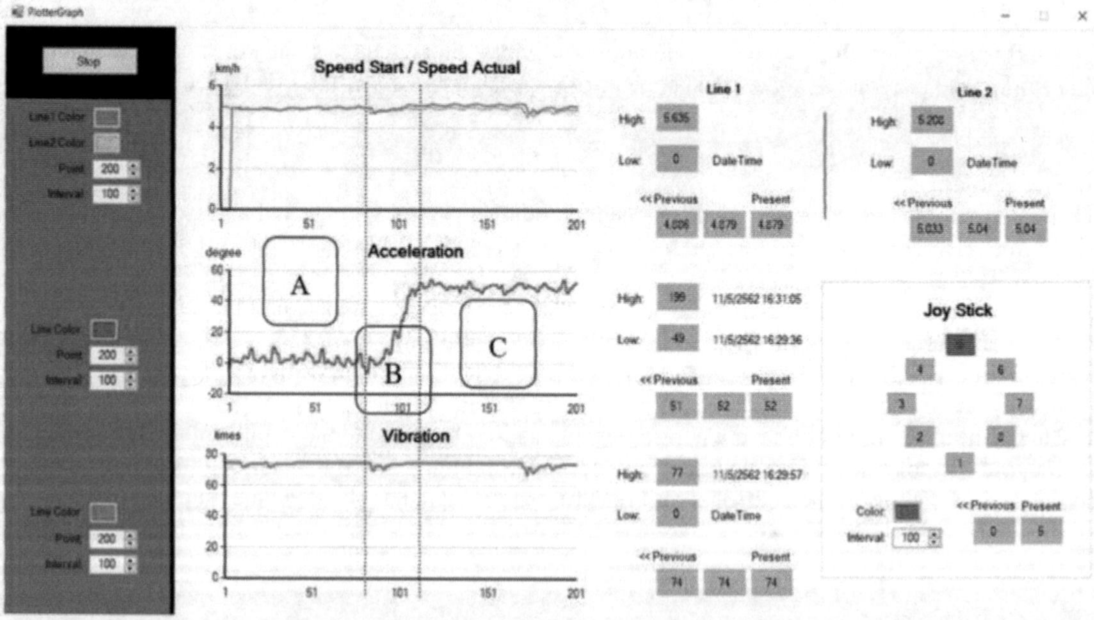

Figure 4. Proportions of the speed, acceleration and vibration while moving upward.

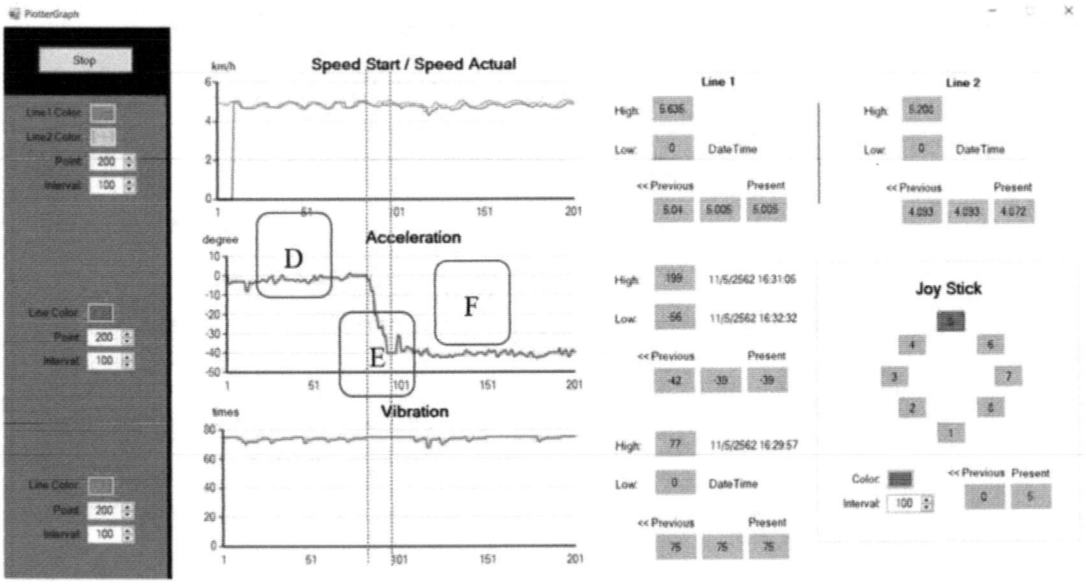

Figure 5. Proportions of the speed, acceleration and vibration while moving downward.

Figure 4 is signal graph for moving upward (forward) on a sloped floor.

Figure 5 is signal graph for moving downward (forward) on a sloped floor. It shows the speed, acceleration and vibration that can be measured from the stretcher movement on a floor with zoning as follows:

- Zone-A and Zone-D is the time when the stretcher moves on a flat area.
- Zone-B is the time period where the movement changes from a flat area to a steep slope in an upward direction. It is found that the acceleration will gradually increase according to the angle of the floor. The control system will maintain a constant speed throughout the movement.
- Zone-C is when the stretcher moves on a steep slope in an upward direction when the acceleration increases to the appropriate value level according to the slope of the floor. Acceleration is constant.
- Zone-E Displays the time zone where the movement changes from a flat floor onto the slope in the downward direction, which indicates that the acceleration will gradually decrease according to the angle of the floor.
- Zone-F is the time zone where the stretcher moves on a steep slope in the downward direction and the acceleration is reduced to the appropriate value level according to the slope of the floor. Acceleration is constant.

5. Conclusion

First conceptual design techniques were used to select the best concept for prototyping power assist systems. We evaluated the operational efficiency of an intelligent control of a modular power assist stretcher operated with an 8-way joystick with hall effect sensors. The PID control of the electric stepping motor (the wheel drive system) has resulted in smooth, non-jerky movement with proper speed and acceleration finely adjusted to the specific floor inclination. The studied power assist technology shows great promise as an effective system with large economic impact for the health care facilities choosing to implement this device. The movement assisted stretcher can operate on half the manpower that would be needed to move a non-assisted stretcher. Our suggestion is for this valuable system to be developed continuously.

References

[1] Tasaki R, Kitazaki M, Miura J and Terashima K 2015 Prototype design of medical round supporting robot 'Terapio' *IEEE International Conference on Robotics and Automation (ICRA)* 829–834

[2] Terashima K. et al. 2014 Medical Round Robot – Terapio *J. Robot. Mechatronics* **26** 112–114

[3] Damoto R and Hirose S 2002 Development of Holonomic Omnidirectinal Vehicle 'Vuton-II' with Omni-Discs *Journal of Robotics and Mechatronics* **14** 186-192

[4] Borgolte U, Hoyer H, Buhler C, Heck H and Hoelper R 1998 Architectural concepts of asemi-autonomous wheelchair *Journal of Intelligent and Robotic System* **22** 233-253

[5] Tadakuma K, Tadakuma R and Berengeres J 2007 Development of Holonomic Omnidirectional Vehicle with 'Omni-Ball': Spherical Wheels *Proc. IEEE/RSJ Int. Conf. on Intelligent Robots and Systems* 33-39

[6] Sarraj AR and Massarelli R 2011 Design History and Advantages of a New Lever Propelled Wheelchair Prototype *Int J Adv Robotic Sy* **8** 12-21

[7] Saaty and Thomas L 2008 Decision making with the analytic hierarchy process *International journal of services sciences* **1** 83-98

[8] Saaty and Thomas L 2000 Fundamentals of decision making and priority theory with the analytic hierarchy process *Rws Publications* **6**

[9] Urbano J, Kitagawa H, Miyoshi T and Terashima K 2004 Haptic feedback control of omni-directional wheelchair considering user's comfort and safety *Proceedings of the 35th International Symposium in Robotics*

[10] Pothi N 2016 Controller design and simulation of a one-degree-of-freedom power assist system for lifting objects *KKU Engineering Journal* **43** 367-369

IOP Conf. Series: Materials Science and Engineering **705** (2019) 012028 doi:10.1088/1757-899X/705/1/012028

Metal shape classification of buried object using multilayer perceptron neural network in GPR data

S N A M Kanafiah*, H Ali, A Z Ahmad Firdaus, M S Zanar Azalan, Y Jusman[3], A A Khairi, M R Ahmad[2], T Sara T Amran[2], I Mansor[2], S A A Shukor, E M Cheng, C Y Fook

[1]School of Mechatronic Engineering, Universiti Malaysia Perlis, 02600 Arau, Malaysia
[2]Agensi Nuklear Malaysia, Bangi, 43000 Kajang, Selangor, Malaysia
[3]Faculty of Engineering, Universitas Muhammadiyah, Yogyakarta, Indonesia

*Corresponding author: aqmariah@unimap.edu.my

Abstract. Ground penetrating radar (GPR) is a non-destructive tool for subsurface investigation becomes an emerging trend in many applications especially in detecting infrastructure utilities. However, interpreting of GPR data in terms of target's geometry is remained a challenging task due to the existence intra/inter-class variations and subtle changes of hyperbolic signature pattern in GPR data. To addressed this issue, this paper proposes to classify the basic shape (i.e. cubic, cylindrical and disc) of buried object using multilayer perceptron neural network. The GPR raw data firstly is pre-processed based on A-scan and B-scan GPR data. Then, the combined statistical features with gray level co-occurrence matrix (GLCM) and Hu moment invariants are extracted from A-scan and B-scan. The extracted features are then fed as input to multilayer perceptron to classify the shapes (cubic, cylindrical and disc) of buried object. Based on the experiment conducted, the highest recognition rate has been achieved is about 83.3%. Thus, the proposed method of using combined statistical with GLCM and Hu moment invariants features to classify shape of buried object using multilayer perceptron is promising.

1. Introduction

GPR is one of the most popular and reliable geophysical devices in shallow subsurface exploration. It successfully used in detecting and locating the subsurface utilities such as pipes and cables [1] and land mines [2]. It produces cross-sectional profile of subsurface without the need of digging or drilling on the ground surveyed. Figure 1 shows the basic principle of GPR system. It transmits short electromagnetic pulses of high frequency electromagnetic wave (EM) into the surveyed ground. These waves propagate with a velocity that depends on the dielectric property of medium. Pulses will be reflected back to the receiver when EM waves hit the materials with different dielectric constant [3]. When antenna is located above a target (i.e. at x_o), the travel time to the target will be minimum (i.e. t_o). Therefore, it creates a hyperbolic signature in GPR images.

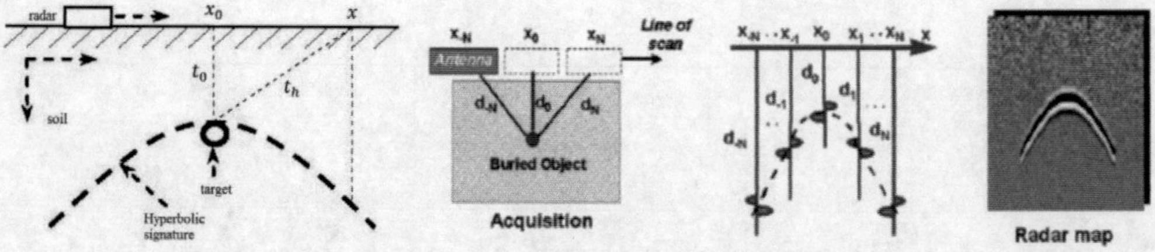

Figure 1. Basic principle of GPR system.

Basically, there are three methods in running the GPR system to acquire the data which are A-scan, B-scan and C-scan as illustrated in Figure 2 [4]. A-scan signal is created from time-series of reflected wave signal that collide with buried object. The consecutive patterns in A-scan can be concatenated to create images of the subsurface, called B-scan. Buried objects exhibit characteristic pattern in B-scan so called radargram. The C-scan is equal to a multiple parallel of B-scan results.

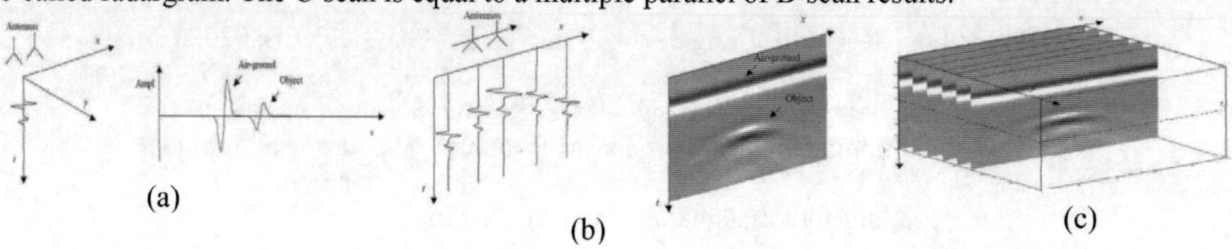

Figure 2. Representation of GPR system: (a) A-scan, (b) B-scan, & (c) C-scan.

In previous work, many techniques have been proposed in pattern recognition in order to detect and recognize the buried objects. Some of them adopted histogram of oriented gradient (HOG) [3, 5], scale invariant feature transform [6], finite different time-domain [7], and Hough transform [8, 9]. For example, Lee & Mokji [3] have used histogram of oriented gradient to automatically detect the hyperbolic signature in GPR data that contains the reflections from target of buried object by narrow down the data into possible reflections with the aid of linear SVM. Syambas [10] proposed decision tree method for predicting the shape and size of a buried object on surface GPR. Carlotto [8] have used Hough transform in GPR images to find the hyperbolic parameter for buried object. Although Hough transform was commonly used method in detecting hyperbolic signature, however this method is time consuming and low accuracy.

This paper presents classifying the metal shape of buried objects in GPR data using combined statistical features derived from A-scan and GLCM and Hu moments invariant features derived from B-scan. The extracted features then served as input to multilayer perceptron neural network to classify the metal shape of buried objects based on hyperbolic signature.

2. Methodology

Figure 3 illustrates the flowchart of the proposed method. It consists of data acquisition, pre-processing, feature extraction and classification. The detailed of each block is described at the subsequent sections.

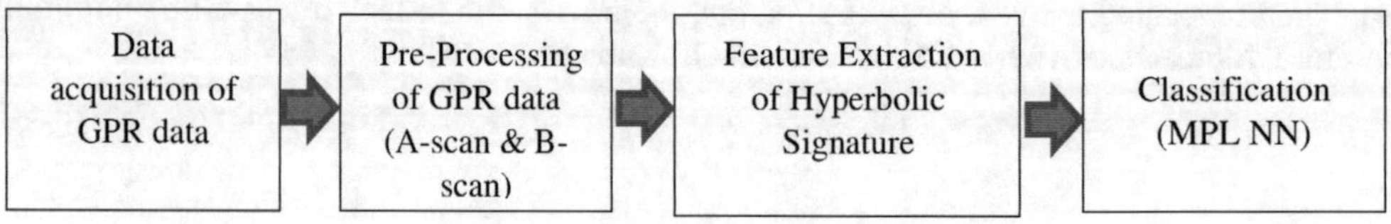

Figure 3. The flowchart of the proposed method.

2.1. Data Acquisition

This research consists of hardware and software development. In hardware development, preparation of samples is required and material of metal specifically aluminum was used in this investigation. Sample fabrication has focused on three basic shapes which are cubic, cylindrical and disc. For GPR system, antenna of 800 MHz transmitter and receiver of Ramac GPR Mala system was used to acquire the GPR images. Figure 4 shows the basic GPR components that consist of antenna, battery pack, control unit, and a portable PC.

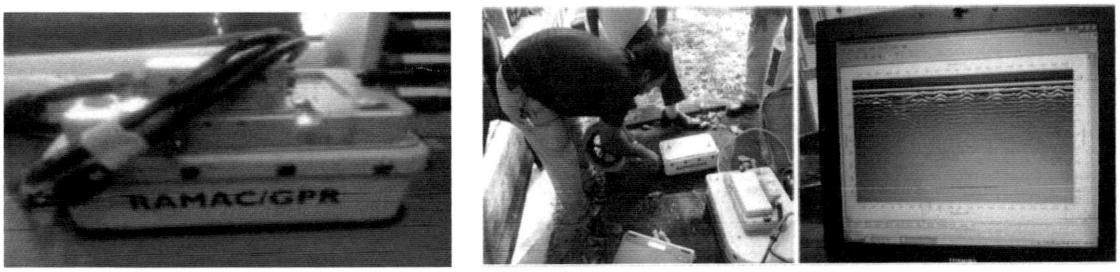

Figure 4. Basic component of GPR system.

The location of buried object is depicted in Figure 5. These samples consist of spherical, disc, cubic and cylindrical shape from material aluminium and perspex. These samples are buried in parallel with depth of 200 cm from the ground surface with 100 cm apart between them. Dry sand was used as soil medium. The operation GPR system was guided by research officer at Non Destructive Test (NDT), Nuclear Agency Malaysia for collecting the GPR data.

In this work, we proposed to investigate the behavior of three different shapes which are cubic, cylindrical and disc for material aluminium (metal) based on GPR images. About 150 GPR images have been recorded in which 50 cubic images, 50 cylindrical images and 50 disc images. Figure 6 illustrates the example of A-scan signal for three different shapes. Meanwhile, Figure 7 shows the B-scan images reveal unique hyperbolic signatures appear in GPR images with respect to the buried object.

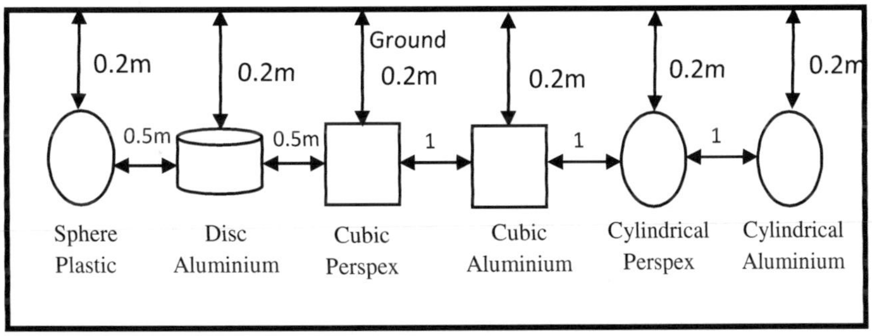

Figure 5. The position of the buried samples.

IOP Conf. Series: Materials Science and Engineering **705** (2019) 012028 doi:10.1088/1757-899X/705/1/012028

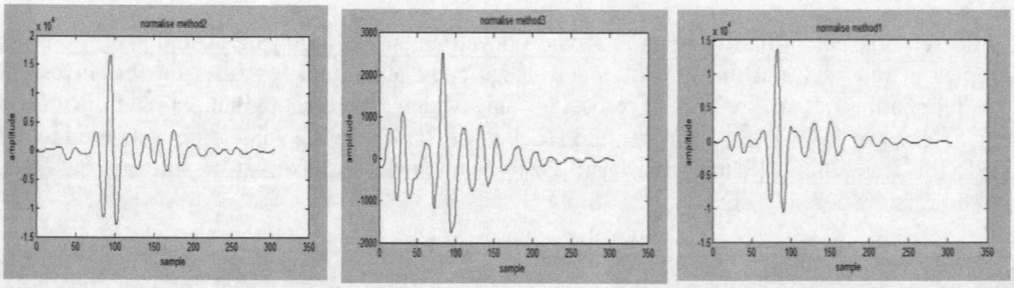

(a) (b) (c)

Figure 6. The example of A-scan signal of the buried objects based on the shape: (a) cubic, (b) cylindrical & (c) disk.

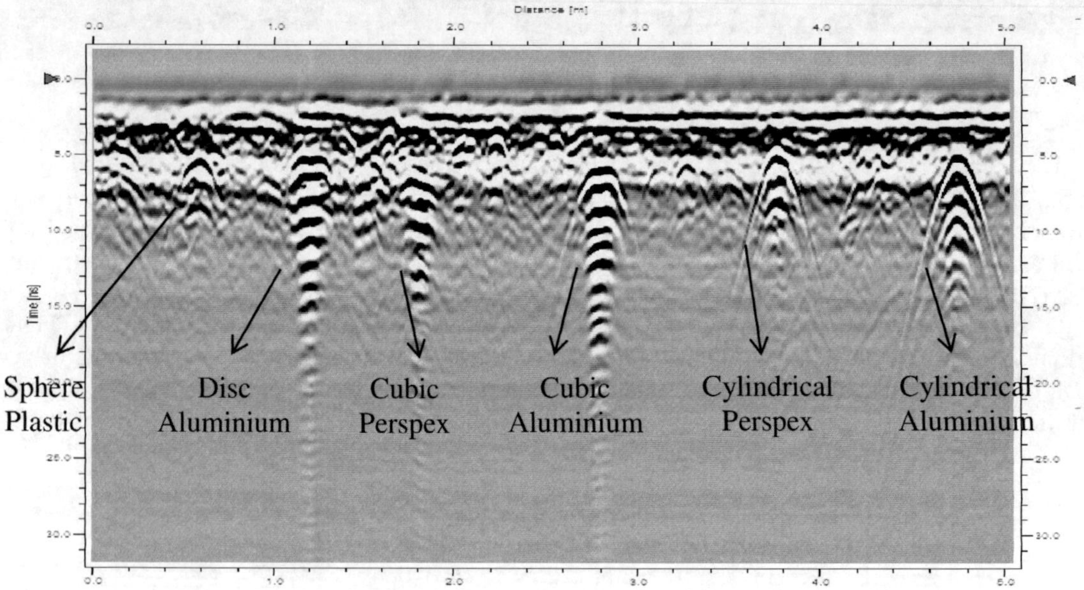

Figure 7. The B-scan of the buried objects reveal the hyperbolic signature pattern.

2.2. Signal & Image Pre-Processing

The acquired signal or image that contains irrelevant parts (noise) should be removed. For the purpose of analysis, the original GPR images needs to be pre-processed. This technique involved for signal pre-processing (A-scan) and image pre-processing (B-scan).

2.2.1. Signal Pre-processing. Figure 8 shows the pre-processed technique for A-scan signal. It consists of cropping, filtering, absolute and normalize of the signal. An example of pre-processed is shown in Figure 9. While for the image, filtering and cropping the specific region to extract the hyperbolic signature was done.

Figure 8. Signal preprocessing technique for the proposed method.

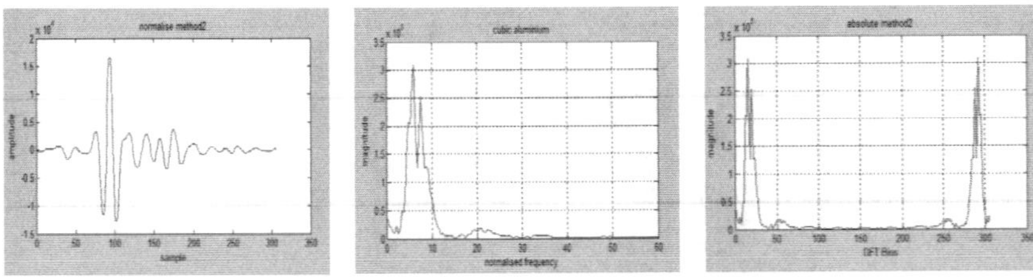

Figure 9. Pre-processing technique for cubic metal shape.

2.2.2. Image Pre-processing. The original GPR image was converted to grayscale image as illustrated in Figure 10. To reduce the noise and preserve useful details in the image, filtering technique was adopted in this work. In this work, we investigate three type of filters which are median filter, average filter and Gaussian filters. The median filter is a non-linear digital filtering technique often used to reduce the noise of an image and preserves useful details in the image, whereas average and Gaussian filter are linear filter. An example of filtered image for cubic metal shape is depicted in Figure 11. Then, the filtered images were further cropping to extract the hyperbolic signature of GPR images. Figure 12 shows the example of the cropped hyperbolic signature image.

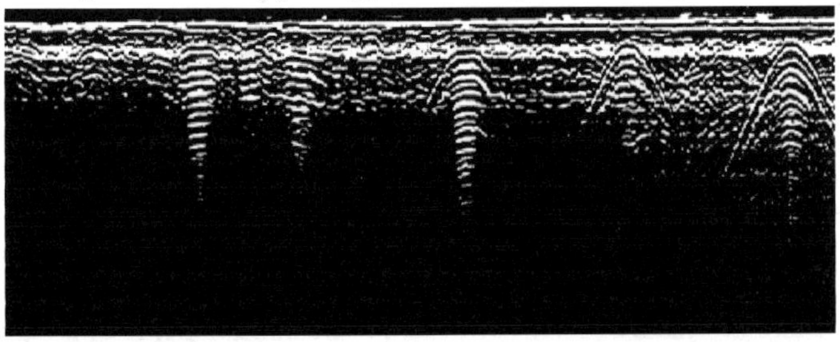

Figure 10. An example of grayscale image of radargram GPR images.

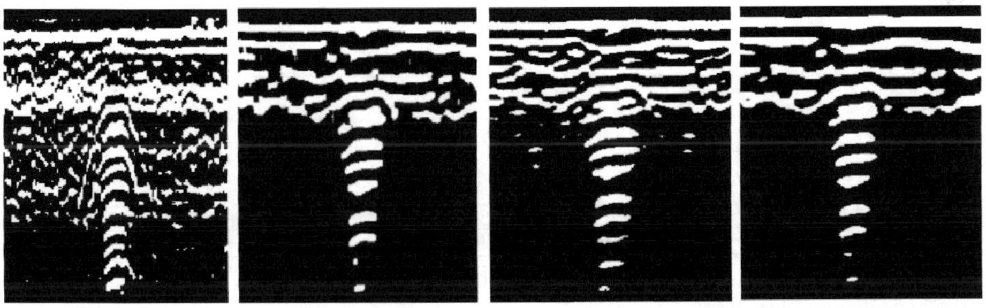

(a) (b) (c) (d)

Figure 11. The example of resultant filter on GPR image of cubic shape: (a) original image, (b) median filter, (c) average filter and (d) gaussian filter.

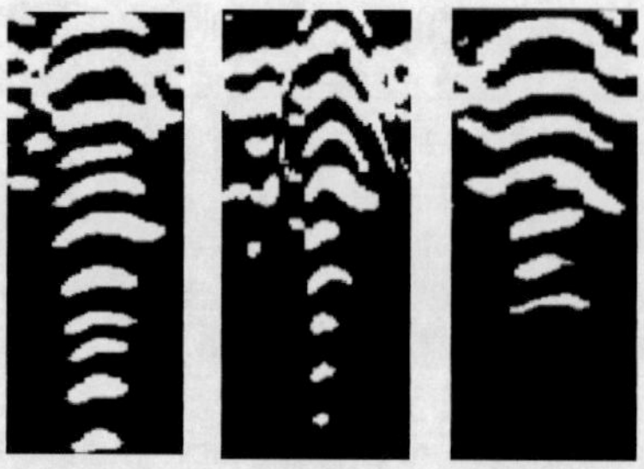

(a) (b) (c)

Figure 12. The radargram for three shapes: (a) cubic, (b) cylindrical, and (c) disc.

2.3. *Feature Extraction*

The hyperbolic signature of buried objects exhibits unique pattern with respect to the shape, size, materials and depth. Feature extraction needs to be done to extract important features. In this work, statistical features such as skewness, standard deviation, maximum value and total number of peaks were extracted from A-scan signal. While, texture and shape features of GLCM and Hu moment invariant were extracted from the hyperbolic signature.

2.3.1. GLCM. Gray Level Co-Occurence Matrix proposed by Haralick [11] consists of 24 parameters from co-occurrence matrix. The parameter such as energy, contrast, local homogeneity, and correlation will be used as features in this work.

2.3.2. Hu Moment Invariant. Moment invariants have been widely used to image pattern recognition in various applications due to its invariant features on image translation, scaling and rotation [12]. Moment can provide characteristics of an object that uniquely represent its shape. Hu introduced seven nonlinear functions which are invariants under object's translation, scale and rotation [12]. The set of seven third order boundary central moments are given by the following equations.

$$M1 = \eta_{20} + \eta_{02} \tag{1}$$

$$M2 = (\eta_{20} - \eta_{02})^2 + 4\,{\eta_{11}}^2 \tag{2}$$

$$M3 = (\eta_{30} - \eta_{12})^2 + (3\,\eta_{21} - \eta_{03})^2 \tag{3}$$

$$M4 = (\eta_{30} + \eta_{12})^2 + (\eta_{21} + \eta_{03})^2 \tag{4}$$

$$M5 = (\eta_{30} - 3\eta_{12})(\eta_{30} + \eta_{12})[(\eta_{30} + \eta_{12})^2 - 3(\eta_{21} + \eta_{03})^2] + 3(\eta_{21} + \eta_{03})(\eta_{21} + \eta_{03})[3(\eta_{30} + \eta_{12})^2 - (\eta_{21} + \eta_{03})^2] \tag{5}$$

$$M6 = (\eta_{20} - \eta_{02})[(\eta_{30} + \eta_{12})^2 - (\eta_{21} + \eta_{03})^2] + 4\,\eta_{11}(\eta_{30} + \eta_{12})(\eta_{21} + \eta_{03}) \tag{6}$$

$$M7 = (3\eta_{21} - \eta_{03})(\eta_{30} + \eta_{12})[(\eta_{30} + \eta_{12})^2 - 3(\eta_{21} + \eta_{03})^2] + (3\,\eta_{21} - \eta_{03})(\eta_{21} + \eta_{03})[3(\eta_{30} + \eta_{12})^2 - (\eta_{21} + \eta_{03})^2] \tag{7}$$

The set of seven boundary moments are treated as feature vectors. The moment 1 (M1) to moment 6 (M6) are moment that invariant to translation, size and rotation. There are analogous to the moment of inertia around the images centroid, where the pixels intensities are analogous to physical density. While moment 7 (M7) is skew invariant, which enables to distinguish mirror images of otherwise identical images. These features are then used for the classification.

2.4. *Multilayer Perceptron Neural Network.*

Artificial neural network is one of the techniques in artificial intelligence (AI) which has a functional imitation of the human brain. It is based on simulated nerve cells or neuron, which joined together to form networks. In general, an ANN is characterized by its architecture, learning algorithm and activation functions. Figure 13 shows the structure of multilayer perceptron neural network used in this project. It consists of an input layer that distributes the input data to the processing elements in next layer. The hidden layers where the nonlinear behavior comes from and the output layer. Both the input and output layers are directly accessible while the hidden layers are not. Each layer contains several processing units which are generally called neurons. In this work multilayer perceptron is used to classify different type of material. The MLP learning algorithm [13] works as follows:

1. Initialize the weights to random values on the interval [-1, +1].
2. Present the first training pattern, and obtain the output
3. Compare the network output with the target output.
4. Propagate the error backwards
   a. Correct the output layer of weights using the following formula.
   $$w_{ho} = w_{ho} + (\eta \delta_o o_h) \tag{8}$$
   where w_{ho} is the weight connecting hidden unit h with output unit o, η is the learning rate, o_h is the output at hidden unit h. δ_o is given by the following.
   $$\delta_o = \mathrm{o}_\mathrm{o}(1 - \mathrm{o}_\mathrm{o})(\mathrm{t}_\mathrm{o} - \mathrm{o}_\mathrm{o}) \tag{9}$$
   where o_o is the output at node o of the output layer, and $t-o$ is the target output for that node.
   b. Correct the input weights using the following formula.
   $$w_{ih} = w_{ih} + (\eta \delta_h o_i) \tag{10}$$
   where w_{ih} is the weight connecting node i of the input layer with node h of the hidden layer, o_i is the input at node i of the input layer, η is the learning rate. δ_h is calculated as follows.
   $$\delta_h = o_h(1 - o_h)\sum_o (\delta_o w_{ho}) \tag{11}$$
   where p is the number of units in the output layer.
5. Calculate the error, by taking the average difference between the target and the output vector. For example, the following function could be used.
   $$E = \frac{\sqrt{\sum_{n=1}^{p} (t_o - o_o)^2}}{p} \tag{12}$$
6. Repeat from 2 for each pattern in the training set to complete one *epoch*.
7. Shuffle the training set randomly. This is important so as to prevent the network being influenced by the order of the data.
8. Repeat from step 2 for a set number of epochs, or until the error ceases to change.

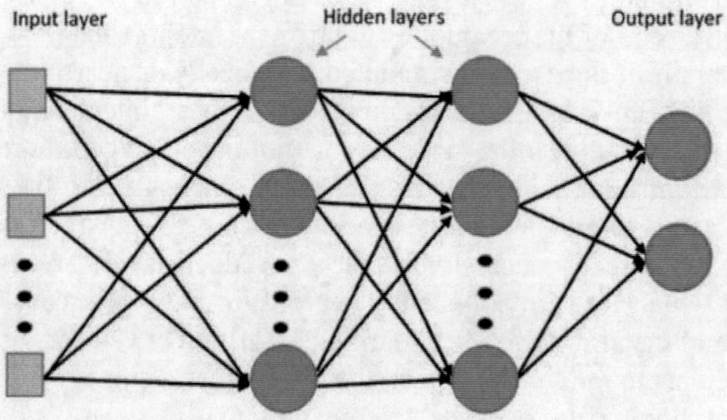

Figure 13. Structure of multilayer perceptron neural network.

3. Experimental results and discussion

To evaluate the effectiveness of the proposed method, 150 recorded GRP data have been used. Three basic shape (cubic, cylindrical and disc) shape from metal specifically aluminum has been adopted. Firstly, the GPR data need to pre-process so that any noise and irrelevant data can be minimized. The statistical features such as skewness, standard deviation, maximum value and total number of peaks were extracted. In addition, GLCM and Hu moment invariant features were extracted from GPR images based on hyperbolic signature. These features then were combined to form one dimensional feature vector that served as input to the multilayer perceptron (MLP). For the MLP NN structure, 50 hidden layers were used at 50 epochs. A series of experiments has been conducted using different learning algorithm such as Bayesian Regulation (BR), Scale Conjugate Gradient (SCG) and Lavernberg Marquardt (LM). In this experiment, the training and testing data was split into 3 strategies which are (90% training, 10% testing), (80% training, 20% testing) and (70% training, 30% testing).

Table 1 shows the average recognition rates based on the combined statistical features with GLCM and Hu moment invariant using MPL NN with different learning algorithms in classifying metal shape of buried object. It can be seen from Table 2 that all training data achieved perfect recognition rates. For testing performance, the proposed method achieved the highest recognition rate which is 83.30% using Scale Conjugate Gradient learning algorithm with (90% training, 10% testing) strategy. When we change the training and testing data to (80% training, 20% testing) the results slightly decrease to 81.30%. Based on the result, it shows that the combined features able to classify the metal shape of buried object in GPR data using MLP NN, thus it shows a promising result of the proposed method.

Table 1. The recognition rates of using combined statistical features with GLCM and Hu moment invariants. using MPL NN with different learning algorithm.

	L A	Data (%,%)	Recognition Rates (%)	
		(Tr, Ts)	*Tr*	*Ts*
MLP NN	BR	(90, 10)	100	66.00
		(80, 20)	100	70.70
		(70, 30)	100	62.20
	SCG	**(90, 10)**	**100**	**83.30**
		(80, 20)	100	81.30
		(70, 30)	100	77.20
	LM	(90, 10)	100	48.70
		(80, 20)	100	40.70
		(70, 30)	100	48.10

Note: LA (Learning Algorithm), BR (Bayesian Regulation), SCG (Scale Conjugate Gradient), LM (Lavernberg Marquardt), Tr (Training), Ts (Testing)

4. Conclusion

This paper has presented the classification of metal shape of buried object using multilayer perceptron based on the combined hyperbolic features extracted from GPR data. The combined of statistical features (skewness, standard deviation, maximum value and total number of peak) with GLCM and Hu moment invariants are extracted and used to classify the metal shape object using MPL NN with different learning algorithms. Based on the results, it shows that, the highest recognition rate which gives 83.30% has been achieved using Scale Conjugate Gradient learning algorithm with (90% training, 10% testing) strategy. Thus, the proposed method shows a promising result in classifying metal shape of buried objects specifically cubic, cylindrical and disc from GPR images. However, further study should be conducted to enhance the performance of the proposed method by utilizing optimization technique of the proposed method.

References

[1] Ni S H, Huang Y H, Lo K F and Lin D C 2010 Buried Pipe Detection by Ground Penetrating Radar Using The Discrete Wavelet Transform *Computers and Geotechnics* **37** 440–448

[2] Ishitsuka K, Iso S, Onishi K and Matsuoka T 2018 Object Detection in Ground-Penetrating Radar Images Using a Deep Convolutional Neural Network and Image Set Preparation by Migration *Hindawi International Journal of Geophysics* 1–8

[3] Lee K L and Mokji M 2014 Automatic Target Detection in GPR Images Using Histogram of Oriented Gradients (HOG) *2nd International Conference on Electronic Design (ICED)* 181–186

[4] Gonzalez-huici M A and Giovanneschi F 2013 A combined strategy for landmine detection and identi fi cation using synthetic GPR responses *J. Appl. Geophys.* **99** 154–165

[5] Torrione P A, Morton K D, Sakaguchi J R and Collins L M 2014 Histograms Of Oriented Gradients For Landmine Detection In Ground-Penetrating Radar Data *IEEE Transactions On Geoscience And Remote Sensing* **52** 1539–1550

[6] Harkat H, Elfakir Y, Bennani S D, Khaissidi G and Mrabti M 2016 Ground Penetrating Radar Hyperbola Detection Using Scale-Invariant Feature Transform *2nd International Conference on Electrical and Information Technologies ICEIT* 392–397

[7] Irving M J and Knight R 2006 Numerical Modelling Of Ground-Penetrating Radar In 2-D Using Matlab *Computers & Geosciences* **32** 1247–1258

[8] Carlotto M J 2002 Detecting Buried Mines in Ground Penetrating Radar Using a Hough Transform Approach *Battlespace Digitization and Network-Centric Warfare II, International Society for Optics and Photonics* **4741** 251–261

[9] Simi A, Bracciali S and Manacorda G 2008 Hough Transform Based Automatic Pipe Detection for Array GPR: Algorithm Development And On-Site Tests *IEEE Radar Conference* 1–6

[10] Syambas N R 2012 An Approach for Predicting the Shape and Size of a Buried Basic Object on Surface Ground Penetrating Radar System *Hindawi Publishing Corporation International Journal of Antennas and Propagation*

[11] Haralick R M, Shanmugam K and Dinstein I 1973 Textural features for image classification *IEEE Transactions on Systems, Man and Cybernetics* **3** 610–621

[12] Vanajakshi B 2012 Classification Of Boundary And Region Shapes Using Hu-Moment Invariants **3** 386–393

[13] Noriega L *Multilayer Perceptron Tutorial* https ://www.academia.edu/ 5142020/ Multilayer_Perceptron_Tutorial

IOP Conf. Series: Materials Science and Engineering **705** (2019) 012029 doi:10.1088/1757-899X/705/1/012029

Development of MOX Gas Sensors Module for Indoor Air Contaminant Measurement

A N Abdullah[1,2], K Kamarudin*[1,2], S M Mamduh[1,2] and A H Adom[1,2]

[1]School of Mechatronics Engineering, Universiti Malaysia perlis (UniMAP), Arau, Malaysia,
Centre of Excellence for Advanced Sensor Technology (CEASTech), Universiti
[2]Malaysia Perlis, Arau, Malaysia

*Corresponding auther : kamarulzaman@unimap.edu.my

Abstract. This paper presents the development of gas sensor module consisting of metal oxide (MOX) gas sensors for indoor air contaminant measurement. The first phase of the studies involves the design of PCBs board for six different metal oxide (MOX) gas sensors particularly CCS803, MiCS5524, GM-402B, GM-502B, GM-702B and MiCS6814. Next, the main board consisting of temperature, humidity and another two MOX gas sensors (i.e TGS2600 and TGS2602) was developed for the acquisition of the sensors. A partially closed chamber was designed and fabricated to fit the main board and allows inflow and outflow of gases. The responses of the gas sensors were measured in a closed room, with the air conditioner turned on and off periodically. The results show that all the MOX gas sensor used are affected by temperature and humidity of the environment. Therefore, the sensor response drift needs to be corrected in order to obtain reliable indoor air contaminant measurement.

1. Introduction

In this era, air contamination is one of the most concern issues that will lead to serious health diseases. According to the report from World Health Organization (WHO), 1.7 million of children died due to the exposure of contaminated environment [1]. The long-term exposure to the polluted environment will cause the life threatening disease. In 2016, Health Effect Institute stated that 6.1 million of people died from lung disease, lung cancer, heart attacks and strokes which caused by the air pollution [2]. Moreover, air pollution is not only limited to just outdoor, but also affects indoor environment, in which the impact is more serious to human.

The indoor air quality (IAQ) issue rise by many factors especially by the human daily activities without noticing it will lead to the pollution of indoor environment. The product that has been used such as air freshener, insect repellent, household cleaning reagent excessively causing the degradation of indoor air quality. Besides, human always neglects indoor air quality issues since it could hardly detected without the help of certain sensors. Some air pollution is odorless and could only be detected by the help of some type of sensing system. Therefore, researcher has been studying on the indoor air quality system that can help in monitoring indoor environment [3].

Volatile organic compound can be listed as one of the main pollutants that exist in atmosphere, which can cause chronic health disease and may also lead to the formation of ozone and mists. Volatile organic compound can be found in many household items and it can easily vaporize at room temperature and pressure which will lower the level of indoor air quality[4]. For example, cleaning reagent that is used in daily activities is the major source of ethanol gas in indoors. It is important to perform indoor air quality monitoring to track on the quality of the air [5].

As technology nowadays has growth exponentially, there is numbers of affordable gas sensor available in market such as resistive sensors, electrochemical sensors, non-dispersive infrared (NDIR) detectors and photoionization detectors (PIDs) [6]. Among all of the gas sensors, resistive metal oxide gas sensor (MOX) has been widely used due to lower in price, smaller in size where offer a lightweight design, low consumption and long-life [7]. The high sensitivity to a large numbers of gases and have a quick responds time increase the market of this type of gas sensor. Due to the ability of the MOX sensor, it has been used in many applications of environmental monitoring [8]. The basic design of MOX sensor is a thin films of metal oxide deposited on a silicon substrate and reaction occur on the surface of the MOX sensors causing a variation of sensing element [9]. This will act as an electrical resistive transducer and the resistance is proportional to the number of molecules absorbed. The MOX gas sensor resistance value will fall when exposed to gas, this fall of resistance is non-linear with increasing gas concentration in the gas measurement [10].

In previous studies, researchers have been discussing on the limitations of MOX sensor, and shows that this type of sensors have limited sensitivity, high power consumption and temporal drift. Kamarudin et al. discussed on the effect of ambient temperature and relative humidity on the gas sensor response and gas distribution map [11]. On another study, he stated that the operating temperature of gas sensor's sensitive layer should be kept between 200-500°C to improve the selectivity and respond rate of the gas sensor. Moreover, the humidity effect on the gas sensor response can be reduced when the temperature is maintained well above 100°C [12]. Therefore, the present study focuses on the development of the sensor's board with chamber for monitoring volatile organic compound (VOC). The aim is to study the respond of different MOX gas sensors in normal indoor environment (i.e. with and without air conditioning) where temperature and humidity fluctuates.

2. Gas Sensor Circuit Development

This section presents the process of developing different MOX gas sensors circuits which include the sensor selection, PCB design and fabrication.

2.1 Selection of Gas Sensors

Most of the gas sensors are not suitable for this experiment due to their lack of volatile organic compounds (VOCs) detection and their concentration limits, where some others can fit such as metal oxide (MOX). There are several reasons why MOX gas sensors are the best selection for this study, which is relevant to (VOCs) and their detection ranges. Table 1 shows eight MOX sensors selected for this study and their features.

These types of sensors were chosen due to their ability to give the desired response to the set parameters; temperature, humidity and gases type, in a way to meet the scope limitation of this study. These gas sensors were proven to provide a reliable result and are widely used due to their stability and relatively high sensitivity compared to other models. According to the datasheet, all the sensors are able to detect VOCs and therefore is deemed suitable for the requirement. The detection range of the sensors is also low (i.e below 1000ppm) which is suitable for indoor air contaminant measurement application.

IOP Conf. Series: Materials Science and Engineering **705** (2019) 012029 doi:10.1088/1757-899X/705/1/012029

Table 1. Selecting a gas sensors for the PCB board

Sensor type	Target gases	Detection range	Features
GM-502B [12]	Carbon monoxide (CO) Nitrogen dioxide (NO3) Ethanol (C2H5OH) Hydrogen (H2) Propane (C3H8) Methane (CH4)	1-1000ppm 0.005-10ppm 10-500ppm 1-1000ppm >1000ppm >1000ppm	• Low power consumption. • High sensitivity. • Fast response. • Simple drive circuit.
GM-402B [14]	Methane (CH4) Propane (C3H8)	1-1000ppm 1-5000ppm	• Low power consumption. • High sensitivity. • Fast response. • Simple drive circuit.
MiCS5524 [15]	Carbon monoxide (CO) Ethanol (C2H5OH) Hydrogen (H2) Methane (CH4)	1 – 1000ppm 10 – 500ppm 1 – 1000ppm >1000ppm	• Smallest footprint for compact design. • Robust MEMS sensor for harsh environments. • High-volume manufacturing for low-cost applications.
MiCS6814 [16]	Carbon monoxide (CO) Ethanol (C2H5OH) Hydrogen (H2) Methane (CH4) Propane (C3H8)	1 – 1000ppm 10 – 500ppm 1 – 1000ppm 1-500ppm >1000ppm	• Smallest footprint for compact design. • Robust MEMS sensor for harsh environments. • High-volume manufacturing for low-cost applications.
GM-702B [17]	Methane (CH4)	1-500ppm	• Low power consumption. • High sensitivity. • Fast response and resume. • Simple drive circuit.
CCS803 [17]	Ethanol (C2H5OH)	5 – 10ppm	• low power consumption. • High sensitivity to Ethanol.
TGS2600 [19]	Methane Carbon Butane Ethanol Hydrogen	1-100ppm	• Low power consumption. • High sensitivity to gaseous air contaminants. • Long life and low cost. • Uses simple electrical circuit. • Small size.
TGS2602 [20]	Methane Carbon Ethanol Hydrogen	1-50ppm	• High sensitivity to VOCs. • Low power consumption. • High sensitivity. • contaminants Long life. • Use simple electrical circuit. • Small size.

2.2. Gas Sensor's Board Design

2.2.1 Schematic design and PCB layout design

Figure 1 and Figure 2 shows schematic diagram and the PCBs layout of six PCB circuits for CCS803, MiCS552, GM-402B, GM-502B, GM-702B and MiCS6814. The reason of designing individual board for each sensor is because in the future if any of the sensors mounted onto the board is damaged there will be a difficulty replacing it since it is applied to the board using surface-mount technology (SMT). Furthermore, if it was designed without modularity in mind, losing one sensor causes the whole board to be unusable. To avoid such case, it is preferable to build and develop separate boards for each gas sensor then connect them all into one main prototype. This procedure helps in different aspects in addition to the one mentioned earlier. Tests on each individual sensor can be performed to ensure its functionality and efficiency. In this case, if any of the sensors shows malfunction after the test then it can be replaced with relative ease. The arranging of the small PCB gas sensors on main board, after knowing all sensors are checked, is easier where all sensors are set in two lines near each other in order to ensure that they are exposed to the gas and each sensor gets to detect it simultaneously.

.

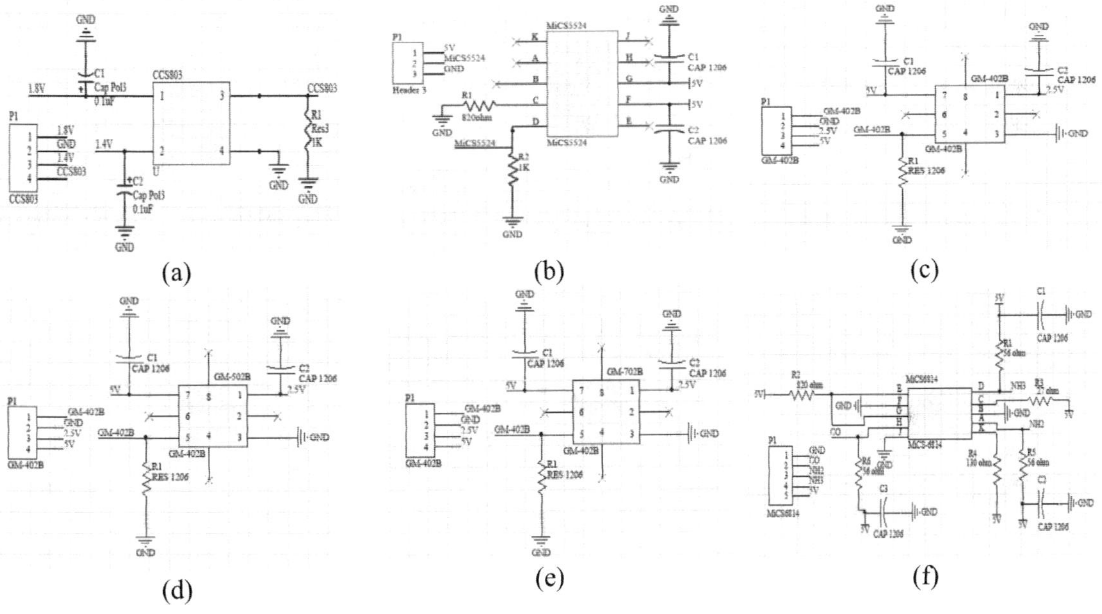

Figure 1. Schematic circuit of several types of the gas sensor's (a) CCS803 (PCBV1), (b) MiCS5524 (PCBV2), (c) GM-402B (PCBV3), (d) GM-502B (PCBV4), (e) GM-702B (PCBV5) and (f) MiCS6814 (PCBV6).

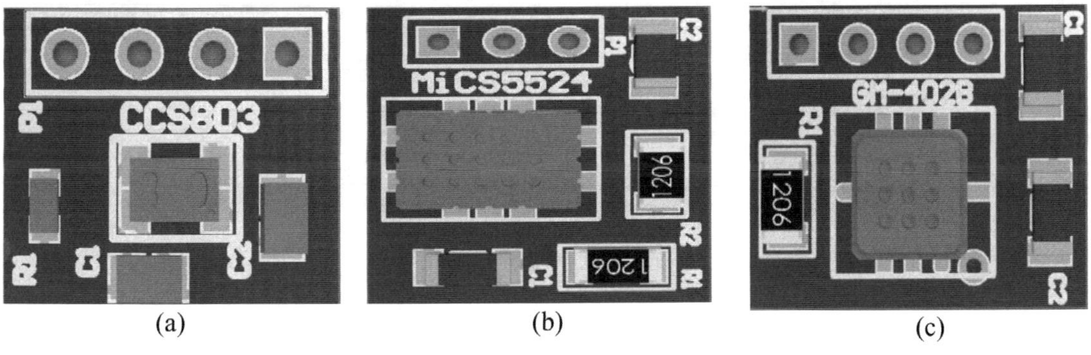

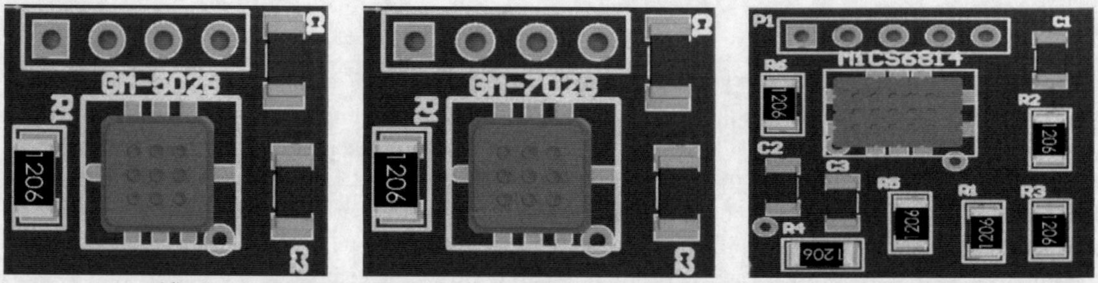

(d) (e) (f)

Figure 2. PCBs layout design of several types of the gas sensor's (a) CCS803 (PCBV1), (b) MiCS5524 (PCBV2), (c) GM-402B (PCBV3), (d) GM-502B (PCBV4), (e) GM-702B (PCBV5) and (f) MiCS6814 (PCBV6).

2.2.2 Main PCB board design

This section describes the development of main PCB board that includes TGS2600, TGS2602 gas sensor, temperature sensor and humidity sensor. Figures 3 and 4 show the schematic diagram and PCB design for the main board respectively. To ensure that the flow of gas crosses smoothly on sensors' sensitive part, the SMD gas sensors (refer Section 2.2.1), TGS2600 & TGS2602 sensors were arranged at similar height and close to each other to ensure they are induced with similar gas concentrations Temperature and humidity sensors were added to enable other parameters related to the environment in proximity to the gas sensors to be captured/recorded. Multiple voltage regulators are used to provide the sensors with correct DC Voltages. The circuit is designed on a double-sided PCB. A double-sided PCB consists of two copper layers on both sides of the substrate layers. The double-sided PCB improves the component density used on the board and minimizes the circuit traces needed in a single sided PCB.

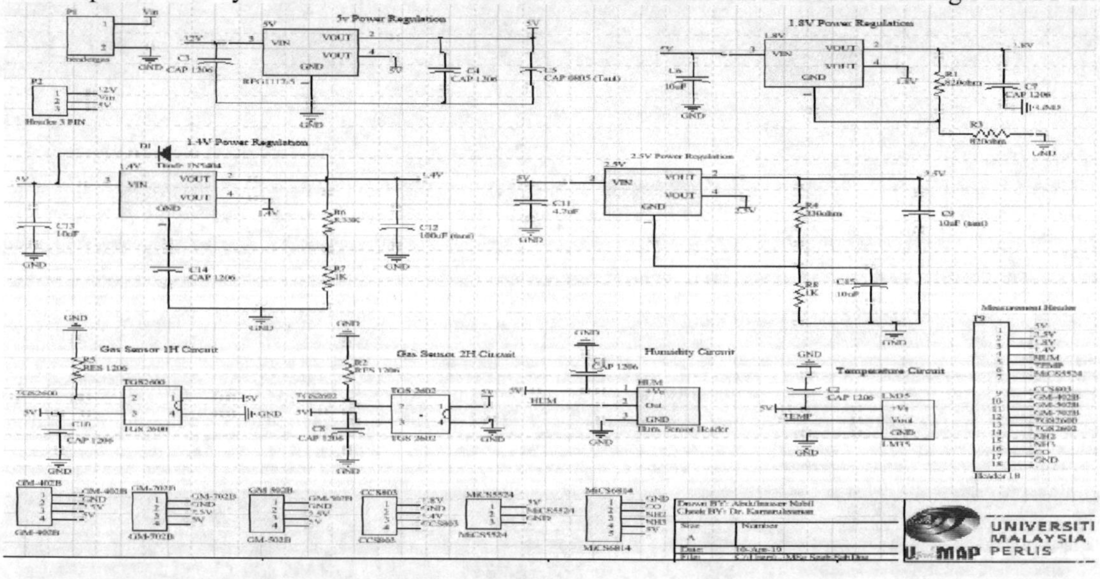

Figure 3. Complete main Schematic design of the gas sensors.

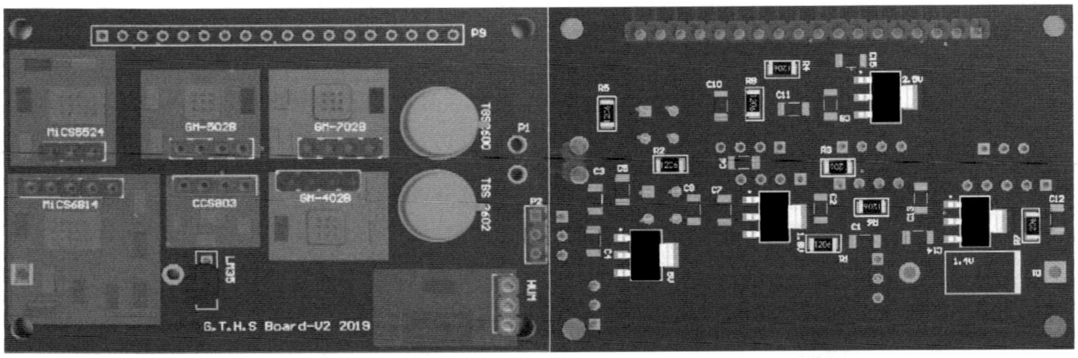

(a) (b)

Figure 4. (a) top layer view (b) bottom layer view of the complete PCB board with sensors.

3. Closed Chamber design

Figure 5 shows a partially closed chamber that will be used to control the inflow and the outflow of the gas. It allows the measurement of low flow rates accurately since; the gas is less disturbed by the environment. Additionally, the use of a chamber will eliminate unwanted effects from the environment. Note that the chamber consists of air strips at the top cover that allows the exchange of the air in the chamber and the air outside the chamber. This design will ensure that the temperature will not rise beyond the desired temperature due to the built-in heater of the gas sensors.

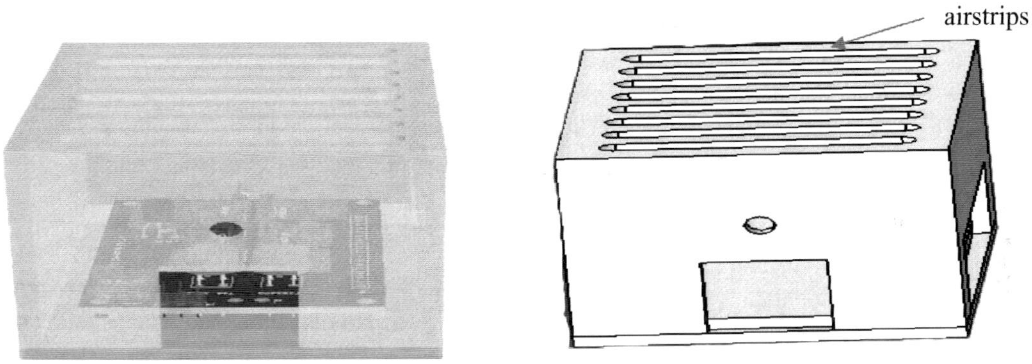

(a) **(b)**

Figure 5. A partially closed chamber was designed to allow inflow and outflow of gas in and out: (a) shows the placement of the PCB board inside the chamber, and (b) shows the external chamber

4. Experimental Results

This section describes the results of the two experiment conducted using the integrated gas sensors board. The gas sensor board was tested placed inside the chamber as well as without the chamber. The experiment was conducted in a closed room where the air conditioner was switched on and off periodically for every 2 to 3 hours. The NI USB-6211 was used to collect the data and displayed on LabVIEW front panel. Figure 6 shows temperature, humidity and gas sensors resistance with respect to time for both experiments.

IOP Conf. Series: Materials Science and Engineering **705** (2019) 012029 doi:10.1088/1757-899X/705/1/012029

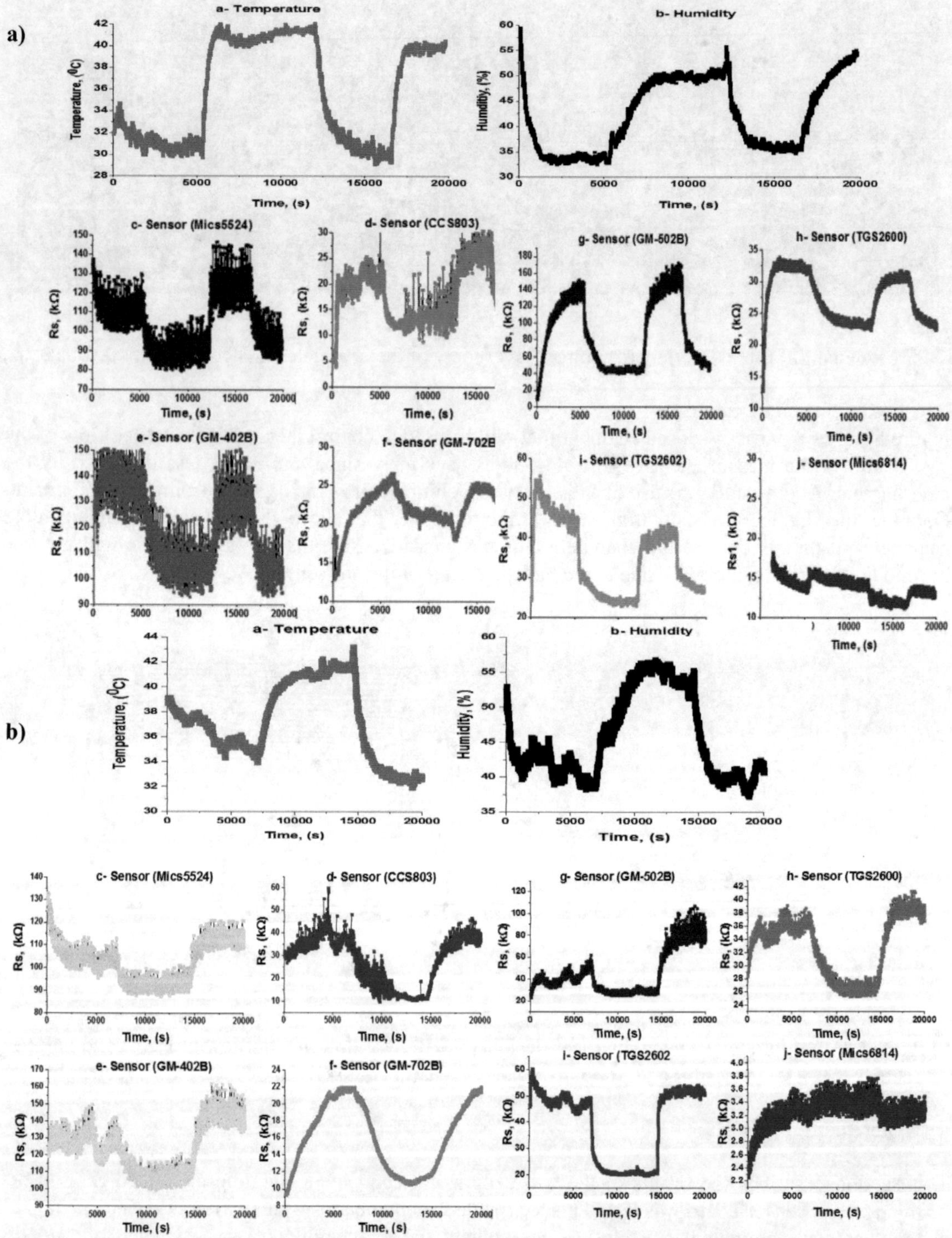

Figure 6. The plots for temperature, humidity and gas sensors response (Rs) with respect to time where a) the experiment was conducted without chamber b) the experiment was conducted using the chamber.

Referring to Figure 6, the temperature and humidity for both experiments fluctuate according to the air-conditioning. When turned on, the temperature and humidity drops and opposite behavior is observed while it is turned off. The temperature and humidity for both approaches, with or without the chamber shows similar levels since it has airstrips (refer Figure 5) s. All sensors are affected by temperature and humidity where the combined increase in temperature and humidity generally cause the sensor response to significantly decrease (though there is no gas release) and vice versa. It is important to note that the plots obtained using the chamber generally contains less noise than without the chamber. Moreover, the GM-502B, GM-702B, TGS2600 and TGS2602 gas sensors are seen to have lesser noise for both experiments conducted with and without chamber. Overall, the outcomes suggest that the correction to temperature and humidity drift is required to improve the reliability of the metal-oxide gas sensor response.

5. Conclusions

In this paper, the integrated gas sensor module was developed for indoor air contaminant monitoring applications. The metal-oxide (MOX) gas sensing technology has been reported as the most widely used technology in monitoring. These sensors are however cross-sensitive to the temperature and humidity of the environment. Therefore, to verify and study the significance of the cross-sensitivity problem, PCB boards containing different MOX gas sensors have been developed together with partially closed chamber. The experiment on the system was conducted in closed room with and without chamber under different air-conditioning control. It was found that all the MOX gas sensors are significantly affected by temperature and humidity of the environment, where the responses are more stable (i.e less noise) when the board is placed inside the chamber. Therefore, it is recommended that these drifts are corrected in order to obtain reliable indoor air contaminant measurement.

6. Acknowledgement

The author would like to acknowledge the support from the Fundamental Research Grant Scheme (FRGS) under a grant number of FRGS/1/2018/TK04/UNIMAP/02/15 from the Ministry of Education Malaysia.

References

[1] Anon WHO | World Health Organization
[2] Niall McCarthy Air Pollution Contributed To More Than 6 Million Deaths In 2016 [Infographic]
[3] Abraham S and Li X 2014 A cost-effective wireless sensor network system for indoor air quality monitoring applications *Procedia Comput. Sci.* **34** 165–71
[4] Joshi S 2008 The sick building syndrome *Indian J. Occup. Environ. Med.* **12** 61
[5] Burgués J and Marco S 2018 Low power operation of temperature-modulated metal oxide semiconductor gas sensors *Sensors (Switzerland)* **18**
[6] Aleixandre M and Gerboles M 2012 Review of Small Commercial Sensors for Indicative Monitoring of Ambient Gas **30**
[7] Huerta R, Mosqueiro T, Fonollosa J, Rulkov N F and Rodriguez-Lujan I 2016 Online decorrelation of humidity and temperature in chemical sensors for continuous monitoring *Chemom. Intell. Lab. Syst.* **157** 169–76
[8] Peterson P, Aujla A, Grant K, Brundle A, Thompson M, Vande Hey J and Leigh R 2017 Practical Use of Metal Oxide Semiconductor Gas Sensors for Measuring Nitrogen Dioxide and Ozone in Urban Environments *Sensors* **17** 1653
[9] Romain A C and Nicolas J 2010 Long term stability of metal oxide-based gas sensors for e-nose environmental applications: An overview *Sensors Actuators B Chem.* **146** 502–6
[10] Fine G F, Cavanagh L M, Afonja A, Binions R, Fine G F, Cavanagh L M, Afonja A and Binions

IOP Conf. Series: Materials Science and Engineering **705** (2019) 012029 doi:10.1088/1757-899X/705/1/012029

R 2010 Metal Oxide Semi-Conductor Gas Sensors in Environmental Monitoring *Sensors* **10** 5469–502
[11] Kamarudin K, Mamduh S M, Md Shakaff A Y, Saad S M, Zakaria A, Abdullah A H and Kamarudin L M 2015 Flexible and autonomous integrated system for characterizing metal oxide gas sensor response in dynamic environment *Instrum. Sci. Technol.* **43** 74–88
[12] Anon 2017 *MEMS VOC Gas Sensor GM-502B*
[13] Combustible M and Sensor G 2015 MEMS Combustible Gas Sensor GM-402B
[14] Sheet D The MiCS-5524 is a compact MOS sensor 1–5
[15] SGX sensortech 2017 MiCS-6814 Data Sheet, 1143 rev 8 1–5
[16] Sensor G 2017 MEMS Carbon Monoxide Gas Sensor GM-702B
[17] Anon 2015 Datasheet for CCS803 Ultra-low power gas sensor for Ethanol detection 0–3
[18] Figaro 2005 Tgs 2600 *Datasheet* 1–2
[19] FIGARO PRODUCT INFORMATION TGS 2600 - for the detection of Air Contaminants *Power*

IOP Conf. Series: Materials Science and Engineering **705** (2019) 012030 doi:10.1088/1757-899X/705/1/012030

Orientation Correction of Kinect's 3D Depth Data for Mapping

A A A Mosed[1], K Kamarudin[*1,2] and S Puveneswari[1]

[1]School of Mechatronics Engineering, Universiti Malaysia Perlis (UniMAP), Arau, Malaysia.
[2]Centre of Excellence for Advanced Sensor Technology (CEASTech), Universiti Malaysia Perlis, Arau, Malaysia.

[*]Corresponding author: kamarulzaman@unimap.edu.my

Abstract. Simultaneous localization and mapping (SLAM) is considered to be one of the primary tasks for autonomous robot navigation process. The first element required by the robot to start the navigation process is to create and assemble an accurate map of the surroundings or ground plane to envision its location. Mobile robots that use 1D or 2D sensors such as laser range finder, ultrasonic or sonar for the mapping process cannot provide enough information on the obstacle locations while 3D sensors like Kinect can. However, if the 3D sensor is not mounted correctly, the orientation of the 3D image obtained will also be affected and thus producing inaccurate maps. In this project, Microsoft Kinect was used to scan the environment. RANSAC algorithm was implemented to detect the ground plane. The orientation of the ground plane was corrected, and the depth data acquired by Kinect was then converted into 2D map. It was found that the methods applied have successfully mapped the obstacles detected.

1. Introduction

Simultaneous localization and mapping (SLAM) is considered to be one of the primary tasks to allow autonomous robot navigation. This mobile robotics field was extensively studied in the last twenty years. Even though a number of researches have been done on this particular application, various problems and challenges still exist. Researchers have been doing sustained efforts to make sure that the navigation system more reliable as any failure can cause harm to human beings [1],[2],[3].

Navigation is a technique applied on a robotic or a gadget for the purpose of allowing it to navigate and move from an area to any other targeted place. Although this problem looks to be straightforward, it is a bit difficult in reality due to the sensors and actuators are behaving non-ideally [1].

To start navigation, the first element a robot usually needs is to create and assemble a map of the surroundings or ground plane. To envision its location, the robot requires an accurate map meaning that both approaches need to be held at the same time. The map is vital to a success navigation procedure, in which the generated map ought to seize enough obstacles to save the robot from collision. The most common sensors that have been used for obstacle detection and avoidance were single range measurement sensors like laser range finders or sonar, or 3D ones like Xbox Kinect.

The possibility of missing an obstacle or misinterpret its location tends to be high with the usage of one dimensional or two-dimensional sensors for the mapping process. It has been proven that the generated maps using those sensors were not efficient as obstacles vary in shape and size. To elaborate more, Figure 1 shows a case where object size needs to be considered during mapping process. For

IOP Conf. Series: Materials Science and Engineering **705** (2019) 012030 doi:10.1088/1757-899X/705/1/012030

this case, the 1D or 2D sensors will not be able to detect the front obstacle, which will lead to a collision during the robot navigation. To overcome this limitation, usage of 3D sensor is believed to be ideal because both shape and size of the object can be obtained. Microsoft Kinect is one of the available 3D sensors that can be used for this purpose [1].

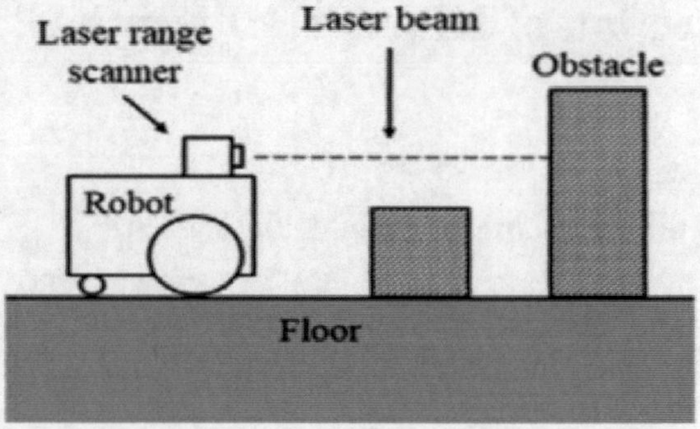

Figure 1. Effect of using a robot equipped with 1D or 2D sensors [1].

Kinect has been used instead of the laser range finder to extract object data in indoor environments. Having a three-dimensional view of the surrounding can still be done by rotating laser range finder; however, Kinect sensor is preferable due to its simple usage and advantages. Besides, RGB image and skeleton data, Kinect sensor can also provide the depth image which then can be manipulated to extract helpful information like the point cloud, object shape and laser scan data [4].

2. Previous Work

The conversion of 3D depth data to 2D obstacle location has been done by developing and implementing several algorithms and methods, however problems like detecting the ground plane and correcting the 3D depth data orientation is still a problem that is faced during the conversion process. Building maps during the navigation process was done by other researchers using several types of sensors. Those sensors are 2D ones like Ultrasonic, sonars and laser scan finder or 3D ones like Kinect. Some studies have utilized ultrasonic and sonars to perform 2D mapping. The results were not efficient because both systems that used ultrasonic and sonars can detect the objects that are only on the same level as the sensor. This limitation is undesired since any collision can happen at any instance due to the existence of the obstacles that are either above or below the sensor level [5],[6],[7]. Another study utilized Kinect and Laser scanner to perform 2D SLAM [8]. Although the combination of Kinect and laser has improved the accuracy of detecting obstacles, there are still some deficiencies and cost issues. Even though a 3D sensor was used, the system still cannot detect negative obstacles like holes or stairs. In addition, things become somehow difficult when the mobile robot is installed with Kinect sensor. Sometimes, the mobile robot has to encounter environments with a bumpy road or slope surface. This will cause the depth image captured not align with the floor which will lead to obtain an inaccurate reading and maps of the environment. Economically, the use of both laser scan and Kinect tends to be a bit expensive [3],[8]. 2D SLAM was also performed using Kinect only in several studies. Despite the efficient results obtained, negative obstacles as well as bumpy roads or slope surfaces that a robot may encounter in the environment were not considered [2],[4],[9],[10].

3. Methods

3.1. Ground Plane Detection and Orientation Correction

Plane detection is crucial for numerous vision-based tasks. As objectives of the study focus on detecting positive obstacles as well as correcting the orientation of Kinect's point cloud data, detecting the ground plane will always be the first step that should be taken. Once the ground plane is successfully detected, both positive obstacles detection and Kinect's point of cloud data orientation correction can be achieved.

One of the well-used algorithms in the field of computer vision is Random Sample Consensus (RANSAC). Its principle is based on finding the best plane among a 3D point cloud. The algorithm randomly elects three points and calculates the parameters of the corresponding plane. Then it detects all points of the original cloud belonging to the calculated plane, according to a given threshold. Afterwards, it repeats these procedures N times; in each one, it compares the obtained result with the last saved one. If the new result is better, then it replaces the saved result by the new one [11].

Correcting the orientation of the depth image acquired depends on identifying two main planes. One is the ground plane of the image and another one is the real plane where the ground plane of the depth image must lay on. By the time those planes are identified, a rotation matrix and a rotation angle can be computed by the mean of using the normal vectors of the two planes,

$$\overrightarrow{n_2} = (A_2 + B_2 + C_2) \tag{1}$$

$$\overrightarrow{n_1} = (A_1 + B_1 + C_1) \tag{2}$$

where n1 and n2 represent the normal vector of the plane where the depth image must lay on the normal vector of the ground plane detected by RANSAC algorithm respectively.

The rotation angle is then found by the following equation

$$\theta = \tan^{-1}\left(\frac{A_1 \cdot A_2 + B_1 \cdot B_2 + C_1 \cdot C_2}{\sqrt{A_1^2 + B_1^2 + C_1^2} \cdot \sqrt{A_2^2 + B_2^2 + C_2^2}}\right) \tag{3}$$

The extracted plane is then rotated according to the following rotation matrix.

$$Rot(z,\gamma) = \begin{pmatrix} \cos\gamma & -\sin\gamma & 0 \\ \sin\gamma & \cos\gamma & 0 \\ 0 & 0 & 1 \end{pmatrix} \tag{4}$$

$$T(P) = \begin{pmatrix} Rot(z,\gamma) & T_{3\times1} \\ 0 & 1 \end{pmatrix}\begin{pmatrix} P_{3\times1} \\ 1 \end{pmatrix} \tag{5}$$

3.2. 3D Depth data to 2D Obstacle Locations

The conversion of 3D depth data to a 2D obstacle location depends directly on the possibility of detecting the floor surface. Once the floor surface is detected, it can be removed to start the conversion process. For this study, the Kinect sensor will be placed at a height of 30 cm from the floor. After the orientation correction algorithm is being applied, the floor surface is filtered out by considering all point cloud between 3 cm to be removed. Moreover, the pixels above the height of the Kinect will be eliminated since they are not representing an obstacle that can be faced by the mobile robot.

3.2.1. Nearest Point Method
A method that can be used to detect the positive obstacles is by implementing the concept of getting the nearest point in each column of the array that represents the depth image [1]. To do that, the 11-bit raw depth image of the environment that is represented as an array of 640 × 480 has to be retrieved from Kinect and to be converted into real depth. Then, X and Y coordinate that are corresponding to each pixel in the real depth image are calculated using Equations 6 and 7 resulting in three 640 × 480 arrays that represent X, Y and Z. Once the three are obtained, the minimum element in each column in Z array is extracted out to get 640 ×1 array as shown in the Equation 8,

$$X_{i,j} = \left(J - \frac{w}{2}\right) \times \frac{320}{w} \times M \times Z_{i,j} \tag{6}$$

$$Y_{i,j} = \left(I - \frac{h}{2}\right) \times \frac{240}{h} \times M \times Z_{i,j} \tag{7}$$

$$Z_j' = min(Z_{0,j}, Z_{1,j}, \dots, Z_{479,j}) \tag{8}$$

where I is the pixel's row number, J is pixel's column number, h and w is the height and width obtained from Z array respectively and M is the NUI camera depth image to skeleton multiplier constant.

The X locations for the corresponding Z-elements were also selected giving another 640x1 array. These two arrays indicate the Z and X coordinates of the nearest pixels and can also be regarded as 2D obstacle locations as obtained when using a laser scanner. The minimum-selection method in equation 8 was implemented to avoid false detection of the obstacle and minimize processing power and time. Also note that the Y-Coordinate was ignored as the robot can only move in X and Z direction. The arrays have already represented the nearest obstacle.

4. Results and Discussion
This Section presents and discusses the result after undergoing ground plane detection, depth image orientation correction and 3D to 2D depth data conversion. The results are divided into several parts for further discussion. The first part focuses on the results obtained with the implementation of 3D to 2D depth data conversion algorithm by showing several depth images. The second part focuses on the results obtained with the implementation of the RANSAC algorithm to detect the ground plane of the scanned environment and finding the rotation angle where the last part focuses on the results obtained with implementation of orientation correction algorithm.

4.1. Results Obtained with The Implementation of 3D to 2D depth data Conversion algorithm

Figure 2 shows the RGB and depth Image of the surrounding environment that was obtained by Kinect. The purpose of including RGB image is just to show view of the Kinect during the scanning process. The depth image shows dark and bright areas. The bright areas represent the parts of the environment that were accessible by the Kinect where the brighter the object is the closer it is to the Kinect. On the other hand, the black areas represent an invalid data of the environment.

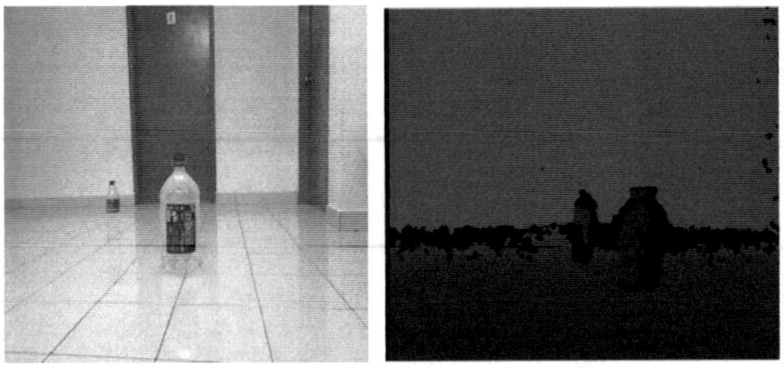

(a) (b)

Figure 2. Image of the scanned environment: (a) is the RGB image, (b) is the depth image.

The invalid data comes from three main sources which are the sensor itself, the measurement setup and the properties of the object surface such as smooth and shiny surfaces. These surfaces that appear overexposed in the infrared impede the measurement of disparities and result in gaps which are represented in Fig 4 as black pixel with value of 0.

The algorithm starts by rotating the ground plane detected by RANSAC so that it is aligned with the horizontal plane. Then, the ground plane was removed by applying the vertical limit filter where all features that are higher than 27 cm and below than 8 cm were removed. In addition to the vertical limit filter, the horizontal limit filter was also applied to remove all pixels that are further than or equal to 2.5 m from Kinect resulting the image shown in Figure 3.

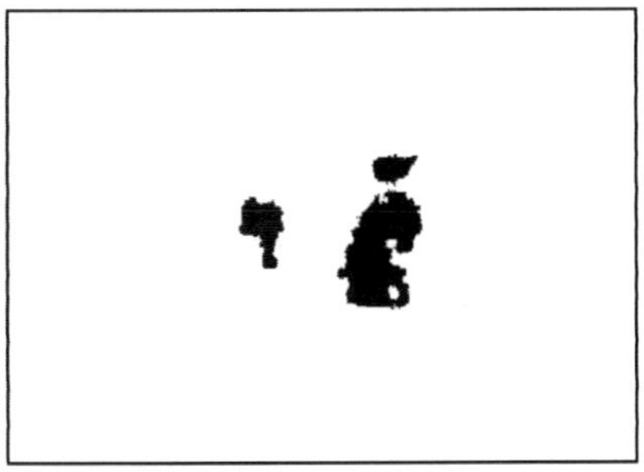

Figure 3. Depth image after filtering process.

The conversion process was continued by extracting the smallest value stored in each column of the filtered image with its location along the X axis (see Section 3.2.1). The obstacles were then plotted based on the extracted data and the result of the conversion is shown in Figure 4.

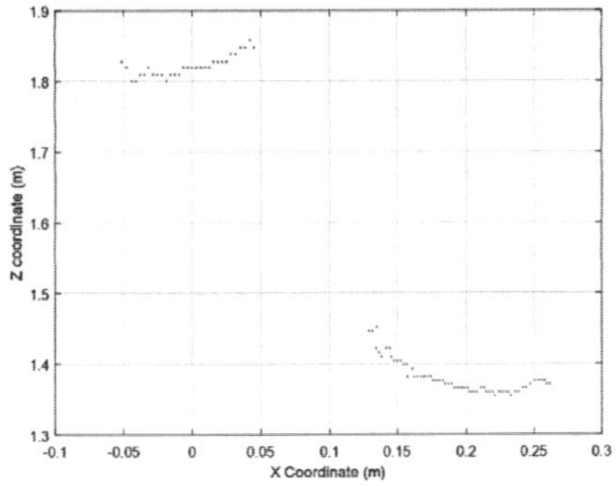

Figure 4. 3D depth data are represented in 2D obstacle location.

4.2. Results Obtained with The Implementation of RANSAC Algorithm to Detect the Ground Plane

It is known that the results of RANSAC algorithm directly depend on two main variables. Those variables are the number of iterations which represent how many times RANSAC will be implemented on an image to extract the ground plane and the threshold distance that is set to test how far points are from the fitted plane. Results obtained with the implementation of RANSAC algorithm to detect the ground plane with different iterations and different threshold distance are represented in Table 1 and Table 2 respectively.

Table 1. Ground plane detection with different iterations and threshold distance of 0.3cm.

Number of iterations	80	50	10
Result		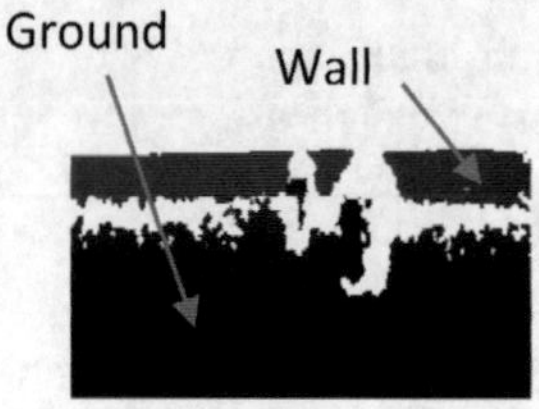	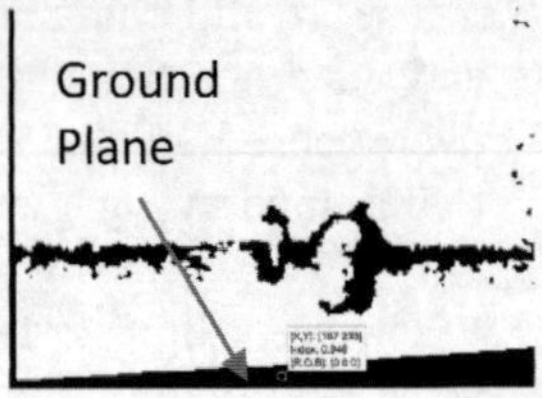

Table 2. Ground plane detection with different threshold distance.

Distance	0.3cm	0.4cm	0.5cm
Result	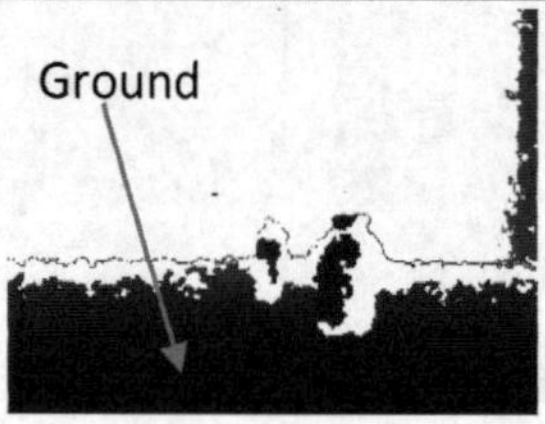	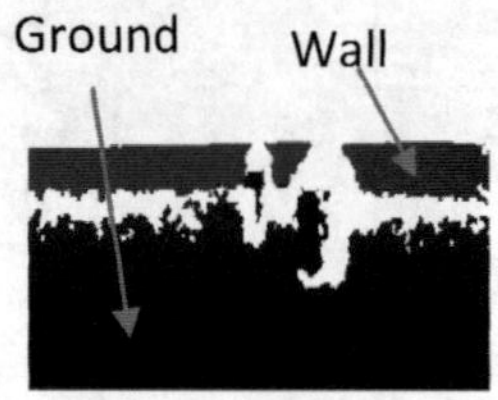	

The results obtained showed that a threshold distance of 0.3cm and 80 iterations are the best for RANSAC to detect the ground plane. It was also noticed that with 50 iterations, RANSAC successfully detected the ground plane; however, it is not efficient due to the detection of some pixels that represent the wall of the environment. This leads to the fact that the higher the number of iterations, the more accurate the RANSAC is. The reason behind this is when the number of iterations is high, it allows RANSAC to test more points and compare the results obtained to choose the best one. This does not mean that RANSAC will not test the point of the 3D point cloud set when the number of iterations is low. Testing the points of the 3D point cloud set is made in both cases but the higher the number of testing times the more accurate the results obtained are.

Trying different threshold values also influences RANSAC results. Table 2 shows the results obtained by implementing RANSAC with 0.3 cm, 0.4 cm and 0.5 cm as a threshold distance. It was found that 0.3 cm was the best threshold distance for this scenario while 0.5 cm was the worst as the output image was the same as the input image. This is because when the distance is increased, more points will be added to the set of points that represent inliers of the ground plane and the opposite happens when the distance is decreased.

4.3. Results Obtained with The Implementation of Orientation Correction Algorithm

Figure 5 shows a comparison between the point cloud image before and after applying the orientation correction algorithm. Image (a) is the same image shown in previous Figures but represented in terms of 3D point cloud. RANSAC result has clearly shown that there is an angle between the detected ground plane and the real plane where the ground plane must lay on. As a result, the orientation correction algorithm was implemented, and the result presented in image (b) shows that the orientation of the plane was successfully corrected.

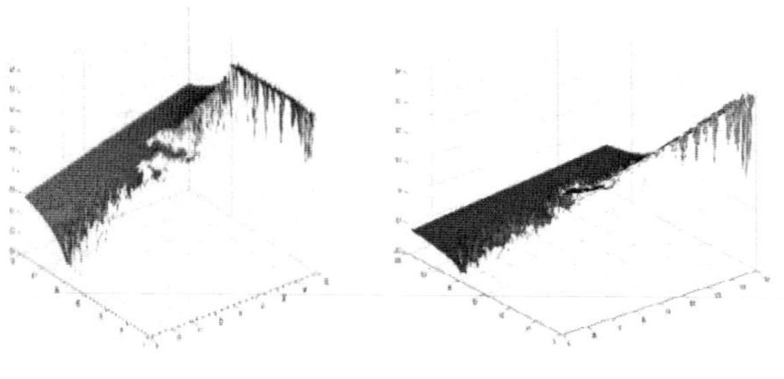

(a) (b)

Figure 5. Orientation correction result: (a) is the depth image represented in terms of 3D surface, (b) is the depth image.

5. Conclusion

The project was mainly focusing on correcting the orientation of Kinect so that 3D depth data acquired by the sensor can be properly converted to 2D obstacle location. To achieve that, ground plane had to be detected using RANSAC and checked either it lies on the horizontal plane of the Kinect. Orientation correction had to be applied to the detected ground plane in case it did not lie on the horizontal plane of the Kinect. The orientation correction method is inspired by the robotic theory and geometric coordinate system. Transformation matrix is used to solve the orientation problem. The 3D to 2D conversion method is then utilized to represent the depth data as 2D obstacle location.

As referring to the results obtained, ground plane detection, ground plane orientation correction and 3D depth data to 2D obstacle locations conversion were successfully achieved. The method used to detect the ground plane is affected by number of iterations and distance between points of plane while other methods have shown their reliability.

As future improvement for 3D depth data conversion to 2D obstacle location, RANSAC algorithm could be implemented and enhanced with estimated surface normal. The orientation correction is implemented on X-Z axis plane and for further improvement, the orientation correction can be implemented on X-Y plane.

The current work focussed on the detection of the positive obstacles. The detection of negative obstacles would make the system more reliable during the robot navigation. A suggested method to detect the negative obstacles is to utilize both farthest point method, instead of the nearest one, and virtual floor projection method.

6. Acknowledgement

The author would like to acknowledge the support from the Fundamental Research Grant Scheme (FRGS) under a grant number of FRGS/1/2018/TK04/UNIMAP/02/15 from the Ministry of Education Malaysia.

References

[1] Kamarudin K et. al. 2013 Method to convert Kinect's 3D depth data to a 2D map for indoor SLAM *Proc. - 2013 IEEE 9th Int. Colloq. Signal Process. its Appl. CSPA 2013* 247–251

[2] Kamarudin K, Mamduh S M, Md Shakaff A Y and Zakaria A 2014 Performance analysis of the microsoft kinect sensor for 2D simultaneous localization and mapping (SLAM) techniques *Sensors (Switzerland)* **14** 23365–23387

[3] Graham B 2013 Obstacle Mapping with a Mobile Robot using the Microsoft Kinect Sensor *The UNSW Canberra at ADFA journal of Undergraduate Engineering Researchers* **11** 1–13

[4] Ghani M F A, Sahari K S M and Kiong L C 2014 Improvement of the 2D SLAM system using Kinect sensor for indoor mapping *Jt. 7th Int. Conf. Soft Comput. Intell. Syst. SCIS 2014 15th Int. Symp. Adv. Intell. Syst. ISIS 2014* 776–781

[5] Barak N, Gaba N and Aggarwal S 2016 Two Dimensional Mapping by using Single Ultrasonic Sensor *International Journal of Advanced Research in Computer Science Research* **7** 254–257

[6] Kalmegh S K 2010 Obstacle avoidance for a mobile exploration robot using a single ultrasonic range sensor *International Conference on Emerging Trends in Robotics and Communication Technologies* 8-11

[7] Avirup Basu S S, Ghosh S K 2016 Autonomous Navigation and 2d Mapping Using Sonar *5th International Conference on Wireless Networks and Embedded Systems (WECON)* 1–5

[8] Md Shakaff A Y, Kamarudin K, Visvanathan L M, Mamduh S M, Kamarudin A Y M S and Zakaria A 2015 Feasibility Analysis Of 2d-Slam Using Combination Of Kinect And Laser Scanner *J. Teknol.* 1–7

[9] Peasley B and Birchfield S 2013 Real-time obstacle detection and avoidance in the presence of specular surfaces using an active 3D sensor *IEEE Work. Robot Vision, WORV 2013* 197–202

[10] Aichele F, Schenke B, Eckstein B and Groz A 2016 A framework for robot control software development and debugging using a real-time capable physics simulation *47th Int. Symp. Robot. ISR 2016* 173–180

[11] McDaniel M W, Nishihata T, Brooks C A and Iagnemma K 2010 Ground plane identification using LIDAR in forested environments *Proc. - IEEE Int. Conf. Robot. Autom* 3831–3836

IOP Conf. Series: Materials Science and Engineering **705** (2019) 012031 doi:10.1088/1757-899X/705/1/012031

Facial expression recognition in JAFFE and KDEF Datasets using histogram of oriented gradients and support vector machine

S K Eng*, H Ali, A Y Cheah, Y F Chong

School of Mechatronics Engineering, Universiti Malaysia Perlis (UniMAP), Kampus Pauh Putra, Arau, 02600, Perlis, MALAYSIA

*Corresponding author: kheng@unimap.edu.my

Abstract. This paper presents the used of histogram of oriented gradient (HOG) for facial expression recognition using support vector machine (SVM). In this work, the facial expression images are firstly preprocessed by face detection and cropped images. Then, HOG method is adopted as feature extraction on facial image. The ability of HOG to preserve the local information and orientation density distribution in facial images suitable as shape descriptor for facial expression. It divides the image into cell or patch that has magnitude and orientations. The extracted HOG was then concatenated into histogram bin to form one feature vector before feed into SVM classifier. Both JAFFE and KDEF datasets were employed to evaluate the performance of proposed method. Based on results, the average recognition rates of JAFFE and KDEF datasets are 76.19% and 80.95% respectively. The results show that the performance of expression surprise has outperformed compared to others expression while expression fear contributes the lowest recognition rate. Thus, utilization of HOG features with SVM classifier have shown the promising results in recognizing facial expression.

1. Introduction

Facial expression refers to the explicit transformation of human face due to the automatic response of the emotional states in fact it is induced by the activations of facial muscles that temporarily deformed both permanent and transient features. Temporal dynamics of facial muscle typically brief normally ranging between 5 seconds [1]. Facial expression plays an important role in human communication since 55% of facial modality involve in human-to-human interaction [2]. Nowadays, with advanced technology, facial expression recognition can be applied to education field, where the emotion state of the students in the classroom can be detected. The facial expression recognition might help in establishing a healthy learning and teaching environment.

Over the past decades, facial expression recognition becomes the prominent research in the field of computer vision, pattern recognition and image analysis. Many approaches constantly proposed by previous researchers. Some of them utilized facial components [3], facial points [4], facial landmarks [5], shape and texture [6], empirical mode decomposition [7], higher order spectra [8], neural-Adaboost [9] and local binary pattern [10] in recognizing facial expression.

Even though facial expression recognition has reached a certain level of success, however its development of a robust facial expression recognition is still ongoing and challenging as there are still many unsolved aspects due to various unpredictable facial variations and complicated exterior environment conditions. Therefore, this paper proposes histogram of oriented gradient of facial images using support vector machine to recognize the seven facial expressions. Although this paper similar to [11], however the pre-processed facial image, HOG [12] framework and the datasets used (JAFFE and KDEF) in the proposed method are varies one with [11]. In this work, face detection using Viola Jones method [13] is adopted and the detected face region is automatically cropped. Then, the pre-processed images were subjected to HOG as feature extraction. The rationale of using HOG is that it has ability to preserve the local information using orientation density distribution and gradient of the edge in facial images. Thus, extracted HOG features will used as input before fed to SVM classifier.

The outline of the paper is organized as follows: Section 2 presents brief methodology used in facial expression. It consists of pre-processing of facial image, extracting facial features and classifier. Section 3 provides the experimental results and discussion of facial expression recognition using HOG technique and SVM in classifying seven facial expressions. Section 4 concludes the proposed method.

2. Methodology

The flowchart of the proposed method is depicted in Figure 1. The details of each block are described in subsequent sections.

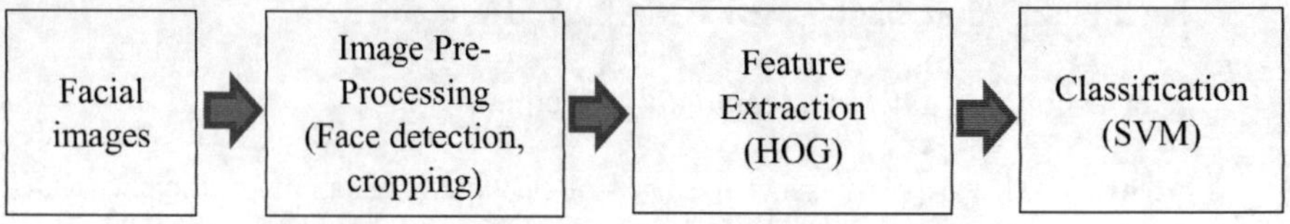

Figure 1. The flowchart of the proposed method

2.1. Facial Expression Database

Two benchmark databases were used in this work: JAFFE and KDEF database.

2.1.1. JAFFE Database. Japanese Female Facial Expression (JAFFE) database was taken from publicly available data which consists of 213 facial expression images of 10 subjects of Japanese female. Each subject performed six basic emotions plus neutral (30 angry, 29 disgust, 33 fear, 30 happiness, 31 sad, 30 surprises and 30 neutral) in which each expression contains 3 to 4 images per subjects. The image has the resolution in grayscale. All the facial images have been taken under strict controlled conditions of similar lighting and no occlusion such as hair or glasses. All the expression in frontal view and the resolution of the original image are 256 x 256 pixels.

2.1.2. KDEF Database. Karolinska Directed Emotional Faces (KDEF) database is another publicly available dataset consists a set of 4900 facial expression images. It contains 70 individuals, each displaying 7 different emotional expressions, each expression being photographed (twice) from 5 different angles.

2.2. Image Pre-Processing

Pre-processing of the facial image is considered as an important step in facial expression recognition. The image needs to be preprocessed, so that the final image will have extracted face region with uniformity of size and shape.

2.2.1. Face Detection. Face detection is a method used to determine the region of the face. In this work, Viola Jones face detection method [13] has been adopted. Figure 2 shows an example of face detection. Then, the original image is automatically cropped based on the face region obtained. Figure 3 shows the results of cropped images for both datasets. Further, the KDEF images which in RGB format, need to convert into grayscale images for further process. Finally, histogram equalization is performed on the images to enhance contrast of grayscale images and have uniform distribution of intensities that will aid in the feature extraction process.

Figure 2. Example of face detection using Viola Jones method [13]

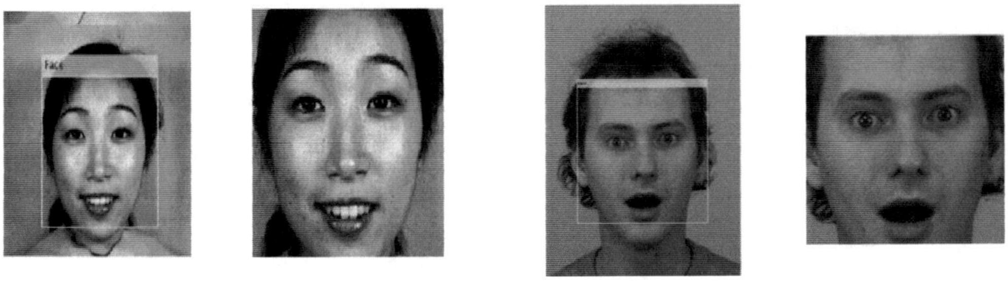

(a) JAFFE dataset (b) KDEF dataset

Figure 3. Results of face detection and cropped images

2.3. Feature Extraction Based on Histogram of oriented Gradient

HOG [12] descriptor is widely used for human detection, object detection and pedestrian identification. HOG is computed using magnitude and orientation. It is based on the accumulation of gradient directions over the pixel of a small spatial region referred as "cell" and in the subsequent construction of a 1D histogram whose concatenation supplies the features vector to be considered for further purposes. Let L be an intensity (grayscale) function describing the image to be analyzed. The image is divided into cells of size $N \times N$ pixels and the orientation $\theta_{x,y}$ of the gradient in each pixel is computed by means of the following equation:

$$\theta_{x,y} = \tan^{-1} \frac{L(x, y+1) - L(x, y-1)}{L(x+1, y) - L(x-1, y)} \quad (1)$$

Successively, the orientations $\theta_i^j i = 1 \dots N^2$ belonging to the same cell j are quantized and accumulated into a M-bins histogram. Finally, all the achieved histograms are ordered and concatenated into a unique HOG histogram that is the final outcome of this algorithmic step, i.e. the features vector to be considered for the subsequent processing. Figure 4 shows the HOG features extraction process based on cell size $N*N$ pixels:

IOP Conf. Series: Materials Science and Engineering **705** (2019) 012031 doi:10.1088/1757-899X/705/1/012031

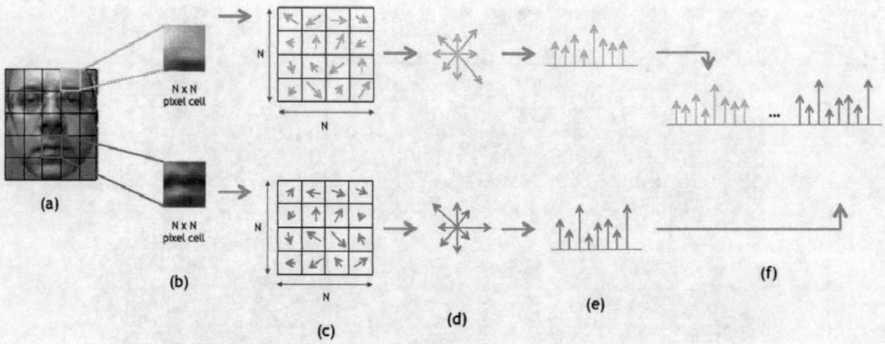

Figure 4. Histogram of oriented gradient [14]

In this HOG framework, the preprocessed facial image is divided into cells of size $N \times N$ pixels. The orientation of all pixels is computed and accumulated in an M-bins histogram of orientations. Then, all the cell histograms are concatenated to construct the final features vector. Figure 5 shows the illustration for different cell size of 32 x 32 orientations with cell size 16 x16 orientations for the cell histograms.

Normally two main parameters are used to describe characteristics of the HOG descriptor i.e. orientation bines and cell size. The dimension of the patch involved in the single histogram computation is represented by cell size. Appropriate selection of parameters plays a vital role and should be selected with care for better performance of classification algorithm. Features extracted using HOG are affected by varying cell size. In the situation of using larger cell size, spatial information of certain region from facial image is squeezed into the unit cell histogram, and the contribution turn into less significant. High resolution analysis on the other hand can be carried out by using smaller cell size, but it results in more detailed information [11].

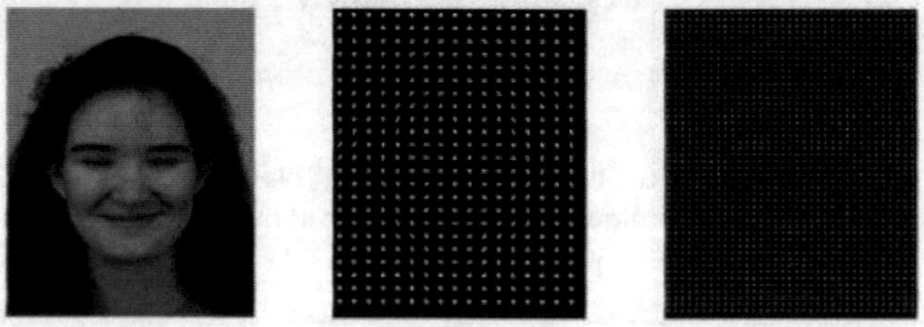

(a) (b) (c)

Figure 5. Illustration of different cell size with 32 x 32 orientations and 16 x 16 orientations

2.4. Support Vector Machines

The foundation of SVMs have been developed by [15] and are gaining keen interest due to attractive features and competitive performance. SVMs are based on structural Risk minimization (SRM) principal, which has been shown to be superior, to traditional empirical risk minimization. SRM minimizes an upper bound on the expected risk as opposed to ERM that minimizes the error on training data. It gives better generalization abilities which are the ultimate goal in the statistical learning. SVMs have been used widely in classification problems such as object recognition, speaker identification and face recognition. Given a set of labelled training data (supervised learning), the algorithm computes an optimal hyperplane (the trained model) which categorizes new examples in the right class. Given the training vectors $x_i \in \mathbb{R}^n, \dot{i} = 1, \dots, l$ and the corresponding set of l labels $y_i \in \{1, -1\}$, the following primal optimization problem is solved:

$$\min_{w,b,\xi} \frac{1}{2} \mathbf{w}^T \mathbf{w} + C \sum_{i=1}^{1} \xi_i \quad (2)$$

subject to $$y_i(\mathbf{w}^T \phi(\mathbf{x}_i) + b) \geq 1 - \xi_i, \xi_i \geq 0, i = 1, \ldots, l \quad (3)$$

where ξ_i is the misclassification error for the ith training vector; ξ is the total misclassification error, $\boldsymbol{w}$ is the normal vector to the hyperplane. Training vectors are mapped into higher dimensional space by the function $\emptyset$. Furthermore, $K(x_i, x_j) = \Phi(x_i)^T \Phi(x_j)$ is called the kernel function. The choice of kernel function is also significant when designing SVM-based classifier. The 2 basic kernel function are usually used:

$$\text{Linear function: } K(x_j, xj) = x_i^T x_j \quad (4)$$

$$\text{Radial basis function: } K_{(x_i - x_j)} = (\gamma x_i^T x_{j+r})^d, \gamma > 0 \quad (5)$$

The classical SVM approach is suitable only for a two classes problem but, unfortunately, FER involves multi-class handling. Multi-class problems can be addressed by using "one-against-one" method as proposed by [16], the approach is based on the construction of an SVM classifier for each pairwise of classes and a voting system aided to elect the predicted class when an unseen item is tested. More specifically, let k be the number of classes, then $k(k-1)/2$ classifiers K_{ij} are trained from the available data of the ith and jth classes. The final prediction is returned by a voting system: the unseen example is analysed by each classifier K_{ij} that gives a vote to one or to the other class. Finally, the class with the largest number of votes is chosen.

Let considered 3-class problem. In this way, three classifiers came out, example, $C_i(x), (i = 1,2,3)$. Calculate the testing data sample one by one with three classifiers, and pick the maximum value $C_i(x), (i = 1,2,3)$, so the testing data sample belongs to the ith class. Figure 6 shows the examples of SVM classifier for one-against-one and one-against-all.

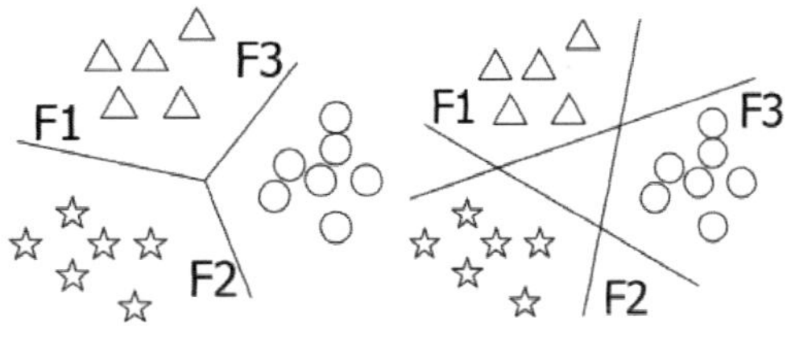

(a)OAOSVM (b)OAASVM

Figure 6. Examples of SVM classifier for one-against-one and one-against-all [17]

3. Results & Discussion

The proposed method is evaluated using two publicly available datasets, JAFEE and KDEF. In JAFFE, a total of 213 images has been employed. While, in KDEF, a total of 980 images have been adopted. All the images have been subjected to image preprocessing such as face detection and facial cropped. Then, the HOG features were extracted and used as input to predict the expression using SVM classifier.

In this experiment, the data was split into training and testing set based on conventional method with 70% training and 30% testing strategy. Figure 7 shows the recognition rates based on extracted HOG features using SVM classifier on JAFFE and KDEF datasets. Based on Figure 7, it can be seen that the expression surprise achieved the highest recognition rate in both datasets which are 88.89% (JAFFE) and 97.62% (KDEF). This can be inferred that, the HOG as shape descriptor able to counts occurrences of gradient orientations in localized portions of an image and able to model the shape of the surprise facial muscles via edge information effectively compared to others facial expression. In contrast, expression fear shows the lowest recognition rate in both JAFFE and KDEF datasets which is 55.56% and 52.38% respectively. On the other hand, the expression disgust in JAFEE and KDEF datasets shows the recognition rate of 66.67% and 76.19%, respectively. While, expression happy, neutral and sadness are comparable with recognition rates of 77.78% in JAFFE dataset.

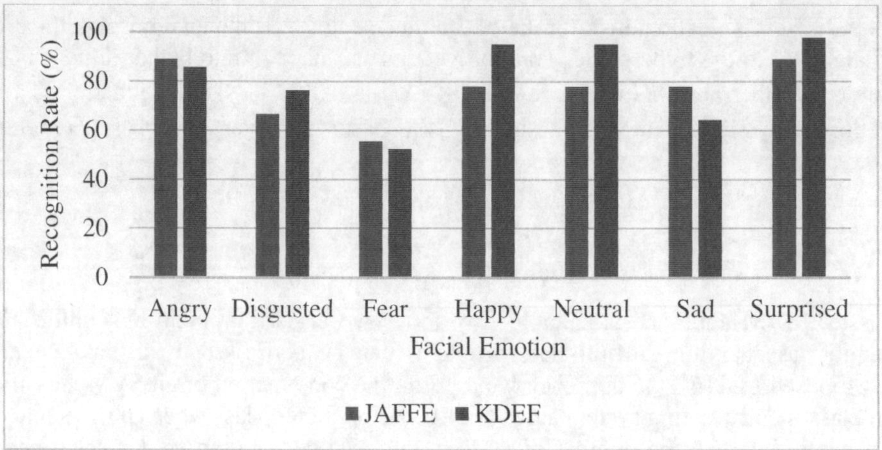

Figure 7. Recognition rates of seven facial expression using SVM classifier in JAFFE dataset and KDEF dataset

As summary, the recognition rate of expression surprise gave the best results in both JAFFE and KDEF datasets. The features of expression surprise may be easily distinguishable. Meanwhile, the recognition rate of expression fear shows the lowest results in both datasets. It can be inferred that; the expression fear is difficult to recognize due to the existence of intra-class variations of different expression between the subjects. Thus, the system confused to classify the expression fear correctly.

4. Conclusion

This paper presented the used of histogram-oriented gradient as feature descriptor of facial image using SVM classifier to classify the seven facial expressions. In this work, the facial expression images are firstly preprocessed by face detection and cropped images. Then, HOG method is adopted as feature extraction on facial image. The ability of HOG to preserve the local information and orientation density distribution in facial images able to distinguish the characteristic of facial expression. The extracted HOG was then concatenated into histogram bin to form one feature vector before subjected as input to SVM classifier. Both JAFFE and KDEF dataset were employed to evaluate the proposed method. Based on results, average recognition rates of JAFEE and KDEF datasets are 76.19% and 80.95%, respectively. It can be seen that the expression surprise has shown the highest recognition rate while fear provide the lowest recognition rate. Further, the average recognition rates of KDEF dataset (80.95%) is slightly higher than JAFFE dataset (76.19%). The results show that the performance of expression surprise has outperformed others expression while expression fear slightly hard to recognize thus contribute the lowest rate. As overall, utilization of HOG features with SVM classifier have shown the promising results in recognizing facial expression.

However, further study needs to be conducted to improve the recognition rate of expression fear by investigating the several HOG features for different magnitude and orientation as well as utilizing different machine learning classifiers to enhance the performance of the proposed method.

References

[1] Fasel B and Luettin J 2003 Automatic facial expression analysis: A survey *Pattern Recognition* **36** 259–275

[2] Mehrabian A 1968 Communication without words *Psychological Today* 8–9

[3] Ilbeygi M and Shah-Hosseini H 2012 A novel fuzzy facial expression recognition system based on facial feature extraction from color face images *Engineering Applications of Artificial Intelligence* **25** 130–146

[4] Pantic M and Rothkrantz L J M 2000 Expert system for automatic analysis of facial expressions *Image and Vision Computing* **18** 881–905

[5] Aifanti N and Delopoulos A 2014 Linear subspaces for facial expression recognition *Signal Processing: Image Communication* **29** 177–188

[6] Kotsia I, Zafeiriou S and Pitas I 2008 Texture and shape information fusion for facial expression and facial action unit recognition *Pattern Recognition* **41** 833–851

[7] Ali H, Hariharan M, Yaacob S and Adom A H 2015 Facial emotion recognition using empirical mode decomposition *Expert Systems with Applications* **42** 1261-1277

[8] Ali H, Hariharan M, Yaacob S and Adom A H 2015 Facial Emotion Recognition Based on Higher Order Spectral using SVM *Journal of Medical Imaging and Health Informatics* **5** 1272-1277

[9] Owusu E, Zhan Y and Mao Q R 2014 A neural-AdaBoost based facial expression recognition system. *Expert Systems with Applications* **41** 3383–3390

[10] Luo Y, Wu C and Zhang Y 2013 Facial expression feature extraction using hybrid PCA and LBP *The Journal of China Universities of Posts and Telecommunications* **20** 120–124

[11] Nazir M, Jan Z and Sajjad M 2018 Facial expression recognition using histogram of oriented gradients based transformed features *Cluster Computing* **21** 539-548

[12] Dalal N and Triggs B 2005 Histograms of oriented gradients for human detection *IEEE Computer Society Conference on Computer Vision and Pattern Recognition* **1** 886-893

[13] Viola P and Jones M 2001 Rapid Object Detection using a Boosted Cascade of Simple Features *IEEE Computer Society Conference on Computer Vision and Pattern Recognition (CVPR'01)* **1** 511-518

[14] Del Coco M, Carcagnì P, Leo M and Distante C 2015 A Minimax Framework for Gender Classification Based on Small-Sized Datasets *International Conference on Advanced Concepts for Intelligent Vision Systems* 415-427

[15] Vapnik V 1998 Statistical Learning Theory *John Wiley and Sons Inc New York*

[16] Knerr S, Personnaz L and Dreyfus G 1990 Single-layer learning revisited: A stepwise procedure for building and training neural network *Neurocomputing: Algorithms, Architectures and Applications, NATO ASI, Berlin: Springer-Verlag* 41-50

[17] Ying Z and Zhang G 2009 Facial Expression Recognition Based on NMF and SVM *International Forum on Information Technology and Applications, Chengdu* 612-615

IOP Conf. Series: Materials Science and Engineering **705** (2019) 012032 doi:10.1088/1757-899X/705/1/012032

Preliminary Analysis of Organic Wireless Positioning System in an Indoor Environment

A H Ismail[*1,2], J S Ngo[1], M I Taib[1], M S M Hashim[1], M S M Azmi[1], M J A Safar[1] and M H Marhaban[2]

[1]School of Mechatronic Engineering, Universiti Malaysia Perlis, Pauh Putra Campus, 02600 Arau Perlis, MALAYSIA.
[2]Department of Electrical and Electronics Engineering, Faculty of Engineering, Universiti Putra Malaysia, 43400 Serdang Selangor, MALAYSIA.

[*]Corresponding author: ihalim@unimap.edu.my

Abstract. The massive deployment of wireless infrastructure especially in the trending 4th Industrial Revolution (IR4.0) has opened the opportunity for wireless-based positioning and localization. The works on wireless positioning system revolves on two techniques, the geometrical technique employing either triangulation or trilateration, and mapping technique employing signal fingerprint method. The later is more favoured since the accuracy was superior to geometrical technique. However, the signal fingerprint method required enormous efforts in order to record and stored the signal fingerprint. This paper investigates the new approach in constructing the signal fingerprint using organic approach. The signal is recorded using the developed apps, and then the user sends the signal data into the cloud. The organic fingerprinting data were compared to the conventional fingerprinting. The preliminary results show promising accuracy in the experimented indoor environment.

1. Introduction

Positioning of an object of interest in an indoor environment typically uses integrated sensors on the objects, for instance the proximity sensors and sometime visual sensors to locally map the targeted indoor environment [1]. From the information of theses sensors, the position of the object of interest can be computed, recently at great accuracies. The other alternative of indoor positioning could be realized by manipulating wireless signals detectable by the object of interest. The later positioning approach are more cost facilitative with ubiquitous wireless infrastructure deployment in recent years [2][3]. It is cheaper solution compared to integrated sensors and could be extended into numerous other applications.

The indoor positioning technique employing wireless solution such as WiFi can be categorized by two different approaches, the geometrical method, and fingerprinting technique. The geometrical method uses mathematical computation for example the triangulation and trilateration technique. This method works by converting the wireless signal parameter i.e. the Received Signal Strength (RSS) into distance metric, usually by suitable signal propagation models. However, there are no generic signal propagation models that suit any environment, thus limit the practicality of the triangulation and trilateration technique. Often, the accuracy studies of triangulation and trilateration method were unacceptable [4][5].

The fingerprinting technique on the other hand, works by comparing the present RSS signal with the constructed radio map (also known as the signal fingerprint) beforehand. This technique has been with proven significant accuracy by many [4][6]. The use of more complex fingerprinting algorithm [5], and state-of-the-art filtering technique such as Kalman Filter has further increased the positioning accuracy [6][7]. Figure 1 depicts the fundamental difference between the triangulation/trilateration and fingerprinting technique.

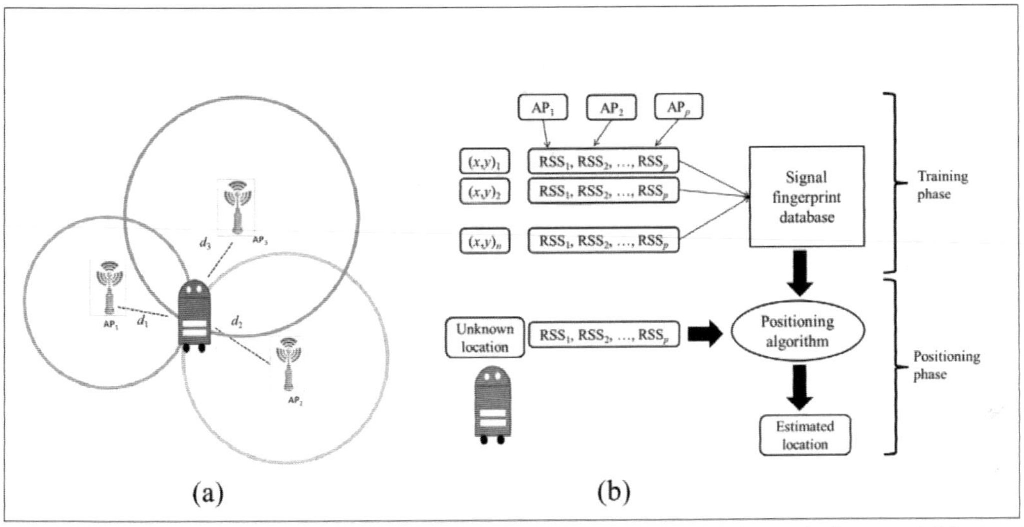

Figure 1. Wireless positioning system, (a) triangulation and trilateration technique and (b) signal fingerprinting technique.

The principal problem with the fingerprinting method is the database creation. In our earlier works, a considerable amount of efforts was required to record all the signal data at all required locations which has led to the development of automatic database creation by mean of interpolation [8][9]. However, such method although significantly improve positioning accuracy is also prone to other errors for instance when the arrangement of the environment has changed. Thus, the database must be re-made. This issue is troublesome and not sustainable for longer time, and not suitable particularly in the fourth industrial revolution (IR4.0) application.

Hence, a new approach of database creation for signal fingerprinting is proposed. This new method takes anonymous participants to voluntarily upload their obtained data to the databased stored in the online cloud. This method is called organic fingerprinting method or crowdsourcing fingerprinting [10]. This paper explored the potentials of organic fingerprinting by comparing the initial result between the conventional fingerprinting technique and the proposed organic fingerprinting technique. To realize the organic fingerprinting technique, we developed a simple android apps, and used the signal recorded by the apps to update our database regularly. In the same time, the signal data acquisition using conventional data recording software is also conducted. Finally, the accuracy between the two are compared by using the KNN matching algorithm.

2. Methodology

2.1. Overview

In order to conduct the experiment and data collection, the environment must be firstly identified. The environment must be indoor, easy access, less people wandering around to minimize the signal fluctuations, as well as large enough to cover a there WiFi Access Point (AP). The Mechatronic System Design (MSD) Laboratory located at School of Mechatronic Engineering, Universiti Malaysia Perlis satisfies all the requirement.

The Android apps were later developed for organic data collection. A user is asked to stay at pre-determined locations, and then recorded the data before uploading them into an excel file stored in the cloud. On the other hand, the conventional data collection were recorded using Homedale® software [11]. The WiFi RSS data are collected over 2 minutes time with 2 seconds interval. Finally, the average of RSS were calculated for the fingerprinting database for both conventional and organic method for all reference locations.

The K-Nearest Neighbour (KNN) matching method was used to evaluate the positioning accuracy of both methods. Euclidean distance between the test data and database are computed. Then, the location is the one that has the most minimum value among the error matrix of the Euclidean distance between the test data and database.

2.2. *Indoor Environment*

Figure 2 shows the 2-dimensional layout of the experimented indoor environment. The experiments were conducted at the MSD Lab. The lab area is 171×90 metre with typical arrangement of teaching facilities such as tables, chairs and projector. These facilities are marked as boxes for easier visualization. The lab was unoccupied during data collection periods, and thus minimizing signal fluctuation due to human movements.

A total of 24 locations are identified as 'reference' locations and marked with '×' symbol, and three WiFi are used and set up as prior database. Therefore, the database vector matrix is 24×5 arranged such as way that depicted in equation (1),

$$\text{db}_{c/o} = \begin{bmatrix} x_n & y_n & s_{AP1} & s_{AP2} & s_{AP3} \end{bmatrix}, \quad (1)$$

where parameter $\text{db}_{c/o}$ is the database for both conventional and organic database, x and y represent the Cartesian coordinates at n reference location, s_{AP1-3} is the average of RSS from AP1, AP2 and AP3 at n location, respectively. As a concern, more WiFi AP is advisable for larger environments, however in our case, these three WiFi AP are enough to serve the purposed.

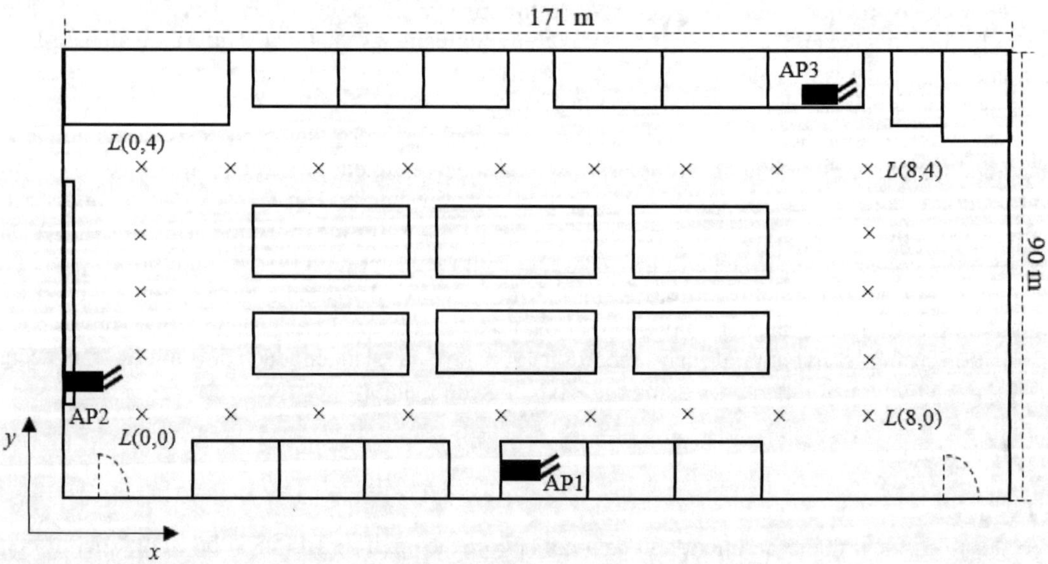

Figure 2. Layout of Mechatronic System Design (MSD) Lab at Universiti Malaysia Perlis, the locations marked with '×' were the reference locations. Also shown the placement of the three WiFi Access Points.

2.3. Android Apps Development

The Android apps is developed in the Android Studio Ver 3.2.1 software. The WiFi connectivity is managed by a dedicated API i.e. Wifimanager at a very low level. The information that can be accessed includes network link and speed, IP address, negotiation state and many other information. With proper programming, permission and syntax, the API can scan, add, save, terminate as well as initiate WiFi connections. The developed Android apps run on Android 6.0 and above since the lower Android version has connectivity problem with WiFi. Figure 3 shows the interface of the developed Android apps.

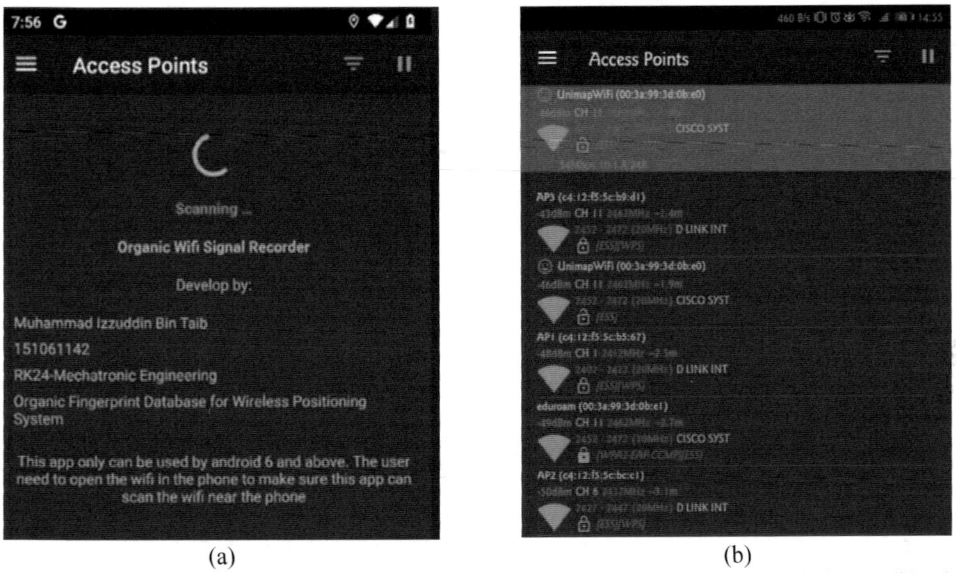

Figure 3. (a) Interface of developed Android Apps running on Android 6.0 and above, (b) List of detectable WiFi Access Points and their wireless properties uploadable into the cloud storage.

2.4. Data Collection

The locations marked with '×' as depicted in Figure 2 were the reference locations. These locations were in fact the locations where the user stays for data collection. For the conventional database collection, the user recorded the WiFi data i.e. WiFi RSS for 2 minutes in the interval of 2 seconds using the Homedale® software installed in a laptop. With 24 locations, the data collection took about one hour to complete including initial setting time. For organic database, the user was asked to install the developed Android apps to his smartphone, then he/she can immediately scan for WiFi and send the required data into the cloud storage (Gdrive). In term of ease of operation, the use of Android apps is simpler and user-friendly. Moreover, the data collection can be extended for multiple users at the same time. Hence, it is convenience.

To compare the positioning accuracy between both approaches, two test locations are taken into consideration. The first location is $L(0,0)$. It is near to AP2, moderate to AP1 and far enough from AP3. Hence, the signal should be linearly separable. The second test location on the other had is chosen at $L(3,4)$ as it bears almost the same distance from all three APs. Figure 4 and 5 shows the signal from these two locations. It can be seen from Figure 4 that the signal obtained from AP2 is much stronger compared to other two APs. Figure 5 shows the WiFi RSS from test location $L(3,4)$ where it shows chaotic signal from the three APs. The complex behavior of WiFi signal such as this example shows the challenges faced in this research topic, and as of our knowledge, there is no common solution to solve this signal fluctuations yet.

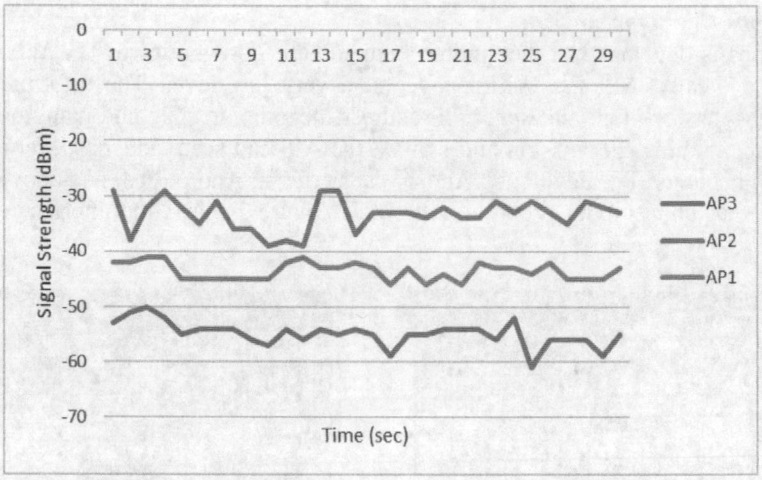

Figure 4. The RSS distribution from test location $L(0,0)$

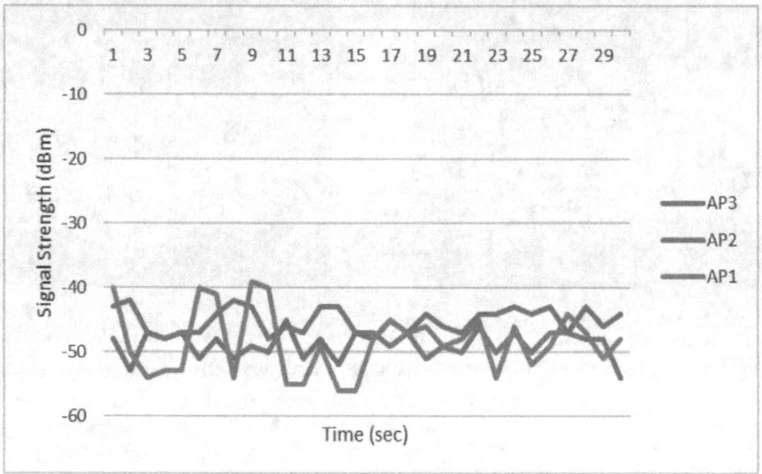

Figure 5. The RSS distribution from test location $L(3,4)$

3. Results and Discussion

The K-Nearest Neighbour (KNN) technique is used to compute the estimated position. In this work, the most nearest neighbour is applied, i.e. K is set to 1. The metric that is used to evaluate the neighbour is Euclidean distance. Then the minimum value of the metric is reflected back to database matrix taking corresponding (x_n,y_n) as the estimated position. Equation (2) and (3) explain this technique.

$$\arg\min_{x \in \mathrm{R}^2} e_i = \sum_{i=1}^{n} \sqrt{\left(s_{t1} - s_{AP1}\right)^2 + \left(s_{t2} - s_{AP2}\right)^2 + \left(s_{t3} - s_{AP3}\right)^2}\ , \tag{2}$$

$$\%acc = \frac{(\hat{L} - L)}{L} \times 100\% \tag{3}$$

where e_i is the error between the test signal s_t and signal in the database $s_{AP1\text{-}3}$ at iteration i. The iteration is computed for all the data in the database of size n, respectively with n reference locations. The parameter $\%acc$ represent the positioning accuracy, by comparing the estimated location $\hat{L}$ and the known exact location L.

For simplicity, two test locations as describe previously are presented in this paper. The first location is taken at $L(0,0)$ which is closer to AP2. The data are linearly separable, and thus matching with the database is fair. The conventional fingerprinting result in 53.33% positioning accuracy while the organic fingerprinting is at 93.33% accuracy. This tremendous different between the two databases is expected since the organic database creation is sustainable as the user uploaded the data in the cloud storage more recently compared to the conventional database. The operation is somewhat easier to the user to upload their data.

At the test location 2 i.e. location $L(3,4)$ where the signal has roughly similar data between each APs due to similar distance from the APs result in poor accuracy. The conventional fingerprinting has only 23.33% accuracy while the organic fingerprinting has 33.33%. The issue of wireless signal fluctuations and instability remained a tough challenge for the researchers in this field.

Table 1 summarises the result of accuracy study between the conventional and organic fingerprinting. Comparing between the two databases, the organic fingerprinting outdone the conventional database at both test locations. This experiment is simple and principle for both signal fingerprinting techniques. Since the organic fingerprinting is proven better than conventional technique, the research will continue to explore relevant algorithms for the indoor positioning system.

Table 1. Accuracy of positioning system on KNN for the two test locations

Test Location	Conventional Database	Organic Database
$L(0,0)$	53.33%	93.33%
$L(3,4)$	23.33%	33.33%

4. Conclusion

This paper studies the accuracy between the conventional database and organic database of the fingerprinting technique for wireless positioning system. Three WiFi Access Point and systematic reference locations for data collection of both databases has been initialized. The conventional database is made by collecting and recording the wireless data using computer's software, while the organic database is made and collecting and storing the wireless data using the developed Android apps, and efficiently uploaded the data into cloud storage. Two test locations have been taken for consideration and finally the position is computed by using the K-Nearest Neighbour algorithm by setting K=1. It has been observed that at both test locations, organic fingerprinting outperforms the conventional fingerprinting. This is due to the fact that the organic database has primary advantage over conventional database which is the ease of operation as well as sustainable to environment arrangement changes. Hence, better quality.

In the future, an enhancement of Android apps is to be made. A more user-friendly apps with better navigational operation will be considered. Furthermore, a more advanced positioning algorithm will be also investigated, especially to suit the 4th industrial revolution.

5. Acknowledgement

The authors would like to acknowledge the support from the Fundamental Research Grant Scheme (FRGS) under a grant number of FRGS/1/2018/ICT05/UNIMAP/02/4 from the Ministry of Education Malaysia. This paper is dedicated to the corresponding author's late son who stood by him while this paper is underwritten, Aysar Muhaimin Bin Abdul Halim. Al-Fatihah.

References

[1] Syaqur W A, Yeon A S A, Abdullah A H, Kamarudin K, Visvanathan R, Ismail A H and Zakaria 2018 Mobile Robot Based Simultaneous Localization and Mapping in UniMAP's Unknown Environment *International Conference on Computational Approach in Smart Systems Design and Applications (ICASSDA)* 1-5

[2] Yang C and Shao H 2015 WiFi-based indoor positioning *IEEE Communications Magazine* **53** 150-157

[3] Liu H 2007 Survey of Wireless Indoor Positioning Techniques and System *IEEE Trans. Systems Man and Cybernetics Part C: Applications and Reviews* **37** 1067-80

[4] Mok Esmond and Günther Retscher 2017 Location determination using WiFi fingerprinting versus WiFi trilateration *Journal of Location Based Services* **1** 145-159

[5] Tian J, Zhou Z, Wu J, Du S, Xiang C and Kuang C 2018 Synthetization of Fingerprint Recognition and Trilateration for Wi-Fi Indoor Localization Through Linear Kalman Filtering *Springer China Satellite Navigation Conference* 373-386

[6] Ismail A H and K Terashima 2018 Prediction of WiFi Signal using Kalman Filter for Fingerprinting-based Mobile Robot Wireless Positioning System *Journal of Telecommunication, Electronic and Computer Engineering* **10** 1-15

[7] Lee S H, Lim I K and Lee J K 2016 Method for Improving Indoor Positioning Accuracy Using Extended Kalman Filter *Mob. Inf. Syst.* 1–15

[8] Jan S S, Yeh S J and Liu Y W 2015 Received Signal Strength Database Interpolation by Kriging for a Wi-Fi Indoor Positioning System *Sensors* **15** 21377-21393

[9] Ismail A H, Mizushiri Y, Tasaki R, Kitagawa H, Miyoshi T and Terashima K 2017 A Novel Automated Construction Method of Signal Fingerprint Database for Mobile Robot Wireless Positioning System *International Journal of Automation Technology* **11** 459-471

[10] Park J, Charrow B, Curtis D, Battat J, Minkov E, Hicks J, Teller S and Ledlie J 2010 Growing an organic indoor location system *Proc. 8th Int. Conf. Mob. Syst. Appl. Serv. MobiSys* **10** 271

[11] T S Development Homedale:Wi-Fi/WLAN Monitor *http://www.thesz.com/products/homedale/*

IOP Conf. Series: Materials Science and Engineering **705** (2019) 012033 doi:10.1088/1757-899X/705/1/012033

Monitoring Oil Palm Tree Health– A Review

J H Hoe*, **S A A Shukor**

School of Mechatronic Engineering, Universiti Malaysia Perlis, Arau, Perlis, 02600 Malaysia.

*Corresponding author: jiahong@studentmail.unimap.edu.my

Abstract. Oil palm is known as one of the most important resources in Malaysia which able to be exported to other countries. Oil palm, also known as "Elaeis guineensis", has many usages and benefits. To ensure the productivity and quality of the oil palm, the health status of the tree needs to be monitored, as they can be prone to various diseases. This paper is going to discuss current methods other researchers have adapted that able to be used in monitoring oil palm trees health. In addition, some analyses and conclusion were drawn to address related issues in the area.

1. Introduction

The oil palm tree, scientifically known as "Elaeis guineensis", are generally planted in the South-East Asia region, especially in Indonesia, Malaysia, and Thailand [1]. Malaysia is known as one of the world's largest palm oil exporters among these countries and currently handles 39% of world palm oil production and 44% of world exports [2], which makes the oil palm tree plantation as one of the important business that is beneficial to other investor countries [3].

Commonly, the oil palm is mainly grown for the use of vegetable oil production, and it has been the most consumed vegetable oil in the world compared to other types of vegetable oil [1]. Figure 1 shows the worldwide consumption of vegetable oil, which includes palm oil, which has been used extensively, beating other well-known types like soybean and rapeseed oil.

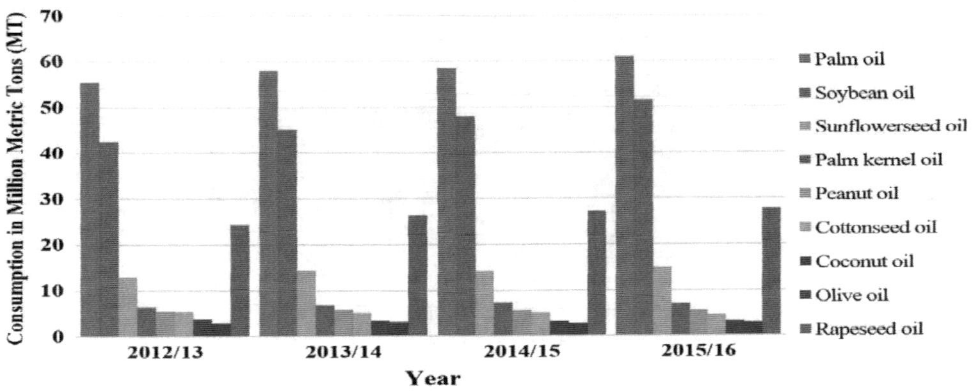

Figure 1. Consumption of vegetable oil throughout the world [1].

IOP Conf. Series: Materials Science and Engineering **705** (2019) 012033 doi:10.1088/1757-899X/705/1/012033

1.1. Introduction to Oil Palm
Oil palm is a type of plant that able to adapt to the tropical humid climate, where it can be planted in a rainy and sunny environment, and with warm temperature. It contains some characteristics like a crown-shaped top and a single-stemmed stalk. The fronds are emerging from the stalk top and extending outward with eight fronds. It has a shape like an eight-pointed star from sky view. The oil palm is known as a perceptual plant, which is more similar to a forest tree than other agricultural crops [1].

1.2. Oil Palm Diseases
Due to a large amount of exportation of palm oil to all over the world, the producers have to ensure the good health level of oil palm and able to supply sustainably. Therefore, all oil palm planters and farmers have to be careful of any diseases that able to destroy their plantation. Generally, pest or plant disease will influence a large number of commercial crops and causes a large economy lost. Previous report has shown that more than 10% of food production of the world is lost due to plant diseases [4]. Many types of plant diseases can be found around oil palm, and one of the most famous plant diseases is basal stem rot disease or Ganoderma Boninense.

Ganoderma is a critical issue to all oil palm plantations in the South-East Asia region, especially in Malaysia and Indonesia. This disease is caused by a type of fungus that is known as Ganoderma Boninense. This is a dangerous disease for oil palms as it can causes at least 50% loss of the palm in a farm [5][6]. It is hard for oil palm farmers and planters to detect the disease at the early stage of infection on the basal stem and leaves by visual inspection, and when they are noticeable, it is too late to be cured [7]. The infection of Ganoderma disease between oil palm trees is spreading by root to root contact, where healthy roots will have the possibility to be infected if contacted with the infected roots. When infected, brown discolouration of oil palm occurred due to the formation of Ganoderma. There is also some noticeable early-stage infection visual symptom from oil palm trees which is the formation of fruiting bodies. At first, it shows white button emerging on the surface of the bark, and then developed into mature fruiting body. Then, the waxy brown will turn into dark brown in colour at the upper surface and then white to creamy white in colour of the bottom surface. For young oil palm trees, disease symptom is shown by observing the yellow and drying of the older leaf. It can start from a certain part of the palm where the infection of disease is started [8]. Figure 2 shows an example of Ganoderma that has affected an oil palm tree.

Figure 2. Ganoderma Boninense [9].

Other than Ganoderma disease, oil palm plantations in Malaysia are also facing serious pest problem caused by bagworms, scientifically known as Psychidae which is a family of the Lepidoptera. It is known as leaf-eating caterpillars that have characteristics of possession of the bag and built from pieces of dried plant materials such as leaves and small twigs [10]. Consequently, yellow rust is also one of the plant diseases that have to be solved in the future to ensure the quality of oil palm exportation [4]. Figure 3 and 4 show the sample of bagworm and yellow rust.

Figure 3. Bagworm, also known as Psychidae [11].

Figure 4. Yellow rust, known as Phragmidium spp. [12].

2. Methods Used in Monitoring Oil Palm Tree Health

In oil palm plantation monitoring, there are many ways that have been suggested by researchers and scientists. A combination of hardware and sensors like the Unmanned Aerial Vehicles (UAV), remote sensing imaging and spectrometer with respective processes involving Artificial Neural Network (ANN) and Deep Learning (DL) algorithm were preferred. In the following sub-section, every method is discussed in sections of hardware and algorithm. Then, their advantages and limitations are observed and discussed in monitoring oil palm diseases.

2.1. Hardware

2.1.1. Remote Sensing Imaging
Remote sensing is known as a technique to deliver temporal and multimodal data representing the Earth's surface. As more sensors and advanced image fusion techniques can be implemented together, it will create the opportunities to process and extract complementary information even better. For example, optical and microwave remote sensing image fusion is a valuable tool to optimize multi-sensor image exploitation [13][14]. Remote sensing imaging techniques are very useful for oil palm studies and many applications can be applied by using it apart for pest and disease detection, such as land cover classification, tree counting, as well as for tree age, carbon production and yield estimation [15].

This method has brought a lot of advantages in monitoring Ganoderma disease because of its ability to obtain accurate information, preventing the spreading of disease and nutrient deficiency sufficiently from a satellite image. Figure 5 shows an example of a satellite image.

Figure 5. An example of satellite image [15].

2.1.2. Unmanned Aerial Vehicles (UAV)
UAV is a type of aircraft that can operate without human pilot control and is used for various purposes. It can perform multi-spectral and multimodal data acquisition by taking orthophotos. In [16], UAV is used in assisting the monitoring of oil palm trees growth and identifying the condition of the trees.

It has the advantages of providing low cost and quickly response information that suits for smallholders. Apart from that, it has a much better resolution up to several centimeters and also has greater flexibility in selecting suitable payloads with appropriate time and spatial resolutions [16].

2.1.3. Spectrometer
A spectrometer is a common tool used by various scientists to obtain information about an object or substance through the analysis of its light properties. In oil palm tree health monitoring, by observing the ultraviolet (UV), visible (blue, green and red bands) and near-infrared (NIR) wavelengths with certain range, it can be used to identify contaminants or levels of various substances in the leaves to detect possible diseases or unwanted toxin [17][18][19][20]. Figure 6 shows an example of the widely-used spectrometer.

IOP Conf. Series: Materials Science and Engineering **705** (2019) 012033 doi:10.1088/1757-899X/705/1/012033

Figure 6. GER 1500 field spectrometer [10].

2.2. *Algorithm*

2.2.1. *Vegetation Index Algorithm*

To determine the health status of the oil palm tree, it is important for the researcher to understand the green substance in a plant which is chlorophyll. The chlorophyll content is a significant biochemical parameter for testing crop status. It can be used to identify crop growing conditions, their physiological status, and health [21]. By investigating the chlorophyll pigment in the plant, researchers can perform image processing to detect vegetation area and then observe the health status of plants from the satellite image [22]. Apart for tree health status, users can also able to perform a count of oil palm trees at the plantation area automatically from this method [23][24][25].

In summary, four basics steps can be performed to detect vegetation area and then check the health status of oil palm trees. Firstly, a satellite image is needed as the input into the software. Secondly, the image will undergoes pre-processing such as noise or disturbance removal from the image. Thirdly, the clean satellite image will be segmented to extract necessary data that will be used in the research. Fourthly, an output graph or result data is produced. For example, the estimated vegetation area and chlorophyll of oil palm trees which can be used to monitor their health status.

2.2.2. *Artificial Neural Network (ANN) Algorithm*

ANN can be applied in the oil palm tree health monitoring to let the computer systems able to perform tasks normally requiring human intelligence, such as infected oil palm tree recognition. By applying the ANN analysis technique, the system can perform discriminating and classifying fungal infections in oil palm trees automatically at an early stage using raw, first, and second derivative spectroradiometer datasets. Besides that, ANN is also able to model the relationship between variables derived from different collected data [26]. An example of the algorithm structure representing ANN can be seen in Figure 7.

IOP Conf. Series: Materials Science and Engineering **705** (2019) 012033 doi:10.1088/1757-899X/705/1/012033

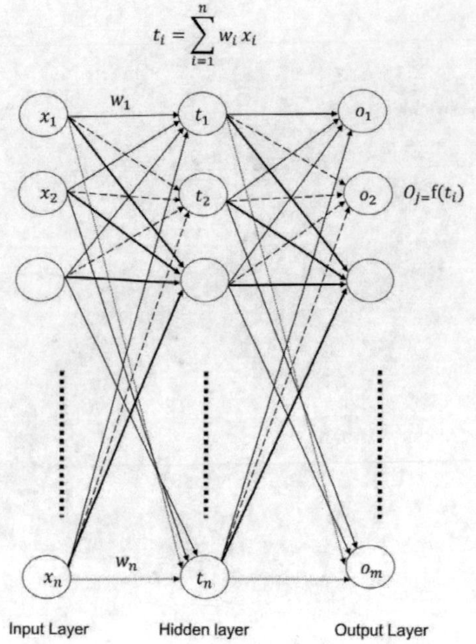

Figure 7. The structure of the artificial neural network (ANN) [7].

2.2.3. Deep Learning (DL) Algorithm

The idea of the DL algorithm is simple – to allow machine in understanding elements by learning the features and it is usually will be very good at decision making. To understand it easily, DL is about a machine using neural networks with more neurons and layers compared to the ANN, and is having interconnectivity with each other to calculate for the result. In other words, DL is a subset of machine learning that enables computers to solve more complex problems depending on different conditions and reasons [27]. By applying DL theory, the system can be designed to check the oil palm tree health. Figure 8 shows a sample of DL model [28].

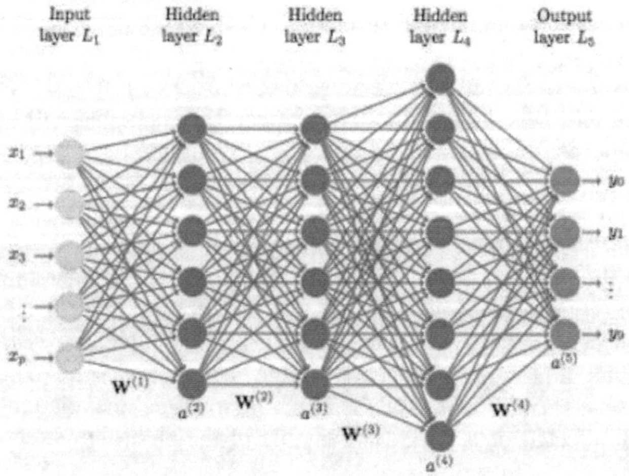

Figure 8. Feedforward deep learning models [28].

3. Discussions

In this section, previously mentioned methods will be discussed and compared to and the best way to monitoring oil palm tree disease can be observed. Comparison is carried out into two groups respectively, which are hardware and data processing or also known as an algorithm.

In here, a scoring system is introduced to compare them. Three levels are assigned to observe the effectiveness of every medium which is 1 represents weak tool, 2 represents moderate tool and 3 represents best tool. By comparing differences of hardware and software and then accumulate all the scores, method with highest score is assumed to be the best tool to be used. Table 1 compares the related hardware used while Table 2 summarizes the algorithm.

Table 1. Comparison of different hardware.

Differences	Remote sensing imaging	Unmanned Aerial Vehicles (UAV)	Spectrometer
Availability	Easy and free to get (3)	Need special tool (1)	Need special tool (1)
Accuracy	High (3)	High (3)	Moderate (2)
Consistency	High (3)	Moderate (2)	Moderate (2)
Sensitivity	High (3)	High (3)	High (3)
Timeliness	Short time (3)	Short time (3)	Long time (1)
Costing	Low (3)	High (1)	High (1)
SCORE	18	13	10

Table 2. Comparison of different algorithm.

Differences	Artificial Neural Network (ANN) Algorithm	Vegetation Index Algorithm	Deep Learning (DL) Algorithm
Completeness	Good (3)	Moderate (2)	Good (3)
Accuracy	Moderate (2)	High (3)	Moderate (2)
Consistency	Moderate (2)	High (1)	Moderate (2)
Sensitivity	High (3)	High (3)	High (3)
SCORE	10	9	10

From Table 1, it can be seen that remote sensing imaging is the best choice. This is because it has the highest score of 18 among all of the hardware. For the availability, remote sensing imaging can provide an easy and free platform for users to download from the internet directly without any costing. For example, Google Maps is one of the platforms that frequently used by researchers to get satellite images immediately. Because of that, remote sensing imaging has involved low expenses in researches from the consideration of costing and also it just takes a short time for researchers to get climate data or geographical images from satellite compared to other hardware. Remote sensing imaging and UAV are mostly used to collect the image of an oil palm farm from Earth's surface, the only difference between them is one is for wider coverage and the other one is for the smallholder. Thus, remote sensing imaging will be the best tool to be used in the system of monitoring the oil palm tree health especially for large scale plantation.

Table 2 has shown three types of the algorithm that are possible to be used. From here, ANN and DL have the same score of 10, whereas the vegetation index algorithm has a score of 9. The operation model of DL is the upgrading version of the ANN algorithm, the difference between them is DL has enabled many practical applications of machine learning and by extension the overall field of artificial intelligence. Moreover, the vegetation index algorithm can be run manually by just inserting data into the software but with the aid of the ANN algorithm and DL algorithm, it is possible to let the monitoring oil palm tree health system to run automatically. This is said to be a contribution to the current research for oil palm disease.

The current program for the vegetation index algorithm that has been used to check the health status of oil palm is suggested to be upgraded which able to show the different healthy levels of oil palms by using percentage, instead of the conventional way of calculating the vegetation index. With this function, users can understand how the infection level of the tree is immediately.

Also, it is good for future research to try to combine the ANN / DL algorithm with the vegetation index into a complete system. This idea can let the monitoring Ganoderma disease system to check for any infection of the oil palm tree automatically and also able the system to self-learning from the experience for upgrading the system to be more complete. It is also suggested to set in the detection system so that it can run automatically to check the damage level of the oil palm and also it can inform the owner immediately by using a message.

4. Conclusion

Oil palm products have become more and more important in our daily lives, especially to the exporter. Due to the high economic potential of oil palm in the future, Malaysia should find out the most effective way to prevent disease attack for the purpose to ensure the good quality of palm oil production. To monitor oil palm tree health, remote sensing imaging can be further applied due to its low costing, short timing to get satellite image, high accuracy, high consistency, high sensitivity, and easy to handle. To process the data, it will be a good idea to combine the ANN / DL and the vegetation index algorithm to build an artificial intelligence monitoring system.

References

[1] Chong K L, Kanniah K D, Pohl C and Tan K P 2017 A review of remote sensing applications for oil palm studies *Geo-spatial Information Science* **20** 184–200

[2] Khairunniza-Bejo S, Yusoff Y, Yusoff N S N, Seman I A and Anuar M I 2015 Identification of Healthy and BSR-Infected Oil Palm Trees Using Color Indices *International Journal of Agricultural and Biosystems Engineering* **9** 785–788

[3] Tugi A, Rasib A W, Suri M A, Zainon O, Yusoff A R M, Rahman M Z A, Sari N A and Darwin N 2015 Oil Palm Tree Growth Monitoring For Smallholders By Using Unmanned Aerial Vehicle *Jurnal Teknologi* **77** 17–27

[4] Huang W, Luo J, Zhang J, Zhao J, Zhao C, Wang J, Yang G, Huang M, Huang L and Du S 2012 Crop Disease and Pest Monitoring by Remote Sensing *Remote Sensing – Application* 32–76

[5] Santoso H, Gunawan T, Jatmiko R H, Darmosarkoro W and Minasny B 2010 Mapping and identifying basal stem rot disease in oil palms in North Sumatra with QuickBird imagery *Precision Agriculture* **12** 233–48

[6] Mohd N and Maris N 2015 Remote Sensing on Ganoderma Disease *18th International Oil Palm Conference* 1–46

[7] Ahmadi P, Muharam F M, Ahmad K, Mansor S and Seman I A 2017 Early Detection of Ganoderma Basal Stem Rot of Oil Palms Using Artificial Neural Network Spectral Analysis *Plant Disease* **101** 1009–16

[8] Alias M, Adnan A M I, Jugah K, Ishaq I and Fizree Z A 2013 Detection of Basal Stem Rot (BSR) disease at oil palm plantation using hyperspectral imaging *5th Workshop on Hyperspectral Image and Signal Processing: Evolution in Remote Sensing (WHISPERS)* 1–4

[9] Benoît C 2017 Selection of Ganoderma-resistant material: identification of 4 resistance loci [online] Available at: https://www.palmelit.com/en/news/selection-of-ganoderma-resistant-material [Accessed 8 May 2019].

[10] Aziz N, Omar W, Kassim R and Kamarudin N 2012 Remote sensing measurement for detection of bagworm infestation in oil palm plantation *MPOB Information Series* 1-8

[11] Lee T 2015 It's In the Bag [online] Available at: https://kentuckypestnews.wordpress.com/2015/06/30/its-in-the-bag/ [Accessed 8 May 2019].

[12] Symptoms of Diseases and Disorders Palm Symptoms [online] Available at: https://idtools.org/id/palms/symptoms/factsheet.php?name=Leaf+Spots+and+Leaf+Blights [Accessed 8 May 2019].

[13] Pohl C, Chong K L and John G 2011 Multisensor Approach To Oil Palm Plantation Monitoring Using Data Fusion and GIS *ACRS 2015 - 36th Asian Conference on Remote Sensing: Fostering Resilient Growth in Asia, Proceedings* 1-10
[14] Nisfariza M N M 2018 A Decade of Remote Sensing Research in Ganoderma Disease in Malaysia *19th International Oil Palm Conference* 1-50
[15] Wang J, Sammis T W, Gutschick V P, Gebremichael M, Dennis S O and Harrison R E 2010 Review of Satellite Remote Sensing Use in Forest Health Studies *The Open Geography Journal* **3** 28–42
[16] Shamshiri R R 2007 A Lecture note on Unmanned Aerial Vehicles (UAV) to Support Precision Agriculture Research in Oil Palm Plantations 1-3
[17] Shafr H Z M and Hamdan N 2009 Hyperspectral Imagery for Mapping Disease Infection in Oil Palm Plantation Using Vegetation Indices and Red Edge Techniques *American Journal of Applied Sciences* **6** 1031–5
[18] Li C, Zhu X, Wei Y, Cao S, Guo X, Yu X and Chang C 2018 Estimating apple tree canopy chlorophyll content based on Sentinel-2A remote sensing imaging *Scientific Reports* **8** 1-10
[19] Rendana M, Abdul R S, Lihan T, Mohd R I W and Ali R Z 2015 A Review of Methods for Detecting Nutrient Stress of Oil Palm in Malaysia *Journal of Applied Environmental and Biological Sciences.* **5** 60–4
[20] Anuar M I 2009 Modified vegetation indices for Ganoderma disease detection in oil palm from field spectroradiometer data *Journal of Applied Remote Sensing* **3** 1-24
[21] Nandibewoor A, Muddebihal S and Hegadi R 2017 Estimation Of Chlorophyll Content In Crop Using Multispectral Satellite Imagery *International Journal of Scientific & Technology Research* **6** 351–4
[22] Pandya A and Priya R S 2015 Classification of Vegetation Area From Satellite *International Journal of Research in IT, Management and Engineering* **5** 48–56
[23] Kalapala M 2014 Estimation of Tree Count from Satellite Imagery through Mathematical Morphology *International Journal of Advanced Research in Computer Science and Software Engineering* **4** 490–5
[24] Bechtel B, Ringeler A and Böhner J 2008 Segmentation for Object Extraction of Trees Using Matlab and Saga 1–12
[25] Emad A A and Noor A A 2018 A count of palm trees from satellite image *Contemporary Engineering Sciences* **11** 3663–72
[26] Collingwood A, Treitz P, Charbonneau F and Atkinson D 2014 Artificial Neural Network Modeling of High Arctic Phytomass Using Synthetic Aperture Radar and Multispectral Data *Remote Sensing* **6** 2134–53
[27] Ball J E, Anderson D T and Chan C S 2017 Comprehensive survey of deep learning in remote sensing: theories, tools, and challenges for the community *Journal of Applied Remote Sensing* **11** 042629
[28] Feedforward Deep Learning Models [online] Available at: http://uc-r.github.io/feedforward_DNN [Accessed 8 May 2019].

IOP Conf. Series: Materials Science and Engineering **705** (2019) 012034 doi:10.1088/1757-899X/705/1/012034

Gas source localization accuracy: A comparison between conventional, weighted arithmetic mean and kernel-based gas distribution mapping methods in small indoor area

A S A Yeon[*1,2], R Visvanathan[1,2], K Kamarudin[1,2], S M Mamduh[1,3], A Zakaria[1,2], L M Kamarudin[1,4]

[1]Centre of Excellence for Advanced Sensor Technology (CEASTech), Universiti Malaysia Perlis (UniMAP), Arau, Perlis 02600, Malaysia.
[2]Pusat Pengajian Kejuruteraan Mekatronik, Universiti Malaysia Perlis (UniMAP), Arau, Perlis 02600, Malaysia.
[3]Pusat Pengajian Kejuruteraan Mikroelektronik, Universiti Malaysia Perlis (UniMAP), Arau, Perlis 02600, Malaysia.
[4]Pusat Pengajian Kejuruteraan Komputer dan Perhubungan, Universiti Malaysia Perlis (UniMAP), Arau, Perlis 02600, Malaysia.

[*]Corresponding author: ahmadshakaff@gmail.com

Abstract. This work focuses on comparing gas source localization accuracy of three types of gas source localization methods which are the conventional method, weighted-averaging mean method and variance map from DM+V method. Gas source localization accuracy in this work means the distance error between the calculated source location and the actual source location. All methods were implemented offline using the measurements collected from real-time gas source localization experiments that were done beforehand. The dataset consists of gas sensor measurements, and mobile robot's location in a 3m by 6m test area. The goal of the work is to shine a light on why recent works in the field are more focusing on improving probabilistic and map-based algorithms. This work also shows how the kernel size of kernel-based gas distribution mapping might affect the gas source localization accuracy in a small indoor area.

1. Introduction

Gas source localization in a way is a problem of calculating the exact gas source location in the environment. Gas source localization using mobile robots is still a fledgling field which started in the early 90s where researchers began to build mobile robots that can find the source of chemical releases [1].

A problem that continuously occurred to researchers in the past when trying to tackle the gas source localization problem is that the gas plume is circulated through wind which is turbulent in nature. This makes it difficult for researchers to accurately calculate gas source location because of the gas concentration is not a smooth gradient from the source towards the open environment. Instead, it is intermittent and often fluctuating. It was often found that the highest concentration patch in the environment is not the actual source location [2].

Gas source localization tasks are usually divided into three sub-tasks [3] as below:

a) Gas plume searching: Searching the environment for the presence of gas plumes.
b) Gas plume tracing: Tracing the found gas plume towards the gas source.
c) Gas source declaration: Declaring that the source is found be it because the mobile robot has reached the vicinity of the source or the final location of the source has been calculated.

As mentioned above, the third subtask which is gas source declaration is quite vague between research groups. The traditional way of declaring the source is by manually observing the mobile robot reached the vicinity of the gas source location [4]. This method is acceptable as early works in the field were mostly focusing on testing gradient-based or bio-inspired algorithms capability to find and trace gas plumes towards the source. But in real life implementation, it is ideal for the mobile robot to be fully autonomous and aware that it has found the gas source or calculate where the source location is.

The focus of this work is to compare the accuracy of gas source localization or in other words, the distance error from the calculated gas source location to the actual gas source location of three methods that can be used to solve gas source localization problem. The three methods are described in Section 3. The main goal of this work is to show the significance developing new methods for gas source declaration and to shine a light on why modern works in this field are now more focusing on improving probabilistic and map-based algorithms.

2. Literature Review

The first impactful work on gas source localization using a mobile robot was done by equipping a mobile robot with 4 gas sensors at 4 different positions. The robot will stop and take gas sensor measurements before taking steps towards the direction of the highest recorded measurement gas sensor [1]. The process was repeated and if the robot managed to get into the vicinity of the gas source, the experimental run was marked as a success and the mobile robot's run is terminated.

Gas source declaration by terminating the robot when the robot gets near the gas source continued for quite some time [4-7]. However, the researchers were not specifically focused on gas source declaration but rather to show that the implemented gas plume searching, and gas plume tracing techniques are working as intended which is to move the mobile robot towards the gas source. Moreover, hardware limitations at the time are preventing researchers from implementing more complex algorithms. With advancements made in terms of hardware in recent years, algorithms or methods with high computing cost are now possible to be implemented on mobile robots.

High computing cost algorithms such as probabilistic-based gas source localization was discussed as an idea in 2007. It was demonstrated that gas source localization task can be performed by mobile robots using an infotactic approach rather than gradient climbing [8]. Infotactic algorithms are typically Bayesian inference methods where the mobile robot holds a probability distribution of where the gas source location is. The probability distribution keeps updating when more information is made available. Typically, the information is used to build a gas distribution map (GDM). The map is updated continuously until the probability of a certain point in the map increases to 1 [9-11]. Moreover, the information can be used to guide the robot towards the gas source instead of map building. This was specifically done in 2011 but with a particle filter-based algorithm instead of Bayesian inference-based algorithm where the work shows that the mobile robot performed a searching and exploratory-like behavior while also calculating the gas source location [12].

Breakthrough in kernel-based GDM method to solve gas source localization were demonstrated in 2004 [13]. More importantly, in 2007 the method was implemented alongside simultaneous localization and mapping (SLAM) [14]. Results from the work shows significant potential as the actual gas source location corresponds to the highest estimated concentration. Improvements were made by the introduction of Kernel DM+V where variance estimation is computed to build a variance map [15]. The work shows that variability lies with the gas sensor measurements which makes the variance map is more accurate on estimating the gas source location. Several work on DM+V were done in the following years to improve it such as Kernel DM+V/W [16] where wind information is added and also a 3-dimensional GDM [17].

2.1. Related Works

This work is heavily influenced by previous works done by the research group in CEASTech. A 6 by 3-meter testbed in an open indoor area were used specifically for real-time gas source localization experiments [18]. A large gas sensor array (LGSA) was built specifically to collect gas sensor measurements during an experiment. The sensor array covered the experimental area and the collected data can act as ground truths. Furthermore, the area is covered by 4 mounted ceiling cameras that can track mobile robot's location in the area [19]. The overview of the testbed and its system is shown in Figure 1. Multiple gas source localization works have been done previously utilizing the testbed to study behavior-based algorithms [20,21,22] and gas dispersion with induced airflow on the testbed [23]. Moreover, CEASTech have also been working on integrating simultaneous localization and mapping (SLAM) with GDM in recent years [24] that have shown potential to localize gas source locations in real-world scenarios.

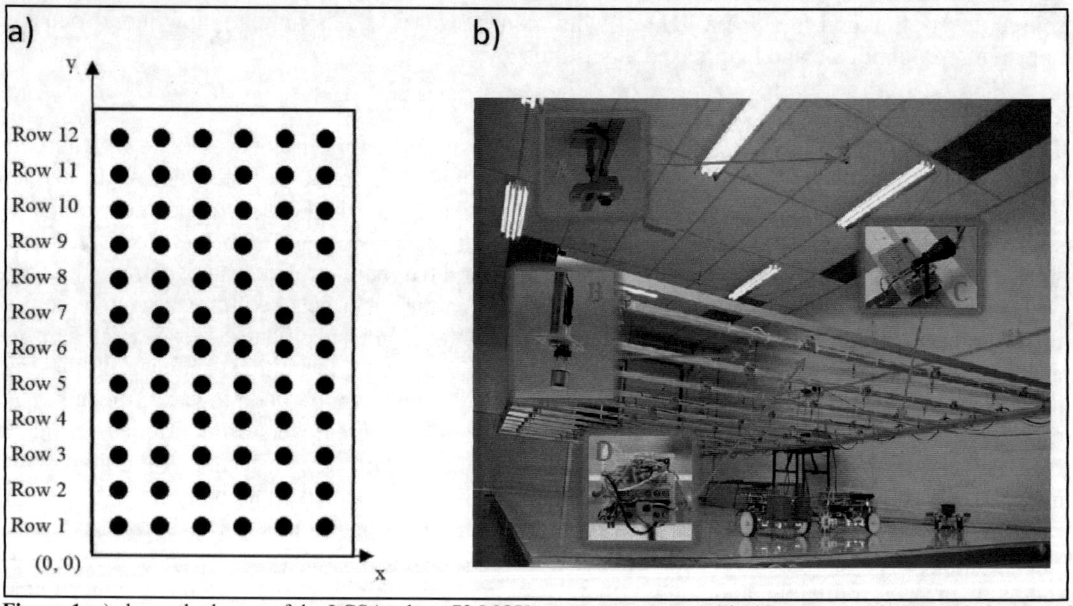

Figure 1. a) shows the layout of the LGSA where 72 MOX gas sensors arranged in 12 rows covering the whole area of the testbed and where each row consists of 6 MOX gas sensors while b) shows the mobile robot testbed consisting of ceiling mounted cameras, gas sensors, sensor boards and wireless modules.

3. Methodology

With the focus of this work is to compare gas source localization accuracy of three methods, this work was based on the dataset from gas source localization experiments on the testbed shown in Figure 1 with no obstacles or induced wind [18] using a mobile robot that implemented behavior-based or reactive gradient climbing algorithms [21] while performing gas source localization tasks. The behavior-based algorithm used for this work is explained in Section 3.1. During the experiment, ethanol gas was released into the test area at location $(x,y) = (150,220)$.

From the dataset, we obtained the mobile robot's location and gas sensor's measurements. Using the data, we performed 3 methods of gas source localization offline to compare its distance error from actual gas source location using the average measurements of the two FIGARO TGS2600 metal oxide gas sensors that were installed on the mobile robot. The three methods are conventional, weighted averaging mean and variance map of DM+V which are thoroughly described in Section 3.2.

3.1. *Zig-Zag and Braitenberg Fusion*

The Zig-Zag and Braitenberg algorithms were implemented on a mobile robot. Each of the algorithm is triggered during the experiment based on the sensor signal value *s* in a state machine manner. Sensor signal value is calculated from Equation (1) where R_0 represents the baseline resistance of the gas sensor taken at the start of an experimental run and R_s is the resistance of the gas sensor calculated at time *t*.

$$s = \frac{\Delta R}{R_0} = \frac{(R_0 - R_S)}{R_0} = 1 - \frac{R_S}{R_0} \quad (1)$$

The mobile robot performed the Zig-Zag behavior when *s* is below than 0.3 and Braitenberg when it's higher which means that the mobile robot will perform a searching like behavior when it is not detecting high gas plume concentration and a gas plume tracing behavior when its sensors measured a high gas concentration. Example of the mobile robot's behavior with the implementation of this algorithm is depicted in Figure 2 where it shows that the mobile robot lingered around the areas where the gas concentration is higher.

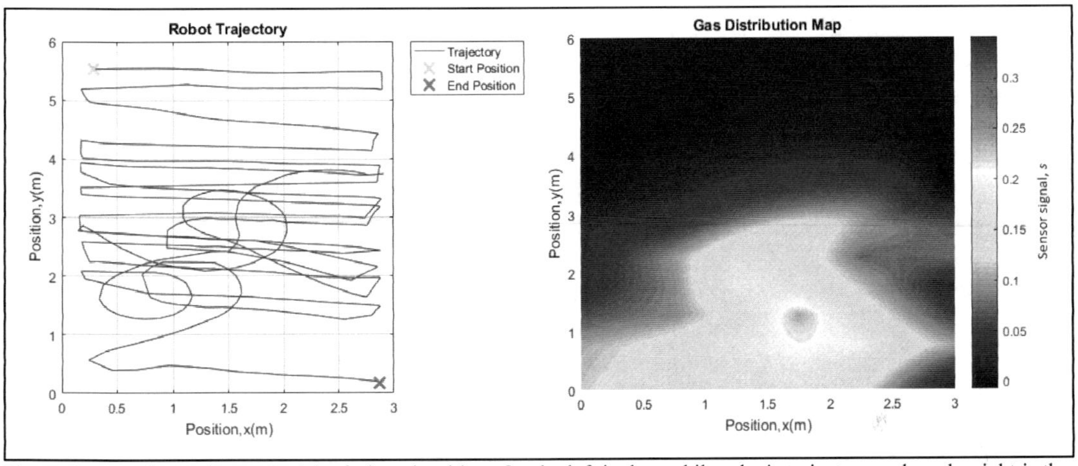

Figure 2. Experimental run #7 of the fusion algorithm. On the left is the mobile robot's trajectory and on the right is the cubic interpolation gas distribution map obtained using the mean measurements from testbed's LGSA.

3.2. *Gas Source Localization Method*

As calculating the gas source location, three methods were chosen for this work to compare its accuracy. The three methods are as follows:

a) Conventional method: Taking the location of the highest recorded gas sensor measurement.
b) Weighted arithmetic mean: Calculate the location of the gas source by applying sensor signal values as weights.
c) DM+V variance map: Taking the location of the highest variance value within the testbed as gas source location.

More details on weighted arithmetic mean method and DM+V variance map method is presented in Section 3.2.1 and 3.2.2.

3.2.1. *Weighted Averaging Mean.* By applying the sensor signal values to act as weights to the robot's location, a centroid is possible to be calculated. This strategy is a simplified version of a gas source localization algorithm based on particle filter in wireless sensor network which was done by calculating the source location using information gained by multiple static gas detectors [25]. A non-empty set of data consists of *x*-axis and *y*-axis coordinates denoted as $\{x_1, x_2, x_3, \ldots, x_n\}$ and

$\{y_1, y_2, y_3, \ldots, y_n\}$. A set of sensor measurements taken at the coordinate such as sensor voltages or signal values act as weights can be denoted as$\{w_1, w_2, w_3, \ldots, w_n\}$. Equation (2) and Equation (3) denotes the calculation of the weighted mean where x_p and y_p is the predicted *x*-axis and *y*-axis location of the gas source. In this work, *w* is sensor signal value *s* where only s≥0 is taken into consideration.

$$x_p = \frac{\sum_{i=1}^{n} s_i x_i}{\sum_{i=1}^{n} s_i} \tag{2}$$

$$y_p = \frac{\sum_{i=1}^{n} s_i y_i}{\sum_{i=1}^{n} s_i} \tag{3}$$

3.2.2. *DM+V Variance Map.* Kernel DM+V [15] models the gas dispersion using a univariate Gaussian function *N* to convolve the sensor measurement r_i at location x_i across a grid map expressed by Equation (4). the *dist(k)* is defined as the Euclidian distance of k_{th} cell in a grid map from point x_i.

$$N(dist(k), \sigma) = \frac{1}{\sigma\sqrt{2\pi}} e^{-(dist(k)^2)/2\sigma^2} \tag{4}$$

A weight map, Ω_k and a weighted reading R_k for *n* sensor readings are computed by Equation (5) and Equation (6) respectively. The σ is the kernel width which means the amount of extrapolation of the gas sensor measurement r_i. The likelihood of r_i to be expanded wider across the grid map is higher with the increase of σ. The confidence of estimation of cell *k* is obtained by normalizing Ω_k to a confidence map, α_k with scaling parameter, σ_α denoted by Equation (7).

$$\Omega_k = \sum_{i=1}^{n} N(dist(k), \sigma) \tag{5}$$

$$R_k = \sum_{i=1}^{n} N(dist(k), \sigma) . r_i \tag{6}$$

$$\alpha_k = 1 - e^{-\frac{(\Omega_k)^2}{\sigma_\Omega^2}} \tag{7}$$

Mean concentration map can be calculated using Equation (8) where r_o *is* the average of all gas sensor measurements. Variance map v_k is computed from weighted variance V_k from Equation (9) and Equation (10) respectively where *k(i)* is the cell of gas sensor measurement and point x_i, $r_{k(i)}$ is the mean value of cell *k* and v_o is the average overall variance computation. v_o represents the sensor variability at every measurement point. The cell with the highest v_o is the predicted gas source location.

$$r_k = \alpha_k \frac{R_k}{\Omega_k} + \{1 - \alpha_k\} r_o \tag{8}$$

$$V_k = \sum_{i=1}^{n} N(dist(k), \sigma) \left(r_i - r_{k(i)}\right)^2 \tag{9}$$

$$v_k = \alpha_k \frac{V_k}{\Omega_k} + \{1 - \alpha_k\} v_o \tag{10}$$

4. Results

A total of 20 experimental runs were done during the experiment where the mobile robot performed gas source localization tasks using the algorithm in Section 3.1. The collected data was then used by to estimate the gas source location with the 3 methods mentioned in Section 3.2. The accuracy of gas source localization or the distance error between the estimated location and actual gas source location were calculated using the 3 methods. The results are shown in Table 1.

Table 1. Distance error of the estimated source location using the three methods.

	Distance Error (cm)		
Run Number	Variance Map (σ=0.1)	Weighted Arithmetic Mean	Conventional
1	31.89	53.43	25.50
2	57.71	62.56	70.08
3	165.53	145.19	186.34
4	72.20	111.45	227.75
5	81.74	81.67	81.42
6	37.12	53.40	40.71
7	53.94	12.45	44.92
8	55.80	68.79	43.27
9	56.30	98.60	82.15
10	33.29	17.65	22.15
11	47.43	15.50	20.91
12	99.13	140.41	49.80
13	166.41	20.34	97.77
14	200.94	107.60	70.78
15	159.14	67.32	57.14
16	81.15	23.39	91.24
17	20.62	42.75	66.36
18	39.20	73.24	38.93
19	37.00	86.46	54.74
20	17.49	19.38	43.83
Average (cm)	75.70	65.08	70.79

From Table 1, it can be deduced that in a less complex small indoor area, weighted arithmetic mean method in general is sufficient enough for accurate gas source localization. This could be best implemented on mobile robots when testing out behavior-based path planning algorithms as it is not computationally taxing. Table 1 also clearly shows the problem that researchers of this field have been facing where the location of the highest average gas measurement is not always near the actual gas source location as in Run 3 and 4, the error is around 2 meters.

Moreover, results have shown that the variance map method is in average the least accurate of the three, but it is the most accurate in 7 out of 20 runs (Run 2, 4, 6, 9, 17, 19 & 20). From experimenting with building the variance map of DM+V, we have noticed that the kernel size played an important part in getting accurate gas source location estimation. Table 2 shows how the kernel size affects the accuracy of gas source localization. From Table 2, it can be seen that the best kernel size is on a case by case basis. Kernel sizes larger than 0.4 is not shown as the interpolation area becomes too large and the highest variance are often out of the test area boundaries.

Table 2. Distance error of the estimated source location using variance map of multiple kernel sizes.

Run Number	Distance Error (cm)			
	σ=0.1	σ=0.2	σ=0.3	σ=0.4
1	31.89	48.33	58.82	72.78
2	57.71	69.12	77.47	28.23
3	165.53	98.29	98.08	230.27
4	72.20	239.24	238.84	265.71
5	81.74	75.03	69.89	64.20
6	37.12	26.93	24.74	16.40
7	53.94	57.72	73.76	74.97
8	55.80	89.47	95.08	90.09
9	56.30	149.16	149.86	154.54
10	33.29	44.05	43.27	37.34
11	47.43	44.55	43.01	36.50
12	99.13	58.52	78.16	74.33
13	166.41	106.51	112.73	139.09
14	200.94	198.81	191.08	53.26
15	159.14	130.10	87.80	76.85
16	81.15	50.99	47.07	39.56
17	20.62	16.76	31.38	47.42
18	39.20	39.62	45.45	55.76
19	37.00	150.56	51.62	49.48
20	17.49	34.99	25.94	26.93
Average (cm)	75.70	86.44	82.20	81.69

5. Conclusion & Future Work

The accuracy of gas source localization via kernel-based gas distribution mapping is influenced by the kernel size. Other works have also shown that the kernel shape and size can be manipulated with additional information such as wind [16], temperature and humidity [26]. The shape of the environment such as walls and obstacles should also be taken into consideration when determining the kernel size as to avoid gas sensor measurements being interpolated out of bounds. It can be said that the gas distribution mapping method has a lot of potential to be further explored.

References

[1] Rozas R, Morales J and Vega, D 1991 Artificial smell detection for robotic navigation *Fifth Int. Conf. on Advanced Robotics' Robots in Unstructured Environments* pp 1730-1733
[2] Fukazawa Y and Ishida H 2009 Estimating gas-source location in outdoor environment using mobile robot equipped with gas sensors and anemometer *SENSORS, 2009 IEEE* pp 1721-1724
[3] Chen X X and Huang J 2019 Odor source localization algorithms on mobile robots: A review and future outlook *Robotics and Autonomous Systems* **112** pp 123-136
[4] Lilienthal A J and Duckett T 2003 Experimental analysis of smelling Braitenberg vehicles *IEEE int. conf. on advanced robotics (Coimbra)* **1** pp 375-380
[5] Ferri G, Caselli E, Mattoli V, Mondini A, Mazzolai B and Dario P 2009 SPIRAL: A novel biologically-inspired algorithm for gas/odor source localization in an indoor environment with no strong airflow *Robotics and Autonomous Systems* **57** pp 393-402
[6] Sandini G, Lucarini G and Varoli M 1993 Gradient driven self-organizing systems *Proc. of Int. Conf. on Intelligent Robots and Systems* **1** pp 429-432
[7] Kuwana Y, Shimoyama I and Miura H 1995 Steering control of a mobile robot using insect

antennae *Proc. Int. Conf. on Intelligent Robots and Systems. Human Robot Interaction and Cooperative Robots* **2** pp 530-535

[8] Vergassola M, Villermaux E and Shraiman B I 2007 'Infotaxis' as a strategy for searching without gradients *Nature* **445** pp 406-409

[9] Blanco J L, Monroy J G, Lilienthal A and Gonzalez-Jimenez J 2013 A kalman filter based approach to probabilistic gas distribution mapping *Proc. of the 28th Annual ACM Symposium on Applied Computing* pp 217-222

[10] Grünbaum D and Willis M A 2015 Spatial memory-based behaviors for locating sources of odor plumes *Movement ecology* **3** 11

[11] Hajieghrary H, Hsieh M A and Schwartz I B 2016 Multi-agent search for source localization in a turbulent medium *Physics Letters A* **380** pp 1698-1705

[12] Li J G, Meng Q H, Wang Y and Zeng M 2011 Odor source localization using a mobile robot in outdoor airflow environments with a particle filter algorithm *Autonomous Robots* **30** pp 281-292

[13] Lilienthal A and Duckett T 2004 Building gas concentration gridmaps with a mobile robot *Robotics and Autonomous Systems* **48** pp 3-16

[14] Lilienthal A J, Loutfi A, Blanco J L, Galindo C and Gonzalez J 2007 A rao-blackwellisation approach to gdm-slam: Integrating slam and gas distribution mapping *3rd European conf. on mobile robots* pp 126-131

[15] Lilienthal A J, Reggente M, Trincavelli M, Blanco J L and Gonzalez J 2009 A statistical approach to gas distribution modelling with mobile robots-the kernel dm+ v algorithm *Int. Conf. on Intelligent Robots and Systems* pp 570-576

[16] Reggente M and Lilienthal A J 2009 Statistical evaluation of the Kernel DM+ V/W algorithm for building gas distribution maps in uncontrolled environments *Procedia Chemistry* **1** pp 481-484

[17] Reggente M and Lilienthal A J 2010 The 3d-kernel dm+ v/w algorithm: Using wind information in three dimensional gas distribution modelling with a mobile robot *SENSORS, 2010 IEEE* pp. 999-1004

[18] Syed Zakaria S, Visvanathan R, Kamarudin K, Ali Yeon A, Zakaria A and Kamarudin L 2015 Development of a Scalable Testbed for Mobile Olfaction Verification *Sensors* **15** pp 30894-30912

[19] Visvanathan R, Mamduh S M, Kamarudin K, Yeon A S A, Zakaria A, Shakaff A Y M and Saad F S A 2015 Mobile robot localization system using multiple ceiling mounted cameras *2015 IEEE SENSORS* pp 1-4

[20] Yeon A S A, Visvanathan R, Mamduh S M, Kamarudin K, Kamarudin L M and Zakaria A 2015 Implementation of behaviour based robot with sense of smell and sight *Procedia Computer Science* **76** pp 119-125

[21] Yeon A S A, Kamarudin K, Visvanathan R, Zakaria S M M S, Zakaria A and Kamarudin L M 2018 Gas Source Localization via Behaviour Based Mobile Robot and Weighted Arithmetic Mean *IOP Conf. Series: Materials Science and Engineering* **318** 012049

[22] Mamduh S M, Kamarudin K, Shakaff A Y M, Zakaria A, Visvanathan R, Yeon A S A and Nasir A S A 2018 Gas Source Localization using Grey Wolf Optimizer *J. of Telecommunication, Electronic and Computer Engineering* **10** pp 95-98

[23] Mamduh S M, Kamarudin K, Visvanathan R, Yeon A S A, Shakaff A Y M, Zakaria A and Abdullah A H 2017 Gas dispersion with induced airflow in mobile olfaction testbed *AIP Conf. Proc.* **1808** 020051

[24] Kamarudin K, Md Shakaff A Y, Bennetts V H, Mamduh S M, Zakaria A, Visvanathan R and Kamarudin L M 2018 Integrating SLAM and gas distribution mapping (SLAM-GDM) for real-time gas source localization *Advanced Robotics* **32** pp 903-917

[25] Li Q, Liu Z and Xiao X 2015 A gas source localization algorithm based on particle filter in wireless sensor network *Int. J. of Distributed Sensor Networks* **11** 874532

[26] Kamarudin K, Bennetts V H, Mamduh S M, Visvanathan R, Yeon A S A, Shakaff A Y M and Kamarudin L M 2017 Cross-sensitivity of metal oxide gas sensor to ambient temperature and humidity: Effects on gas distribution mapping *AIP Conf. Proc.* **1808** 020025

IOP Conf. Series: Materials Science and Engineering **705** (2019) 012035 doi:10.1088/1757-899X/705/1/012035

Tree Crown Density Analysis from Hyperspectral Image

Rina Syazwani Zulkafli, SAA Shukor*

School of Mechatronic Engineering, Universiti Malaysia Perlis, 02600 Perlis, Malaysia

*Corresponding author: shazmin@unimap.edu.my

Abstract. A study was conducted to investigate whether reflectance data of hyperspectral image of an area could be used to extract related physical features to produce mapping of vegetation density. This paper explains on estimating percentage of vegetation coverage based on Normalized Difference Vegetation Index (NDVI). Image segmentation based on thresholding was used to separate different features of the land entities like soil, water and road. From here, NDVI values can be integrated for further segmenting the vegetation features. The colour segmentation method is then able to classify the vegetation according to their density level, which can be used to determine tree crown density. Test conducted towards a hyperspectral image shows that different density level can be extracted, where it contains about 7.5% high level tree crown density, 1.8% medium crown density and 5% low. More tests need to be conducted in order to proof the workability of the developed algorithm in analysing hyperspectral images from tree crown density mapping.

1. Introduction

Forest inventory is an activity that deals with trees and information obtained in a specific circumstance land area. This activity is not new and was carried out since the late 18th century using traditional method [1]. There are several reasons which makes forest inventory important, which includes structural development about the forest management, like evaluating the raw material of wood and acquiring number of individual trees. Apart from that, forest inventory activity would define spatial distribution of urban forest, and can be used to distinguish species composition. An optimal harvesting schedule can also be planned if a large-scale inventory is carried out.

In the early stage, the main method used in forest inventories was totally enumeration. However, they need a long period of time and energy to accomplish the inventories since some of the area is inaccessible. Sometimes, the interested area cannot be estimated completely since the area may be steep and at the same time, they may be facing the threat of wild animals. Furthermore, this method cannot provide complete coverage of large areas.

Due to this limitation, researchers have developed new method to estimate the timber information that can reduce the time consumed and even more effective than the manual method, which is based on image analysis. This would allow foresters or researchers to interpret the image into some valuable information regarding to the electromagnetic radiation from the Earth's surface. This modern method offers increases in speed of analysis, high precision results and high timeliness of forest inventory. Remote sensing has been a valuable source of information in providing suitable images for mapping and monitoring forest activities [2].

Among the available approaches in remote sensing that support forest inventory is through hyperspectral imaging. Hyperspectral is the technology of extracting information from Earth's surface through radiance obtained either from airborne or spaceborne sensors [3]. The image produces by the sensor contains spectral bands due to its wavelength, which is suitable in extracting spatial-based information, like what is being done in this project. Hyperspectral images have been used for various applications in forestry and agriculture, such as for tree species recognition [4] and detection of plant disease and stress [5] [6].

One of the important studies in forest inventory is tree crown density. Data of forest / vegetation area taken from airborne, which is less complex to be collected compared to terrestrial / land, can be used to map tree crown density, which can be used to estimate tree volume, canopy structure and improve tree species classification [7].

This paper concentrates on the work of extracting and evaluating the density level of tree crown from a hyperspectral image. Due to the important role of segmentation in image processing, especially in this case of determining and mapping tree density, several methods of segmentation are tested. The selected methods are then being compared to determine the best in segmenting and extracting the image feature, i.e. tree crown density. Feature extraction results will classify the level of crown density (low, medium or high). In addition, the total pixels for all levels and percentage of tree crown density will also be calculated. To assist the usage of the algorithm in the future, a suitable Graphical User Interface (GUI) is also developed.

Several studies have been performed in image segmentation. Researchers in [8] stated that segmentation can be categorized into three classes, feature thresholding or clustering, edge detection and region extraction, while in [9], there are six segmentation methods which include histogram-based method, neural network-based segmentation, region-based method and clustering. Segmentation in image processing can be defined as a process of assigning a label to every pixel in an image of certain characteristics. The characteristics can be either with respect to colour, intensity or texture. For this project, three methods of segmentation will be tested which are threshold NDVI, k-means clustering and polygon Region of Interest (poly ROI), due to their mostly usage in image processing.

According to [10], thresholding can be classified into two schemes which are global (uses only one threshold for the entire image) and local (threshold that is used to an image that had been partitioned into several subregions) [11]. Meanwhile, the Normalized Difference Vegetation Index (NDVI) is a numerical index that uses the visible and near-infrared bands to analyse remote sensing measurement and used to determine whether the region of interest that being observed contains green vegetation or not. NDVI was first used in 1974 by [12]. Since the characteristic of plants across the electromagnetic spectrum can be obtained, the NDVI information can be extracted by focusing on the satellite bands that are most sensitive to vegetation information. The bigger the difference between near-infrared and red reflectance, the more vegetation there has to be. Theoretically, NDVI values are represented in value ranging from -1 to 1 and negative values always represent water, concrete, paved road and built-up area, values around zero represent soil and values over 0.6 always represent dense green vegetation.

K-means clustering, which is part of clustering-based image segmentation, has been used by many researchers [9] [10]. It works by to partitioning the image into various cluster based on the similarities features such as colour or texture. The algorithm divides the image into k-cluster based on the similarity between the pixels in the cluster. However, they discovered that this method encounters great difficulties when computing the number of clusters that are present in the future space or extracting the suitable feature.

Polygonal Region of Interest (Poly ROI) is refers to a special region in an image that is of particular interest or importance to the user. Poly ROI always deals with two regions which are region of interest and background region. The selected sample region is then being matched with the entire pixels in the image. The region that holds the same features with the selected sample region will be extracted together [13].

2. Methodology

Figure 1 shows the process flow conducted in this project. The first and foremost is to acquire the digital image from a source. As in this project, the image is captured by a sensor and digitized into

IOP Conf. Series: Materials Science and Engineering **705** (2019) 012035 doi:10.1088/1757-899X/705/1/012035

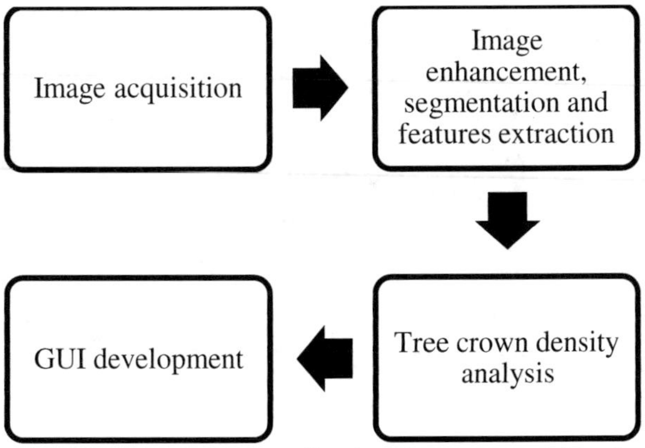

Figure 1. Process flow of the work, from data collection (image acquisition) to the suitable Graphical User Interface (GUI) development in producing tree crown density analysis of an area

a simpler image form so that it can be processed. The image obtained is in the form of visible green, red and near-infrared bands images.

The study site is located in the South of Iran (32°10'21.18'' N- 48°46'54.36'' E) lying at the head of the Persian Gulf and bordering Iraq on the west. It is one of the 31 provinces of Iran that generally hot and occasionally humid while cold and dry during winter. Figure 1 shows the image of the site taken from Google Maps.

Figure 2. The study site image from Google Maps

The respective hyperspectral image was acquired on 20th September 2014 from the Landsat 8 platform and Operational Land Imager (OLI) sensor image. The raw imagery does not need any atmospheric correction since the sky is clear. The image obtained is in the form of visible green, red and near-infrared (NIR) bands images. Figure 3 shows the respective hyperspectral image in their bands used in this study.

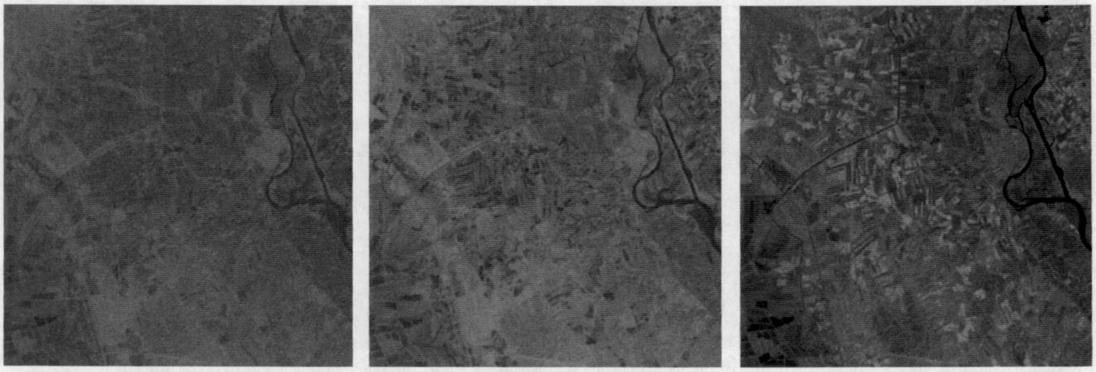

(a) (b) (c)

Figure 3. Hyperspectral image in: (a) visible green band; (b) visible red band; (c) NIR band

The images were then enhanced using histogram manipulation for intensity transformation. This is to allow modification towards original image histogram so that the image contrast is improved. Next, all the images are combined together into an RGB colour image. Then, vegetation area will be determined using NDVI as shown in Equation (1), which calculated by dividing the difference in the near-infrared (NIR) and red colour bands by sum of the NIR and red colour bands for each pixel in the image.

$$NDVI = \frac{(NIR - RED)}{(NIR + RED)} \tag{1}$$

where:
NIR = Near-Infrared
RED = Red visible

As been discussed earlier, k-means clustering and poly ROI could also be used to segment the image, however upon testing, their results were not as accurate as of thresholding NDVI. Figure 4 shows the results of these processes. As can be seen from this figure, the results of thresholding NDVI shows the best compared to the other segmentation methods.

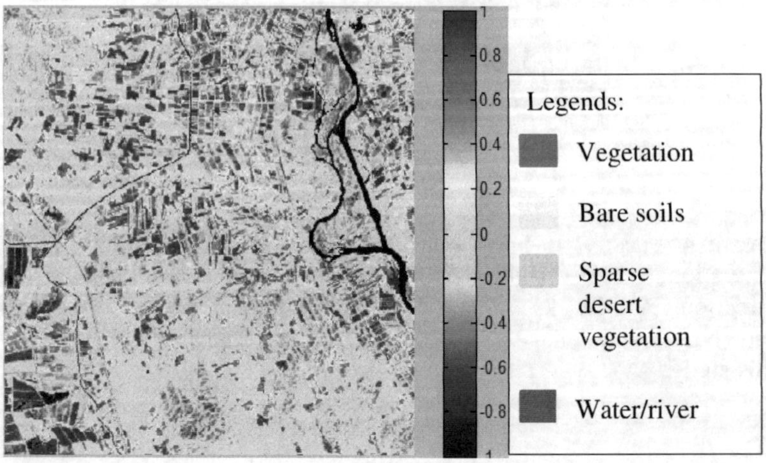

(a)

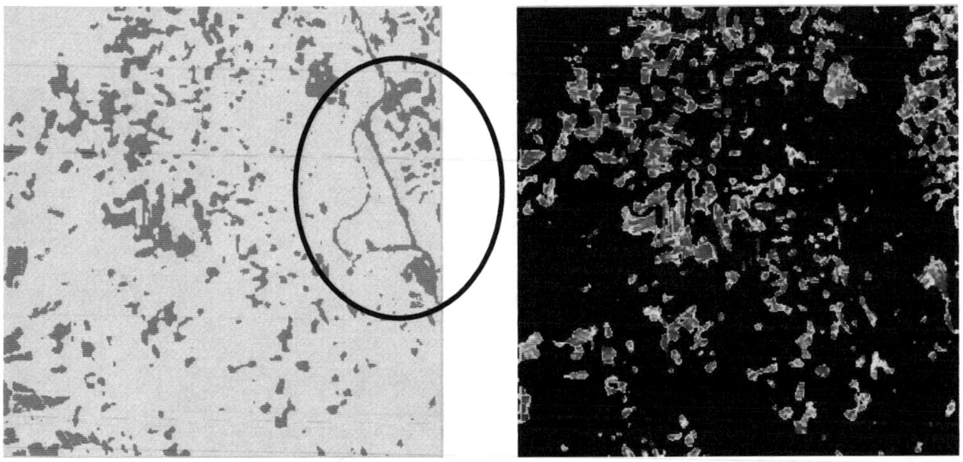

(b) (c)

Figure 4. (a) Regions segmented based on their spectral band values as according to the NDVI; (b) Segmentation based on poly ROI where water/river areas were selected as well (marked in the circle); (c) Results of k-means clustering where non-vegetation regions are also segmented

After thresholding, the image will be converted to grayscale image. As the grayscale image stored 256 different intensities, hence, the pixel can be divided into several ranges and can be used in the next step which is feature extraction. Feature extraction is always required as it can pull out the relevant information from the image. The information is then will be used in the classification stage of image processing data. Therefore, it is important to extract the exact information so that they can be classified correctly. In this project, the feature used as the input is colour ranges. This feature will distinguish the level of vegetation density into low, medium or high. The total area will be calculated and displayed.

A Graphical User Interface (GUI) is also designed in this project to allow better accessibility. It will serve as a better tool in providing a fast result and can demonstrate the overall functionality of the algorithm. User needs to choose the respective three images (green, red and NIR bands) needed to produce an RGB image which will be shown. Next, the NDVI segmented image will be generated and the features according to the level of tree crown density will be calculated and shown.

3. Results and Discussion

Figure 5 shows the respective GUIs for tree crown density analysis of the data used in this project, while Table 1 summarizes the results. It can be seen that for high density level, the total pixels is 12073 with 7.5456 %, 2934 pixels (1.8338%) for medium density and 8082 pixels (5.0513%) for low density tree crown. A total coverage area of 14.4307% representing vegetation / forestry site shows that it does not represents the majority of the area, hence proved by the desert-type location (refer to Figure 2). High value of density indicates that the tree has a large amount of leaf material available for photosynthesis and has growing conditions that enable full and symmetrical growth. While, low density level value may indicates poor amounts of foliage, a thin crown, or missing section of crown or may result from stresses such as insect and disease or any other environmental factors such as drought, wind, competition or soil compaction [14]. Thus, it can concluded that in this area, the total tree crown with high-density level has the highest distribution, followed by the tree crown with low-density level and lastly, the tree crown with medium-density level. From here, further processing like tree species recognition can be done.

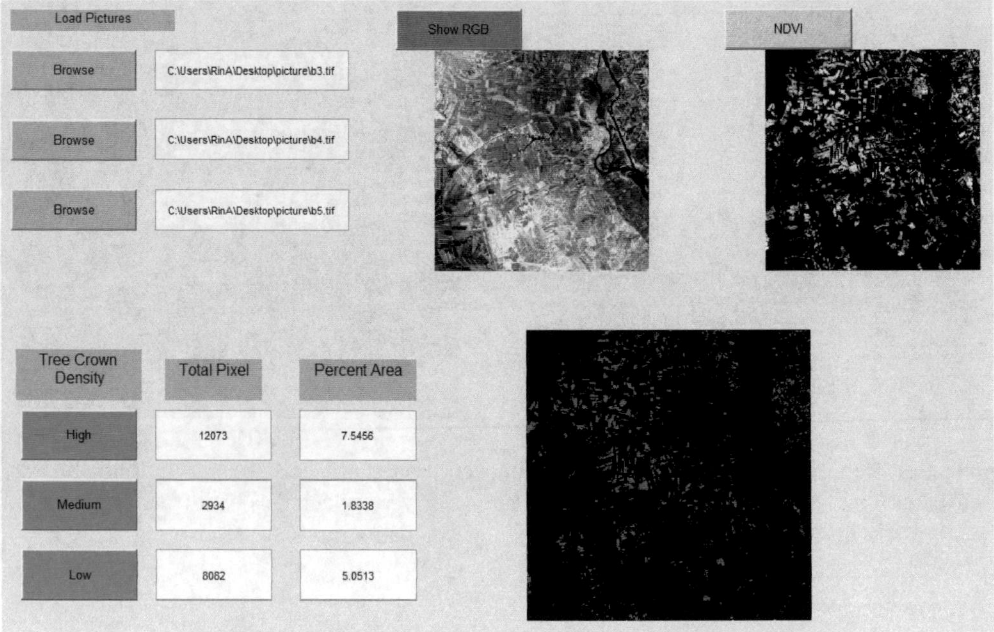

Figure 5. GUI calculating and mapping tree crown density

Table 1. Summary of results representing the tree crown density of the study area

Tree Crown Density	Total Pixel	Percentage Area
High	12073	7.5456%
Medium	2934	1.8338%
Low	8082	5.0513%

4. Conclusion

This project has shown the capability of remote sensing, specifically hyperspectral image in extracting and classifying information regarding tree crown density. Standard image processing approaches are needed and due to its importance, several dominant methods in image segmentation has been used and tested. From here, it has been proven that threshold NDVI gives the best segmentation results in vegetation area arrangement. Further tests are required to prove the workability of the overall methods in analysing tree crown density.

References

[1] Dian Y, Li Z and Pang Y 2013 Forest tree species clssification based on airborne hyper-spectral imagery *MIPPR 2013: Remote Sensing Image Processing, Geographic Information Systems, and Other Applications* **8921** 892107

[2] Krahwinkler P and Rossmann J 2010 Analysis of hyperspectral and high-resolution data for tree species classification *Proc. ASPRS 2010 Annual Conf.* (San Diego) 26-30

[3] Bioucas-Dias J M, Plaza A, Camps-Valls G, Scheunders P, Nasrabadi N and Chanussot J 2013 Hyperspectral remote sensing data analysis and future challenges *IEEE Geoscience and Remote Sensing Magazine* **1** 6-36

[4] Hycza T, Stereńczak K and Bałazy R 2018 Potential use of hyperspectral data to classify forest tree species *New Zealand Journal of Forestry Science* **48** 18

[5] Lowe A, Harrison N and French AP 2017 Hyperspectral image analysis techniques for the detection and classification of the early onset of plant disease and stress *Plant methods* **13** 80
[6] Adão T, Hruška J, Pádua L, Bessa J, Peres E, Morais R and Sousa J 2017 Hyperspectral imaging: A review on UAV-based sensors, data processing and applications for agriculture and forestry *Remote Sensing* **9** 1110
[7] Zhen Z, Quackenbush L and Zhang L 2016 Trends in automatic individual tree crown detection and delineation—Evolution of LiDAR data *Remote Sensing* **8** 333
[8] Fu K S and Mui J K 1981 A survey on image segmentation *Pattern Recognition* **13** 3-16
[9] Dubey S R, Dixit P, Singh N and Gupta JP 2013 Infected fruit part detection using K-means clustering segmentation technique *Ijimai* **2** 65-72
[10] Taxt T, Flynn P J and Jain A K 1989 Segmentation of document images *IEEE Transactions on Pattern Analysis and Machine Intelligence* **11** 1322-1329
[11] Davies E R 2004 *Machine Vision: Theory, Algorithms and Practicalities (Oxford: Academic Press)*
[12] Rouse Jr J, Haas R H, Schell J A and Deering D W 1974 Monitoring vegetation systems in the Great Plains with ERTS *Proc. NASA Goddard Space Flight Center 3d ERTS-1 Symp.* (Texas)
[13] Tabassum M and Shorif Uddin M 2011 Extraction of ROI in Geographical Map Image *Journal of Emerging Trends In Computing And Information Sciences* **2**
[14] United States Department of Agriculture 2014 *Report on Forest Health Monitoring*

IOP Conf. Series: Materials Science and Engineering **705** (2019) 012036 doi:10.1088/1757-899X/705/1/012036

Map-Based Localization Indoor Environment for Object Tracking using RF Trackers

Hing C H[1], K Fong P W[1,2], S M M Syed Zakaria[1,2], L M Kamarudin[2,3], A Zakaria[2,4], E Kanagaraj[2,3], K Kamarudin[2,4], Y A Rahim[2,5] and N Ahmad[1]

[1] School of Microelectronic Engineering, Universiti Malaysia Perlis (UniMAP), Kampus Pauh Putra, Arau, Perlis, 02600, Malaysia
[2] Centre of Excellence for Advanced Sensor Technology (CEASTech), Universiti Malaysia Perlis (UniMAP), Kompleks Pengajian Jejawi II, Arau, Perlis, 02600, Malaysia
[3] School of Computer and Communications Engineering, Universiti Malaysia Perlis (UniMAP), Kampus Pauh Putra, Arau, Perlis, 02600, Malaysia
[4] School of Mechatronic Engineering, Universiti Malaysia Perlis (UniMAP), Kampus Pauh Putra, Arau, Perlis, 02600, Malaysia
[5] School of Manufacturing Engineering, Universiti Malaysia Perlis (UniMAP), Kampus Pauh Putra, Arau, Perlis, 02600, Malaysia

Corresponding author: smmamduh@unimap.edu.my

Abstract. This paper proposed a RF-based mapping-based indoor localization technique to locate their packages and items. The Received Signal Strength Indicator (RSSI) is commonly used in trilateration to allocate the packages or items location. However, several problems need to overcome to increase the accuracy of the RSSI signal propagation for the localization. The problems include the complexity of the indoor environment will affect the radio waves experienced multipath effect such as the signal may be reflected, refracted, absorbed or blocked by other matter. The paper is designed to study the propagation of radio waves in indoor conditions, identify the type of antenna used and determine the accuracy of the localization model. The RSSI is recorded for every 1m until reached the maximum read range of 20m to study the behaviour of radio waves propagation and the relationships between RSSI and distance. The RSSI mapping experiment was carried out to determine the type of antenna used to carry out the experiment. The localization is performed with mapping-based technique to predict the random coordinates in the testbed. The predicted coordinates and actual coordinates were compared to determine the accuracy of the localization model. The error between the actual coordinates and predicted coordinates can achieve an average of 0.95 meter. The RF mapping-based indoor localization technique results shown that the effectiveness of the techniques to be implemented in a company to improve the efficiency in the production line.

1. Introduction

Indoor localization has become one of the main focus in research due to the low accuracy of the GPS localization in indoor environments. The GPS propagates weak signal which is unable to penetrate the building blocks and results in high error location estimation in the indoor environment [1,2]. Many researchers have reported works on solutions for indoor localization by proposing multiple devices such as WiFi, RFID, ZigBee and other devices to be used for indoor localization [3]. The indoor localization positioning approaches can be divided into Time of Arrival (TOA), Angle of Arrival (AOA), Time Difference of Arrival (TDOA) and Received Signal Strength Indicator (RSSI) based system. Among all of the indoor localization positioning approaches, the RSSI has a major advantage over other methods because it does not require extra hardware to support the infrastructure of the algorithm [4]. The most common RSSI technique usage is the RSSI fingerprinting localization. However, the RSSI based indoor localization is prone to fluctuations due to signal strength being non-linear against distance and is affected by the presence of obstacles [5,6]. In this work, map-based RF indoor localization technique is proposed with the RSSI mapping in the experimental testbed. The RSSI mapping is being used in the algorithm to predict the unknown location with the predefined RSSI mapping.

2. Literature Review

RSSI is now widely used for location sensing in many location tracking systems based on the RF signal. The most common classes of RSSI positioning technique are the radio propagation model distance prediction and fingerprinting techniques [7]. The typical radio propagation model distance prediction uses the relationship between distance and RSSI as shown in Equation 1;

$$RSSI = -(10n \log d + A) \tag{1}$$

where A is defined as the absolute energy measured at 1m away from the transmitter, n is the signal transmission constant and d is the distance between the transmitter and the receiver. However, RSSI is a measure of signal strength and it is subjected to the non-linear and non-Gaussian relationship with distance. Therefore, Equation 1 may not satisfy all the unique conditions indoor environment.

On the other hand, the fingerprinting technique consists of two phases: offline training phase and online phases [8]. During the offline phase, a database is built by collecting all of the RSSI for every reference points (RP) in the experimental testbed. Then, in the online phase, the position can be estimated by comparing the real-time RSSI value with the RSSI database built. However, this is a time-consuming process to collect all the RSSI value in every RPs manually and it is subjected to environmental changes [8].

The trilateration technique uses signal strength from at least three different access points to predict the location of the mobile terminal [9]. The trilateration forms circles at the centre of each access points and the radius of the circle can be determined by measuring the signal strength (RSSI) of the mobile terminal. The intersection points between at least three different access points give the estimated location for the mobile terminal [10]. The trilateration technique illustration is shown in Figure 1. Another work stated that the trilateration technique would be prone to errors due to the signal strength can be affected by obstacles or error while converting the RSSI values to distance [11].

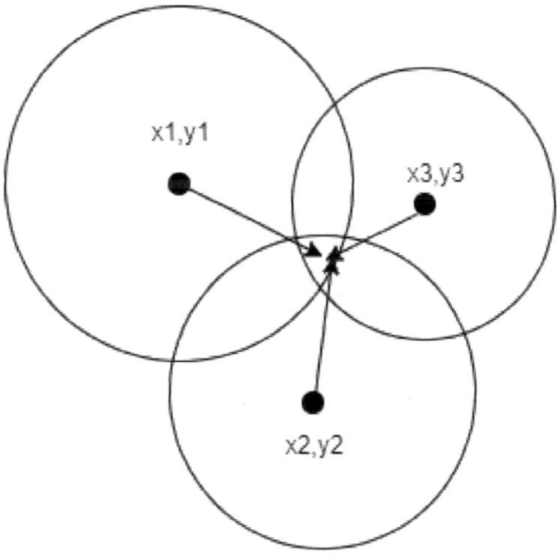

Figure 1. Illustration of trilateration technique

Although many efforts being proposed by researchers to reduce the error in the trilateration method, there are still errors presented in the trilateration technique due to the complexity of the indoor

environment. Another work presents a new error correction algorithm for trilateration technique as an alternative for the standard trilateration algorithm calculations that will reduce the offset error by more than 50% [12].
In this paper, a combination of RSSI mapping method is proposed. The localization is based on the measured distance between the RSSI mapping and the overlapping area for every antenna in the testbed.

3. Methodology

First, the reader's performance is evaluated by performing distance and RSSI measurements throughout the testbed. This is to ensure the reader's placement is able to fully cover all the testbed area. Based on the reader's performance evaluation, the relationship between the RSSI and distance is obtained by curve fitting the RSSI against distance graph with a logarithmic curve.
The experiment is divided into two stages and is summarized in Table 1. At the first stage, the 14m-by-8m testbed is divided into 2m by 2m grid. There are three identical readers placed at locations (0, 0), (0, 8) and (14, 8) respectively. The RSSI values for signal transmission at every grid point is collected based on three readers. The experiment continues until all the RSSI for every point are collected. All of the collected data is plotted in multiple contour map graphs to observe the RF signal propagation in each of the readers. Then, in the second stage, a transmitter is placed at random coordinates. Based on the RSSI value collected from three different location readers, the location of the transmitter can be based on the combined contour graph and pattern matching with the current unknown RSSI location. The error of the location prediction is quantified based on the Euclidean distance between the actual position and the predicted position.

Table 1. Experiment steps for Mapping-based localization

Stage	Experiment Steps
Step 1:	Experimental testbed was divided into 2m by 2m grid and marked.
Step 2:	Antenna 1, antenna 2 and antenna 3 was placed at locations (0, 0), (0, 8) and (14, 8) respectively.
Step 3:	RSSI values of the transmitter were collected for each antenna at each grid point coordinates.
Step 4:	Contour map of the RSSI value for each of the antennas was plotted.
Step 5:	Transmitter is placed at random coordinates.
Step 6:	Transmitter RSSI values at random coordinates were recorded when it was detected by each of the antennas.
Step 7:	Resemblance of transmitter RSSI values at random coordinates were identified based on the intersection between antennas from the previous plotted contour map.
Step 9:	The predicted coordinates and actual coordinates were compared to determine the accuracy of the proposed method.

4. Results and Discussions

4.1. Relationship between RSSI and distance

Due to the complexity of the indoor environment, the RSSI exhibited fluctuations throughout the experiment. It is difficult to model a distance equation model based on only RSSI because the model may not be suitable for all the different indoor environmental conditions. Therefore, the RSSI value for every 1m distance is collected until the transmitter is not detected by the receiver. Due to the fluctuations of the collected RSSI value, the logarithmic curve fitting method was chosen to obtain the relationship equation between RSSI and distance as shown in Figure 2.

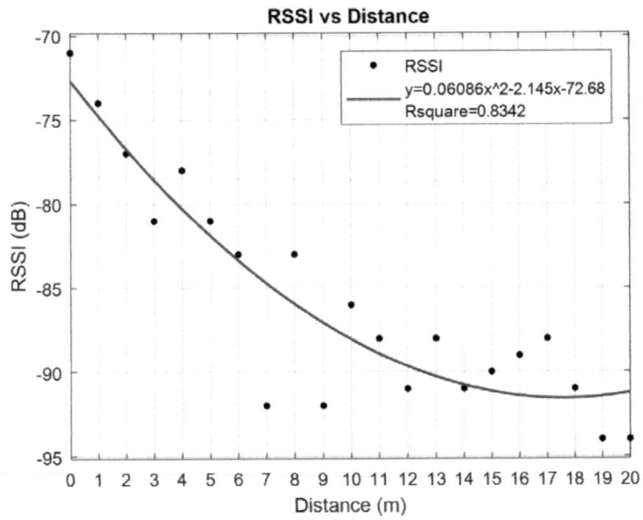

Figure 2. Relationship between RSSI and distance based on logarithmic curve fitting

Based on Figure 2, the logarithmic curve fitting is done with the MATLAB curve fitting tools and obtained the model equation as shown in Equation 2. The R^2 of the curve fitting model is 0.8342. It is approximately 83.42% of the observed variation can be explained by the model's inputs.

$$RSSI = 0.06086x^2 - 2.145x - 72.68 \quad (2)$$

Where x is the distance between the transmitter and the receiver in meter.

4.2. RSSI Mapping

This experiment is designed to determine the antenna range used in the following experiments. The antenna is placed at the centre of the testbed as shown in Figure 3. Based on Figure 3, the omnidirectional antenna is able to cover 14 meters x 8 meters area of the testbed. As the transmitter moves further from the antenna, the RSSI value became more negative and this indicates the signal strength decrease with increasing distance. It is difficult to model a distance model equation due to the non-linear RSSI changes as shown in Figure 3.

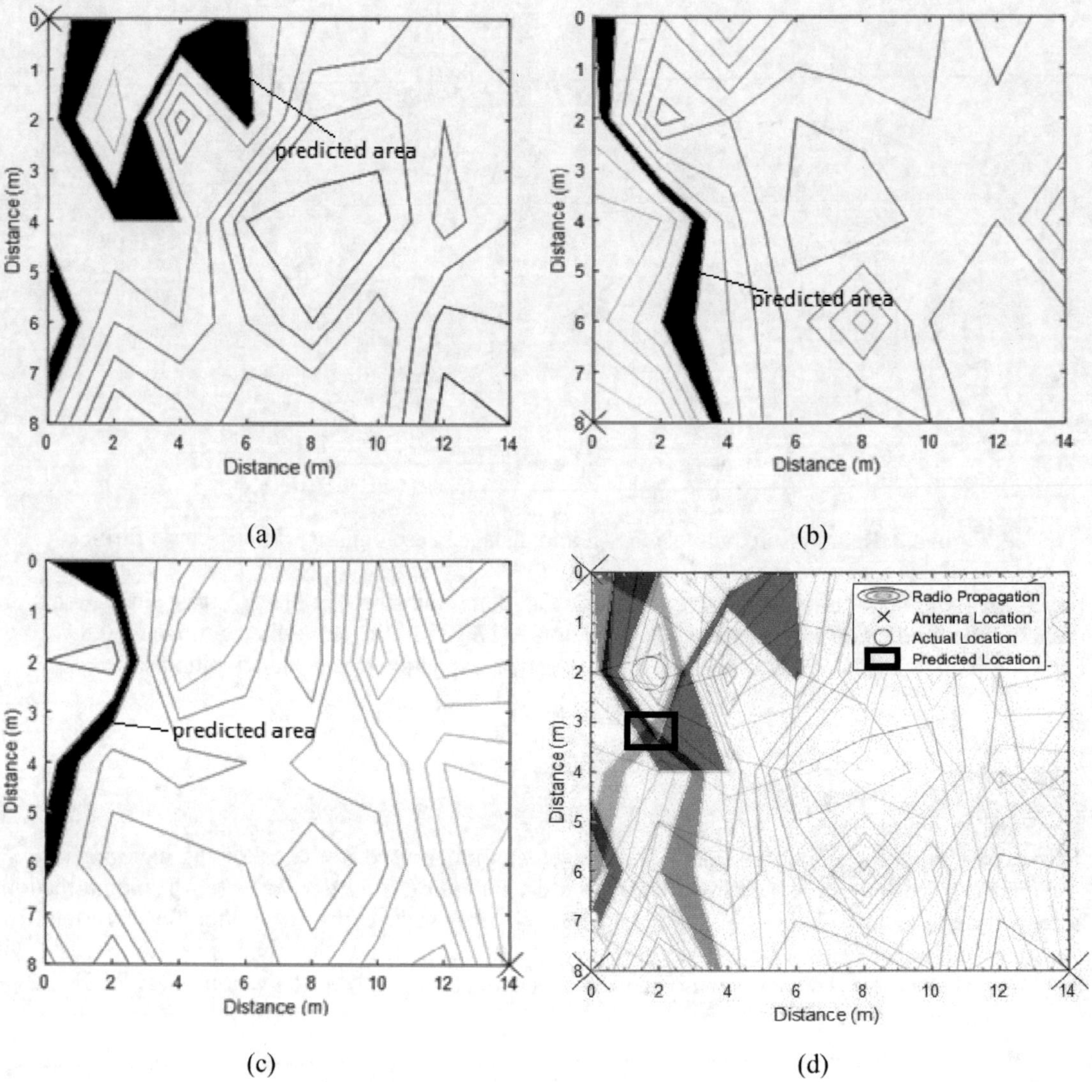

Figure 3. RSSI mapping-based algorithm for localization (a) Antenna 1 coverage (b) Antenna 2 coverage (c) Antenna 3 coverage (d) Comparison actual and predicted location

4.3. RSSI-based Localization

Based on the transmitter's RSSI values at random coordinates, the location can be predicted based on previously recorded RSSI contour map. The centre of the intersection location of three different contour map graphs is the predicted location. The proposed algorithm location prediction is shown in Figure 3. Based on Figure 3, the actual location and the predicted location are different due to the error present in the proposed algorithm. The error between the actual location and the predicted location is calculated by using the Euclidean distance equation as shown in Equation 3.

$$Error = \sqrt{(x_1 - x_2)^2 + (y_1 - y_2)^2} \quad (3)$$

Based on Equation 3 and Figure 3., the actual coordinate of the transmitter is (2, 2) and the predicted coordinate by the algorithm is (2, 3.5). By using Equation 3, the error of the algorithm is 1.5 meters. The experiment is repeated at another random coordinate to observe the reliability of the algorithm. The proposed algorithm's predicted location and the actual location is tabulated in Table 2.

Table 2. Formatting sections, subsections and subsubsections.

	Actual Location (x,y)	Predicted Location (x,y)	Error (m)
Position 1	(2,2)	(2,3.5)	1.5
Position 2	(8,6)	(8,5.6)	0.4

The results represent a promising start for this approach to RSSI map-based localization in an indoor environment. Rather than relying on building a perfect model for RF propagation to execute trilateration algorithm, the mapping-based algorithm is able to produce comparable accuracy compared to previous works reported on RF based location with simple approaches.

5. Conclusion

The proposed method with RSSI measurement metric has an average error of 0.95m. The localization error in the method is expected mostly due to the multipath propagation of the RF signal causing fluctuations and uncertainties in the RSSI. In this explorative work, the RF signal behavior in an indoor environment was studied and a method to localize objects and items based on RSSI maps was proposed. The relationship between the RSSI and distance is proposed with the logarithmic curve fitting method. The accuracy of the algorithm can be improved by increasing the number of antennas in the testbed but this will increase the infrastructure cost. The trade-off between the cost and the accuracy is required to be considered while planning the research. The current method can be further improved with the integration of machine learning algorithm or integration of probabilistic methods in future works. If there is a high amount of data supplied to the machine learning algorithm, the accuracy of the localization is envisaged be improved.

Acknowledgement

This research was supported by Ministry of Education (MOE) through Fundamental Research Grant Scheme (FRGS/1/2018/SKK06/UniMAP/02/1). The authors would also like to acknowledge the support from Industrial Grant (9002-00076) provided by the industry.

6. References

[1] Huang C N and Chan C T 2011 ZigBee-based indoor location system by K-nearest neighbor algorithm with weighted RSSI *Procedia Comput. Sci.* **5** 58–65

[2] Wang X, Gao L and Mao S 2016 CSI Phase Fingerprinting for Indoor Localization with a Deep Learning Approach *IEEE Internet Things J.* **3** 1113–23

[3] Li G, Geng E, Ye Z, Xu Y, Lin J and Pang Y 2018 Indoor positioning algorithm based on the improved rssi distance model *Sensors (Switzerland)* **18** 1–15

[4] Shukri S and Kamarudin L M 2017 Device free localization technology for human detection and counting with RF sensor networks: A review *J. Netw. Comput. Appl.* **97** 157–74

[5] Subaashini K, Dhivya G and Pitchiah R 2012 Zigbee RF signal strength for indoor location sensing-Experiments and results ... *Technol. (ICACT), 2012 ...* **1** 12–7

[6] Shukri S, Kamarudin L M, Cheik G C, Gunasagaran R, Zakaria A, Kamarudin K, Zakaria S M M S, Harun A and Azemi S N 2017 Analysis of RSSI-based DFL for human detection in indoor environment using IRIS mote *2016 3rd Int. Conf. Electron. Des. ICED 2016* 216–21

[7] Xue W, Qiu W, Hua X and Yu K 2017 Improved Wi-Fi RSSI Measurement for Indoor Localization *IEEE Sens. J.* **17** 2224–30

[8] Yiu S, Dashti M, Claussen H and Perez-Cruz F 2017 Wireless RSSI fingerprinting localization *Signal Processing* **131** 235–44

[9] Kodippili N S and Dias D 2010 Integration of fingerprinting and trilateration techniques for improved indoor localization *2010 7th Int. Conf. Wirel. Opt. Commun. Networks, WOCN2010*

1–6
[10] Wang Y, Yang X, Zhao Y, Liu Y and Cuthbert L 2013 Bluetooth positioning using RSSI and triangulation methods *2013 IEEE 10th Consum. Commun. Netw. Conf. CCNC 2013* 837–42
[11] Rusli M E, Ali M, Jamil N and Din M M 2016 An Improved Indoor Positioning Algorithm Based on RSSI-Trilateration Technique for Internet of Things (IOT) *Proc. - 6th Int. Conf. Comput. Commun. Eng. Innov. Technol. to Serve Humanit. ICCCE 2016* 72–7
[12] Zaniani M M, Shahar A M and Azid I A 2010 Trilateration target estimation improvement using new error correction algorithm *Proc. - 2010 18th Iran. Conf. Electr. Eng. ICEE 2010* 489–94

IOP Conf. Series: Materials Science and Engineering **705** (2019) 012037 doi:10.1088/1757-899X/705/1/012037

Analysis of Mobile Robot Indoor Mapping using GMapping Based SLAM with Different Parameter

W.A.S Norzam[1], H.F.Hawari[1], K.Kamarudin[2]

[1]Department of Electrical and Electronic Engineering, Universiti Teknologi PETRONAS 32610 Bandar Seri Iskandar, Perak Darul Ridzuan, Malaysia
[2]School of Mechatronic Engineering, Universiti Malaysia Perlis (UniMAP), Arau, Perlis 02600.

wan_18003530@utp.edu.my, huzeinfahmi.hawari@utp.edu.my

Abstract. Mapping is one of the elemental application of the mobile robot. The map is created using the mobile robot by employing sensors such as camera, sonar and laser sensor. One of the most popular mapping methods is the Simultaneous Localization and Mapping (SLAM). SLAM allows the map to be created while localizing the robot location in the map at the same time. GMapping is one of the widely used algorithms in SLAM which will be used in this project. The mobile robot is equipped with a Hokuyo Laser Range Finder sensor and netbook. The router is used for wireless communication between the mobile robot and the user. The GMapping is done in two different locations of different lab size and amount of features in the area. Three trial is conducted to investigate the effects of different parameters such as robot speed, mapping delay and particle filter on the mapping quality. The results show a significant difference in terms of mapping accuracy and the time taken to complete the process as the parameter changed from the three trial. As a result, the parameter used in the second trial, robot speed 0.1333m/s, mapping delay 1s and particle filter 30 is considered as the best based on the time taken and the map accuracy.

1. Introduction

Mobile Robot is a type of robot that can move around in the environment freely either using the wheel-based or foot-based system. These robots are developed in order to help solve the problems that usually arise because the task is harmful to human. Tasks like repetitive movement would cause ergonomic risk or deal with something unknown, like the exploration of the unmapped area. In the exploration of an unknown area, the mobile robot can be sent first to check the condition and mapping of the area for any danger or hazard that may exist. The mobile robot continuously receiving the attention of many researchers as one of the ways to solve problems arising and continuous research done in order to explore more on mobile robot capability.

Mapping is the process of creating a map that is readable by both human and mobile robot. These maps created using a mobile robot with suitable sensors like visual (camera), sonar and laser sensor. Mapping is considered as one of the interesting contributions of mobile robot since the map can be used for various application. For example, mobile robot autonomous navigation which requires the map of the area in order to navigate safely [1]. Other application includes being able to be used in search and rescue robot which requires the robot to create the map while searching and mapped the location of the potential victim [2]. As such, mapping has become one of the important research in the mobile robotic area with various mapping technique currently available.

One of the most popular mapping methods is Simultaneous Localization and Mapping (SLAM). SLAM is the mapping method that is available in the Robotic Operating Software (ROS) platform. It is able to construct or update the map while at the same time keeping in track the location of the mobile robot [3]. This will reduce the errors in term of mapping accuracy by taking into account the robot location while mapping. As such, many algorithms of SLAM method is developed and used with the mobile robot in other research [4]. There are many algorithms that are available under SLAM, for example, GMapping, Core SLAM, Graph SLAM, and Hector SLAM.

However, most of the study done by others are comparing the algorithms to see which able to construct the most accurate map. Each of the algorithms also has its own parameters which can be changed based on the environment the robot is working. Using different parameters like robot speed and mapping update delay not only able to cause changes in the time taken but also the accuracy of the map. Hence, it is important to know the effect of how different parameter of the algorithm can change the quality of the map which is the focus of this paper.

This paper is organized as follows. Section II discusses other researcher works on SLAM topic. Section III shows the mobile robot setup which is used in this research. Section IV explains the algorithm used for SLAM and the mapping flow process. Section V shows the results and discussion of this research. Section VI is the conclusion of the paper.

2. Related Works

SLAM is continuously being researched in order to create or improve on the algorithm available to create a better quality map. As mapping is considered as crucial especially for navigation purpose which is considered as the beginning of the process many research had been conducted in the past.

The research by Kohlbrecher et al [5] uses Hector open source modules for the map generation using their robot. Their research is to do the mapping and navigation for the rescue robot. Using the Robotic Operating System (ROS) as the platform, they used Hector SLAM in order to solve the SLAM problem and generate a highly accurate metric map for exploration of uneven environment for rescue. Figure 1 shows the hector SLAM used with Hector UGV robot and the quadrotor. The researcher used the proposed method in RoboCup 2012 for a rescue operation, which is the autonomous exploration of the robot in the arena to locate all the four victims.

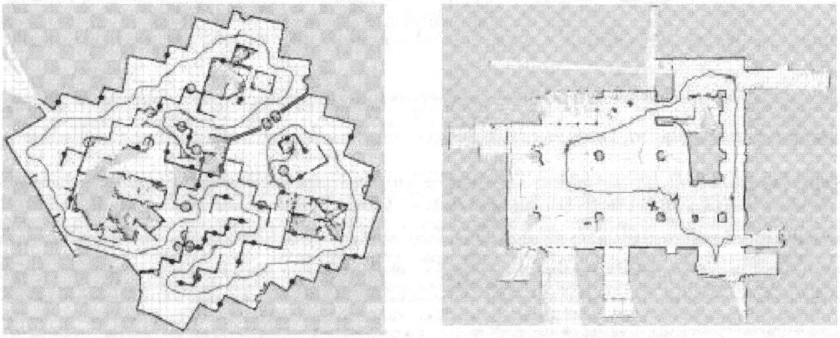

(a) (b)

Figure 1. Map learned using Hector modules, (a) mapping results using hector_slam and hector_exploration_planner with Hector UGV robot, (b) hector SLAM application on ccny quadrotor [5].

Balasuriya et al [6] research on outdoor robot navigation uses the GMapping based SLAM algorithm to generate the map. The robot equipped with a laser scanner with a combination of sonar sensor as the backup. Using the slam_gmapping algorithm will create the grid-based map. Figure 2 shows the results of the GMapping for an indoor and outdoor environment. It can be seen that the walls which should have been a straight line but are not updated in the map as straight. This is due to slippage between the tracks and robot wheel occurring constantly during the mapping. From the results the GMapping algorithm able to cope with both stationary and dynamic landmarks. It keeps the stationary and removed the dynamic landmarks from the created map as the map is updated.

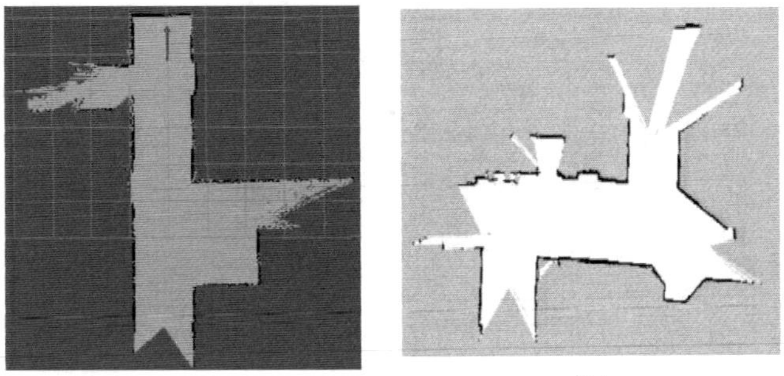

(a) (b)

Figure 2. GMapping mapping results, (a) Indoor, (b) outdoor [6].

Li et al [7] propose the use of FastSLAM with new jacobian free Neural Network (NN). FastSLAM suffers from the limitation on error accumulation when it is introduced with incorrect odometry model and inaccurate linearization of the SLAM. Using the multilayer of NN able to compensate the odometry error and the NN is trained online during the SLAM. The researcher also used the third-degree Cubature rule for Gaussian weighted integral to calculates nonlinear transition density of Gaussian to estimate correctly the SLAM state. The researcher simulated the proposed algorithm using the MATLAB simulator with Ackermann vehicle model. Their algorithm was able to outperform the traditional FastSLAM 2.0.

Based on the works by others, it can be seen all the algorithm GMapping, Hector SLAM and FastSLAM able to provide a good mapping quality. GMapping is also used by other researchers [8-10] in their research making it one of the popular choices for SLAM mapping algorithm. However, besides the differences in the algorithm, there are also parameters in each mapping technique that would be able to affect the mapping quality. Hence, in this research, it is decided to focus on different parameters effect of the GMapping technique.

3. Mobile Robot Setup

The mobile robot used for this research is the IRobot Create. The robot has the total max speed of 500mm/s is equipped with a Hokuyo Laser Range Finder (LRF) and netbook on top of it. The netbook is connected with the robot through serial port wire communication. Figure 3 shows the mobile robot together with the Hokuyo LRF and netbook. The LRF sensor is able to sense the surrounding in range of 0.02m to 5.6m and covering up to 240° angle. The laptop or remote station called as a base station while the robot with the netbook is the workstation. Both of these stations communicated using wireless communication of the Router as shown in figure 4.

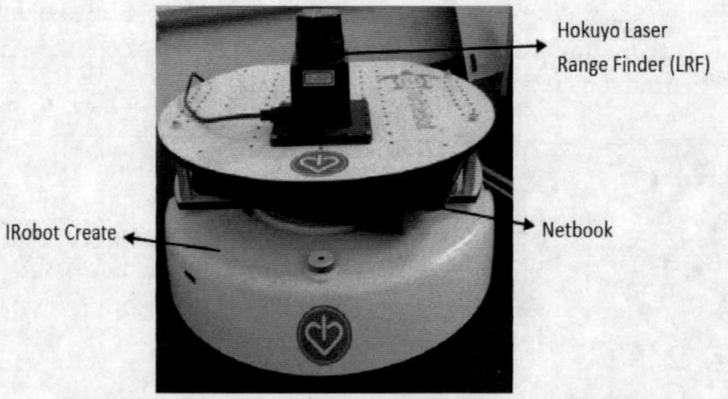

Figure 3. IRobot with Hokuyo LRF and Netbook.

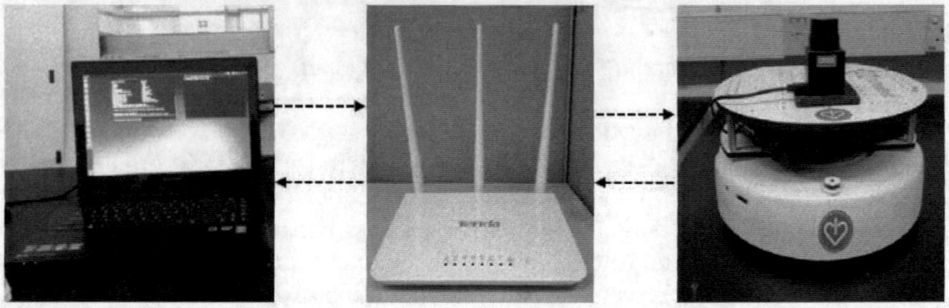

Figure 4. Wireless communication between base stations to workstation using router.

4. Simultaneous Localization and Mapping (SLAM)

SLAM is one of the methods available for the mobile robot mapping using the platform Robot Operating System (ROS) that are available in Ubuntu. The algorithm chosen for this research is GMapping. GMapping used the Rao-Blackwellized Particle Filter (RBPF) and take data from both laser sensor and robot pose to create a 2D grid map [9]. Figure 5 shows the mapping flow process which is used for this research.

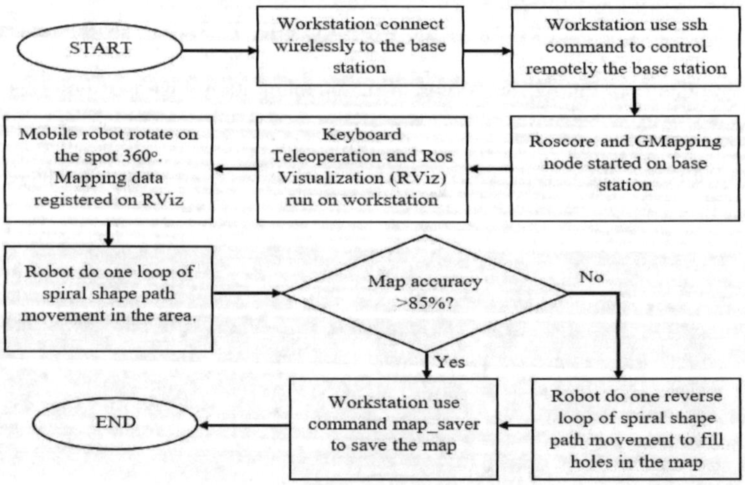

Figure 5. SLAM mapping flow process.

The base station and workstation is connected wirelessly using a router based on the network configuration of [11]. Once the connection is established then the corresponding node for the mapping process is run on the remote pc which able to access the netbook on the robot through ssh command. The robot will move in a spiral path pattern as shown in figure 6.

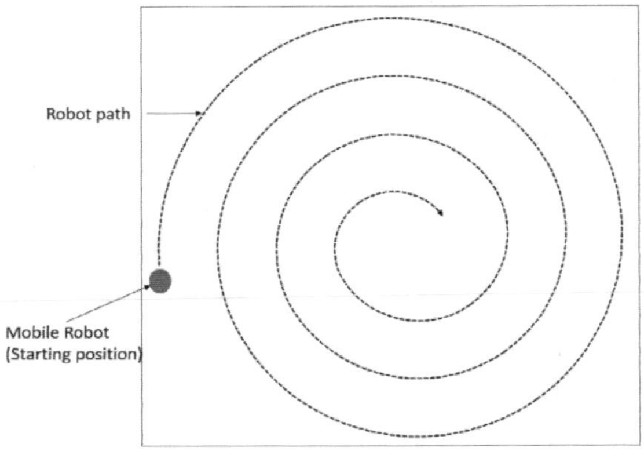

Figure 6. Robot mapping path.

The map accuracy is calculated when the robot finished the mapping. The equation (1) is used to determine the accuracy of the map based on the size of the real map layout. The x represents the total length of the map created by the robot while y is the total length of the real map. If the accuracy is high or more than 85% then the map will be saved if it is not then the robot will do one more loop to fill any missing parts of the map.

$$\text{Accuracy }(\%)=\frac{x}{y}\times 100 \tag{1}$$

5. Result and Discussion

The result is obtained from the SLAM mapping of two different indoor areas. The first mapping location is in Lab A and the second location is Lab B. Both are located indoors. Each of the area mappings is done in three trials with the different parameter of mobile robot speed, mapping delay and particle filter as shown in table 1.

Table 1. Mapping parameter for three trial.

Trial	1	2	3
Robot speed (m/s)	0.1	0.133	0.161
Map update interval (s)	5	1	0.1
Particle filter	30	30	15

The mapping for Lab A is done in three trial as mentioned before. The map area of characterization lab is 108m^2 and the total length is 41.8m. The robot starting area is the same for all three trial of the mapping process. Figure 7 shows the real map layout and three trial mapping of Lab A. Comparing the three results with the real map in visual alone it can be seen that as the trial goes the maps become more incomplete (increasing empty unmapped area). The results of the mapping are shown in table 2. The mapping accuracy decreases even though the time taken for mapping becomes shorter. For Lab A the best map is the second trial. This is due to the time taken to complete the map is almost two times shorter than the first trial even if the accuracy lower by 1%.

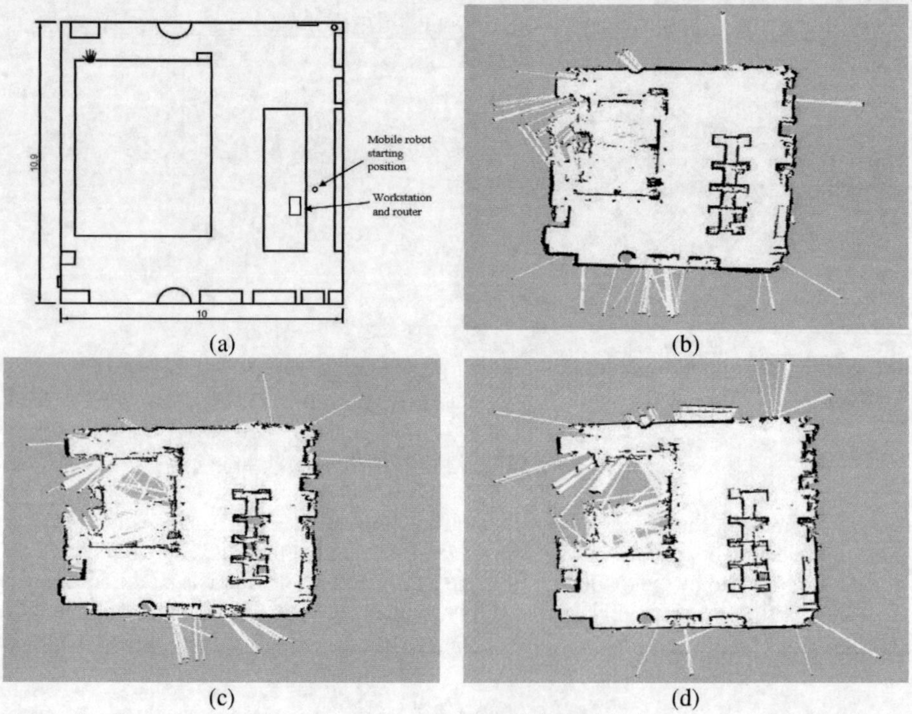

Figure 7. The Lab A mapping, (a) real map layout, (b) first trial, (c) second trial (d) third trial

Table 2. Mapping result for Lab A.

Trial	1	2	3
Map size perimeter (m)	38.8592	38.1266	36.1965
Time taken to complete mapping (min : s)	24 : 33	12 : 19	6 : 14
Map accuracy (%)	92.96	91.21	86.59

Lab B map area is 184m^2 and the total length is 55.6m. The lab consists of two rooms separated by wall and open door in the middle. The notable difference with Lab A is that this lab B is bigger in size but the fewer amount of feature presents in the area. Fig. 8 shows the real map layout and three trial of Lab B mapping. The results of Lab B mapping is shown in table 3.

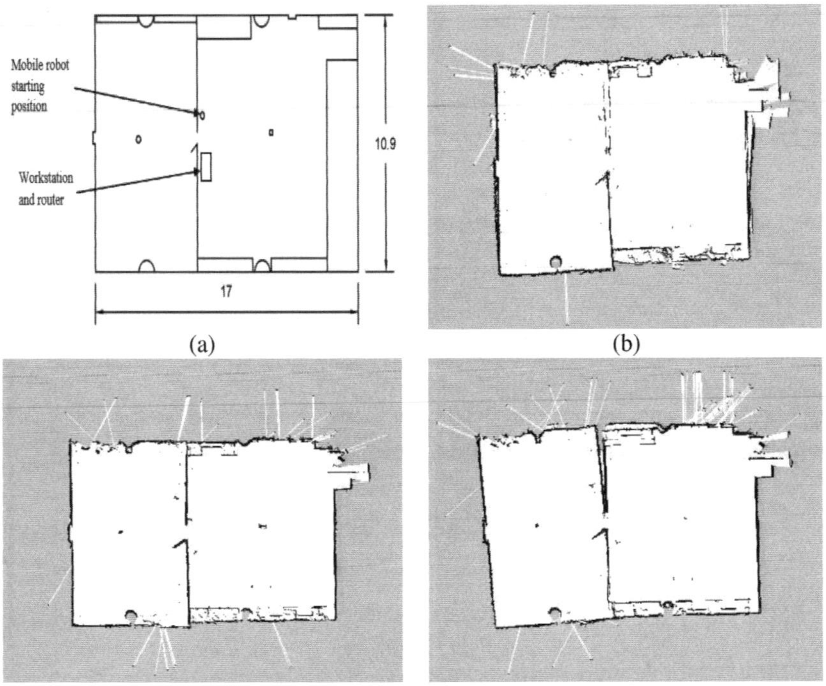

(a) (b) (c) (d)

Figure 8. Lab B mapping, (a) real map layout, (b) first trial, (c) second trial (d) third trial

Table 3. Mapping results for ISDP Lab.

Trial	1	2	3
Map size perimeter (m)	54.5963	54.0231	51.8253
Time taken to complete mapping (min : s)	40 : 22	19 : 35	15 : 01
Map accuracy (%)	98.19	97.16	93.21

The first and second trial is quite similar in terms of the accuracy of the map. However, the first trial took quite a long time to finish the mapping which is 40min compared to the second trial ends at 19min. This is due to the lower speed and long mapping delay causing it takes some time to finish the mapping. Since not only the map size is bigger but also the number of features which is comparatively lesser than in Lab A, it also contributes to the longer time it takes to finish the map. Unlike the first and second trial which shows high-quality map in terms of visual and high accuracy, the third trial, however, the second room on the left becomes slanted. This is due to the shorter map update delay and low particle filter causes high computational load for the robot causing some data loss giving an inaccurate representation of the map. This also causes the robot movement to occasionally move and stop a few times even with the manual command being sent from the base station. Finally, from the results in Table 3, the second trial is considered as the best map with the highest accuracy.

Based on the result from both lab mapping the time taken for mapping decreases as the robot speed increases, map update delay and particle filter decreases. However, it can be seen that the mapping quality also decreases especially in the third trial. This is due to the higher robot speed and much lower mapping delay that resulted in the increases of computational load for the mobile robot. As a result, some of the new data are being sent to the workstation are lost in the process. Not to mention that the faster robot speed means the sensor will not have enough time to scan every area in the lab. Thus, this will induce more holes and errors in generating the map.

6. Conclusion

The SLAM method is used with the GMapping algorithm for the mapping process in this research. Three trials of the different parameter of robot speed, map update interval, and particle filter are set to investigate its effect on the mapping quality. The results show the significant differences in the map quality and time taken as the changes take effect. The robot speed needs to be complemented with the map update delay as the speed increase the delay also should be lower. This is due to faster speed will give not enough time for the map update if the delay is high causing some parts of the map to remain unexplored. The same can be said on the particle filter if too high will cause heavy computational process effect while too low will affects the map accuracy.

Acknowledgment

The project's work was supported by Universiti Teknologi PETRONAS (UTP). We appreciated the funding provided by UTP under the Graduate Assistantship (GA) scheme.

References

[1] C. Gao et al 2019 *Research on a Panoramic Mobile Robot for Autonomous Navigation* in *Proc. of the 3rd Int. Conf. on Mechatronics Engineering and Information Technology* 87 pp. 204–207

[2] F Niroui, K Zhang, Z Kashino, and G Nejat 2019 Deep Reinforcement Learning Robot for Search and Rescue Applications: Exploration in Unknown Cluttered Environments in *IEEE Robotics and Automation Letters* **4** 2 610–7

[3] M Filipenko and I Afanasyev October 2018 *Comparison of Various SLAM Systems for Mobile Robot in an Indoor Environment in 9th IEEE Int. Conf. Intell. System* (Portugal: Madeira)

[4] A R Khairuddin, M S Talib, and H Haron November 2015 *Review on simultaneous localization and mapping (SLAM)* in *Proc. 5th IEEE Int. Conf. on Control System, Computing and Engineering* (Malaysia: George Town) pp 85–90

[5] S Kohlbrecher, J Meyer, T Graber, K Petersen, U Klingauf, and O von Stryk 2014 *Hector Open Source Modules for Autonomous Mapping and Navigation with Rescue Robots* in *RoboCup 2013: Robot World Cup XVII* vol. 8371 pp 624–31

[6] B L E A Balasuriya, B A H Chathuranga, B H M D Jayasundara, N R A C Napagoda, S P Kumarawadu, D P Chandima and A G B P Jayasekara 2016 *Outdoor robot navigation using Gmapping based SLAM algorithm* in *2nd International Moratuwa Engineering Research Conference* (Sri Lanka: Moratuwa) pp 403–8

[7] Q L Li, Y Song, and Z G Hou 2014 Neural network based FastSLAM for autonomous robots in unknown environments in *Neurocomputing* **165** 99–110

[8] H I M A Omara and K S M Sahari 2015 *Indoor mapping using Kinect and ROS* in *Int. Symp. On Agents, Multi-Agent Systems and Robotics* (Malaysia: Putrajaya) pp. 110–116

[9] R N Darmanin and M Bugeja 2016 *Autonomous Exploration and Mapping using a Mobile Robot Running ROS* in *Proc. of the 13th Int. Conf. on Informatics in Control, Automation and Robotics* (Portugal: Lisbon) vol. 2 pp 208-15

[10] D Shen, Y Huang, Y Wang, and C Zhao 2018 *Research and Implementation of SLAM Based on LIDAR for Four-Wheeled Mobile Robot* in *2018 IEEE International Conference of Intelligent Robotic and Control Engineering* (China: Lanzhou) pp. 38–41

[11] AnisKoubaa 2016 Network Configuration *wiki.ros.org* Online Available: http://wiki.ros.org/turtlebot/Tutorials/indigo/Network Configuration. Accessed: 6-May-2019

[12] B Gerkey 2013 GMapping *wiki.ros.org* Online. Available: http://wiki.ros.org/gmapping. Accessed: 10-May-2019

IOP Conf. Series: Materials Science and Engineering **705** (2019) 012038 doi:10.1088/1757-899X/705/1/012038

RSSI-based Localization Zoning using K-Mean Clustering

Kenny Fong Peng Wye[1,2], Ericson Kanagaraj[2,3], Syed Muhammad Mamduh Syed Zakaria[1,2], Latifah Munirah Kamarudin[2,3], Ammar Zakaria[2,4], Kamarulzaman Kamarudin[2,4] and Norhawati Ahmad[1]

[1] School of Microelectronic Engineering, Universiti Malaysia Perlis (UniMAP), Kampus Pauh Putra, Arau, Perlis, 02600, Malaysia
[2] Centre of Excellence for Advanced Sensor Technology (CEASTech), Universiti Malaysia Perlis (UniMAP), Kompleks Pengajian Jejawi II, Arau, Perlis, 02600, Malaysia
[3] School of Computer and Communications Engineering, Universiti Malaysia Perlis (UniMAP), Kampus Pauh Putra, Arau, Perlis, 02600, Malaysia
[4] School of Mechatronic Engineering, Universiti Malaysia Perlis (UniMAP), Kampus Pauh Putra, Arau, Perlis, 02600, Malaysia
[5] School of Manufacturing Engineering, Universiti Malaysia Perlis (UniMAP), Kampus Pauh Putra, Arau, Perlis, 02600, Malaysia

kennyfong@ieee.org

Abstract. This document discusses the novel approach to localize human location based on the current zone via k-mean clustering. A pilot experimental analysis of k-mean clustering to group similar RSSI pattern is compared to user defined zones. The dataset collected has demonstrated the difficulties of deploying trilateration and fingerprinting methods in dynamic conditions, as it introduces fluctuating RSSI measurements. The k-mean clustering was proposed to validate antenna placement to divide the testbed into zones, and to create a baseline accuracy which can be compared with other algorithms in the future. It was found that, for Zones 2, 3 and 4, the k-mean clustering of antennas agree with the planned antenna placement grouping. However, for Zone 1, there are differences, which may be attributed to a large metal obstacle occupying the zone. The k-mean clustering also recorded a peak accuracy of 79% with k=4, which agrees with the number of planned zones.

1. Introduction

The maturity of the Global Positioning System (GPS) has allowed the implementation of various remote time-critical applications. However, GPS is limited in the sense that it requires Line-of-Sight (LOS) between a minimum of 4 satellites and the receiver to achieve accurate localization. This issue is exacerbated further indoors where it is impossible to receive LOS signals from any satellite. The dependence on LOS transmission to the satellite renders indoor localization using GPS inaccurate and unreliable.

Therefore, researchers try to find alternatives for indoor localization where Non-Line-of-Sight (NLOS) conditions are prevalent. Many research works reported use devices such as RFID, ZigBee, WiFi, MEMS and others for RF based localization. The measurements made to locate trackers can be generalized into Time of Arrival (ToA), Angle of Arrival (AoA), Time Difference of Arrival (TDoA), and Received Signal Strength Indicator (RSSI) [1].

Lately, Received Signal Strength (RSS) has been the focus of researchers due as the RSSI functions are built in the devices and require no additional hardware to support the localization technique [2]. Although RSSI location technique does not require additional hardware, the RSS is prone to multipath effect and interferences in the complex indoor environments [3].

Another solution is implementing the RSSI localization measurement that uses fingerprinting. The fingerprinting technique can achieve a high level of accuracy in localization. However, it requires a high amount of time and labour intensive for reference point database construction for each deployment environment. The fingerprinting technique has two stages; offline and online stage. The offline stage is the collection of each unique RSSI value at the reference points to form a database for the online stage. In the online stage, location can be predicted based on the database and matched the new unknown location RSSI value with the predefined RSSI database previously [4].
However in most practical applications, high accuracy localization is not required. Most applications need area-based localization. In these sorts of applications, the actual position of the tracked object is not required, but rather to ensure the objects are within a certain area. Rather than having accurate position of the object, users are more interested to know with absolute certainty that the tracked object is within a certain area. For example, a shopkeeper would want to know if certain products are available in certain zones. Likewise, in a workshop, supervisors only need to know whether a worker is in their work zone rather the accurate position of the worker due to mobile nature of the workshop job. High accuracy position information does not deliver any added value to the monitoring supervisors. Hence, what is needed is a method to locate tracked objects in zones with high certainty and reliability.

2. Literature Review & Related Work

Previously, k-mean clustering algorithm was used to improve the accuracy of the position function in NLOS environment [5]. The position function is established with TOA based localization in LOS environment and eliminating the unwanted NLOS measurement. The author suggested that NLOS error due to the complexity of environment can cause large path loss in a signal propagation model is the main reason of the error reported. To improve the accuracy of the localization technique, k-mean clustering can be used to identify and eliminate unwanted NLOS error value to improve the localization model.
RF signal strength has also been reported to be cost-advantageous compared to other techniques such as AOA, TOA, TDOA [6]. The author uses k-mean clustering algorithm to improve the accuracy of the indoor location sensing similar to fingerprinting method. It was pointed out that if a RSSI value that has not been fingerprinted, location cannot be identified. The RSSI value would be added into the specific cluster which has the lowest Euclidean distance value between the RSSI and the centroid to estimate locations which were not fingerprinted.
In another work, Berz et al. proposed an indoor positioning system with RFID technology to estimate the item-level location of stationary objects. The Support Vector Regression machine learning is integrated into a fingerprinting indoor positioning system. To further improve the accuracy of the model, k-mean approach is applied similar to Subaashini [6]. Again, the RSSI values collected from the same tag suffers from interferences, obstacles and multipath effects that would affect the localization of the tag. Shukri et al. demonstrated the fluctuations of signal strength in RSSI based localization using the IRIS motes [7].
Razavi et al. introduced a solution of time-consuming data collection for fingerprinting RSS technique with k-mean clustering algorithm [8]. The proposed solutions achieved a similar accuracy with Nearest-Neighbor (NN) fingerprinting. The k-mean clustering algorithm is used for floor estimation in indoor mobile localization. Bai et al. suggested that fingerprinting techniques in indoor localization system suffers from heavy computation [4]. The heavy computation could be reduced with k-mean clustering and the average searching cost becomes 1/K. Since fingerprinting requires different access point (AP) to form the fingerprinting database, it could result in too many different dimensions for every anchor's fingerprint. Author suggested that Principal Components Analysis should be performed in those fingerprints and limits the dimension's property when clustering [4].

Previously work has been done on indoor propagation and detection of occupants via RSSI. The nature of indoor propagation has been studied in depth and has demonstrated that RSSI fluctuations are affected by number of occupants and obstacles [9]. A number of methods have been proposed to detect and track occupants based on RSSI variations [7]. Various algorithms have also been proposed for control of transmission to ensure reliability in transmission indoors and outdoors environment [10,11]. While the proposed methods have been able to perform to similar accuracies to methods from other researchers, further probing needs to be done to explore other modalities of RF-based localization. As part of continuing work, an active RFID system was deployed at a warehouse in an industrial environment. The work presented in this manuscript uses this testbed to explore new approaches to tracking and localization.

3. Experimental Setup

The testbed used in this research is a large workshop environment. The workshop is divided into different zones, based on different activities done in the workshop. The placement of the directional Antenna 1-12 in the workshop is shown in Fig. 1; placed as such to cover the areas of interest labelled Zone 1-4. The antennas are placed 30m above ground, directed horizontally across the zones.

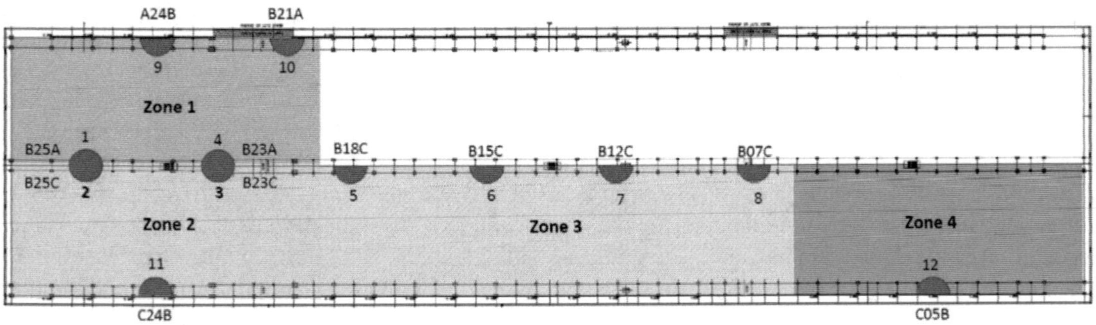

Figure 1. Testbed for the experiment environment

3.1. RSSI Mapping

RSSI mapping was done to ensure the coverage of the antenna in all the zones. The measurement position for Antenna 8 is shown in Fig. 2. From this experiment, the coverage and the strength of the signal received by the antenna from different parts of the area can be observed. The selection of measurement point was affected by the ongoing work in the workshop. Hence some parts of the area were not measured.

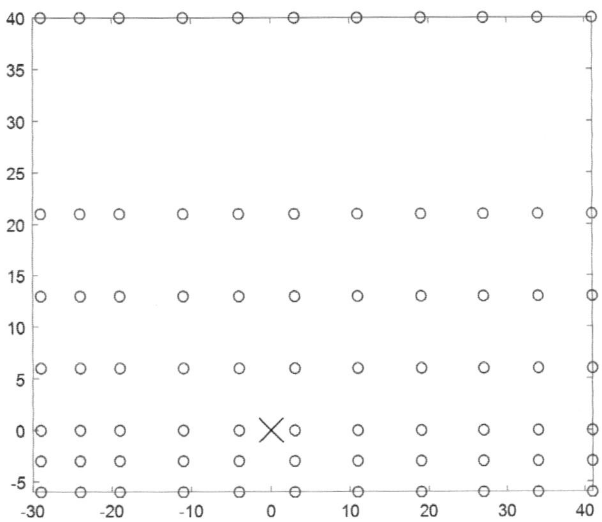

Figure 2. Measurement points for Antenna 8 RSSI mapping

3.2. K-Mean Clustering

IOP Conf. Series: Materials Science and Engineering **705** (2019) 012038 doi:10.1088/1757-899X/705/1/012038

k-mean clustering is an unsupervised machine learning that cluster data into k exclusive clusters. k-mean operates on the observations from the dataset supplied to create clusters. Each of the clusters is created where similar data is partitioned together and other non-similar data are grouped as another cluster. Each of the clusters is defined by the similarity and the centroid of the cluster. The centroid for each cluster is the point at which the sum of distances from all members in that cluster is minimized.
Based on Fig. 1, there is a total of 4 zones in which different activities are conducted. Hence, the placement of the antennas to locate objects in these zones needs to be verified to be adequate, and, the baseline reliability and accuracy of such zone-based localization needs to be established. Hence, k-mean clustering is a suitable method to demonstrate both of these objectives.
The RSSI data stored in the database is divided into user-defined number of clusters in the k-mean clustering. To determine the optimal number of clusters, k-mean model was built with initial cluster from 0 to 13 and observing the elbow pattern. Then, the accuracy of the clusters in the elbow region was studied and compared.
The RSSI of 4 transmitters was collected while it is moved within and between the different zones. The actual time and zones of the transmitter were recorded. This data is fed into the k-mean clustering method to evaluate whether the RSSI data can be used to distinguish transmission from the 4 different zones. By clustering the recorded RSSI, the natural coverage of zones by the antenna can be determined to be suitable to its intended purpose. The grouping of the k-mean as reflected by the centroids derived for each antenna would demonstrate whether the zones are suitably covered by its corresponding antenna.

3.3. Min-Max Normalization
Before the data is fed into the k-mean clustering, the data is normalized using Min-Max method. The Min-Max Normalization used is as shown in Equation 1.

$$RSSI' = 1 - \frac{RSSI - RSSI_{min}}{RSSI_{max} - RSSI_{min}}(newMax - newMin) + newMin \tag{1}$$

The normalized RSSI value is represented by RSSI' as shown in Equation 1. The maximum and minimum value of RSSI from 12 antennas is used for the normalization. The new maximum value after the normalization is 1 and the new minimum value is 0. In this Min-Max normalization, the stronger signal strength should be more approaching to 1 and vice versa.

4. Results and Discussions
The objective of this paper is to propose a novel approach for zoning-based localization with unsupervised learning. The unsupervised learning model used in this paper is k-mean clustering. As mentioned, fingerprinting required time consuming and labour-intensive data collection to build the database. Based on the measurement shown in Fig. 2, a contour graph is plotted in Fig. 3.

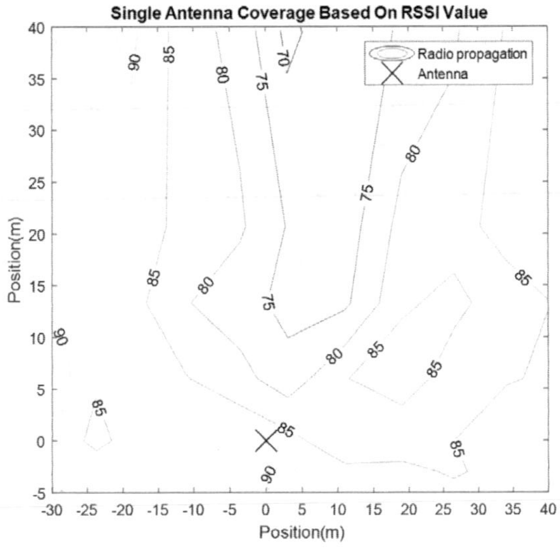

Figure 3. Measurement for Antenna 8 RSSI

The collected data shows some unexpected peaks and troughs in the distribution, due to the fluctuation observed during the measurement. Since there were many metal obstacles in the environment, some measured areas, for example (20, 10), drops in RSSI was recorded. The results demonstrate that in dynamic environments, especially with metal obstacles, RSSI based localization techniques such as trilateration or fingerprinting may not be feasible.

Then, data was collected while transmitters were moved in and between the different zones. The collected data was fed into k-mean model as training data. The elbow graph for k cluster is plotted in Fig. 4(a). Based on Fig. 4(a), the elbow pattern can be observed when k=3 until k=6. The accuracy of the model is further analysed and compared to the reference positioning as plotted in Fig. 4(b).

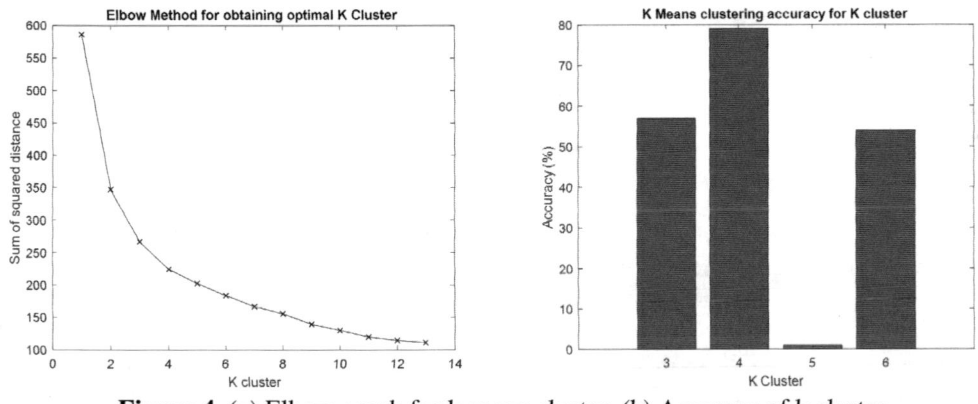

Figure 4. (a) Elbow graph for k-mean cluster. (b) Accuracy of k cluster.

Based on Fig. 4(b), the highest accuracy is observed when k=4 which is similar to the zoning division based on Fig. 1. The accuracy for k=4 is 79% which fulfils the antenna objective of zoning into 4 divisions.

The model when k = 4 is further examined and tabulated in Table 1. Here, the main features; which are the antenna reading, from the dataset that was taken to group the data can be observed in the centroids. As demonstrated in Equation (1), a higher value for centroid means that the RSSI reading for the corresponding antenna in the corresponding zone is expected to be higher.

Table 1. Tabulated centroid for Zoning division in 12 antennas.

Z	A1	A2	A3	A4	A5	A6	A7	A8	A9	A10	A11	A12
1	0.0573	0.5587	0.0749	0.0148	0.0109	0.0118	0.0109	0.0109	0.0305	0.0336	0.0263	0.0109
2	0.0372	0.4990	0.1802	0.0151	0.0111	0.0129	0.0109	0.0109	0.0550	0.0451	0.4532	0.0109
3	0.0184	0.0129	0.0396	0.0139	0.0469	0.1086	0.0800	0.1294	0.0261	0.0468	0.0450	0.0159
4	0.0109	0.0109	0.0109	0.0109	0.0109	0.0109	0.0120	0.0262	0.0109	0.0109	0.0109	0.4059

The centroid of the clustering when k = =4 is summarized in Table 1. The unsupervised zoning grouping can be observed by finding the higher value centroids for each of the zones. This means that transmitters are considered to be in the corresponding zone if the RSSI readings of the corresponding antenna are higher. This way, the unsupervised zoning localization can be compared with the testbed zoning plan to determine whether the antenna placement is suitable for zone based localization.
Based on the centroids in Table 1, the zonal grouping is summarized in Table 2.

Table 2. Zones Based Antenna Grouping

Zones	No. of Antennas	Planned Zone Grouping (Antenna)	Unsupervised Zone Grouping (Antenna)
1	4	1,4,9,10	1,2,3,10
2	3	2,3,11	2,3,11
3	4	5,6,7,8	5,6,7,8
4	1	12	12

As shown in Table 2, the unsupervised k-mean zone grouping agrees with the planned zone grouping for zones 2, 3 and 4. However, there are some differences while the grouping of antenna for zone 1. This is because both zone 1 and zone 2 are near to each other and the antennas in both zones overlap with each other hence causing different features being taken into consideration. Furthermore, it was noted that there was a large metal object almost entirely occupying Zone 1; which may have affected the measurements made in the area.
The findings in this work demonstrated that zone based localization has the potential to be used reliably as localization technique in obstacle laden indoor environments. Furthermore, it was observed that fluctuations in the recorded data may render trilateration methods unreliable in real world environments without extensive computation strategy.

5. Conclusions

The work presented in this document has managed to demonstrate the feasibility of zone based localization in obstacle present environment. Furthermore, the dataset collected has demonstrated the difficulties of deploying trilateration and fingerprinting methods in dynamic conditions, as it introduces fluctuating RSSI measurements. The k-mean clustering was proposed to validate antenna placement to divide the testbed into zones, and to create a baseline accuracy which can be compared with other algorithms in the future. It was found that, for Zones 2, 3 and 4, the k-mean clustering of antennas agree with the planned antenna placement grouping. However, for Zone 1, there are differences, which may be attributed to a large metal obstacle occupying the zone. The k-mean clustering also recorded a peak accuracy of 79% with k=4, which agrees with the number of planned zones.

Acknowledgment

This research was supported by Ministry of Education (MOE) through Fundamental Research Grant Scheme (FRGS/1/2018/SKK06/UniMAP/02/1). The authors would also like to acknowledge the support from Industrial Grant (9002-00076) provided by the industry.

References

[1] Xiao H, Zhang H, Wang Z and Gulliver T A 2017 An RSSI based DV-hop algorithm for wireless sensor networks *2017 IEEE Pacific Rim Conf. Commun. Comput. Signal Process. PACRIM 2017 - Proc.* **2017-Janua** 1–6
[2] Shukri S and Kamarudin L M 2017 Device free localization technology for human detection and counting with RF sensor networks: A review *J. Netw. Comput. Appl.* **97** 157–74

[3] Azmi N, Kamarudin L M, Mahmuddin M, Zakaria A, Shakaff A Y M, Khatun S, Kamarudin K and Morshed M N 2011 Interference issues and mitigation method in WSN 2.4GHz ISM band: A survey *2014 2nd Int. Conf. Electron. Des. ICED 2014* 403–8

[4] Bai S and Wu T 2013 Analysis of K-Means algorithm on fingerprint based indoor localization system *2013 5th IEEE Int. Symp. Microwave, Antenna, Propag. EMC Technol. Wirel. Commun. MAPE 2013* 44–8

[5] Cheng L, Wu X and Wang Y 2017 A non-line of sight localization method based on k-means clustering algorithm *Proc. 2017 IEEE 7th Int. Conf. Electron. Inf. Emerg. Commun. ICEIEC 2017* 465–8

[6] Subaashini K, Dhivya G and Pitchiah R 2012 Zigbee RF signal strength for indoor location sensing-Experiments and results ... *Technol. (ICACT), 2012 ...* **1** 12–7

[7] Shukri S, Kamarudin L M, Cheik G C, Gunasagaran R, Zakaria A, Kamarudin K, Zakaria S M M S, Harun A and Azemi S N 2017 Analysis of RSSI-based DFL for human detection in indoor environment using IRIS mote *2016 3rd Int. Conf. Electron. Des. ICED 2016* 216–21

[8] Razavi A, Valkama M and Lohan E S 2015 K-means fingerprint clustering for low-complexity floor estimation in indoor mobile localization *2015 IEEE Globecom Work. GC Wkshps 2015 - Proc.* 1–7

[9] Turner J S C, Kamarudin L M, Ndzi D L, Harun A, Zakaria A, Shakaff A Y M, Saad A R M and Mamduh S M 2011 Modelling indoor propagation for WSN deployment in smart building *2014 2nd Int. Conf. Electron. Des. ICED 2014* 398–402

[10] Kamarudin L M, Ahmad R B, Ndzi D, Zakaria A, Ong B L, Kamarudin K, Harun A and Mamduh S M 2012 Modeling and simulation of WSNs for agriculture applications using dynamic transmit power control algorithm *Proc. - 3rd Int. Conf. Intell. Syst. Model. Simulation, ISMS 2012* 616–21

[11] Mamduh S M, Shakaff A Y M, Saad S M, Kamarudin K, Kamarudin L M, Zakaria A, Kamarudin H, Ezanuddin A M M, Sa'Ad F S A, Nooriman W M and Abdullah A H 2012 Odour and hazardous gas monitoring system for swiftlet farming using wireless sensor network (WSN) *Chem. Eng. Trans.* **30** 331–6

IOP Conf. Series: Materials Science and Engineering **705** (2019) 012040 doi:10.1088/1757-899X/705/1/012040

Human Identification through Kinect's Depth, RGB, and Sound Sensor

P Shunmugam[1], K Kamarudin[1,2,a], A A A Mosed[1,2] and S A A Shukor[1]

[1]School of Mechatronics Engineering, Universiti Malaysia Perlis (UniMAP), Arau, Malaysia,

[2]Centre of Excellence for Advanced Sensor Technology (CEASTech), Universiti Malaysia Perlis, Arau, Malaysia

[a]kamarulzaman@unimap.edu.my

Abstract. Human identification is a very important subject in computer field and has been researched widely. This paper proposes human identification system based on Kinect's individual sensor as well as combination of all its available sensors. In the first part of the project, each sensor on the Kinect (i.e IR depth sensor, RGB camera and microphones) was used for skeleton recognition, face recognition and speech recognition respectively. Then, these individual recognition methods are combined as step-by-step process into a multi-sensor recognition system. Few experiments were carried out to test the reliability of the developed human identification systems. The results show that multi-sensor based human identification system is highly efficient compared to the single-sensor system. This is because, multi-sensor recognition system involves many recognition stages as each recognition stage needs particular biometric information of the user.

1. Introduction

Human identification has good application prospect and potential economic value. Human identification is a systematic process, involves biometrics technology using human biological characteristics. Human identification is beneficial in many fields mainly on authentication and security system [1].

One of the most significant inventions in the year 2010 is the Microsoft Kinect because of the high-resolution depth, four arrays of microphones, and visual (RGB) features that it provides for a relatively much lower cost when compared to other 3D cameras such as stereo cameras and Time-Of-Flight cameras[2]. Kinect is used for applications like playing a virtual violin, to applications in health care and physical therapy, retail, education, and training [3]. Figure 1 shows the various biometric technologies [1][4][5][6] and components of the Kinect sensor. The followings are the specification of the Kinect; an RGB camera with a 640 x 480-pixel, an infrared (IR) emitter and an IR depth sensor to capture depth images an array of four microphones with 16-kHz, 24-bit mono pulse code modulation (PCM) and a 3-axis accelerometer to determine the orientation of the Kinect [2].

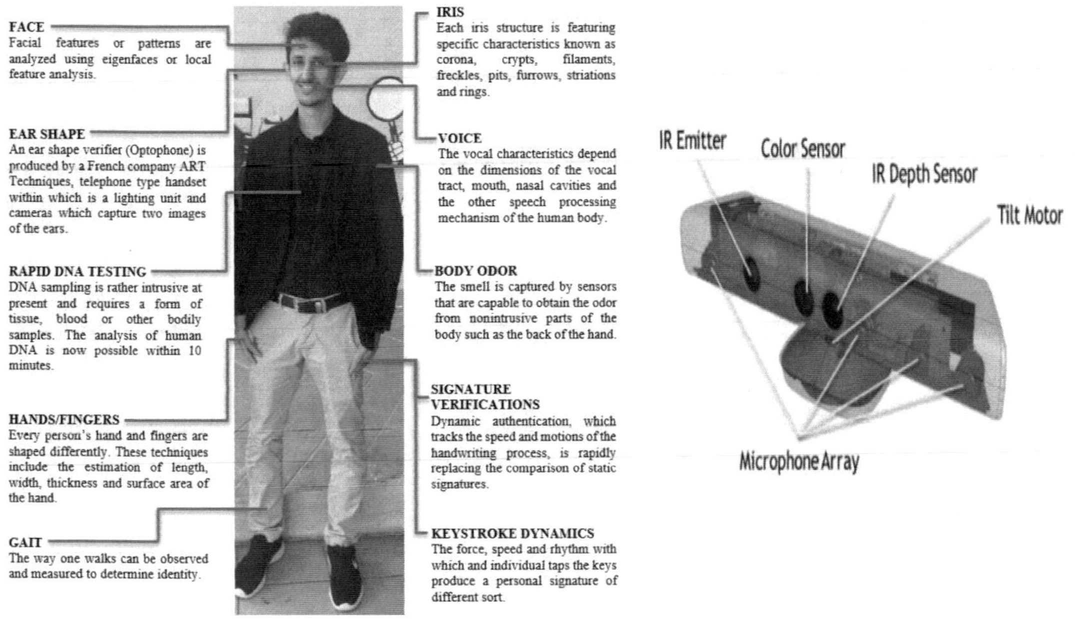

(a) (b)

Figure 1. (a) Various biometric technologies [1][4] [5] [6]; (b) Components of the Kinect sensor [2].

The availability of various sensors available on the Kinect can be utilized for multi-typed sensor data based human identification purpose where more data can be obtained and collected, which resulting in more accurate and efficient identification. Although, Kinect consists of many sensors, but the price of Kinect much lower compared to other multi sensor devices [7]. By using Kinect, low cost, highly efficient and accurate human identification system can be developed. In this research, speech recognition system using microphone, skeleton recognition system using the IR depth sensor and face recognition system using the RGB camera was developed for the single-typed sensor data based human identification system. Speech recognition recognizes speech command, skeleton recognition recognizes user's 19 skeleton mean information while face recognition recognizes user's face, eye, nose and lips template. For multi-typed sensor data based human identification system, speech, skeleton and face recognitions are combined into one system where the process will run step-by-step. The reliability of the developed human identification methods was tested and analyzed

2. Related Work

In previous study, the Mel-frequency cepstrum coefficients, the logarithmic power and their related values are calculated from the personal voice [8]. Another study in [9] was on facial depth data of a speaking subject, captured by the Kinect device, as an additional speech informative modality to incorporate to a traditional audiovisual automatic speech recognizer. They presented feature extraction algorithm for both visual and accompanying depth modalities, based on a discrete cosine transform of the mouth region-of-interest data, further transformed by a two-stage linear discriminant analysis projection to incorporate speech dynamics and improve classification. An autonomous voice and motion-controlled video camera system was developed by Jody (Pritchett) Shu [10]. The Kinect data is processed in real time by the software system on the laptop to track the lecturer's movement and automatically change focus to the board, projector screen, or other visual aids. In addition to posture and voice-command recognition, the system filters out background sounds and acts as a directional microphone to pick up the lecturer's speech in an optimal fashion for voice recognition [10].

3. Methodologies

3.1. Single-typed sensor data based human identification

3.1.1. Speech Recognition System (Microphone)

This system uses automatic speech recognition method which is speech to text technique. The flow chart of the system is shown in Figure 2. The Kinect's microphone array collected the 24-bit audio data and pre-processed to remove the background noise and echoes of the signal using automatic echo cancellation (AEC) algorithm [11]. When there is a multiple set of microphones, the time that sound arrives from an audio source to each microphone is slightly different. Audio data captured from Kinect microphones is fed to the preliminary processor and at that section Beam forming and Sound localization techniques determine the direction of the sound source and the set of microphones is used as a directional microphone.

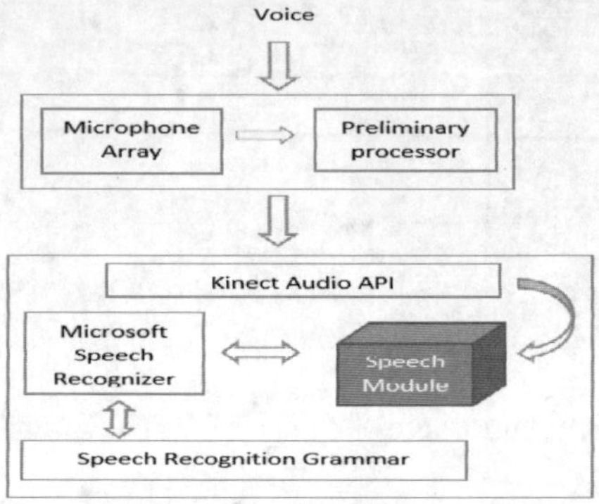

Figure 2. Speech Recognition system flow chart

Speech module section consists of two parts which are Microsoft Speech Recognition Engine [12] and Microsoft speech recognition grammar. Microsoft speech recognition engine matches vocal inputs with words and phrases, as defined by grammar rules. Two grammar rules exist in the speech recognition grammar. First is the simple rule which recognize small words and commands. Another rule can recognize and organize semantic content with various user accents. In this system, we had implemented the simple rule where the user needs to add the word into the command list and then the word will be initialized. This system built by grammar builder and speech recognizer from the "System.Speech(4.0.0.0)" of NET Constructor with LabVIEW software [13].

3.1.2. Skeleton Recognition System (IR Depth sensor)

Depth sensor is used to recognize human skeleton in this research using Kinesthesia toolkits. It utilizes the "Distance and Displacement between Joints VI" which obtain 20 joint coordinates of the skeleton [14] starting from head until feet. When the Microsoft Kinect starts operating, the subject to be recognized must stand at a minimum distance of 1.5 meters from the Kinect. 20 joints will be detected, and the coordinates are obtained. All the joints are connected, where 19 pairs of adjacent joints make up the nones for the whole skeleton. The distance between joint pair (i.e bone) calculated using Euclidean distance algorithm (refer Equation 1) [15]:

$$d = \sqrt{(x_2 - x_1)^2 + (y_2 - y_1)^2 + (z_2 - z_1)^2} \quad (1)$$

where d is the distance between 2 joints, (x^1, y^1, z^1) are the coordinates of the first joint and (x^2, y^2, z^2) are the coordinates of the second joint. By applying this algorithm, again it is possible to obtain 19 absolute distances between the 20 adjacent joints. The data then is written in one spreadsheet file continuously. After Euclidean distances have been calculated, the mean for each of 19 absolute distances is computed using Equation 2:

$$\bar{x} = \frac{1}{n}\left(\sum_{i=0}^{n} x_i\right) = \frac{x_1 + x_2 + \ldots + x_n}{n} \tag{2}$$

Where n is the total number of reading. The data saved as the subject's name and will be registered as user. During the recognition, the 19 mean distance data stored in the file (.csv) will be compared $\pm$ 1 range with real time data of the 19 Euclidean distance data of subject to be recognized. The 19 outputs are either logic 1 or logic 0. Then all of these are summed using "Compound Arithmetic" function and choosing the "Add" option. If the score is equal to or more than 15 thresholds, then the subject is recognized as the registered user.

3.1.3. Face Recognition System (RGB camera)

One of the most remarkable abilities of human vision is that of face recognition [16]. RGB camera was used to recognize face in this section. This recognition was developed using IMAQ Vision Development Toolkit [17]. The system uses built-in colour and pattern matching capabilities to perform face recognition on a Kinect camera stream. Given right templates and adequate settings it does perform well with respect to its simplicity. The whole approach can be split into two main parts; Face Detection and point following. Face detection was accomplished by matching eye, nose and lips templates against the camera stream image. Figure 3 shows the sample of the eye, nose and lips template in .png file format.

(a) (b) (c) (d)

Figure 3. Sample of pattern matching templates: (a) face (b) eye (c) nose (d) lips.

3.2. Multi-sensor based Human Identification

This system developed for human identification based on multiple types sensor data particularly depth, RGB, sound sensor. It is the combination of all the above speech recognition, skeleton recognition and face recognition approaches which are executed step by step. Firstly, the user must set up the data of his speech command, skeleton information and face template in the system. To be recognized by the system, the subject must speak the speech command. If the speech command is recognized, then the subject will undergo second step which is the skeleton recognition. When the skeleton is recognized, then the subject will go through face recognition as the last step. The user will be identified and authorize the system only if all 3 stages of recognitions are passed. If any of the stage failed to be recognize, he or she will not be recognized, and the system will end the process.

4. Experimental Result

Few experiments were carried out on eight different subjects to test the reliability of the method for the single and multi-typed sensor data based human identification system. The results of the experiments are shown and discussed in this chapter.

4.1. Single-typed sensor data based human identification system

4.1.1. Speech Recognition System (Microphone)

The efficiency of the speech recognition system was tested by using "measure" and "scan" commands. Results are shown in Table 1. Each test consists of two rounds where in each round, subject is allowed to say the command for a of maximum of 10 attempts to be recognized. If more than 10 attempts, the subject will not be recognized, and the system ends the recognition operations.

Table 1. "Measure" command test on speech recognition system

Subjects	Command "Measure"		Command "Scan"	
	1st round	2nd round	1st round	2nd round
Osamah	Fail	Fail	1	1
Fiqa	Fail	3	8	7
Izyan	6	2	1	1
Ling	4	3	4	2
Khei	Fail	1	3	2
Raj	1	2	2	1
Teven	2	1	1	1
Puva	3	2	1	1

From Table 1, we can observe that Osama failed to pronounce "measure" command as he has Arabic mother tongue and he faced difficulties in pronouncing the command correctly. But Raj, Teven and Puva was able to pronounce the "Measure" command correctly and easily in both rounds. Also, we can observe Osamah, Izyan, Teven and Puva were able to easily pronounced the "Scan" command and was recognised by the system at the very first attempt. But still few subjects like Fiqa unable to be recognised early by the system as she has very soft voice and very low sound. This shows that the subject must say the command bold and confident. The weakness of this system is, any subject with the right command will be recognized as the user by the system. To increase the efficiency of this system, the command must be unpredictable word, difficult to be pronounced and must kept secretly by the user.

4.1.2. Skeleton Recognition System (IR Depth sensor)
Few experiments were conducted on eight subjects to test the reliability of this skeleton recognition method. Firstly, the matching score (threshold) is fixed to 15. Each subject tested two random users' skeleton information. The results are shown in Table 2.

Table 2. Tesing random user data on random subject.

Subjects	*User*	Match score	Recognized?	*User*	Match score	Recognized?
Osamah	*Puva*	7	NO	*Khei*	10	NO
Fiqa	*Puva*	10	NO	*Osama*	16	YES
Izyan	*Puva*	8	NO	*Fiqa*	17	YES
Ling	*Puva*	9	NO	*Izyan*	14	NO
Khei	*Ling*	15	YES	*Fiqa*	5	NO
Raj	*Khei*	11	NO	*Fiqa*	15	YES
Teven	*Raj*	11	NO	*Osama*	12	NO
Puva	*Ling*	8	NO	*Fiqa*	12	NO

From Table 2, we can observe misrecognition happened when subject (Khei) tried to access the user (Ling) skeleton data. Also, we can conclude that few misrecognitions occurred for Fiqa, Izyan and Raj. This happened because subject and user have the similar same height and body build. The weakness of this system is, few subjects were misrecognized by the system where the subject and the user have similar body built.

4.1.3. Face Recognition System (RGB camera)
Efficiency of the system was determined by testing random user's face template data on random subject. Each subject tested the system twice. The results are shown in Table 3.

Table 3. Result of testing random user data on random subject.

	Experiment 1		Experiment 2	
Subjects	User	Recognized?	User	Recognized?
Osamah	Puva	No	Khei	Yes
Fiqa	Puva	No	Osama	No
Izyan	Puva	No	Fiqa	No
Ling	Izyan	Yes	Fiqa	No
Khei	Ling	No	Fiqa	No
Raj	Fiqa	No	Khei	No
Teven	Raj	No	Osama	No
Puva	Osama	Yes	Teven	No

From Table 3, we can observe misrecognition occurred while testing subject (Ling) with user (Izyan) face template data. Also, we can observe that Osamah was misrecognized by Khei data and Puva was misrecognized by Osamah data. This happened because both the user and subject have fair skin, and have similar face features such as eyes, nose and lips. From the experiment result we can conclude that 700 is the best threshold as user can be recognized easily. But still misrecognition occurred when tested with random user data on random subjects. This happened because the subject and the user have the similar face features such as same type of nose, eyes and lips.

4.2. Multi-sensor based human identification system

For multi sensor based human identification system, two experiments were carried out to test the reliability of the system. First experiment used eight subjects to test user (Vishnu) system. User chooses "measure" as the voice command password as it's difficult to pronounce by others and he can pronounce it correctly and easily. User's skeleton information and face template also set up in the system. The results are shown in Table 4.

Table 4. Eight subject tested user (Vishnu) system.

Subject	*User*	Stage 1: Speech Command: Measure	Stage 2: Skeleton Match Score:15	Stage 3: Face Threshold 700	Authorised?
Osamah	*Vishnu*	Failed	X	X	X
Fiqa	*Vishnu*	Failed	X	X	X
Izyan	*Vishnu*	Yes	Failed	X	X
Ling	*Vishnu*	Yes	Failed	X	X
Khei	*Vishnu*	Failed	X	X	X
Raj	*Vishnu*	Failed	X	X	X
Teven	*Vishnu*	Yes	Failed	X	X
Puva	*Vishnu*	Yes	Yes	Failed	X
Vishnu	*Vishnu*	*Yes*	*Yes*	*Yes*	*Yes*

From Table 4, we can observe that only 4 subjects (Izyan, Ling, Teven and Puva) able to pass first stage (speech recognition). Out of four subjects only one subject (Puva) able to pass second stage (skeleton recognition). But this subject (Puva) too failed to pass the last stage (face recognition) of the system. At the end, this multi-typed sensor based human identification system can only be authorized by the user (Vishnu). Second experiment was conducted using cross user method where five persons act as user as well as subjects tested the system 25 times. The results are shown in Table 5. While Figure 4 shows an example of the process where the user (Puva) has been successfully recognized by the multi-sensor data based human identification system.

IOP Conf. Series: Materials Science and Engineering **705** (2019) 012040 doi:10.1088/1757-899X/705/1/012040

Table 5. Five user/subject cross test the system 25 times.

	Users				
Subjects	Osamah	Fiqa	Izyan	Ling	Puva
Osamah	Recognised	Not	Not	Not	Not
Fiqa	Not	Recognised	Not	Not	Not
Izyan	Not	Not	Recognised	Not	Not
Ling	Not	Not	Not	Recognised	Not
Puva	Not	Not	Not	Not	Recognised

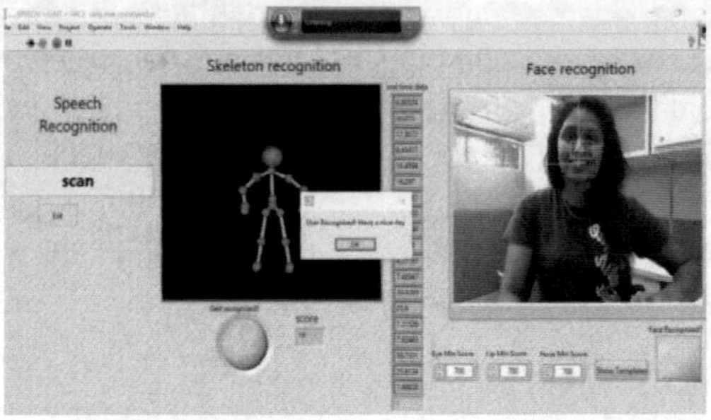

Figure 4. Puva was recognized by the multi-sensor data based human identification system.

From Table 5, we can observe that the system did not misrecognize any subjects. All these experiments proved that multi-sensor human identification system is more efficient compared to the single sensor human identification system as only the correct user was recognized and authorized by the system.

5. CONCLUSION

Multi-typed sensor human identification algorithm was developed, analyzed and implemented successfully by a detailed evaluation for real life scenarios such as highly efficient and low-cost user authentication system. From all the experiments carried out, we can conclude that single-typed sensor data based human identification system are less efficient compared to the multi-typed sensor data based human identification system.

One of the limitations are, voice command password must be kept secretly. Secondly, the person to be identified and tracked must not move his/her body for 10s. Thirdly, for the face recognition the person should maintain the appearance, for example if he or she used glasses in the template image, then he or she must wear glasses during the recognition process. Finally, since Microsoft Kinect uses infrared, camera and microphone, this system only can be used indoors where there is no direct sunlight, and under controlled brightness.

Kinect originally can detect 6 people and track 2 at the same time, but with Kinesthesia toolkit in LabVIEW, only one person is recognized and tracked which reduces the practicality of the algorithm developed. As future work, the Kinesthesia toolkit can be improved to recognize all the six detectable people and track two, which will enhance the practicality of the system.

6. ACKNOWLEDGMENT

The authors would like to acknowledge the support from the Fundamental Research Grant Scheme (FRGS) under a grant number of FRGS/1/2018/TK04/UNIMAP/02/15 from the Ministry of Education Malaysia.

IOP Conf. Series: Materials Science and Engineering **705** (2019) 012040 doi:10.1088/1757-899X/705/1/012040

References

[1] Bhattacharyya D, Ranjan R, Alisherov F and Choi M 2009 Biometric authentication: A review *Int. J. u-and e-Service, Sci. Technol.* **2** 13–28

[2] Recognition H 2013 Human Recognition, Identification and Tracking using Microsoft Kinect Interfaced with DaNI Robot Author: Diyar Khalis Bilal 2012–3

[3] Tashev I 2013 Kinect development kit: A toolkit for gesture-and speech-based human-machine interaction [Best of the Web] *IEEE Signal Process. Mag.* **30** 129–31

[4] Kumar M S N and Babu R V 2013 Human gait recognition using depth camera 1–6

[5] Anon 2013 Face Recognition And Registration For Home Surveillance System Using Iterative Closest Point And Haar Cascade Lgorithm By Kranthi Kumar Kandi , B . tech Presented to the faculty of The University o f Houston Clear Lake In Partial Fulfillment of the Requir

[6] Engineering C 2013 Person Authentication Using Face And Voice 71–9

[7] Rahman M W, Zohra F T and Gavrilova M L 2018 Rank level fusion for kinect gait and face biometric identification *2017 IEEE Symp. Ser. Comput. Intell. SSCI 2017.*1–7

[8] Kita E, Zuo Y, Saito F and Feng X 2017 Personal Identification with Face and Voice Features Extracted through Kinect Sensor *IEEE Int. Conf. Data Min. Work. ICDMW* 545–51

[9] Galatas G, Potamianos G and Makedon F 2012 Audio-Visual Speech Recognition Incorporating Facial Depth Information Captured By The Kinect Institute of Informatics and Telecommunications, NCSR " Demokritos ", Athens , Greece Heracleia Lab ., Dept . of Computer Science and Engin ., University of Tex 2714–7

[10] State C 2013 Autonomous Voice and Motion Controlled Video Camera System for Instructional Technology by

[11] Rathnayake K A S V, Diddeniya S I A P, Wanniarachchi W K I L, Nanayakkara W H K P and Gunasinghe H N 2017 Voice operated home automation system based on Kinect sensor *2016 IEEE Int. Conf. Inf. Autom. Sustain. Interoper. Sustain. Smart Syst. Next Gener. ICIAfS 2016* 1–5

[12] Rami M, Svitlana M, Lyashenko V and Belova N 2017 Speech Recognition Systems : A Comparative Review *IOSR J. Comput. Eng.* **19** 71–9

[13] Ertugrul N 2000 with LabVIEW ™ *Simulation*

[14] Anon 2017 Copyright by Sanobar Kadiwal 2017

[15] Lip C J, Yeon A S A, Kamarudin L M, Kamarudin K, Visvanathan R, Zaidi A F A, Mamduh S M, Zakaria A and Nooriman W M 2018 Human 3D Reconstruction and Identification Using Kinect Sensor *2018 Int. Conf. Comput. Approach Smart Syst. Des. Appl. ICASSDA 2018* 1–7

[16] IEEE 1998 Face Recognition: Features versus Templates *IEEE Trans. Pattern Anal. Mach. Intell.* **15** 1–11

[17] Adiloğlu S 2016 We are IntechOpen , the world ' s leading publisher of Open Access books Built by scientists , for scientists TOP 1 % *Heavy Met. Remov. with Phytoremediation* **i** 13

IOP Conf. Series: Materials Science and Engineering **705** (2019) 012041 doi:10.1088/1757-899X/705/1/012041

Design and Development of a Reconfigurable Screw-wheeled Omnidirectional Mobile Robot

M.J.A. Safar*, M.H. Samsol, A.H. Ismail, S.N. Basah, K.S. Basaruddin, A.F. Salleh

School of Mechatronic Engineering, Universiti Malaysia Perlis, Pauh Putra Campus, 02600 Arau, Perlis, Malaysia

Email: *juhairi@unimap.edu.my

Abstract. Omnidirectional mobility plays an important role for mobile robots to provide flexibility in their motion. However, most of the existing omnidirectional mechanisms are only superior in indoor applications, but inappropriate for outdoor purposes. This paper introduces the design and development of an omnidirectional mobile robot with reconfigurable screw-wheeled mechanism. The orientation of the screw wheels is adjustable in order to enable flexibility in traction control to provide more propulsion drive in certain conditions. The experimental results are given to demonstrate the performance of the omnidirectional mobile robot on different ground surfaces and different wheel's orientation.

1. Introduction

Standard wheel with tyre is one of important invention in human history to provide good traction and fast locomotion system on the ground. However, the performance of the standard wheel may weakening in rough and soft terrains. There are many other alternatives provided by the researchers to solve the locomotion on these kind of terrains such as screw-wheels propelled vehicles [1]. The screw wheel or screw propelled wheel actually are develop by following the concept of the Archimedes screw conveyor that consists of a helix surface surrounding a central cylindrical shaft which are located inside a hollow pipe. The screw-wheel vehicles are good in the operation on the extreme terrains such as muddy, sandy, snowy, clay and grassy area. It can give a better performance of the movement on those kind of terrains compared to the wheel that can be seen mostly at normal vehicles such as cars, buses and motorcycles. The screw-wheels vehicles move based on the direction of rotation that are applied to the wheel with helical edges on its surface [2]. The helical surface on the wheel will produce a resisting force which is able to push enough of the material to attain motion. Besides, the different rotation of the screw wheel can cause the vehicles to move with the direction whether moving forward or backward, it depends the sequence of rotation of the wheel in clockwise or counter-clockwise direction [3,4]. By utilizing this principle and proper design, a new mechanism that can work in omnidirectional motion is developed [5]. This paper introduces a new improvement to the developed reconfigurable screw-wheeled omnidirectional mobile robot (RSOMR) by introducing a Variable Wheel Angle Mechanism (VWAM). The robot design and the principles of the VWAM is discussed in Section 2. Meanwhile, the performance of the new improvement is experimentally tested and evaluated in Section 3.

2. Robot Design and Kinematics

The kinematics model of the reconfigurable screw-wheeled omnidirectional mobile robot (RSOMR) is shown in Figure 1. The RSOMR is constructed by four screw wheels which are arranged in diagonal structure. The helical angle of each wheel is designed at the angle of 45 degrees in order to produce equivalent driven power at the lateral and longitudinal direction of the wheel. Each screw wheel is independently driven and attached to the steering mechanism, Variable Wheel Angle Mechanism (VWAM). By utilizing this VWAM, the orientation of the wheel can be steered at any angle between 0 and 45 degrees as shown in Figure 2.

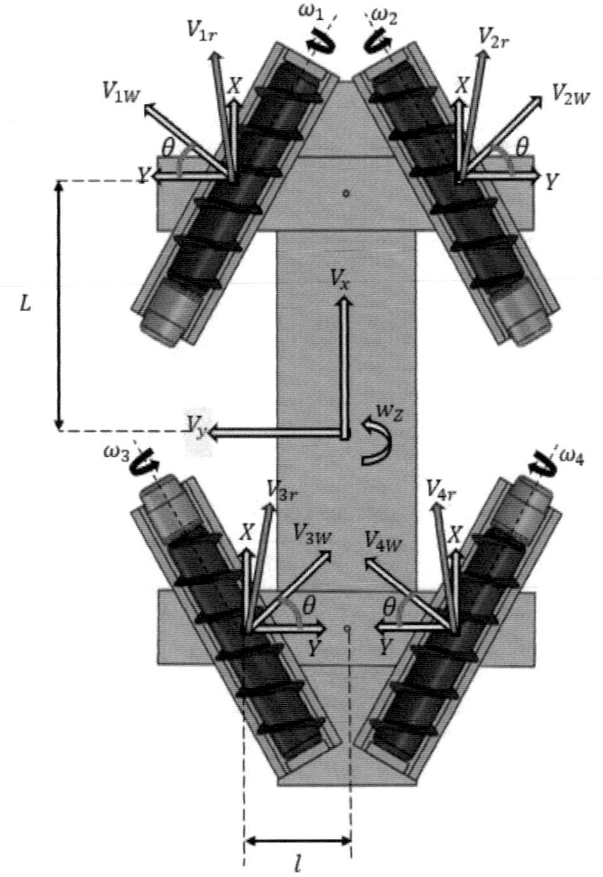

Figure 1. Kinematics Model of RSOMR

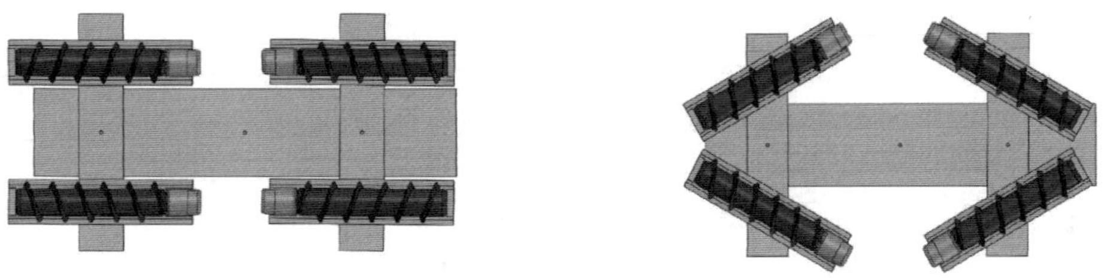

(a) Standard Configuration (0 degree)

(b) Variable Angle Configuration (<45 degree)

Figure 2. Screw wheel's arrangement for RSOMR

The omnidirectional motion is produced at the centre of the RSOMR as the result of the resultant force induced from each individual wheels. This relationship is described by the forward kinematics shown in Equation 1 which is the extended equation presented in [5].

$$\mathbf{v} = A\boldsymbol{\omega} \tag{1}$$

where, $\mathbf{v} = [v_x, v_y, \omega_z]^T$ is the robot velocity which includes the linear velocities at $xy-$axes direction and angular velocity about $z-$axis and $\boldsymbol{\omega} = [\omega_1, \omega_2, \omega_3, \omega_4]^T$ is the angular velocities of the wheels. The matrix A is given by

$$A = \frac{r_w}{4k}\begin{bmatrix} -1 & -1 & -1 & -1 \\ 1/\tan a & -1/\tan a & -1/\tan a & 1/\tan a \\ 1/b & -1/b & -1/b & 1/b \end{bmatrix}$$

where, r_w is the radius of the wheel, $a = \theta + 45°$, θ is the angle of wheel arrangement, $k = \cos\theta \tan a - \sin\theta$ and $b = L\tan a + l$.

Table 1 shows the three basic motions of the omnidirectional mobile robot. Using the same magnitude of angular velocity at each wheels, ω, the RSOMR can be controlled to move in the direction of longitudinal, lateral or rotation just by changing the direction of the angular velocity.

Table 1. Basic motion of RSOMR

Movement	Angular Velocities			
	Wheel 1	Wheel 2	Wheel 3	Wheel 4
Longitudinal (Forward)	$-\omega$	ω	ω	$-\omega$
Lateral (Right Sideway)	ω	ω	ω	ω
Angular (Clockwise)	ω	ω	$-\omega$	$-\omega$

3. Result and Discussion

3.1. Prototype Development

Figure 3 shows the developed prototype of the RSOMR. The omnidirectional mobile robot is controlled by using Arduino microcontroller. It has four wheels and each of the wheel is driven by a DC geared motor. Meanwhile, the VWAM steering mechanism is actuated by a stepping motor which enables a steering angle between 0 to 45 degrees. The helical part of the screw wheels is a 3D printed PLA plastic and it is wrapped to a hollow PVC pipe to construct a screw wheel. Figure 4 shows the RSOMR in 0 degree and 45 degrees orientation, respectively.

3.2. Performance Evaluation

An experimental work is conducted to evaluate the performance of the RSOMR in different ground conditions. There are four angle of screw wheel arrangements tested in the experiment which are 0, 15, 30 and 45 degrees. Each screw wheels arrangement are tested on three ground conditions which are tiles, asphalt and soft soil. The desired velocity of RSOMR based on the angle are shown in Table 2.

Table 2. Desired velocity of the RSOMR

Wheel Angle (deg)	Desired Velocity (m/s)	
	Longitudinal	Lateral
0	0.048	0.772
15	0.044	0.140
30	0.032	0.135
45	0.033	0.066

Figure 3. The developed prototype of RSOMR

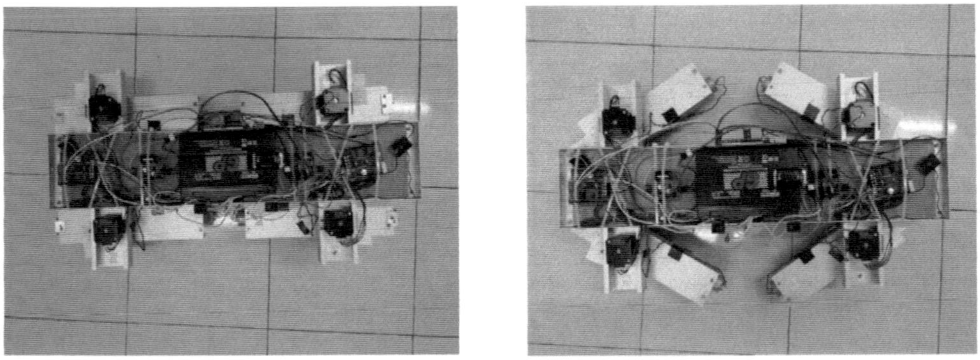

(a) 0 degree (b) 45 degrees

Figure 4. RSOMR in different configurations

Figure 5 shows the performance of the RSOMR for longitudinal movement in three different ground conditions. In general, the RSOMR is possible to produce a longitudinal movement, except for the wheel arrangement of 45 degrees. In most cases the produced velocity is lesser than the theoretical value. The RSOMR also is found to operate better on the soft-soil condition with comparison to another two ground conditions. The soft-soil which flows at the screw wheels is expected to provide more traction force. Figure 6 shows the performance of the RSOMR for lateral movement in three different ground conditions. The velocity of the screw wheel during on the tiles surface is nearer towards the theoretical value. It is because during on the tiles, the rolling force of the wheel has less resistance to attain the obstacle so that the less resistance of obstacle let the rolling forces stay high during the movement compare to the asphalt and soft-soil surface. Therefore, screw wheel has a good locomotion in horizontal on the tiles in any type of angle of wheel arrangement.

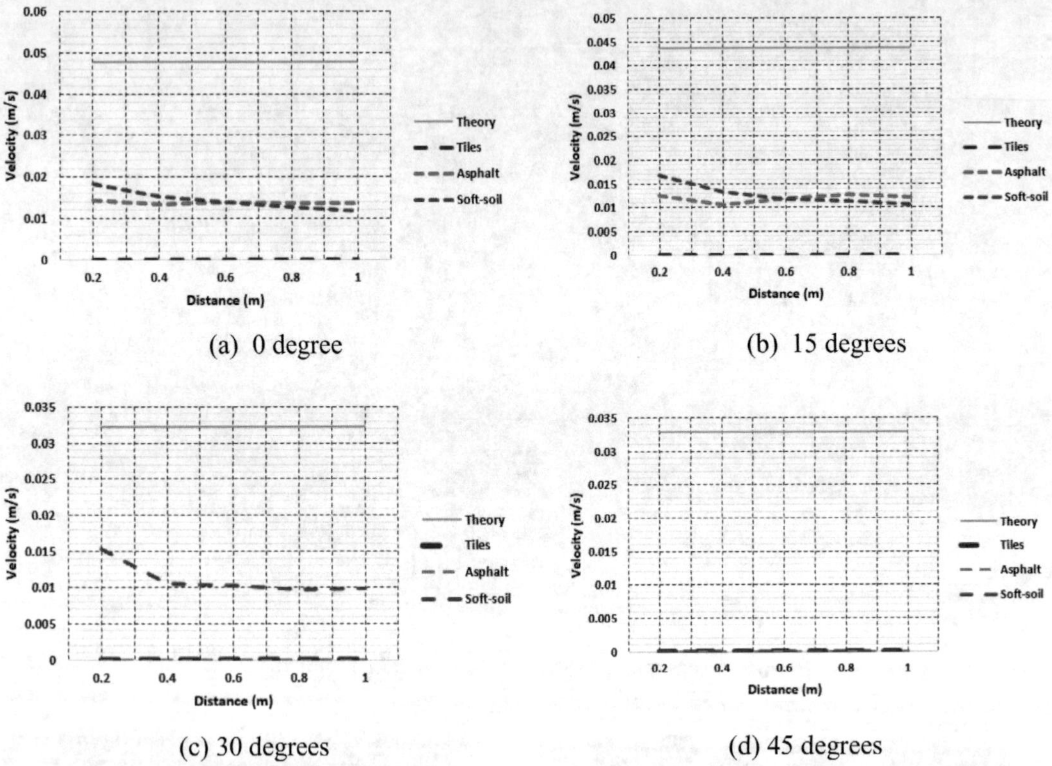

Figure 5. Result for longitudinal movement on different types of surface with different angle of wheel arrangement.

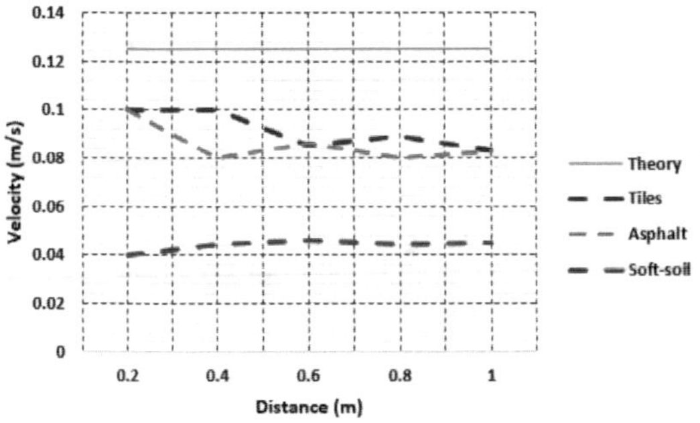

(a) 0 degree

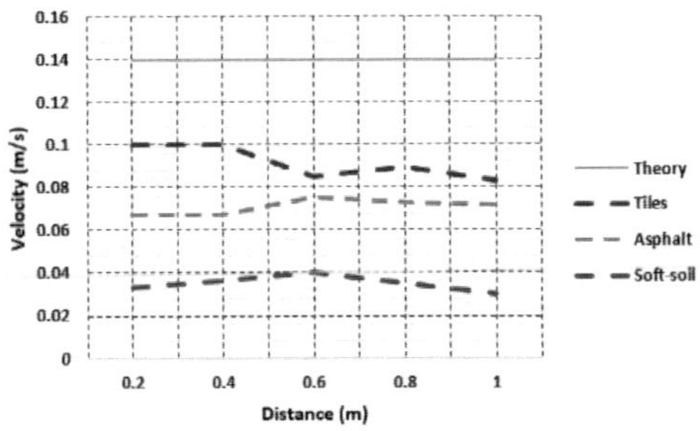

(b) 15 degrees

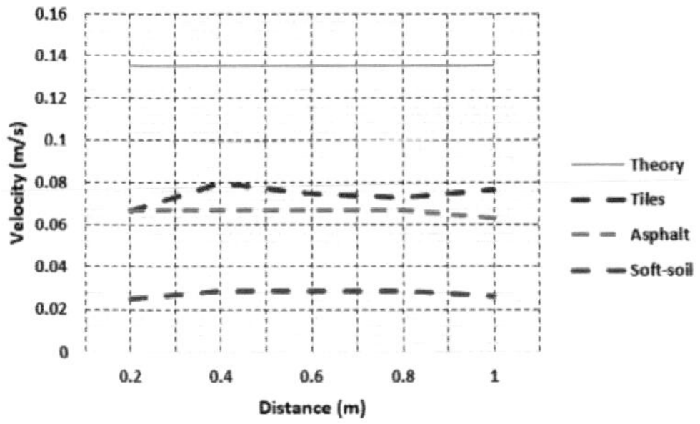

(c) 30 degrees

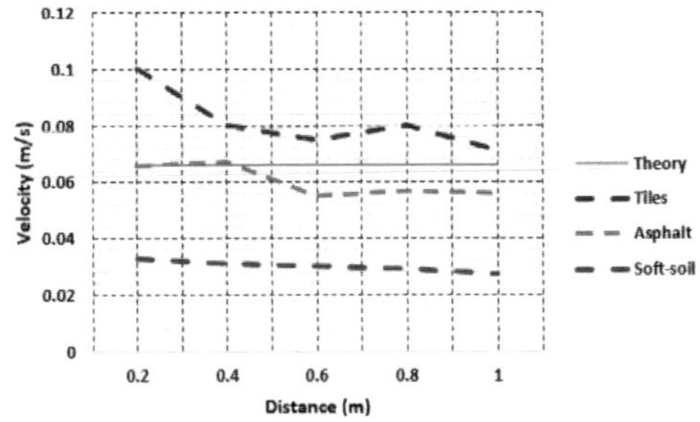

(d) 45 degrees

Figure 6. Result for lateral movement on different types of surface with different angle of wheel arrangement.

4. Conclusion

In this study, an omnidirectional mobile robot with screw wheels is designed and developed. The experimental results shows that the omnidirectional mobile robot is capable to produce the desired motion at the right direction, but the performance are poor and varies according to the ground conditions and wheel's orientation. This performance issue may be caused by uneven contact in rough ground and insufficient traction due to helical design or materials.

5. References

[1] Morath JA 1899 Agricultural Machine *US Patent 635501*

[2] Nagaoka K, Otsuki M, Kubota T, and Tanaka S 2010 Terramechanics-based Propulsive Characteristics of Mobile Robot Driven by Archimedean Screw Mechanism on Soft Soil *Proc. IEEE/RSJ Int. Conf. on Intell. Robots and Systems* 4946–4951

[3] Osiński D and Szykiedans K 2015 Small Remotely Operated Screw-Propelled Vehicle Progress in Automation, Robotics and Measuring Techniques *Advances in Intelligent Systems and Computing* **351** 191-200

[4] Liedke J, Winkler L and Wörn H 2013 An Alternative Locomotion Unit for Mobile Modular Self-reconfigurable Robots Based on Archimedes Screws *9th International Symposium on Mechatronics and its Applications (ISMA)* 1-6

[5] Safar MJA, Chandradekaran Y, Basah SN, Basaruddin KS, Hashim MSM 2016 Kinematics Analysis of a Screw Wheeled Omnidirectional Mobile Robot **10** 1-15

IOP Conf. Series: Materials Science and Engineering **705** (2019) 012042 doi:10.1088/1757-899X/705/1/012042

Development of Web-based Mobile Olfaction System for Confined Space Hazardous Gas Monitoring

A.H. Abdullah[1], M.A.A. Bakar[1], F.S.A. Saad[2], S. Sudin[2], N.H.A. Latif[2], Z.A. Ahmad[3], M. Elshaikh[3], K.A. Ismail[4]

[1] *Faculty of Tech. Eng., Universiti Malaysia Perlis, Perlis, Malaysia*
[2]*School of Mechatronic Eng., Universiti Malaysia Perlis, Perlis, Malaysia*
[3]*School of Computer & Communication Eng., Universiti Malaysia Perlis, Perlis, Malaysia*
[3]*School of Manufacturing Eng., Universiti Malaysia Perlis, Perlis, Malaysia*
abu.hassan@unimap.edu.my

Abstract. Hazardous gas in the confined space is an issue that often causes danger to the worker while doing the tasks i.e. installation, cleaning and maintenance. The nature of the area may exposure the worker to hazardous gas leakage. It is important to monitor the hazardous gas in the surrounding area before the worker start their task inside the confined space. This paper proposes a study of developing a web-based mobile olfaction system for confined space hazardous gas monitoring. The system is integration between a mobile robot and an electronic nose (e-nose). The system will reduce the danger faced by the worker from the hazardous gas. The e-nose consists of a sensor array, embedded controller, electronic components and wireless GSM communication. The instrument sensors i.e. hydrogen sulphide, oxygen, carbon monoxide and methane are used to acquire the confined space air sample. It will also acquire the environment temperature and humidity. The robot which piggyback the instrument will manoeuvre through the environment while acquires data. The data acquired were transmitted and monitored wirelessly to the web-based information system. The website layout is designed using HTML and JavaScript. Data received from e-nose will be stored into a database, MySQL server before being process by the website. The data will go through the conversion process from digital value to PPM. Then the data will be classified using ANN to obtain the Hazardous Gas Concentration Index (HGCI). The website will display the HGCI mapping of the area in real-time. Results show that the system performance is good and can be used to monitor the hazardous gas in the confined space.

Keywords—Electronic Nose, Mobile Olfaction, Confined Space, Hazardous Gas, Web-based Information System.

1. Introduction

The maintenance worker that required to work in confined space i.e. pipeline inspection, repairing utility holes, silo, manholes, tunnels, cave, sewers etc. is dangerous when expose to hazard environment. The area limited space and poor ventilation will affect the worker health while perform the specific task such as installation, cleaning and repairs. The place also vulnerable to serious injuries from expose to hazardous gas i.e. carbon monoxide, hydrogen sulphide lack of oxygen methane and flammable gas. These may arise from the work that being carried out by the worker such as welding or machine operation. So, it is useful to have a system that able to monitor the hazardous gas in the confined space for the safety of the worker.

Olfaction is also being used by animal to avoid dangerous situations or warn the presence of predators. Currently olfaction technique that being used by dogs to detect drug, explosive, escape prisoner, victims of earthquake and gas leakages. Dogs required extensive training and care by expert handlers. Dog attention also short, can easily divert and suffers from fatigue [1]. It's also exposing to the hazardous environment that that will affect it safety and health. Existing research on olfaction are mainly on navigation algorithm on odour detection, source localization and its distribution. The techniques that had been used to detect odour source are such as trail following, plume tracking and localization [2]. To date, the research on olfaction have concentrated mainly in the application of environmental monitoring [3].

The combination of electronic nose (e-nose) and mobile robot called mobile olfaction. In indoor environments, the system could be used to prevent toxic fumes poisoning caused by fire or to mark the areas that are already covered in floor cleaning applications. Metal-oxide (MOS) gas sensors were used in most of the system because of their high sensitivity, compact size and fast response time [4]. The system can be used as the alternative to perform similar tasks with less supervision and more reliability. The system's flexibility and ability to manoeuvre in difficult and isolated areas is important for the safety of humans in the area especially without adequate safety procedure [5]. Therefore, this project has aimed to develop a web-based mobile olfaction system to monitor hazardous gas for confined space application [6]. The web-based information system is used to remotely monitor the hazardous environment for the worker safety and health. The system will able provide safety measures for the worker while perform specific task in the confined space.

2. System Description

The web-based information system architecture is shown in Figure 1 where the e-nose is piggyback on a mobile robot. The e-nose data sampling process is most effective when its front-end position facing the hazard environment [7]. The e-nose acquired data will be transmitted wirelessly to a using a GSM communication module. The system will process and analyse the acquired data, then it will display the finding in website.

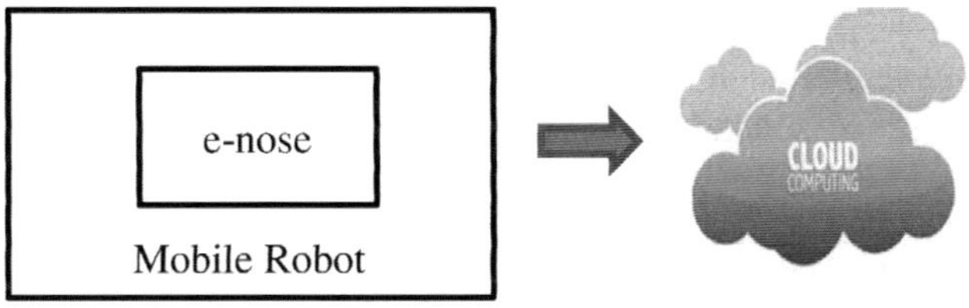

Figure 1. The Mobile Olfaction System.

2.1 Electronic Nose

The e-nose is developed to mimic the human sense of smell, comprises of an array of sensors, sensing chamber, embedded controller, electronic component and pattern recognition system. The instrument is capable of discriminating the sensor signals into classes corresponding to the pollutant i.e. hazardous gas *fingerprint* [8]. The instrument is piggyback on a mobile robots that have advantages over conventional methods [9]. The instrument as shown by Figure 2 others components are sensing chamber, embedded controllers board as well as electronic circuit and DC power supply. The instrument is using selected MOS gas sensors i.e. oxygen, toxic gas and explosive gas as shown in Table 1. The sensors were used because of their high sensitivity, compact size and fast response time [10]. A humidity & temperature sensor was also being attached to the instrument enclosure to monitor the environment condition. The instrument uses a dsPIC33 microcontroller as its embedded controller [11]. The instrument sensing chamber material and enclosure was selected to withstand sensor heater high temperature. The headspace sampling system will deliver the air sample to flow properly into the sensing chamber which ensures all sensors exposure to the air sample effectively. An air-pump was used to draw-in the sample to chamber during *Sniff* cycle and for *purging* cycle it will flush out sample from the sensing chamber. The embedded software was also used to communicate with the web-based information system through GSM communication.

Table 1. The Electronic Nose Sensor

Gas Type	Human Limit	Sensor Sensitivity
Oxygen	15 – 19 %	0 – 30 %
Carbon Monoxide	35 ppm	10 – 300 ppm
Hydrogen Sulphide	5 ppm	1 – 100 ppm
Methane	1000 ppm	300 – 10000 ppm

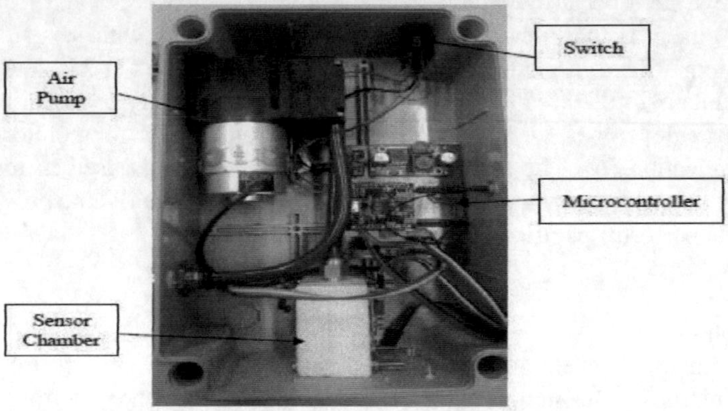

Figure 2. The Electronic Nose

2.2 Mobile Robot

The mobile robot shown in Figure 3 is used for the e-nose to move across the area. The chassis is made from aluminium plate and its size is 40 cm wide, 50 cm long and 30 cm high. It's uses PIC18 microcontroller and embedded software was developed to control the navigation. It's also equipped with remote control for manual mode. Three ultrasonic sensors were situated at the front-end to determine the distance between robot and obstacles. Its will used information from these sensors to manoeuvre inside the confined space.

Figure 3. The Mobile Robot

2.3 Web-Based System

Internet of Things (IoT) system is a wireless communication network that has consisted of sensor that has connected to the Internet. Whereas, Web of Things (WoT) is defined as create IoT that is built with website technology [12]. The web-based information system which used the WoT is being applied by collecting, process, analyse and display the finding about data from the environment by using website [13]. The system website will monitor the confined space atmospheric environment and alert the user for any hazard condition. The website development can be divided into four parts as show in Figure 4. Data collection from the e-nose is transmitted wirelessly using GSM module to web server and saving it in database. The data will be converting from digital value to Part per Million (PPM). Then the data will be classified using Artificial Neural Network (ANN) to obtain the location Hazardous Gas Concentration Index (HGCI). After that the system will display the environment HGCI on the website as graphic representative.

This mapping system will show the location points in the area with the HGCI. The area hazard condition is being displayed and monitor by the website in real time. The web-based information system is also trained to recognize, by different sets of measurements, hazard patterns for different polluting factors acting in the monitored environment, as well as identify accidental patterns of the hazardous gas.

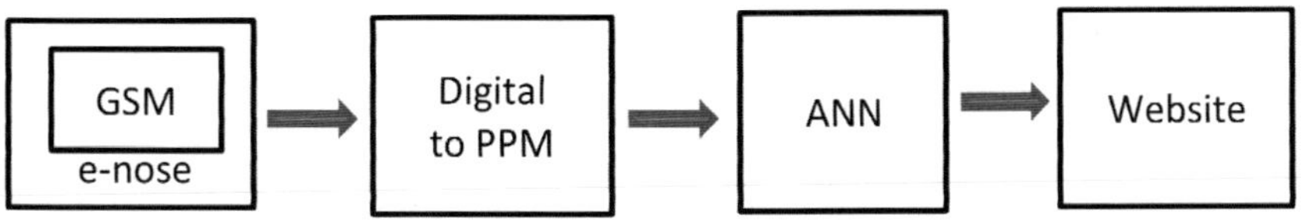

Figure 4. The Web-based Information System

The coding for develop the web-based information system is by using XAMPP software. It contains HTML it is a standard mark-up language to create web pages and web applications [14]. The PHP language an open source general-purpose scripting language for the web development is also embedded into HTML [15]. The MySQL software is an open source relational database management system (RDBMS) based on Structured Query Language (SQL) is used for acquired data storage[16].

Then the data will be converts from digital values to PPM for standard gas measurement unit. Every sensor has different formula for sensor resistance, Rs and the value of sensor resistance in clean air, Ro according to the respective sensor. First step is to determine the value of gradient, m by using a formula in equation (1). The number for y and x is taken from the graph in a datasheet according to the respective sensor. The value for k can be found by choosing any number of y and x to put inside the formula in equation (2). To find the value for y, need to determine the value of Rs by using a formula and the value of Ro that has been given in a datasheet. The value of PPM has been calculated according to the formula given in equation (3) and equation (4). Lastly, the PPM in percentage is calculated by using equation (5).

$$m = \frac{\Delta(\log y)}{\Delta(\log x)} \quad (1)$$

$$k = \frac{y}{x^m} \quad (2)$$

$$y = \frac{R_s}{R_o} \quad (3)$$

$$x = 10^{\left[\frac{\log(\frac{y}{k})}{m}\right]} \quad (4)$$

$$x_{\%} = \frac{x_{ppm}}{10000} \quad (5)$$

IOP Conf. Series: Materials Science and Engineering **705** (2019) 012042 doi:10.1088/1757-899X/705/1/012042

Where, Rs is sensor resistance at gas various concentration, Ro, sensor resistance in clean air, k,y value that cross line x and m, slope of the line.

Afterthat the data were analysing by using ANN for its good adaptability property (learning, generalisation and noise tolerance) which is suitable for the non-linear data. It has the ability to learn the complex interaction between the malodour sample variables [17]. The Multiple Layer Perceptron (MLP) ANN model is used to generate the HGCI which being test for classifier success rate. It will generate the HGCI based on air sample of each location. All of the results indicate that the tests classifications correlate with sample location. The results show that the mobile olfaction system was able to generate the HGCI according to their location. The system performs good and fast measurement, processing and transmission of the sample data. It is very useful that the web-based information system able to monitor hazardous gas in the confined space.

3. Methodology

An experiment was conducted to verify the e-nose sensitivity and detection limit that correlate with the hazardous gas concentration. The samples were collected in a hospital basement room in Malaysia. The process was conducted on-site at four different locations in the environment. The samplings process is illustrated in Figure 5.

Confined Space Mobile Olfaction Sensor Response

Figure 5. The Sampling Process

The e-nose sampling process utilise the Purge *Sniff and Hold* technique whose parameter settings are shown in Table 2. Initially the instrument is being *Purge* using active carbon to clean the sensors before start to *sniff* the air sample. Then it will acquire the baseline air sample data before sniffing at four specific locations in the area. The robot will move to per-set location before the sniffing process being repeated. This technique will *pre-concentrate* and *Hold* the air sample to enhance the instrument sensitivity. The duration of the instrument process cycles is based on the MOS sensor characteristics and *trial and error* [10]. Then the instrument will send the acquire air sample wirelessly to the web server through GSM module.

Table 2. The Electronic Nose Sampling Parameter

Operation	Time (Sec)	Air Pump
Purge Cycle	10	ON
Baseline	30	OFF
Sniff Cycle	10	ON
Steady State	30	OFF

4. Results and Discussion

4.1 Sensor Response

The sensor response was a time series of waveform profiles as shown in Figure 6. The sampling technique was suitable because the acquired data were within the *measurement range*. This indicated that the e-nose functions accordingly to different sampling location. The acquired data was being pre-process by applied filtering and baseline manipulation. Relative baseline is used by comparing the ambient with the sample air. This process will improve the system sensitivity which will enhance the classification result.

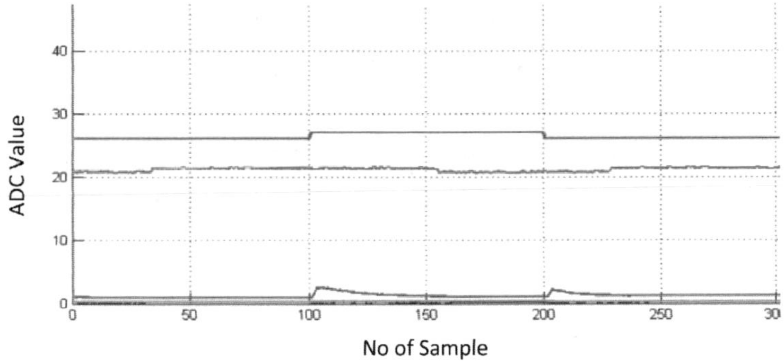

Figure 6. The Electronic Nose Sensor Response

4.2 Web-Based System

The data acquired form the location is transmitted wirelessly by GSM to the web server. MySQL software is used to build a database to store the received data. Hypertext Mark-up Language (HTML) and JavaScript have been used in order to design the website and analyse data from MySQL server. The digital data will be retrieved and convert to PPM as shown in Table 3.

Table 3. Conversion from Digital to PPM

Hazardous Gas	Loc. P1	Loc. P2	Loc. P3	Loc. P4
Oxygen	21%	20.9%	20.9%	21.3%
Carbon Monoxide	0.29	0.25	0.26	0.27
Hydrogen Sulphide	2.13	2.97	3.37	3.32
Methane	114.27	114.50	114.02	114.33

The Multilayer Perceptron (MLP) is used to generate the HGCI ANN model. The model used 70% of the data for training and 30% for testing. The output will be set in between 0.0 to 1.0. Table 4 shows the HGCI colour coding.

Table 4. Hazardous Gas Concentration Index Colour

Colour	Concentration Index	Condition
Green	0 – 0.20	Very low - excellent
Brown	0.21 – 0.30	Low – good
Light Blue	0.31 – 0.39	Medium - normal
Yellow	0-40 – 0.59	High - caution
Red	0.60 – 0.70	Very high - danger

The developed system website is shown in Figure 7. HGCI at each location will be arranged in graphic representation as shown by webpage in Figure 8. The website will show oxygen, carbon monoxide, hydrogen sulphide, methane and overall in rotational order for every 10 second interval. The HGCI location is represented by five different colour code that give different information about hazardous gas in the area as shown by Table 4.

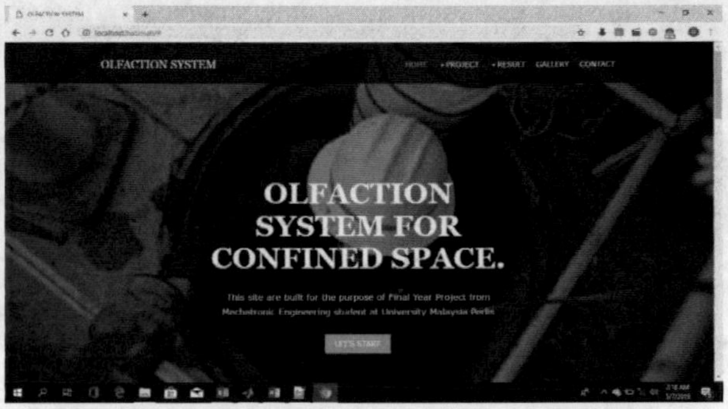

Figure 7. Mobile Olfaction System Website.

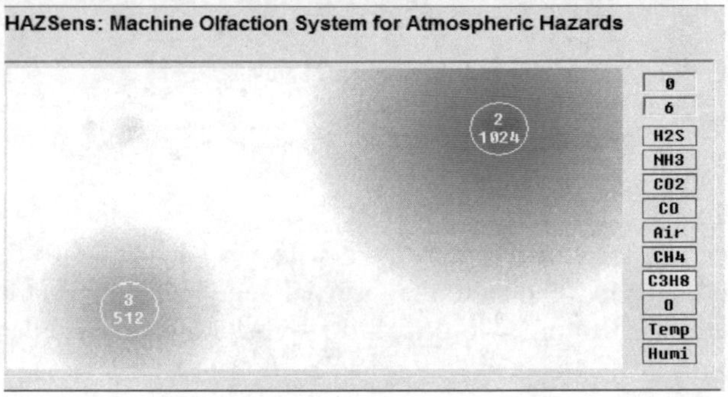

Figure 8. The Webpage Layout.

5. Conclusion

A mobile olfaction system prototype using an e-nose integrate with mobile robot for navigating inside the confined space has been successfully developed The system was tested to acquire air samples and transmitted wirelessly to website using GSM communication module. The web-based information system for monitoring the hazardous gas was developed by using HTML and JavaScript. The real time data collection is MySQL, the conversion from digital to PPM and the HGCI was generate using ANN model. The HGCI of oxygen, carbon monoxide, hydrogen sulphide and methane were displayed by the website. The environment parameter i.e. temperature and humidity also being display. The system also able to send alert signal once the HGCI exceed the safety value. Future work should consider the used of fuzzy logic for the uncertainties in generating the HGCI.

Acknowledgement

This research work is partly funded by the Ministry of Science, Technology & Innovation of Malaysia under the Science Fund Grant Scheme (Project No: 9005-00066). The author also gratefully acknowledges to Universiti Malaysia Perlis (UniMAP) for the opportunities given to do the research.

IOP Conf. Series: Materials Science and Engineering **705** (2019) 012042 doi:10.1088/1757-899X/705/1/012042

References

[1] Elise Hancock 1996 *A Primer on Smell: A Special Issue on the Senses*, (Johns Hopkins Magazine, Baltimore, USA)

[2] Hayes A T, Martinoli A and Goodman R M 2001 *IEEE. Intl Con. Intelligent Robots and Systems* 1073

[3] Gardner J W, Persaud K C, Gouma P and Osuna R 2012 *Sensors Journal* vol. 12(11) 3105

[4] Sandini G, Lucarini G, and Varoli M 1993 *IEEE/RSJ Int. Conf. on Intel. Robots System Proceeding* 429

[5] Pearce T C , Schiffman S S, Nagle H T and Gardner J W 2003 *Handbook of Machine Olfaction: Electronic Nose Technology* (Wiley-VCH Germany)

[6] Ishida Wada and Matsukura, 2012 *IEEE Sensors Journal* 12(11) 3163

[7] Loutfi A 2006 *Odour Recognition Using Electronic Noses In Robotic And Intelligent Systems* (PhD Thesis Örebro University Sweden)

[8] Gardner J and Bartlett P 1994 *Actuators B* 18(1) 211

[9] David M 2014 *Robot, Control, and Sensing* (Craver Dissertation)

[10] Figaro Engineering 2005 *General Information for TGS Sensors* Online: http://www.figarosensor.com/ product/common(1106).pdf

[11] Microchip Technology 2007 *High-Performance 16-bit Digital Signal Controller,* Online: http://freedatasheets.com /downloads/dsPIC33FJXXXMCX06X08X10dsPI CDSC

[12] Diaconescu I M and Wagner G 2015 *Con. On Multiagent System Technologies* 9433 137–153

[13] Sim S and Choi H 2017 *Journal of Cluster Computer* 21(1) 1151-1160

[14] Yadav P and Barwal P N 2014 Int. Journal Sci. Tech. and Research 3(11) 152-155

[15] Bolboacă S, Jäntschi L, and Cadariu A A 2003 *Leonardo Electronic Journal of Practices and Technologies* 2(2) 2 37-52

[16] Ya N, Zainuddin A, Fatinah I, Abidin Z, and Yusof A L 2016 *IEEE Int. Con. on Control System, Computing and Eng. (ICCSCE)* 25

[17] Khong M D, Gale T J, Jiang D, Olivier J C, and Ortiz-Catalan M 2013 *Biomed. Eng. Int. Conf.* 1

IOP Conf. Series: Materials Science and Engineering **705** (2019) 012043 doi:10.1088/1757-899X/705/1/012043

Oil Palm Plantation Monitoring from Satellite Image

Andrew Foong Weng Sum, S.A.A. Shukor[1]

School of Mechatronic Engineering, Universiti Malaysia Perlis, 02600 Perlis, Malaysia

[1]shazmin@unimap.edu.my

Abstract. Malaysia is one of the largest palm oil producers in the world. To meet the production, its plantation needs to be monitored accordingly. Manual monitoring is no longer feasible due to energy, cost and time consumption. Remote sensing technology such as satellite images can be utilized for palm oil monitoring as they are cost effective, time and energy saving, and able to enhance the possibility of classifying the vegetation through spectral and texture analyses. Due to its impact towards Malaysia economy, an effective monitoring of palm oil plantation can be done by using satellite image to maintain the sustainability of palm oil in Malaysia. This project utilized the computation of Normalized Difference Vegetation Index (NDVI) and Soil Adjusted Vegetation Index (SAVI) to determine the plant health. Relationship of NDVI and SAVI with respect to temperature and precipitation is studied based on the data obtained from Malaysian Meteorology Department. By using regression model technique, the relationship of the plant health based on NDVI analysis, nitrogen content based on SAVI analysis and temperature is found out to be highly correlated while plant health and precipitation is found out to be no or low correlated. The results can be utilized by the plantation operators, especially independent smallholders in monitoring the health status of the palm oil trees to ensure high productivity.

1. Introduction

Palm oil and palm kernel oil are among the most consumed vegetable oil in the world [1], where their production has recorded to reach 75.17 million tonnes in 2017 from the plantation area of 19.04 million hectares worldwide. Malaysia alone has its planted area of 5.81 million hectares, where more than 3 million hectares of the area are managed by the smallholders [2]. To meet demands and production, a lot of land has been converted to palm oil plantations, which could bring towards endangers biodiversity and degrades the environment [3]. Thus, there is a need to increase oil yield with lesser new plantations establishment [4]. To achieve this, help is needed to monitor the plantation, as manual monitoring is no longer feasible due to its costing and energy consuming [1].

Remote sensing has benefited a lot of areas including agriculture as to improve farming practices [5], such as for tree detection and counting [6], habitat monitoring in mapping pest and disease [7] and nutrient detection [8], where satellite images are among the favoured source of data. Satellite images provide a huge amount of information about our earth's surface. It is beneficial for researchers especially to get more information on inaccessible areas. Apart from that, satellite-based research has a lot of advantages such as cost effective, time saving in obtaining data without having to go to the site, as well as having the possibility of classifying the vegetation through spectral and texture analyses [9].

Hence, this project will utilize publicly available satellite data to monitor palm oil plantation, in terms of the plants' condition to ensure good productivity. Researchers in [10][11] have shown that oil palm yields higher productivity under high temperature of around 29°C to 33°C maximum. Features like nitrogen content and soil moisture are also playing important parameters that could determine the condition of oil palm, and all these can be determined by applying suitable indexes from images data. As nutrient deficiency could affect the growth of the tree, thus, these features can be utilized to indicate good condition of trees to ensure good productivity. Researchers in [12] utilized UAV data to collect the images, but there is a need to hire a UAV to collect the data and this may lead to additional cost. Furthermore, UAV is not sufficient to cover a large scale area of plantation due to its short flying time.

This paper is separated into several sections – Section 1 on the introduction, where some background on palm oil and the importance to monitor their plantations are highlighted. Methodology is highlighted in Section 2 covering data collection and processing, while results are discussed in Section 3 and finished with a conclusion in Section 4.

2. Methodology

2.1 Area of Interest and Data Collection

As this project is purely educational, a suitable oil palm plantation is chosen randomly where the data is obtained from a free-source public domain. Kamuning Estate is chosen to be the testing palm oil plantation for this project which is located at Sungai Siput, Perak, with 37,241 hectares. The satellite image representing the plantation is obtained from Landsat-8 image data through the United States Geological Survey (USGS) website [13]. Landsat-8 satellite data has been chosen due to its high resolution and consist of more wavebands compared to the other Landsat satellite.

2.2 Data Processing

The image is being processed using a geographic information system (GIS) software called ArcMAP 10.5 of ArcGIS. The satellite image obtained is having 11 bands stored in 11 files, hence it must be stacked into 1 image file in order to obtain the result based on the band requirements. However, only 9 bands are selected as Band 10 and 11 are more suitable for land surface temperature study since they contain data from the thermal infrared sensor. As the data covers neighbouring area as well, validation of the correct area needs to be done. Thus, upon checking the area of interest in Google Earth Pro, the same place is determined in the ArcMAP software manually for further processing. For more detailed analysis, a subset image was created to focus on the specific area of interest with lesser smaller size compared to the original image. Next, classification is done through training and testing to determine the entity exists in the subset image. For example, to train data containing water, multiple polygonal shapes is drawn on the image of water to represent the area. Similar processes are done to other classes of entities like road, soil, urban area, oil palm plantation, cloud and hill. Training of sample is carried out by comparing the image from Google Earth. An accuracy assessment is carried out to determine whether the classification has being done correctly using confusion matrix, with a minimum requirement for the classification to be more than 80% in order to have a success results.

To further analyse the image in monitoring the plants' condition, Normalized Difference Vegetation Index (NDVI) and Soil Adjusted Vegetation Index (SAVI) were used. NDVI can be used to separate the background with respect to the tree image in [1], and furthermore, as healthy vegetation absorbs most of the visible light and reflects most of the near infrared (NIR), they will have higher NDVI values [14][15]. Meanwhile, SAVI is a vegetation index used to restrain the influence of soil on the vegetation data which is proposed to increase the sensitivity to chlorophyll concentration [16]. Both NDVI and SAVI can be calculated using Equation (1) and (2), where L is the soil adjustment factor:

$$NDVI = \frac{Near\ Infrared\ Band - Red\ Band}{Near\ Infrared\ Band + Red\ Band} \tag{1}$$

$$SAVI = \frac{Near\ Infrared\ Band - RedBand}{Near\ Infrared\ Band - Red\ Band + L}(1 + L) \tag{2}$$

IOP Conf. Series: Materials Science and Engineering **705** (2019) 012043 doi:10.1088/1757-899X/705/1/012043

For the purpose of finding the best correlation factor in determining the condition of oil palm trees, regression model were developed for NDVI, SAVI, precipitation and temperature data. Temperature and rainfall data were collected from the Malaysian Meteorology Department of the same time when the satellite data is obtained.

3. Results and Discussion

Figure 1 shows the classification results obtained from the ArcMAP software in segmenting the image into classes of entity exist in the image. A reference map from Google Earth (Figure 2) is used to check the results. It can be seen here that the image has been classified into 7 classes which is water, road, soil, urban area, oil palm plantation, cloud and hill.

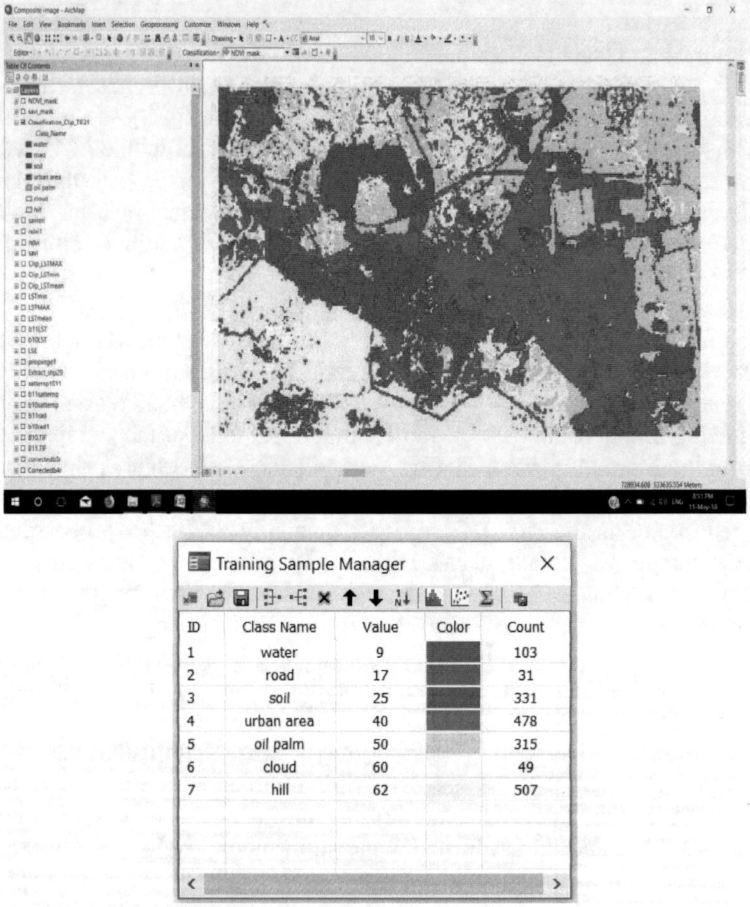

ID	Class Name	Value	Color	Count
1	water	9		103
2	road	17		31
3	soil	25		331
4	urban area	40		478
5	oil palm	50		315
6	cloud	60		49
7	hill	62		507

Figure 1: Result of supervised classification on Kamuning estate

Figure 2: Google Earth image of Kamuning estate

Next, the calculations of the vegetation indices were done. Figure 3 shows the mapping of the plantation with respect to the NDVI distribution and Figure 4 is for SAVI. It can be seen that the healthy vegetation is range between 0.45 and 0.57. Healthy vegetation shows higher NVDI values due to the light is absorbed and reflected by the plants at the leaves. However, soil and urban area show low values of NDVI as both do not produce chlorophyll and no absorption of light and reflectance or infrared band.

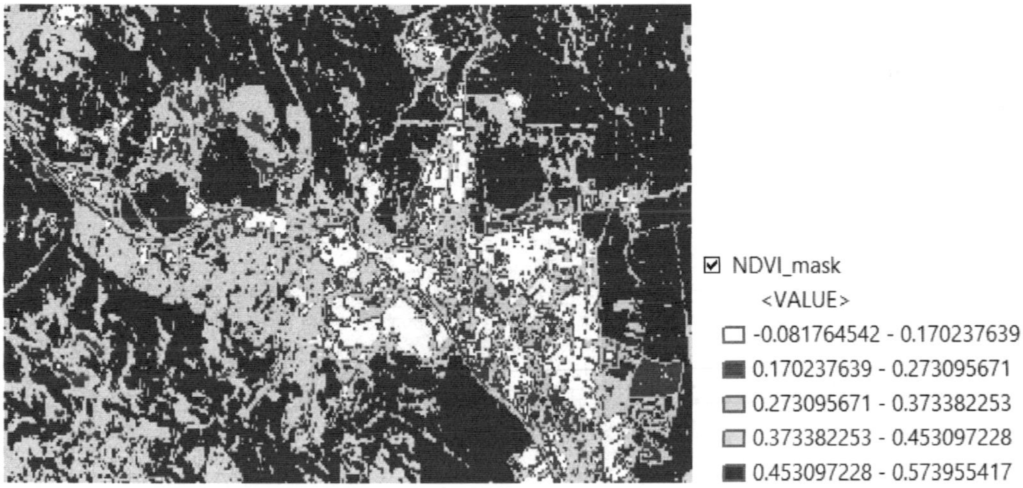

Figure 3: Results of NDVI

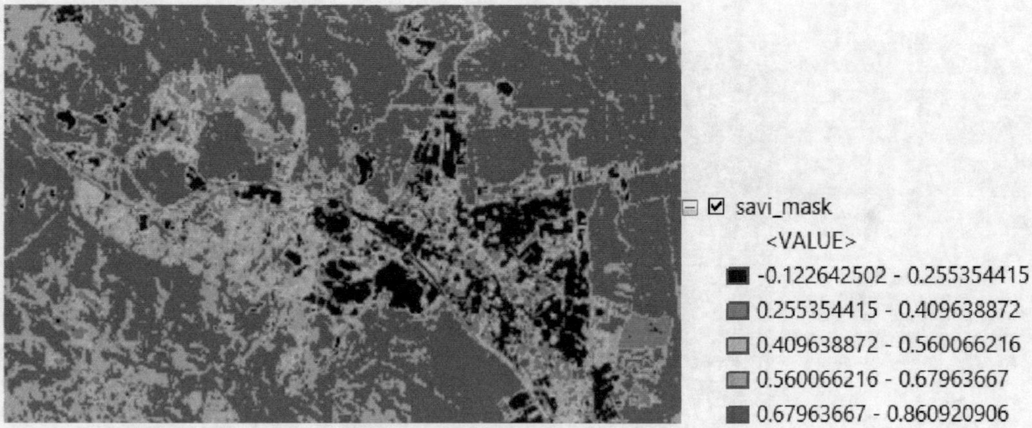

Figure 4: Results of SAVI

SAVI can be used to determine nitrogen content in a plant. Figure 4 shows the distribution of colour based on the values for SAVI. The higher the SAVI value, the more nitrogen contains in the plant vegetation. Based on the results obtained, dark brown area shows the high nitrogen contains from range value of 0.68 to 0.86.

Regression model is used to study the relationship between NDVI and temperature, NDVI and SAVI as well as NDVI and precipitation. Figure 5 shows the relationship between NDVI and the maximum temperature recorded. The coefficient determination, R^2 in this model is 0.8711 and correlation, r = 0.9333 which show strong relationship between these two variables. The temperature data obtained during this time is between 29°C to 34°C (obtained from Malaysian Meteorological Department) which could indicates that there are sufficient sunlight for the oil palm to carry out photosynthesis [17], which is why the NDVI is highly correlated with the temperature.

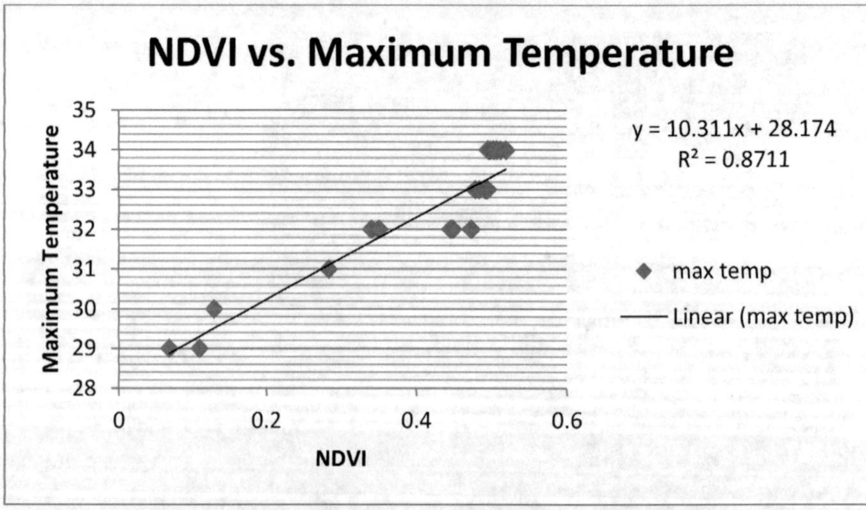

Figure 5: Correlation between NDVI and temperature

Meanwhile, Figure 6 shows the correlation between NDVI and SAVI. From here, the coefficient determination, R^2 in this regression model is 0.9953. The correlation calculated is equal to 0.997 where it shows strong correlation between NDVI and SAVI since it is close to 1. This means that the relationship between NDVI and SAVI are strong. i.e. the plant health depends on the nitrogen content. The higher the SAVI means higher in nitrogen content where it will leads towards higher in NDVI value where it shows healthy plants.

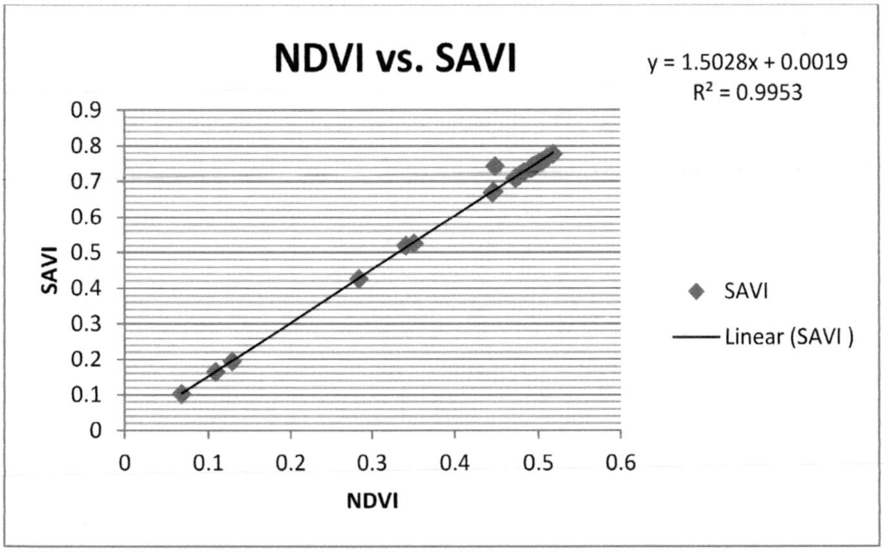

Figure 6: Correlation between NDVI and SAVI

Figure 7 shows the correlation between NDVI and precipitation. From here, the coefficient determination, R^2 in this regression model is 0.0218 and the correlation calculated is equal to 0.148 where it shows weak correlation. Although the average of rainfall receive in March is 8.2 mm, but it does not have much impact on the palm oil growth. The plantation has been took care well by the workers since climate started to change to hot and dry in March to June. Suitable amount of water has been applied on the plantation to maintain the condition of plantation. Besides that, continuous of rainfall may affect the growth of palm oil. Thus, palm oil plantation may not depends on the precipitation.

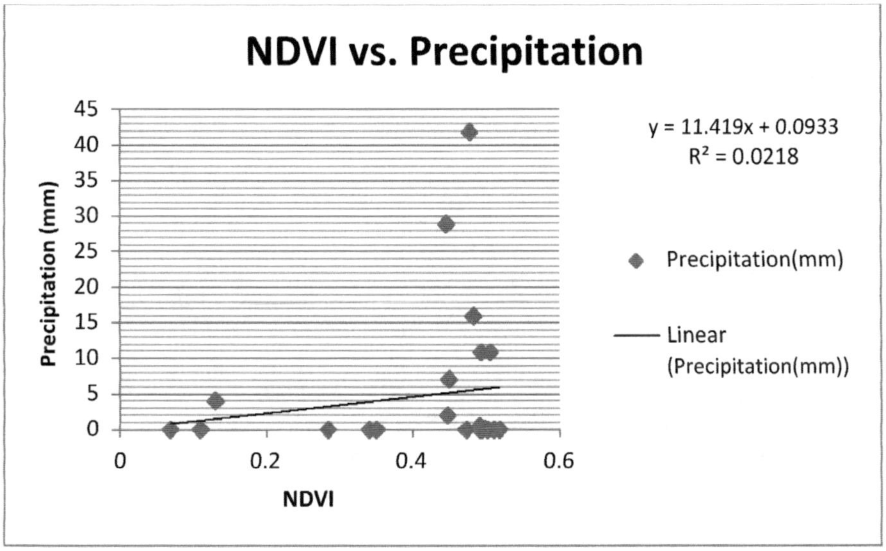

Figure 7: Correlation between NDVI and precipitation

4. Conclusion
This project has shown the capability of satellite image for environmental monitoring in agricultural, especially in oil palm plantation. From the results, it can be seen that NDVI, an index that can be used to determine the health condition of a plant, is having strong correlations with SAVI and temperature, but not with precipitation. Thus, it can be concluded that the image obtained from the satellite can be used for plantation monitoring with appropriate analysis. Further tests especially with suitable laboratory experiments can be done to investigate further the results of this project.

References
[1] Chong, K.L., Kanniah, K.D., Pohl, C. and Tan, K.P., 2017. A review of remote sensing applications for oil palm studies. *Geo-spatial Information Science*, *20*(2), pp.184-200

[2] Kushairi, A., Loh, S.K., Azman, I., Hishamuddin, E., Ong-Abdullah, M., Izuddin, Z.B.M.N., Razmah, G., Sundram, S. and Parveez, G.K.A., 2018. Oil palm economic performance in Malaysia and R&D progress in 2017. *Journal of Oil Palm Research*, *30*(2), pp.163-195

[3] Koh, L.P. and Wilcove, D.S., 2008. Is oil palm agriculture really destroying tropical biodiversity?. *Conservation letters*, *1*(2), pp.60-64

[4] Murphy, D.J., 2014. The future of oil palm as a major global crop: opportunities and challenges. *J Oil Palm Res*, *26*(1), pp.1-24

[5] Khanal, S., Fulton, J. and Shearer, S., 2017. An overview of current and potential applications of thermal remote sensing in precision agriculture. *Computers and Electronics in Agriculture*, *139*, pp.22-32

[6] Li, W., Fu, H., Yu, L. and Cracknell, A., 2016. Deep learning based oil palm tree detection and counting for high-resolution remote sensing images. *Remote Sensing*, *9*(1), p.22

[7] Yuan, L., Bao, Z., Zhang, H., Zhang, Y. and Liang, X., 2017. Habitat monitoring to evaluate crop disease and pest distributions based on multi-source satellite remote sensing imagery. *Optik*, *145*, pp.66-73

[8] Rendana, M., Rahim, S.A., Lihan, T., Idris, W.M.R. and Rahman, Z.A., 2015. A Review of Methods for Detecting Nutrient Stress of Oil Palm in Malaysia. *J. Appl. Environ. Biol. Sci*, *5*(6), pp.60-64

[9] Purnamasayangsukasih, P.R., Norizah, K., Ismail, A.A. and Shamsudin, I., 2016, June. A review of uses of satellite imagery in monitoring mangrove forests. In *IOP Conference Series: Earth and Environmental Science* (Vol. 37, No. 1, p. 012034). IOP Publishing

[10] Yamada, T., Watanabe, K., Okuda, T., Sugimoto, T. and Azlin, Y.N., 2016. Growth and survival of trees planted in an oil palm plantation: Implications for restoration of biodiversity. *Journal of Tropical Forest Science*, pp.97-105

[11] Oettli, P., Behera, S.K. and Yamagata, T., 2018. Climate based predictability of oil palm tree yield in Malaysia. *Scientific reports*, *8*(1), p.2271

[12] Tugi, A., Rasib, A.W., Suri, M.A., Zainon, O., Yusoff, A.R.M., Rahman, M.Z.A., Sari, N.A. and Darwin, N., 2015. Oil palm tree growth monitoring for smallholders by using unmanned aerial vehicle. *Jurnal Teknologi*, *77*(26)

[13] US Geological Survey [online]. Available at: https://www.usgs.gov/ [Accessed 16 March 2017]

[14] Marzukhi, F., Elahami, A.L. and Bohari, S.N., 2016, June. Detecting nutrients deficiencies of oil palm trees using remotely sensed data. In *IOP Conference Series: Earth and Environmental Science* (Vol. 37, No. 1, p. 012040). IOP Publishing

[15] Shashikant, V., Shariff, A.R.M., Nordin, L. and Pradhan, B., 2012. NDVI of oil palm trees by LANDSAT-5 Imagery. In *The 3rd Asian Conference on Remote Sensing* (pp. 1-4)

[16] García, P. and Pérez, E., 2016. Mapping of soil sealing by vegetation indexes and built-up index: A case study in Madrid (Spain). *Geoderma*, *268*, pp.100-107

[17] Van den Besselaar, E.J.M., Sanchez-Lorenzo, A., Wild, M., Klein Tank, A.M.G. and de Laat, A.T.J., 2015. Relationship between sunshine duration and temperature trends across Europe

since the second half of the twentieth century. *Journal of Geophysical Research: Atmospheres*, *120*(20), pp.10-823

IOP Conf. Series: Materials Science and Engineering **705** (2019) 012044 doi:10.1088/1757-899X/705/1/012044

Analysis the Awkward Posture Ergonomic Risk and Workstation Improvement Simulation in Mechanical Assembly Manufacturing Industry using DelmiaV5

Ruzy Haryati Hambali[1,2], Siti Aishah Abdul Rahim[1], Noorain Azizan[1], Zainuddin Zali[1], Suriati Akmal[1,3] and Muhammad Hanafee Zin[4]

1 Centre of Smart System and Innovative Design, Faculty of Manufacturing Engineering, Universiti Teknikal Malaysia Melaka, Hang Tuah Jaya, 76100 Durian Tunggal, Melaka, Malaysia
2 Lancaster Product Development Unit, Department of Engineering, Lancaster University, Lancaster, UK
3 Center for Advanced Computing Technology, Institut Pengurusan Teknologi dan Keusahawanan, Universiti Teknikal Malaysia Melaka, Hang Tuah Jaya, 76100 Durian Tunggal, Melaka, Malaysia
4 Aerospace Malaysia Innovation Centre (AMIC), AMIC-AAC R&T Centre, No. 03-06-01, Jalan Ilmiah, German-Malaysian Institute, 43000 Bangi, Selangor, Malaysia

Email: ruzy@utem.edu.my

Abstract. Ergonomics knowledge is crucial in any industries, without knowledge applied, one easily exposed towards the risks of the working environment. Based on the survey conducted showed awkward posture and heavy lifting were the most highlighted issues occurred. Aim of this study is to analyse the working posture in mechanical assembly manufacturing industry and analyse the effect of the improvement workstation. The use of Rapid Upper Limb Assessment (RULA) in order to analyse the most level of risk of workers during working, develop human manikin following the anthropometric data collected from industry real population, as well to replicate the posture of the workers, simulate the working process via simulation using DELMIA V5. Results reported one of the highest ergonomic risk occurred in this department is drilling activity. The manikin successfully developed with 105 measurements for 685 male workers. Simulations effectively simulated using DELMIA V5 in virtual environment. Step ladder suggested as virtual and this improved the value of RULA score. In conclusion, the software and method used were powerful tool in designing the virtual improvement and this research expected to create a design guideline for the industry that help workers with an ergonomic way at the workstation before and during working.

1. Introduction

Awkward position indicated as the body posture that unusual or hardly to perform by the body. This awkward position caused stress to the muscle in the body that resulted pains that lead to fatigue and injury. Health issue due to awkward position is common especially to workers in the heavy industry includes mechanical assembly manufacturing industry. This awkward posture influenced the performance of the worker which has a greater impact to the company itself. This problem as well mostly related with the design of the workstation, which reported not ergonomically designed. Ergonomic are human-centered and focusing on evaluation of task performance and human performance when there is interaction of user with complex systems which be achieved by applying a task analysis procedure [1]. Ergonomic also a task or activity intended to analyse and solve the injuries of the human due to working problem. Besides, ergonomic is a continuous improvement effort to design or redesign and propose an improvement to the workplace or workstation layout in. The guidelines on ergonomic deficiencies are derived from the accident reports which commonly name employees as direct causes of accident. The main cause of accidents found of significance for prevention purposes include the misalignment of workstations with the psychophysical capabilities of workers [2].

Other ergonomics issues are related with the body posture itself. In order to curb the problem with ergonomics issue related, a well study of the body posture of the worker or human is important to obtain a good workstation. The non-success or inaccurate of the human body posture examination can lead to risk and injury. Most result from the previous studies or researches showed that working with the hands above the shoulders or using repetitive movements and working in awkward posture were associated with shoulder and neck pain, rotator cuff syndrome, and epicondylitis. Other ergonomic risks include vibration, both sitting and bending prolonged sitting without support, carrying out repetitive tasks with hands and fingers, frequent lifting activity, and standing with awkward posture were associated with a high prevalence of shoulder and neck pain [3]–[5].

This project covered the analysis of ergonomic risk specifically to the awkward posture analysis using physical and simulation analysis. This project significantly has potential to help the workers and industry where occupational health of the workers can be improved and avoid any occupational injuries as this work. Results could be extended to the society to enhance the awareness of ergonomic analysis before and during working activities.

2. Methodology

This study took place among the production workers in mechanical assembly department at ABC Company Sdn. Bhd. Working posture worksheet is used to identify and observe the working posture whereas Nordic survey is used as structured interview to screening the MSD problem. Besides, this research is focusing the manikin development with own population measurements where the anthropometric measurements data of workers in aerocomposite industry are taken from data collected in previous research. DELMIA (Digital Enterprise Lean Manufacturing Interactive Application) is among the well-known software that is used to simulate the dynamic simulation aiming to observe the motion of the worker in the workstation virtually by using the manikin developed in the CATIA V5. Numerous studies simulation in DELMIA V5, the ergonomic risks or awkward posture can be analysed based on result obtained. CATIA V5 software is used for the manikin development and DELMIA V5 is used to simulate the design. RULA analysis, energy expenditure and biomechanics single action analysis are used to analyse the working posture of the worker and DELMIA V5 is used to simulate and prove the improvement and effectiveness of the model produced. DELMIA V5 software only focusing a simulation of the model since no fabrication and real implementation can be used to improve the posture in DELMIA V5.

3. Results and discussion

There are seven working postures observed and selected precisely related to awkward posture in localised drilling process. From these seven observed postures, one posture selected according to the result of Employee Assessment Worksheet with highest RULA score and that posture was required by the industry for further critical analysis. The simulation in DELMIA V5, the ergonomic risks or awkward posture can be analysed based on result obtained as in numerous previous researches [3], [6].

For this study, improvement related with workstation design, 2 concepts were suggested; 2 steps ladder and 3 steps ladder.

3.1. RULA Analysis

RULA is a method for the assessment of risk of the body posture by maintaining inadequate by the poor design of working area. The RULA method provides an important and useful tool for evaluating working postures of the worker that may create a risk tool. RULA score starts from 1 to 7 which 1 is the least risky whereas 7 is high risk. Figure 1 showed the results of RULA assessment analysis for current workstation and workstation with improvement.

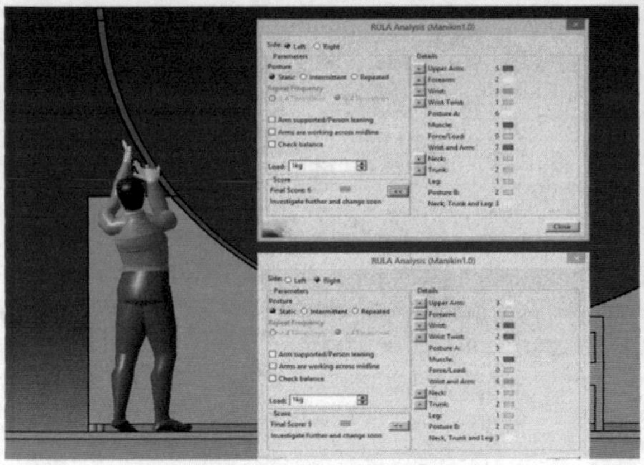

RULA Final Score Left = 6
RULA Final Score Right = 5
(a)

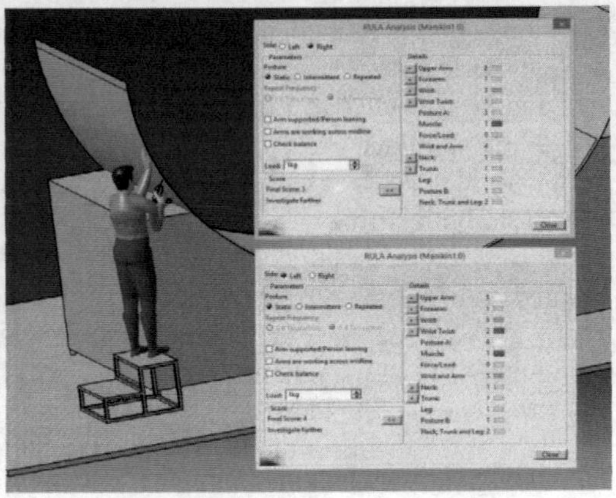

RULA Final Score Left = 3
RULA Final Score Right = 4
(b)

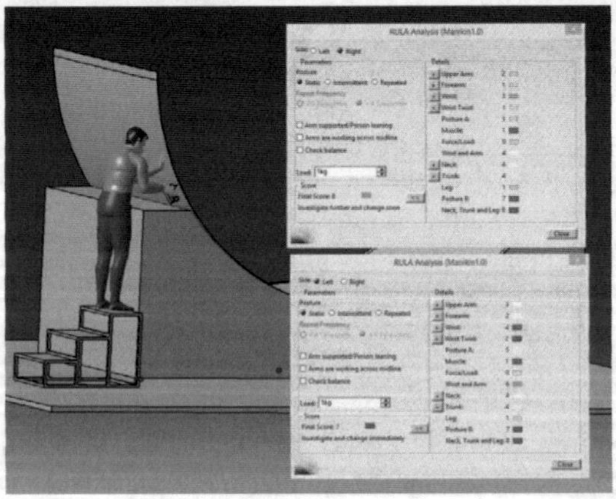

RULA Final Score Left = 6
RULA Final Score Right = 7
(c)

Figure 1. RULA Analysis result for:
(a) current workstation design (b) 2 steps ladder workstation improvement (c) 3 steps ladder workstation improvement

3.2. Energy expenditure analysis

Energy expenditure was analysed after the worker complete performing the task which from the first position of the worker before the process until the last position of the worker after complete all the tasks. Result from energy expenditure analysis indicates that the higher the number of energy, the higher the calories burnt which significantly indicate that the task carried by the worker is heavy and requires large energy to perform it. Thus, worker will easily to get fatigue due to huge energy required to complete the task. Table 1 showed the results of energy expenditure analysis for current workstation and workstation with improvement.

Table 1. Energy expenditure analysis for:
current workstation design (b) 2 steps ladder workstation improvement (c) 3 steps ladder workstation improvement

Energy Expenditure Summary		
Total Energy	1.510	Kcal
Total Time	26.156	S
Total Walk Distance	12117.961	Mm
Average Rate	0.058	Kcal/s
End of Report		

Average rate energy expenditure = 0.058 kcal/s
(a)

Energy Expenditure Summary		
Total Energy	2.187	Kcal
Total Time	44.270	S
Total Walk Distance	18728.784	Mm
Average Rate	0.049	Kcal/s
End of Report		

Average rate energy expenditure = 0.049 kcal/s
(b)

Energy Expenditure Summary		
Total Energy	2.528	Kcal
Total Time	47.975	S
Total Walk Distance	19806.755	Mm
Average Rate	0.053	Kcal/s
End of Report		

Average rate energy expenditure = 0.053 kcal/s
(c)

3.3. Biomechanics single action analysis

This ergonomic tool measures a worker's biomechanical data in a pose. The Single Action Analysis tool calculates and produces information such as the lumbar spinal loads (abdominal force, abdominal pressure, body movements) and the forces and moments on manicinal joints from the current manicinal posture. All the output incorporated in the model is based on the scientific community's research results and algorithms. Table 2 showed the results of biomechanics single action analysis for current workstation and workstation with improvement.

IOP Conf. Series: Materials Science and Engineering **705** (2019) 012044 doi:10.1088/1757-899X/705/1/012044

Table 2. Biomechanics single analysis for:
(a) current workstation design (b) 2 steps ladder workstation improvement (c) 3 steps ladder workstation improvement

Biomechanics Summary		
L4 – L5 Moment	-36	Nxm
L4 – L5 Compression	1132	N
Body Load Compression	464	N
Axial Twist Compression	48	N
End of Report		

L4 – L5 Compression: 1322 N
Axial Twist Compression: 48 N
(a)

Biomechanics Summary		
L4 – L5 Moment	42	Nxm
L4 – L5 Compression	1214	N
Body Load Compression	487	N
Axial Twist Compression	0	N
End of Report		

L4-L5 Compression: 1214 N
Axial Twist Compression: 0 N
(b)

Biomechanics Summary		
L4 – L5 Moment	106	Nxm
L4 – L5 Compression	2649	N
Body Load Compression	443	N
Axial Twist Compression	52	N
End of Report		

L4-L5 Compression: 2649 N
Axial Twist Compression: 52 N
(c)

For improvement related with workstation design, two concepts were suggested; 2 steps ladder and 3 steps ladder. From the RULA analysis, both right and left showed the lowest scores compared to current workstation design and 3 steps ladder workstation design which 3 for right and 4 for left. For energy expenditure result, 2 steps ladder showed the least energy used to burn the calories since the working posture only needs less energy to perform the task which is 0.049kcal/s compared to current workstation design and 3 steps ladder workstation design. Whereas for biomechanics single action analysis, L4-L5 compression is 1214 N that indicates the bending is not more than 45° which better than current workstation design. The axial twist compression is 0 N that indicates the posture is comfort since the worker does not have to twist their arm during drilling process. Thus, an improvement of the workstation with 2 steps ladder for this process was the best improvement that should be suggested to the industry.

4. Conclusion

In conclusion, the working posture of the worker at mechanical assembly workstation is completely simulated using RULA and DELMIA analysis software. A conceptual design of 2 steps ladder workstation design is suggested as an improvement to the current workstation design based on the results of analysis obtained with the lowest value of RULA score, energy expenditure and biomechanics single action analysis. In addition, this software was powerful predictive tool to measure the ergonomic risk value and improve the workstation design of the industry. This simulated results as well is an excellent work platform in order to integrate and compare physical with virtual augmented reality system.

Acknowledgments

The authors would like to acknowledge collaboration to industrial for funding this research under Grant Luar refer GLuar/CTRM-AMIC/2016/FKP-AMC/100018, Faculty of Manufacturing Engineering of UTeM and Centre for Research and Innovation Management (CRIM) for providing assistance throughout this study. Gratitude is also extended to Aerospace Manufacturing Innovation Centre (AMIC) and Composites Technology Research Malaysia Sdn. Bhd for the assistance and permission to carry out this study at their location.

References

[1] K. G. Gómez-bull, J. L. Hernández-arellano, and G. Ibarra-mejía 2015 A proposed methodology for task analysis in ergonomic evaluations, vol. 3, no. Ahfe, pp. 4756–4760

[2] G. Dahlke 2015 Ergonomic criteria in the investigation of indirect causes of accidents, vol. 3, no. Ahfe, pp. 4868–4875

[3] L. E. Charles, C. C. Ma, C. M. Burch, and R. G. Dong 2018 Vibration and Ergonomic Exposures Associated With Musculoskeletal Disorders of the Shoulder and Neck, vol. 9

[4] M. Y. Mustafa , A. Prof , and R. Sutan 2013 Work Related Neck Pain and Its Associated Factors among Registered Female Nurses Who Are Computer Users in Universiti Kebangsaan Malaysia Medical Centre, vol. 1, no. 2, pp. 41–56

[5] A. Chandrasakaran, H. L. Chee, K. G. Rampal, and G. L. E. Tan 2003 The Prevalence of Musculoskeletal Problems and Risk Factors Among Women Assembly Workers in the Semiconductor Industry, vol. 58, no. 5

[6] P. Polášek, M. Bureš, and M. Šimon 2015 Comparison of Digital Tools for Ergonomics in Practice, vol. 100, pp. 1277–1285

IOP Conf. Series: Materials Science and Engineering **705** (2019) 012045 doi:10.1088/1757-899X/705/1/012045

A Study of Ultrasonic Sensor Capability in Human Following Robot System

W. W. Tai*, B. Ilias **, S.A. Abdul Shukor, N. Abdul Rahim, MA Markom

School of Mechatronic Engineering

University Malaysia Perlis

*Email: taiwwei@gmail.com

**Email: bukhari@unimap.edu.my

Abstract – Human following robot is very common in this technology era. Human following is a technique used by robot and autonomous vehicles to follow a human within a specific range. In this case, communication between the human and the robot is the most significant factor where sensor is needed to ensure its successfulness. This paper discussed a human following robot system that utilized ultrasonic sensor. Ultrasonic sensor is preferred for human following robot due to its wide detection area, less light dependency, the ability to detect glass and shining wall, smaller in size, lightweight, use a very low memory, cheaper than Laser Range Finder (LRF) or camera and lower power consumption. However, being a specular type of sensor and not narrowly focused, this can cause wrong estimation and recognition of human, especially at the legs. Therefore, necessary accompanying algorithms should be developed to encounter this issue. Here, three pieces 40 kHz ultrasonic sensors are mounted on the four-wheeled mobile robot platform. Ultrasonic sensors used to detect the human's leg, and the data collected from the sensors were used upon the design of sensor platform. In order to detect human's leg, the sensor platform was tested repeatedly to obtain optimum data. The data which have been collected from the ultrasonic sensor will then be interfaced with Arduino software. Thus, this paper outlines a set of benchmarks for human following and briefly evaluates its performances. The human following robot was tested in a real laboratory environment. The results were good, where the implementation of this algorithm is able to produce an accurate decision.

1. Introduction

In this high technology era, the rise of robotics plays an important role for human beings. Robots proved us to be always a very helpful thing as a more intelligent robot are able to cooperate with human for supporting the human. In order to support the human's works, a robot must be able to detect and follow the human [1]. A robot that able to detect and follow human is called human following robot. A human following robot needs interaction between itself and human in order to give aid / assistance. For example, an intelligent cargo transportation robot helps by carrying luggage in an airport [2-6]. The development of mobile cleaning robot to detect and track the sound source by using three microphones [7]. Besides, one of the domestic applications for a following robot is that it can carry stuffs along with human in shopping malls, home, office or golf courses.

IOP Conf. Series: Materials Science and Engineering **705** (2019) 012045 doi:10.1088/1757-899X/705/1/012045

There are a variety of advantages of this robot for human beings as it can also be used in civil application and industrial area. However, the robot itself cannot track and detect an object. So, the robot needs a mechanism which can make decision for it to take action accordingly so that the task can be performed correctly [8]. The involved mechanism is sensor that can detect obstacles or object around the robot itself. A lot of sensors can be utilized like laser, radio frequency, cameras and others, that able to follow the human in a constant distance with the mobile robot itself. In this project, ultrasonic sensor is used. Ultrasonic sensor is a device that able to measure the distance to an obstacle by using sound waves. The sound wave is travelled out at a specific frequency and returned when its sense a thing to measure distance in real time. It has its faults and this could be an issue for the sensor to recognize an object because the sensor works based on specular basis which mean it is not a narrow range scanning angle. Hence, a precision and accuracy work need to be implemented in order to track an object.

The ultrasonic sensor is low cost device that able to operate in any environment with low visibility where vision or laser systems are not really efficient [9-19]. The ultrasonic sensor is able to measure distance from itself to the certain object by using pulse-echo in real time. Unlike other sensors, ultrasonic sensor provides data about object existence in front of the sensor. Only few researches used ultrasonic sensor in human following system even though it is a very superior sensor in maneuvering and scanning application in various fields and environment conditions [20]. Here, the system utilizes ultrasonic sensor for legs recognition. The position and number of sensor was designed to recognize both human's leg at the same time and follow the human. Three sensors were used to detect both human's left and right leg and their interception is used to differentiate left leg and right leg. Each ultrasonic data is evaluated by using probability and the algorithm that can detect and track human legs is developed. From here, the robot able to follow as it could recognize both human's leg. The performance of human following robot was then analyzed.

The paper is organized as follows – Section 2 explains the method of sensor validation and the proposed human following procedures. In Section 3, the results of the proposed human following algorithm are presented. Finally, the work is concluded in Section 4.

2. Methodology

2.1. Design and Development of Sensor Platform

Three pieces of ultrasonic sensors were used to perform tracking on the both human's legs. The concept is that the sensors are arranged horizontally with a specific distance, where this distance will be the average width of human legs when the human is walking normally. Figure 2.1 shows the basic concept of the arrangement of ultrasonic sensor where the human legs are being sensed within the area of interception. There are two situations in the sensors placement, the first is when there is an intercept signal between sensor 1 and sensor 2, the systems inform the robot that is sense the human's left leg, while when there is an intercept signal between sensor 2 and sensor 3, the system is sensing the right leg.

The dimension of the platform has been determined and the process of cutting Perspex is then carried on. Then, the combination of all cut Perspex is carried on according to the design as Figure 2.2 and connected using screw and nut. One sensor is located in the middle of the platform where another two are located at the right and left of the middle sensor. The purpose of using screw and nut is to aid the users in adjusting the angle while performing calibration on three sensors that used to track human's left and right leg. By adjusting the angle, calibration on an algorithm would be easier to develop.

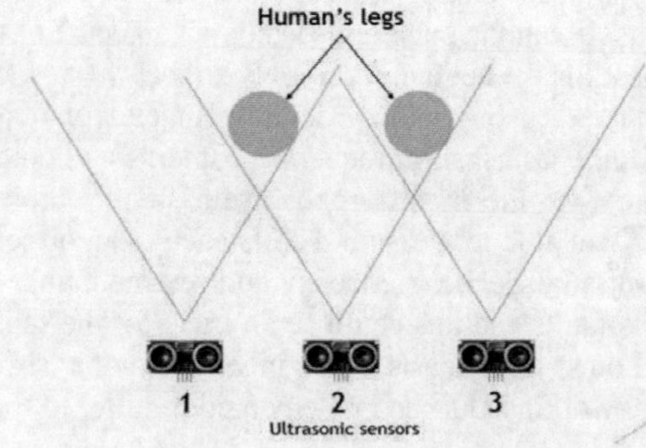

Figure 2.1. Concept of Sensors Works

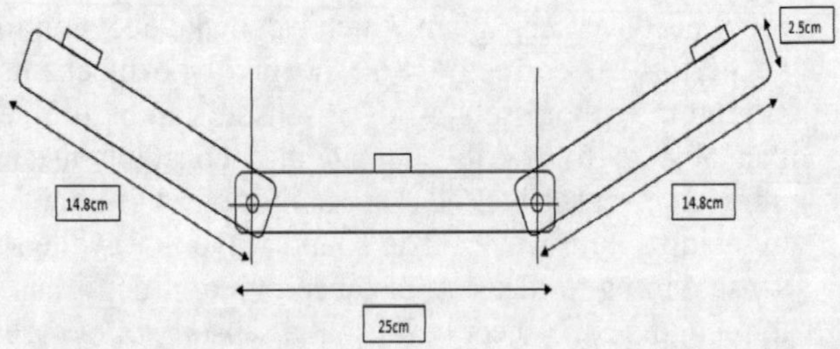

Figure 2.2. Design of Sensor Platform

2.2. Calibration of Sensor Platform

The calibration of sensor platform involved the angle between both sides. After setting the sensor platform, both the right and left of the platform was adjusted to a specific angle of 0°, 10°, 20° and 30° to obtain the intercepted areas. For each angle being set, the areas of detection of combination of all sensors were visualized using 3D drawing. Flowchart in Figure 2.3 shows the steps of calibrating the sensor platform. Figure 2.4 and Figure 2.5 showed each adjusted angle of the sensor platform in order to test and calibrate the platform.

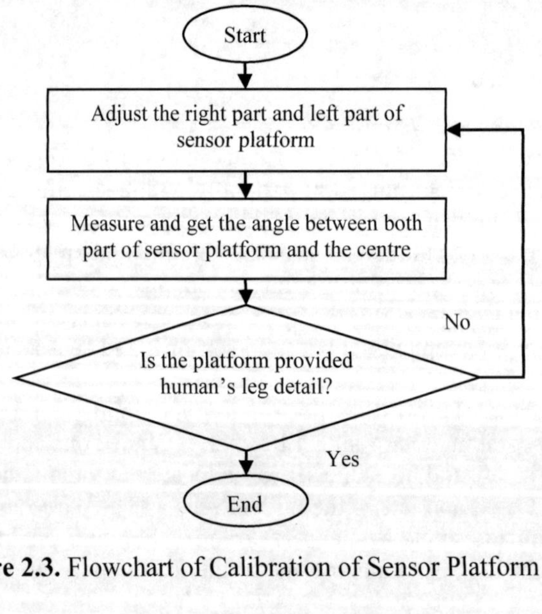

Figure 2.3. Flowchart of Calibration of Sensor Platform

IOP Conf. Series: Materials Science and Engineering **705** (2019) 012045 doi:10.1088/1757-899X/705/1/012045

Adjustments of sensor platform were done as follows:

a) Angle of 0° and Angle of 10°

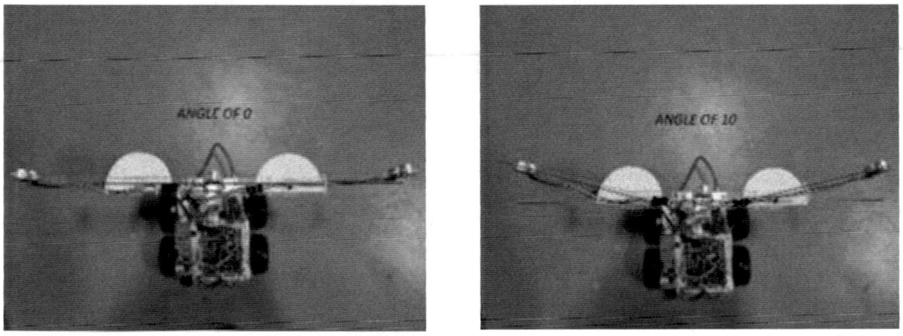

Figure 2.4. Angle of 0° and Angle of 10°

b) Angle of 20° and Angle of 30°

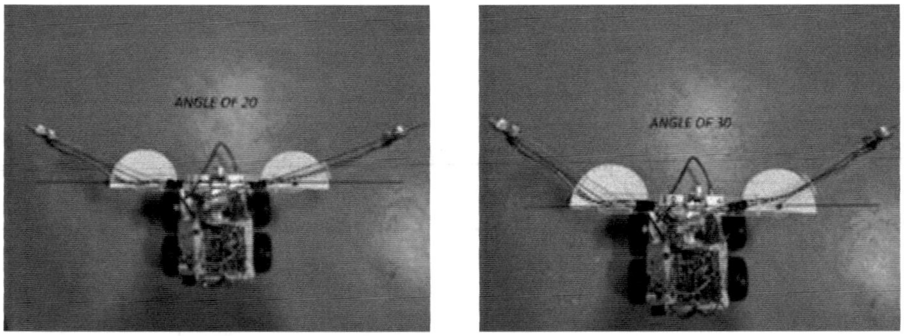

Figure 2.5. Angle of 20° and Angle of 30°

2.3. Position Validation of Sensor Platform

A cylindrical PVC pipe with a diameter of approximately same as human's leg was used. The PVC pipe is placed within an area of one position, where there are three positions to be tested. Each position was randomly tested for 30 times to obtain average results. For 1st position, the tested area is the detection area of sensor 1 and sensor 2, while for 2nd position; the tested area is the detection area of sensor 2 and sensor 3. Last, the 3rd position was tested for both area of detection of 1st and 2nd position.

Besides, the sensor platform also tested with human's leg as shown in Figures 2.6 – 2.8. Left leg of human was placed and tested in the area of 1st position, right leg of human was placed and tested in the area of 2nd position and 3rd positon was used to test both legs in position. The testing of sensor platform take place in indoor lab environment and the human being tested was wearing long pants. The calibration of sensor platform involved;

a) 1st position: Left leg, 1st sensor & 2nd sensor detected, yellow LED lights up, robot in standby mode.

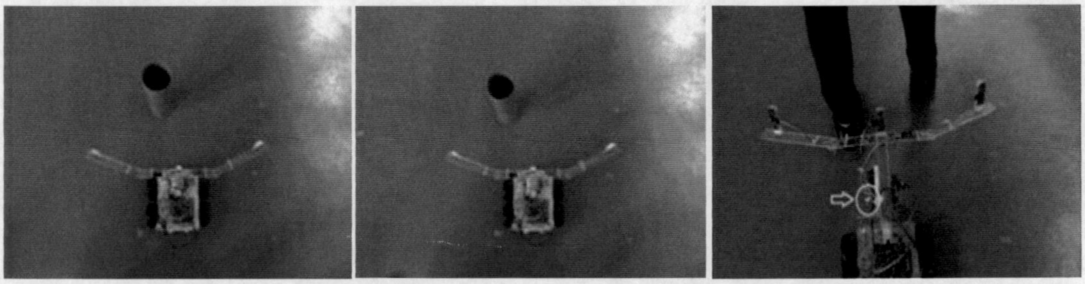

Figure 2.6. Position validations for the 1st position

b) 2nd position: Right leg, 2nd sensor & 3rd sensor detected, yellow LED lights up, robot in standby mode.

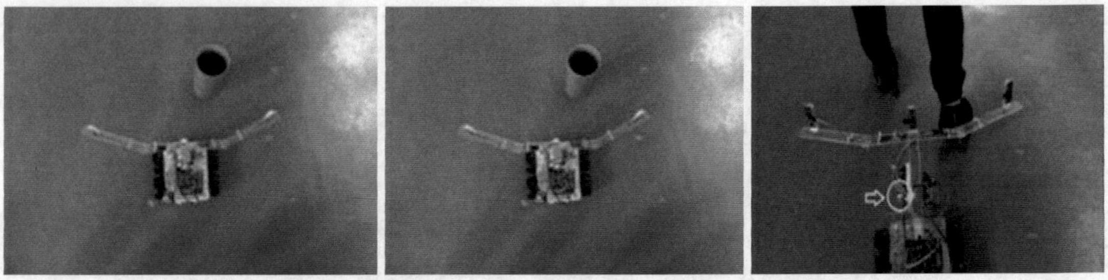

Figure 2.7. Position validations for the 2nd position

c) 3rd position: Left & Right legs. All sensors detected, Red LED lights up, robot in stop mode.

Figure 2.8. Position validations for the 3rd position

2.4. Testing the Ultrasonic Sensor on a Human Following Robot

The human following algorithm is tested and integrated with four different conditions in order to follow the human's legs in a straight line. Robot has to check the human's legs in front of it, then first is for the forward action, second is for stop action, third is the standby mode and last is backward action. Beneath are the algorithms of all condition:

a) Forward: IF (S1 & S2 & S3)>=60

b)	Stop:	IF (25>S1>40) & (25>S2>40) & (25>S3>40)
c)	Standby:	IF ((S1>50) & (25>S3>40)) or ((S3>50) & (25>S1>40))
d)	Backward:	IF (S1 or S2 or S3) <=25

3. Results & Discussion

3.1. Results of Calibration of Sensor Platform

The calibrations of sensor platform were done for the angles of 0°, 10°, 20° and 30°. Figures 3.1 and 3.2 visualized the area of detection between sensors for these angles which were done in 3D to illustrate the area of detection between sensors.

For the angle set to 0°, the area of detection between two sensors are small where this could make the robot hard to make decision in order to detect the human leg. While, for the angle set to 10°, a larger area of detection between the sensors shows that it is able to detect human leg within the area.

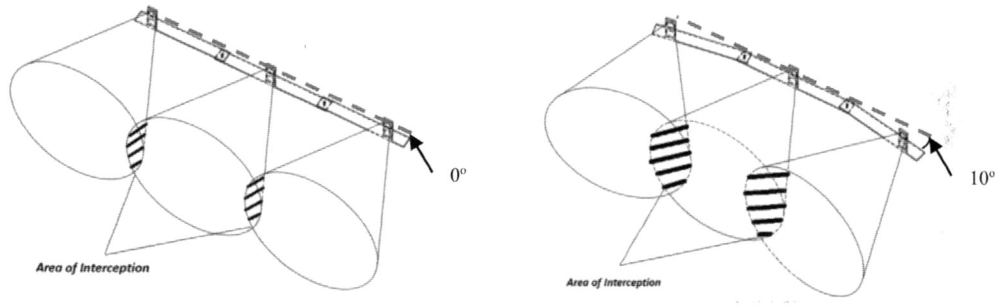

Figure 3.1. Results of Area of Detection for 0° & 10°

Meanwhile, for the angle set to 20°, the area of detection was visualized to be slightly larger than the angle at10°. It could make the robot unable to recognize legs when distance between legs and robot increases. Thus, accuracy could be decreases. The area of detection of angle 30° is too complicated where it could confuse on decision making where accuracy will extremely decreases. Overall, the angle of 10° is said to be the suitable angle for the sensor platform.

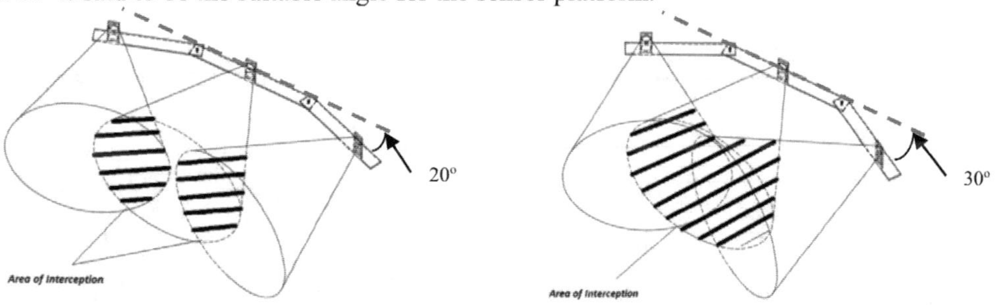

Figure 3.2. Results of Area of Detection for 20° & 30°

IOP Conf. Series: Materials Science and Engineering **705** (2019) 012045 doi:10.1088/1757-899X/705/1/012045

3.2. Results of Position Validation

Table 3.1 shows the results of position validation of sensor platform. Since the result from calibration of sensor platform illustrated that the sensor platform is suitable to work with the angle of 10°, position validation were also conducted at the same angle. For the 1st position, the results show that sensor 1 and sensor 2 able to detect the object which located on the left side of sensor platform. For the 2nd position, the results show that sensor 2 and sensor 3 able to detect the object which located on the right side of sensor platform while sensors 1, 2 and 3 able to detect the object which located on the left and right side of sensor platform. All of the 30 readings for each position were successfully recorded and it can be concluded that the sensor platform is able to detect human left and right legs within the area of detection.

Table 3.1. Results of Position Validation

Angle of 10									
		Position 1			Position 2			Position 3	
No.	S1	S2	S3	S1	S2	S3	S1	S2	S3
1	/	/	-	-	/	/	/	/	/
2	/	/	-	-	/	/	/	/	/
3	/	/	-	-	/	/	/	/	/
4	/	/	-	-	/	/	/	/	/
5	/	/	-	-	/	/	/	/	/
6	/	/	-	-	/	/	/	/	/
7	/	/	-	-	/	/	/	/	/
8	/	/	-	-	/	/	/	/	/
9	/	/	-	-	/	/	/	/	/
10	/	/	-	-	/	/	/	/	/
11	/	/	-	-	/	/	/	/	/
12	/	/	-	-	/	/	/	/	/
13	/	/	-	-	/	/	/	/	/
14	/	/	-	-	/	/	/	/	/
15	/	/	-	-	/	/	/	/	/
16	/	/	-	-	/	/	/	/	/
17	/	/	-	-	/	/	/	/	/
18	/	/	-	-	/	/	/	/	/
19	/	/	-	-	/	/	/	/	/
20	/	/	-	-	/	/	/	/	/

3.3. Results of Ability Test

The ability test was conducted to clarify the performance of the human following robot. The results were obtained in the indoor lab environment. Figures 3.3 – 3.4 show the results of the human following robot that has successfully followed the targeted human for a distance of 360cm.

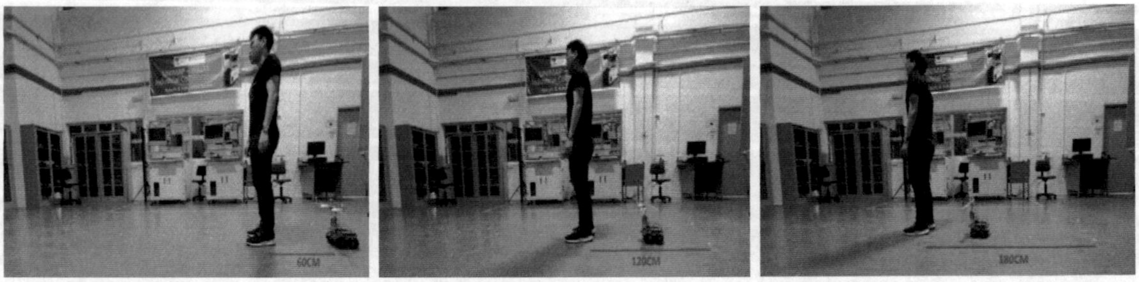

Figure 3.3. Ability Test for 60cm, 120cm & 180cm

Figure 3.4. Ability Test for 240cm, 300cm & 360cm

Table 3.2 shows the results being recorded when the ability tests were constructed. The total movement of robot for 300cm has been verified by tested for every 60cm.

Table 3.2. Results of Ability Test

No.	Human Walking Distance (cm)	Robot Following Distance (cm)	Success (/) / Fail (x)
1	60	0	/
2	120	60	/
3	180	120	/
4	240	180	/
5	300	240	/
6	360	300	/

4. Conclusions

A solution to human following robot has been presented in this paper. Three ultrasonic sensors have been integrated with the sensor platform with an objective of following a human. A study on the suitable number of ultrasonic sensor has been carried out. Besides that, a methodology of verification on the ultrasonic sensor is constructed. Ideally, the ultrasonic sensor has a specification of angle of view with ±30°, while the data obtained in the verification of the sensor is ±60°. The collected data are simply different to the ideal result due to multi-curved surface, tested in indoor environment and single sensor tested at one time. It has been proven that the ultrasonic sensor has a wide angle detection area which visualized like a bell shape. The design of sensor platform has been constructed along with the robot. The sensor platform was constructed to test the performance of human following robot. The human following algorithm was tested repeatedly in order to obtain the best result. The human following robot was tested in a real laboratory environment. The results were good, where the implementation of this algorithm is able to produce an accurate decision. Finally, comparison of different sensor shows that ultrasonic sensor is capable in a human following robot.

References

[1] Morioka K, Lee J H and Hashimoto H 2004 Human Following Mobile Robot in a Distributed Intelligent Sensor Network *IEEE Trans. Ind. Electron* **51**(1) 229-237

[2] Kawarazaki and Noriyuki 2015 Development of Human Following Mobile Robot System using Laser Range Scanner *Procedia Computer Science* **76** 455-460

[3] Burgard W, Cremers A B, Fox D, Hahnel D, Lakemeyer G, Schulz D, Steiner W and Thurn S 1998 The Interactive Museum Tour-guide Robot *in AAAI/IAAI* 11-18

[4] Scheutz M, McRaven J and Cseret G 2004 Fast, Reliable, Adaptive, Bimodal people Tracking for Indoor Environments *in Intelligent Robots and Systems*

[5] Fotiadis and Efstathios 2013 Human Detection from a Mobile Robot using Fusion of Laser and Vision Information *Sensors* **13**(9) 11603-11635

[6] Ismail R, Omar Z and Suaibun S 2016 Obstacle-avoiding robot with IR and PIR motion sensors *IOP Conf. Series: Materials Science and Engineering* **152** 1-6

[7] Jin T S and Han J H 2015 Sound Source Based Mobile Robot Control for Human Following in a Network Intelligent Space *International Journal of Control and Automation* **8**(7) 67-74

[8] Matsumoto Y and Zelinsky A 1999 Real-time Face Tracking System for Human-robot Interaction *IEEE International Conference on Systems, Man and Cybernetics* **2**(2) 830-835

[9] Akshay C 2017 Follower Robotic Cart using Ultrasonic Sensor *International Journal of Engineering Science and Computing* **7**(3) 5794-5797

[10] Aroul K 2016 Ultrasonic Sensor Based Wireless Luggage Follower *Journal of Chemical and Pharmaceutical Sciences* **9**(4) 3263-3265

[11] Ciarcia S 1980 Home In on the Range! An Ultrasonic Ranging System *Byte Publication Inc*

[12] Ye C and Borenstein J 2002 Characterization of a 2-D Laser Scanner for Mobile Robot Obstacle Negotiation *Proceeding of the 2002 IEEE International Conference on Robotics and Automation, University of Michigan, Washington DC, USA* 2512-2518

[13] Salih Y and Malik A S 2011 Comparison of stochastic filtering methods for 3D tracking *Pattern Recognition* **44**(10) 2711-2737

[14] Muñoz-Salinas R, Aguirre E and García-Silvente M 2007 People detection and tracking using stereo vision and color *Image Vision Computing* **25**(6) 995-1007

[15] Jung S, Kim J and Kim S 2009 Simultaneous localization and mapping of a wheel-based autonomous vehicle with ultrasonic sensors *Artificial Life Robotics (ISAROB 2009)* **14**(2) 186-190.

[16] Khoshgozaran A and Shahabi C 2010 A taxonomy of approaches to preserve location privacy in location based services *International Journal of Computational Science and Engineering*, **5**(2) 86-96

[17] Afghani S 2013 Follow Me Robot using Infrared Beacons *SAVAP International* **4**(3), 67-72

[18] Purvi A 2017 Human Follower Robot Using Kinect *International Research Journal of Engineering of Technology (IRJET)* **4** 1635-1637

[19] Chen J 2015 An Intelligent Human-Tracking Robot Based on Kinect Sensor *[Online].Available:https://oaktrust.library.tamu.edu/bitstream/handle/1969.1/156510/CHEN-THESIS-2015.pdf?sequence=1*

[20] Mojaev A and Vorst P 2016 Mobile Service Robots *[Online]. Available: http://www.ra.cs.uni-tuebingen.de/forschung/terminiert/service/welcome_e.html*

IOP Conf. Series: Materials Science and Engineering **705** (2019) 012046 doi:10.1088/1757-899X/705/1/012046

Shape classification of ground penetrating radar using discrete wavelet transform and principle component analysis

Hasimah Ali, Ahmad Firdaus Ahmad Zaidi, Cheng Ee Meng, [1]Mohamad Ridzuan Ahmad, [1]Tengku Sarah Tengku Amran, Siti Nurul Aqmariah Mohd Kanafiah, Chong Yen Fook, Shazmin Aniza Abdul Shukor, [2]Mohamed Elshaikh

School of Mechatronic Engineering, UniMAP, 02600, Arau, Perlis, Malaysia.
[1]NDT-MSI Group, Industrial Technology Division, Malaysia Nuclear Agency, 43000 Kajang, Selangor, Malaysia
[2]School of Computer Engineering and Communication, UniMAP, 02600, Arau, Perlis, Malaysia.

hasimahali@unimap.edu.my

Abstract. Ground penetrating radar is one of the promising non-destructive investigation for shallow subsurface exploration in locating buried utilities. However, interpreting hyperbolic signature of buried objects in GPR images remains a challenging task since the GPR signals are easily corrupted by environmental noise and cause misinterpretation of the size and geometry of subsurface object from the GPR raw profile. Therefore, this paper proposes Discrete Wavelet Transform (DWT) and principal component analysis (PCA) to classify geometry of buried object using k-nearest neighbour. (k-NN). The GPR images firstly being pre-processed. Then, the GPR images are decomposed using DWT into four sub-bands which are LL (Low-Low), LH (Low-High), HL (High-Low) and HH (High-High). The sub-bands LL or a coarse approximation coefficients was extracted as DWT features in order to classify the shape of buried objects. Since DWT features do contain high dimensional data, thus PCA is used to reduce the dimensional features from higher to lower space by linear transformation. The new projected features were then classify using k-NN classifier into four shapes which is cubic, cylinder, disc and sphere. A series of experiments have been conducted on extracted DWT and PCA features from hyperbolic signature of buried objects having different shapes. Based on the results, the proposed method had achieved the average recognition rate of 99.41%.

1. Introduction

The ground penetrating radar utilizes high frequency electromagnetic (EM) wave propagation which transmits waves into the ground by means of moving the antenna along the surface. Whenever wave strikes the object or boundary interface of contrast dielectrics, a portion of the wave reflects back to the surface, and receiving antenna. Figure 1 shows the imaging process of GPR data. It is widely used in a variety of applications such as locating buried utilities infrastructure such as pipes and cables [1], road inspection, land mines [2] and potential collapse area [3].

In previous work, many techniques have been attempted to analyse the hyperbolic signature for detecting and locating the buried object. Carlotto [4] have used Hough transform in GPR images to find the hyperbolic parameter for buried object. Although Hough transform was commonly used method in detecting hyperbolic signature, however this method is time consuming and low accuracy.

In [3], they used of the logarithmic transformed ensemble empirical mode decomposition for subsurface GPR imaging able to assess the risk of potential collapse area in urban environments. Irving & Knight [5] used numerical modeling (finite-difference time-domain) of ground penetrating radar by exploring the link between subsurface properties and GPR data in two dimensions using MATLAB environment. Ni et al. [6] used discrete wavelet transform to detect underground buried pipe based on hyperbolic signature. They used multiresolution analysis of DWT to filter and enhance the GPR data for obtaining more information on the DWT profile. Further, they have used sample of buried pipe made of plastic and metal in their design. Lee & Mokji [7] have used histogram of oriented gradient to automatically detect the hyperbolic signature in GPR data that contains the reflections from target of buried object by narrow down the data into possible reflections with the aid of linear SVM. Syambas [8] proposed decision tree method for predicting the shape and size of a buried object on surface GPR. One of the issues of GPR signals is that they easily corrupted by noisy environment, which cause misinterpretation of the size and geometry of subsurface objects from GPR raw profile.

Therefore, this paper proposes the used of DWT to decompose the GPR images into four sub-bands. Due to advantages of DWT which analyzes the signal at different resolutions by decomposing the signal into a coarse approximation and detailed information, thus generated wavelet coefficients able to reveal signal characteristic enhancement by means of applying a low pass filter to obtain approximation coefficients and high pass filter for detailed coefficients [6]. In this work, LL sub-bands or approximation coefficients at first level decomposition is extracted. Since DWT features have higher dimensional space, thus PCA is adopted to reduce the dimension from higher to lower space. The new projected features were then used to classify the geometry of buried objects using kNN.

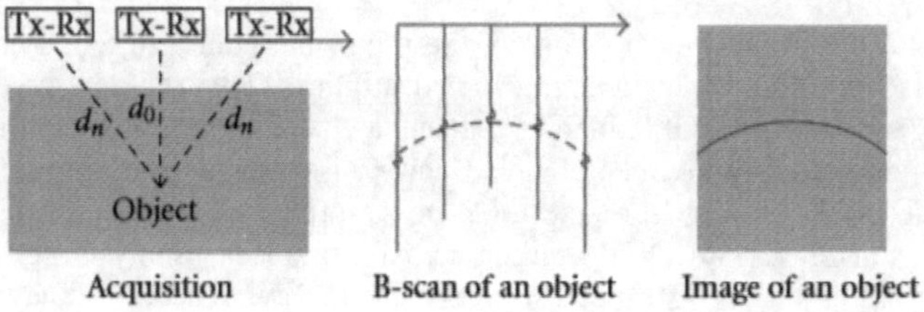

Figure 1. Imaging process of GPR data (left to right) [8].

2. Methodology

The flowchart of the proposed method is depicted in Figure 2. It consists of data acquisition, pre-processing, feature extraction, feature reduction and classification. The detailed of each block is described at the subsequent sections.

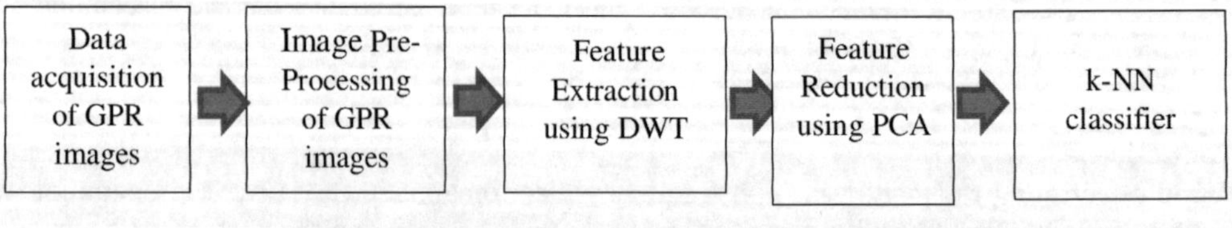

Figure 2. Block diagram of the proposed method

2.1. Data Acquisition

Totally, 340 GPR images are collected from 87 cubic images, 82 cylinder images, 86 disc images and 85 sphere images from aluminium material (metal) using Ramac GPR Mala system with antenna of 800 MHz transmitter and receiver. Figure 3 shows the basic GPR components that consist of antenna, battery pack, control unit, and a portable PC. These samples are buried in parallel with depth of 20 cm from the top surface in the rig test setup at Non Destructive Test (NDT), Nuclear Agency Malaysia as shown in Figure 4, with 40 cm apart between them. Sand is used as soil medium. The operation of GPR system was guided by research officer of NDT, Nuclear Agency Malaysia.

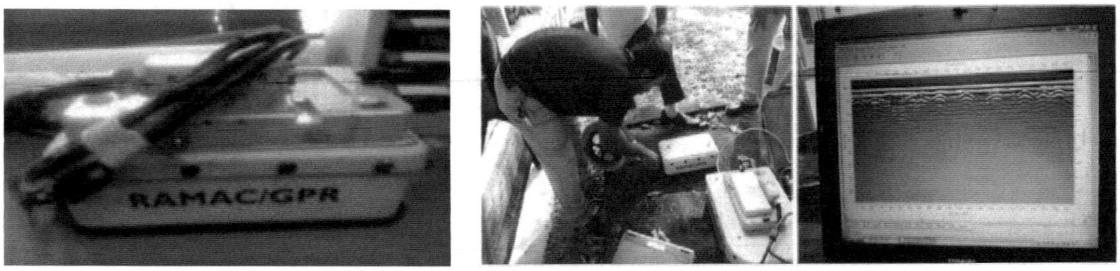

Figure 3. Basic component of GPR system

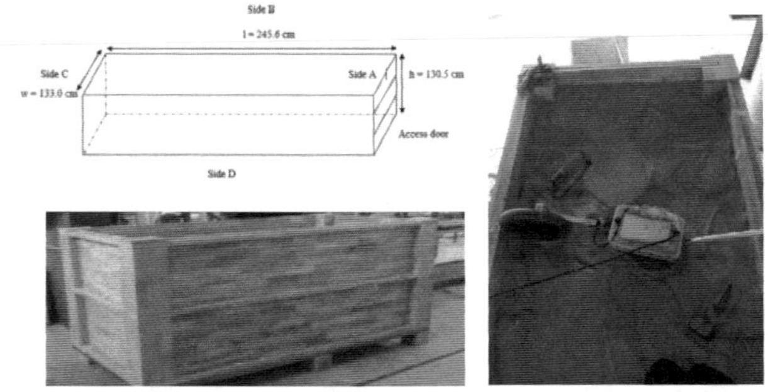

Figure 4. The rig test setup by NDT, Nuclear Agency Malaysia.

2.2. Image Preprocessing
Pre-processing is an important stage in every image analysis which can boost the overall performance of the system. In this work, the original image is firstly scaled and cropped into a fixed size in order to have uniformity in size is achieved. Figure 5 shows the original binary images and the preprocessed GPR images. The preprocessed images are then subjected to the DWT for extracting the GPR features.

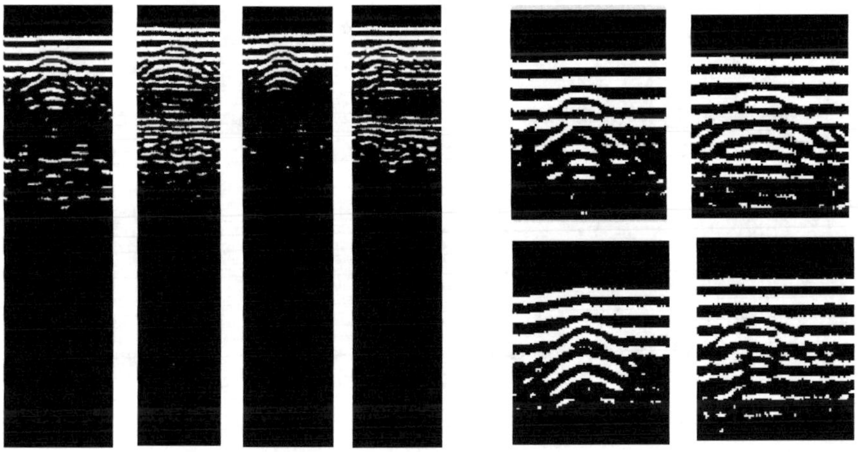

(a) (b)

Figure 5. (a) Binary images of cubic, cylinder, disc and sphere (left to right), (b) cropped hyperbolic signature images of cubic, cylinder (top) and disc, sphere (bottom).

2.3. Feature Extraction: Discrete Wavelet Transform

Discrete Wavelet Transform (DWT) offers advantage of using variables size time-scale for different frequency subbands. Thus, it is a powerful tool for analyzing nonstationary signals like GPR signals which reveals subtle changes of temporal variations in low frequency and abrupt temporal changes in high frequency [6]. The DWT involves some signal processing for example filtering, downsampling and upsampling. The DWT analyzes the signal at different resolutions by decomposing it into a coarse approximation and detailed information. The approximation coefficients are resulted by allowing the signal through a low pass filter, which removes the high frequency component. The process was repeated same for detailed coefficients resulted by allowing the signal through a high pass filter. The extracted wavelet coefficients can then be adopted for signal behavior enhancement. This method has advantages over generating the signal behavior the, especially for a GPR signal with noise and/ or a masked target. Based on this rationale, this study aims to link the behavior of GPR signal with DWT to classify the shape of buried objects based on hyperbolic signature pattern.

In 2D image framework, DWT decomposes images by first applying 1D filtering along rows of the image and then along columns [9]. The output of the analysis decomposition is a set of four *N/2* x *M/2* subimages so-called LL (low low), LH (low high), HL (high low) and HH (high high) sub-bands which correspond to different spatial frequency band in the image [9]. The decomposition of binary GPR image into such four sub-bands is shown in Figure 6. It can be observed that, the LL sub-bands is a coarse (low resolution) version of the original image, and that the HL, LH and HH sub-bands, respectively contain details with vertical, horizontal and diagonal orientation. The total number of pixels in the four sub-bands is equal to the original number of pixels, *NM*.

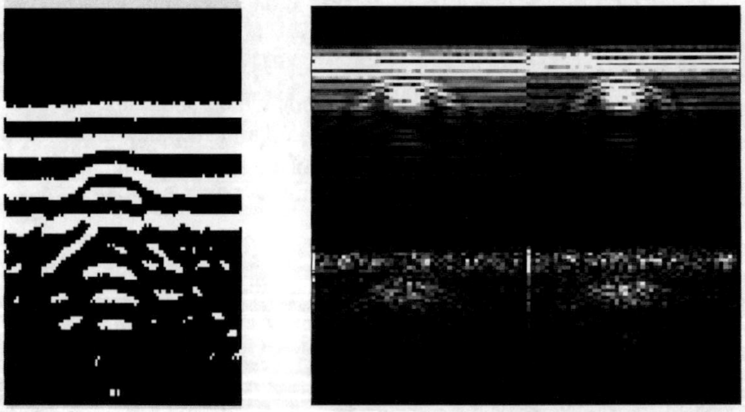

(a) (b)

Figure 6. Decomposition of *N x M* image into four *N/2 x M/2* sub-bands, (a) the original image of cubic shape, (b) first level DWT decomposition on GPR image of cubic shape.

2.4. Feature Reduction: Principal Component Analysis

The goal of PCA is to mapped higher dimensional space into a lower dimensional space by means of linear projection that maximizes the scatter of all projected samples [10]. Let X be a random vector with observations $x_i \in R^d$:

Step 1: Determine the mean of the data matrix μ.

Step 2: Determine the covariance matrix S.

Step 3: Calculate both the eigenvalues λ_i and eigenvectors v_i of S,

Step 4: Sort the eigenvectors in descending order associated with largest eigenvalues. The projected sample in the new space is defined as $y = W^T * x$. Figure 7 illustrates the number of components (no of eigenvectors) with respect to the variance accounted. In this work, the first 250 principal components with 98.37% of variability were investigated.

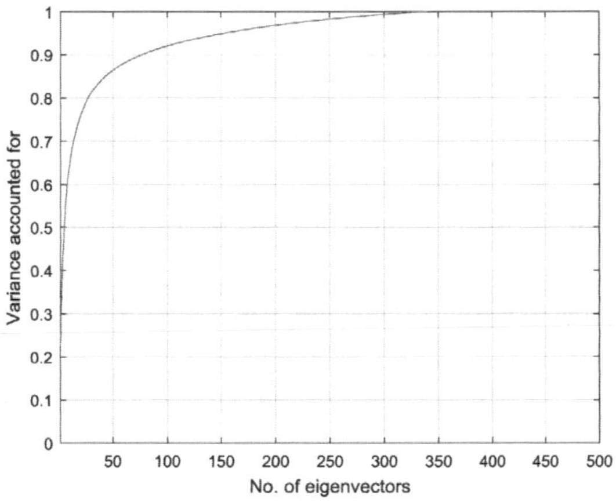

Figure 7. The plot of principal components analysis.

2.5. Classification: K-Nearest Neighbour

The k-Nearest Neighbor algorithm (k-NN) [11] is a nonparametric method used in pattern recognition for both classification and regression by assigning the class label to the input pattern based on the class labels that represented by k-closest neighbour vector. The input pattern is classified by a majority vote of its neighbours.

3. Results and Discussion

In this section, the effectiveness of the proposed method was evaluated based on DWT features and PCA using k-NN. About 340 GPR images of buried objects are collected from 87 cubic images, 82 cylinder images, 86 disc images and 85 sphere images from aluminium material (metal). The GPR images firstly being preprocessed. Then, the GPR images are decomposed using DWT into four sub-bands which are LL, LH, HL and HH. The sub-bands LL or a coarse approximation coefficients were extracted as DWT features in order to classify the shape of buried objects. Since DWT features do contain high dimensional data, thus PCA is used to reduce the dimensional features from higher to lower space by linear transformation. The new projected features were then classify using k-NN classifier into four shape which is cubic, cylinder, disc and sphere. In general, the performance of the DWT features using PCA method varies with the number of principal components. Thus before evaluating the DWT, we firstly performed an experiment using PCA. In this work, the first 250 principal components with 98.37% of variability were investigated.

Figure 8 shows the recognition rates of different shapes which are cube, cylinder, disc and sphere of GPR images based on DWT and PCA features using k-NN classifier. One can observe that both cube and cylindrical shapes have achieved the highest recognition rates which is 100% using DWT and PCA features. In addition, both disc and sphere shapes have achieved the recognition rates which are 98.84% and 98.82%, respectively. Further, Table 1 shows the confusion matrix of DWT features and PCA using k-NN classifier. As observed in Table 1, about 1 out of 86 disc shape was misclassified with cylindrical shape, whereas about 1 out of 85 sphere shape was misclassified with cylindrical shape. This can be inferred that, the geometry of cylindrical shape has tendency to be confused with geometry of disc and sphere shapes since they possess similar hyperbolic pattern in GPR images. As overall, the average recognition rates for four different shapes (cubic, cylinder, disc, and sphere) using

DWT and PCA features on GPR images produced 99.41% recognition rates. Thus, the proposed method shows a promising results.

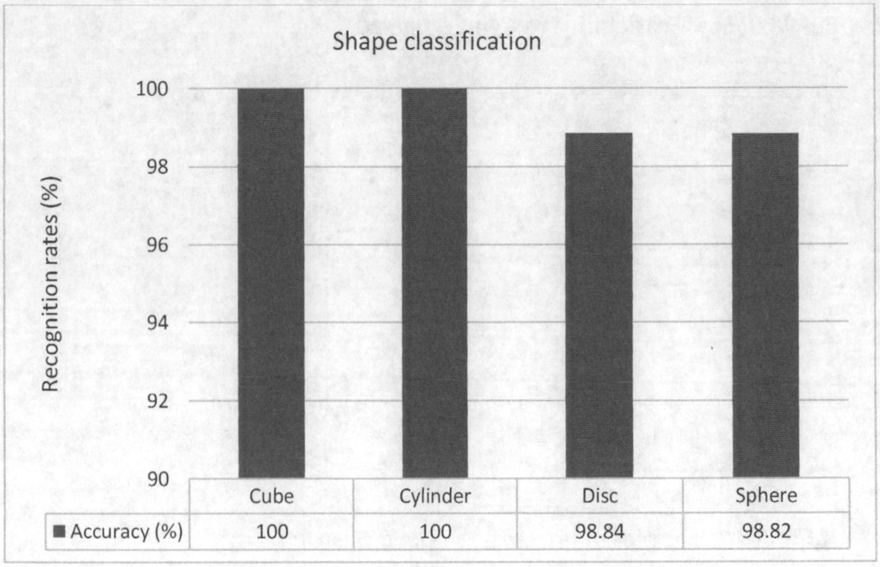

Figure 8. The recognition rates of different shape (metal) based on DWT and PCA features using k-NN classifier.

Table 1. Confusion matrix of recognition rates of cube, cylinder, disc and sphere.

	Cubic	**Cylindrical**	**Disc**	**Sphere**	*Average*
Cubic	87	0	0	0	
Cylindrical	0	82	1	1	
Disc	0	0	85	0	
Sphere	0	0	0	84	
Total	87	82	86	85	**99.41%**

4. Conclusion

This paper has presented the used of discrete wavelet transform and principal component analysis in classifying different shapes such as cubic, cylindrical, disc and sphere using k-NN classifier. Initially, the original image is preprocessed to have uniform size and shape. DWT is adopted to decompose the GPR images into four sub-bands. The LL sub-bands which is a coarse approximation coefficients are extracted and used as features. To reduce the redundancy of the features, PCA was adopted. Then, the reduced features is classify using k-NN classifier. Based on the results obtained, the average recognition rate achieved is 99.41% which cubic and cylinder shapes gave perfect recognition rates of 100%. While, the disc and sphere were achieved 98.84% and 98.82%, respectively. This shows that the DWT and PCA features able to discriminate as well as exhibit the signal characteristic of hyperbolic signature of buried object in GPR images. Thus, the proposed method shown a promising results in classifying the GPR images. However, further study needs to be conducted to automatically recognize the GPR images by applying artificial intelligence so that it can be utilized in real environments investigation.

References

[1] Simi A, Bracciali S & Manacorda G 2008 *Hough Transform Based Automatic Pipe Detection for Array GPR: Algorithm Development And On-Site Tests* IEEE Radar Conference

[2] Ishitsuka K, Iso S, Onishi K, and Matsuoka T 2018 *Object Detection in Ground-Penetrating Radar Images Using a Deep Convolutional Neural Network and Image Set Preparation by Migration.* Hindawi International Journal of Geophysics, pp.1-8

[3] Jeng Y & Chen C S 2012 *Subsurface GPR imaging of a potential collapse area in the urban environments.*Engineering Geology, 57-67

[4] Carlotto M J 2002 *Detecting Buried Mines in Ground Penetrating Radar Using a Hough Transform Approach* Proceedings Volume 4741, Battlespace Digitization and Network-Centric Warfare II;

[5] Irving M J, Knight R 2006 *Numerical Modelling Of Ground-Penetrating Radar In 2-D Using Matlab.* Computers & Geosciences **32**, pp1247–1258

[6] Ni S H, Huang Y H, Lo K F, Lin D C 2010 *Buried Pipe Detection by Ground Penetrating Radar Using the Discrete Wavelet Transform.* Computers and Geotechnics **37**, pp. 440–448

[7] Lee K L, Mokji M M 2014 *Automatic Target Detection in GPR Images Using Histogram of Oriented Gradients (HOG).* 2nd International Conference on Electronic Design (ICED), August 19-21, Penang, Malaysia

[8] Syambas N R 2012 *An Approach for Predicting the Shape and Size of a Buried Basic Object on Surface Ground Penetrating Radar System.* Hindawi Publishing Corporation International Journal of Antennas and Propagation Volume 2012

[9] Bovik Al 2009 *The Essential Guide to Image Processing*, Burlington, MA: Academic Press (Elsevier)

[10] Belhumeur P N, Hespanha J P, & Kriegman D J 1997 *Eigenfaces vs. Fisherfaces: recognition using class specific linear projection.* IEEE Transactions on Pattern Analysis and Machine Intelligence, ***19***(7), 711–720

[11] Altman N S 1992 *An introduction to kernel and nearest-neighbor nonparametric regression.* The American Statistician, ***46(3),*** 175-185

Human Location Classification for Outdoor Environment

M.T.M Talib[1,2*], M.H.F Rahiman[1,2#], L.M. Kamarudin[1,3], Hiromitsu Nishizaki[4], A. Zakaria[1,2]

[1] Center of Excellence for Advanced Sensor Technology (CEASTech), Universiti Malaysia Perlis, 02600 Arau, Perlis, Malaysia.

[2] School of Mechatronics Engineering, Universiti Malaysia Perlis, 02600 Arau, Perlis, Malaysia.

[3] School of Computer and Communication Engineering, Universiti Malaysia Perlis, 02600 Arau, Perlis, Malaysia.

[4] Department of Mechatronics, Faculty of Engineering, University of Yamanashi, Japan.

*masturah@unimap.edu.my, #hafiz@unimap.edu.my

Abstract. Outdoor localisation can offer great capabilities in security and perimeter surveillance applications. The localisation of people become more challenges when involving with the non-linear environment. GPS and CCTV are two localisation techniques usually use to localise human in an outdoor environment. However, they have weaknesses which result in low localisation accuracy. Therefore, the application of Device-free localisation (DFL), together with the Internet of things (IoT) is more appropriate due to their capability to detect the human body in all environmental conditions, and there is no problem losing signals as faced by GPS. This system offers excellent potential in humans localisation because humans can be detected wirelessly without any tracking device attached. In developing the DFL system, the main concern is the localisation accuracy. Although the existing DFL system gives significant result to the localisation, the accuracy is still low due to the large variation in RSSI values. Hence, a Radio Tomographic Imaging-based ANN classification (RTI-ANN) approach is proposed to increase the localisation accuracy. This Artificial Neural Network (ANN) is designed to learn the Radio Tomography imaging (RTI) input for classification purpose. Even though the RTI gives a good result to the localisation, however, it suffers from smearing effect. To eliminates this smearing area and background noise, pre-processing of the RTI image is required. Thus, extracting the valuable information technique from the RTI image has been proposed. By extracting the valuable information data from the RTI image, about 61% to 66% of the smearing noise is removed depending on the size of the RTI image. Only data directly associated with human attenuation used for training and learning of ANN. The experimental results show ANN system can localise human in the right zone for a given dataset.

Keywords: device-free localisation, radio tomographic imaging, noise elimination, neural network

IOP Conf. Series: Materials Science and Engineering **705** (2019) 012047 doi:10.1088/1757-899X/705/1/012047

1. Introduction

Device-free localisation (DFL) has attracted remarkable attention, especially in the security monitoring system in the past few years. Internet of thing (IoT) with wireless technologies in DFL system offers great capability, especially in collecting and transmitting data to processing units. The implementation of IoT and wireless technology in DFL (e.g. Finger-print and Radio Frequency Tomography) makes human localisation easier without any touch with the target (human). Based on wireless sensor networks (WSN), the target is free from wearing a tracking device that makes human localisation easier without any contact with the target (human) [1].

Various wireless localisation techniques have been developed to localise the human position, which enables the transducers to send and receive data without any physical connection. These approaches characterised the differences of Radio-frequency (RF) response by exploiting the radio frequency attenuation due to obstruction of the human body. However, the identification of the human position become inaccurate due to a large variation in radio signal strength (RSS) [1]. Meanwhile, if there have small changes in multipath components, e.g. node position changes and environmental factors, the value of RSSI measurements will be affected [2][3]. Based on the fundamental of radio propagation, the received signal power varies as a function of space, frequency and time. The variation of the received signal in small free space outdoor environment can be classified as small scale fading because it deals with a small distance between transmitter and receiver [4]. The main cause that affects the signal strength is the multipath propagation of the electromagnetic wave. As the distance between transmitter and receiver increase, the local average of the received signal power decreases gradually.

Unlike the Finger-print approach, the Radio-frequency tomography (RFT), usually known as Radio Tomographic Imaging (RTI) considers all the multipath problems in the measurement of the attenuation field. However, this technique suffers two major problems. First, an ill-posed problem happens when the number of pixel in the reconstruction image is greater than the number of projection data [2]. Second, the overlapping of images due to the summation of the back-projected signals for each pixel causes the smearing effect. Due to these problems, the conventional RFT system applies regularisation techniques. However, if the parameter of regularisation selected is not suitable, it will affect the image result [5][6][7][8].

The problem in classification is caused by the DFL problem, in which location and RSS relationships are not directly accessible. Researchers start to tune their focus to solve the classification problem by introducing several machine learning approaches to extract the features for classification purpose. The popular machine learning algorithm used include K-Nearest-Neighbor (KNN), Support Vector Machines (SVM) and Neural Networks. Among this three, Neural Networks give a significant impact on localisation accuracy [9][10]. However, it still has a problem of low accuracy because of data source influenced by environmental noise. Therefore, this paper presents RTI-ANN classification approach as an alternative solution to overcome the localisation problem. In this RTI-ANN classification approach, the pre-processing of the RTI image has been introduced.

2. RSSI ATTENUATION

In this DFL-RTI system, each sensor will act as a transceiver, which can transmit and receive the data. The number of link's measurements (M) can be obtained by

$$M = \frac{n(n-1)}{2} \tag{1}$$

Where n is the number of the transceiver allocate around the monitoring area. By assuming that the measurement is reciprocal due to two-way communication of sensors, therefore, total measurement is divided by two. The average of RSS values is calculated to minimise the variation of RSS data for a set of samples, for each link. Before we convert the RSS data into a matrix form, the data need to be normalised first. This normalisation process is important to ensure that all the data is visible and read in the same way across all records. Mathematically, the normalisation of measurement data, $N_{(Tx,\,Rx)}$ can be defined as

$$N_{(Tx,Rx)} = \frac{M_{(Tx,Rx)} - M_{ref}}{M_{ref}} = \Delta y \tag{2}$$

where $M_{(Tx,\,Rx)}$ is attenuation data for current human position, and M_{ref} is reference data during an empty area. Next, we set the normalised ratio to a certain threshold value. If the normalised ratio $N_{(Tx,\,Rx)}$ is less than the threshold value, normalized data is set to be zeros. Thus, we can generate the unique dataset of each location.

$$M = \begin{cases} 1 & if\ N_{(Tx,Rx)} > threshold\ value \\ 0 & if\ N_{(Tx,Rx)} < threshold\ value \end{cases} \tag{3}$$

Figure 1 illustrates the type of projection used in this research. Fan-beam projection is chosen to maximise the coverage areas of detection.

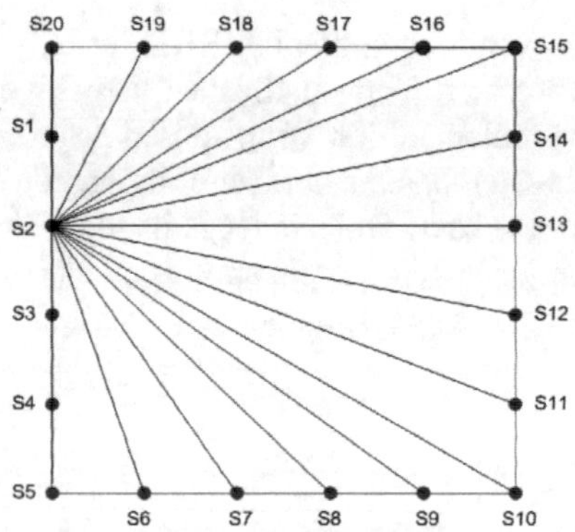

(a) Single layer of Fan-beam projection

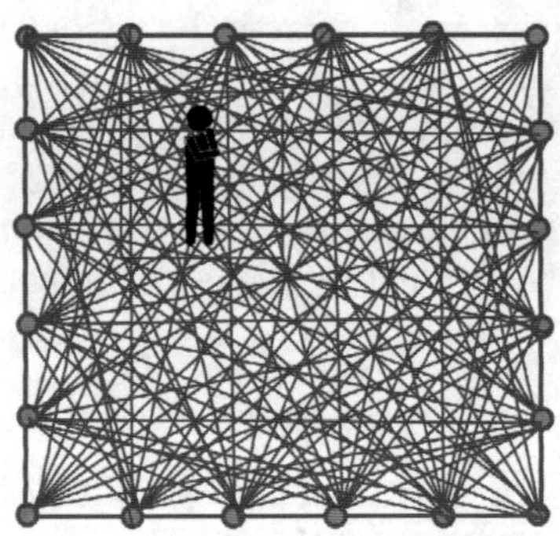

(b) Multiple Fan-beam projection form a DFL network

Figure 1. Fan-beam projection for data.measurement

The attenuation (S) can be modelled as

$$S_i(t) = \sum_{j=1}^{N} \omega_{ij} x_j(t) \tag{4}$$

where $x_j(t)$ is a single link attenuation, and ω_{ij} is the weight for each link. All parameters in (4) compiled into matrix form. Vector RSS (y) can be obtained by multiply the weight of each link (W) with attenuation vector (x) and can be expressed as:

$$\Delta y = W\Delta x + n \tag{5}$$

where Δy is the vector of length M (contain with RSS for all link) and Δx is the image of attenuation to be estimated. The weighting (W) can mathematically describe as

$$\omega_{ij} = \begin{cases} \frac{1}{\sqrt{d_{ij}}} & if\ d_{ij}1 + d_{ij}2 < d_{ij} + \lambda \\ 0 & otherwise \end{cases} \quad (6)$$

where λ is a parameter that determines the ellipse's width, dij1 and dij2 are distances between the centre of a pixel to both points of Tx and Rx, respectively. Thus, the pixels will have non-zero weighting values if and only if their centroid falls within the ellipse [11] as shown in Figure 2.

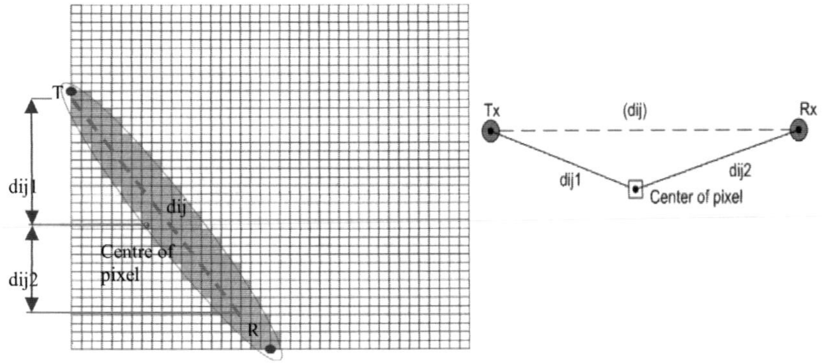

Figure 2. Ellipse model for weighting calculation

3. Localisation-Based –RTI-ANN Classification Approach

The block diagram in Figure 3 shows the Localization System using RTI with ANN classification approach, called an RTI-ANN technique. First, the calibration process is conducted to ensure the data is valid before we measure the attenuation and is saved as a reference dataset. All measured data were compared with the reference dataset to obtain the attenuation data. This approach provides the measurement vector for image reconstruction.

An Artificial Neural Network (ANN) is designed to learn the Radio Tomography imaging (RTI) input. This ANN classification approach is used to increase the localisation accuracy. In this research, the RTI data has been used as an input dataset to this ANN system. Since RTI image contains with un-informative data, extracting valuable information from radio tomography imaging (RTI) is required to eliminate background noise and smearing effect. This extracting information is known as RTI data. The background noise can be removed using image filtering techniques. Training and test ANN is conducted to evaluate the system performance. The ANN will execute all data and classify it according to the designated position class, as stated in Table 1.

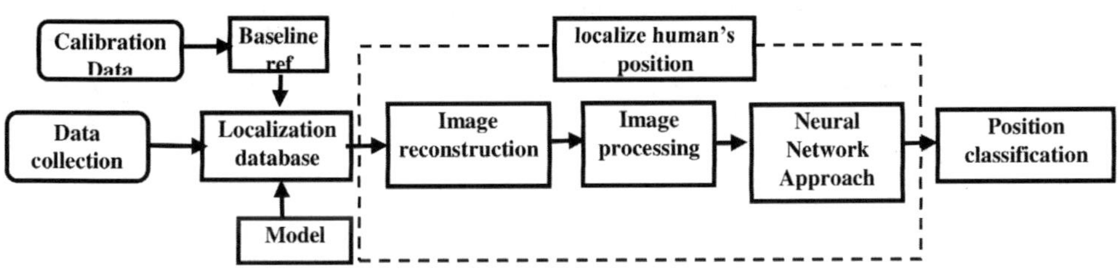

Figure 3. Localization System using RTI - ANN classification approach

3.1. Experimental Design

An experiment has been conducted to gather information about the human body attenuation. The information about RSS data for both stage (calibration and attenuation) help in designing the system model. By taking the average diameter of the human body is 50cm, monitoring area of 5m x 5m has been divided into nine grid to represent a zone, as shown in Figure 4(a). Hence, each zone can contain up to four positions of human. The reason for zoning the monitoring area is to fix the human position in a specific location to ease classification work.

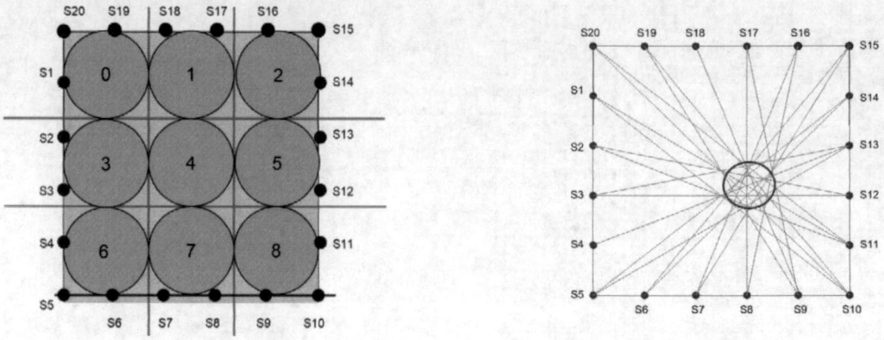

(a) 9 zone of monitoring area (b) Effected pair of links due to human presence

Figure 4. Experimental design for human localization

Figure 5. Sensor array follow experimental design

Figure 4(b) shows the affected pair of the link, which gives high attenuated data. Where the area in the circle contains high attenuated data which presents the human position in the monitoring area. Figure 5 shows the experimental setup for monitoring area of 5m x 5m. The RSSI value of the baseline dataset collected under the calibration period (without the presence of human). Based on the experiment, the range of RSSI without human body attenuation is around -40 to -60 dBm, while the human body attenuates about -61 dBm to -80 dBm.

3.2. Pre-processing of RTI image

Linear back projection (LBP) technique is used to reconstruct the tomographic image. As can be seen from Figure 6(a), the image is slightly blurred, which is due to the smearing effect that can affect localisation accuracy. Technically, the colour in the RTI image represents the intensity or permittivity of the subject. If we focus on the outer layer of the RTI image, we can see the smearing effect on it. Therefore, the elimination of background noise is needed. Now, let us focus more on the centroid of the RTI image. The image filtering technique has been applied to eliminate the smearing area; By assuming that the centre of the RTI (blue) image has a high permittivity value that may represent human, undesirable data can be automatically removed as shown in Figure 6(b). This approach can be proposed to classify human location.

(a) RTI image (b) Extract the valuable information of RTI

Figure 6. the RTI image before and after eliminating the background noise

4. Classification Approach

Figure 7 illustrates the frameworks of RTI-ANN classification approach. Since the RTI data contain the specific pattern associated with the human position, the classification approach will match the similarity of the data. ANN used as a training and learning tool for high accuracy detection. As usual, Multi-Layer Perceptron (MLP) of ANN calculates the position based on the RTI. There are three layers of MLP, the input layer, the hidden layer, and the output layer. Where in most practical application, one hidden layer is sufficient for all input nodes [12].

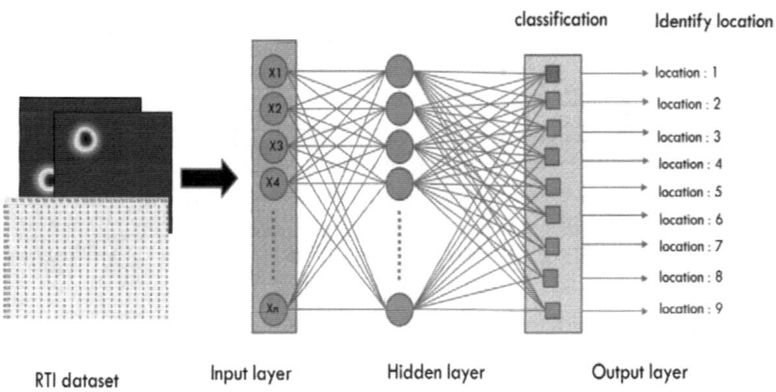

Figure 7. Frame work of RTI-ANN Classification Approach

IOP Conf. Series: Materials Science and Engineering **705** (2019) 012047 doi:10.1088/1757-899X/705/1/012047

The Forward formulation for each hidden node's input can be expressed as:

$$a_k^h = \sum x_n w_{k,x}^h + b_k^h \quad (7)$$

where $x_n w_{k,x}^h$ are weight's connection from input layer's node and b_k^h is a bias node.

Next, this hidden node's input will be transformed by Rectified Linear Unit (ReLU) activation function to compute the MLP output. Comparing to sigmoid and other activation functions, ReLU model is more simple and straightforward. The activation function is formulated in (8), and it works under specific conditions, as stated in (9).

$$h(a) = \max(0, \mathrm{a}) \quad (8)$$

$$S = \begin{cases} 1 & if h(a) > 0 \\ 0 & otherwise \end{cases} \quad (9)$$

The MLP will learn indirect tomography images from RTI data. This approach will allow the ANN system to recognise the unique pattern of the entire dataset by calculating the probability distribution for each node and selecting the highest probability node to represent the human position in a given localised area. ANN system classifies human position based on a specific zone, as mentioned in Table 1. Each zone contains four positions of human. The idea is if a human move from one location to another location in the same zone, ANN will classify it as in the same zone.

Table 1. Human position for different zone area and their training class

Zone	Position	Class
0	1,2,3,4	0
1	5,6,7,8	1
2	9,10,11,12	2
3	13,14,15,16	3
4	17,18,19,20	4
5	21,22,23,24	5
6	25,26,27,28	6
7	29,30,31,32	7
8	33,34,35,36	8

5. Results of Classification Model

This section presents the results of ANN classification for given RTI dataset. The performance of ANN is achieved 100% for training and testing, as shown in Figure 8. On the other hand, loss of validation for training and testing decreases gradually towards zero. A different set of RTI data is used for validation to verify the ANN model. A summary of the results is as listed in Table 2. From this validation stage, the result shows that the classification is successful where the model can localise human correctly in the right zone for a given dataset.

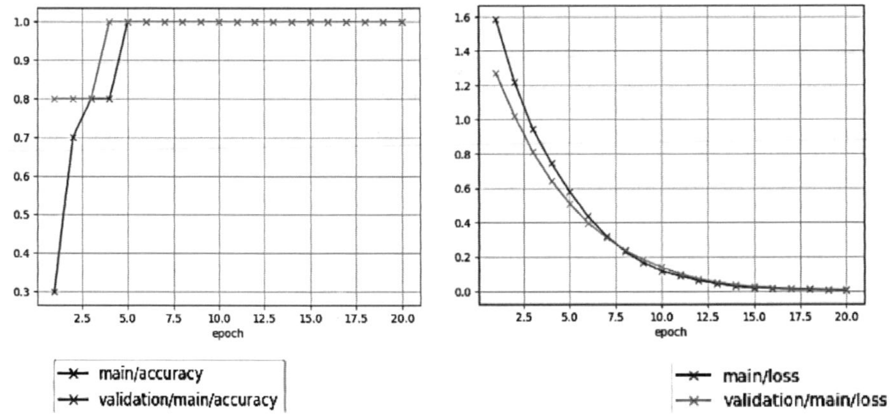

Figure 8. Training (main) and Test (validation) result of ANN system.

Table 2. Human localisation for different zone area

No.	Position	Zone
1	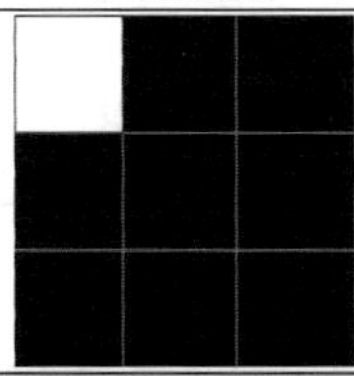	0
2	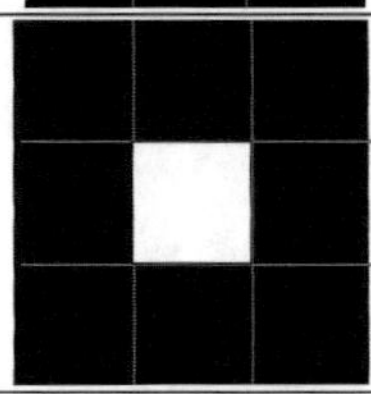	4
3	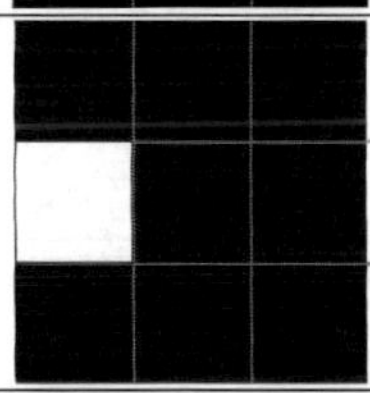	3
4		1,3,5,7
5		0,2,6,8

6. Conclusion

This paper presents the RTI-ANN classification approach. From these results, we can say that RTI performance improved in terms of accuracy of localisation by integrating RTI system with the ANN approach. Removing the smearing area and background noise can help to improve localisation accuracy. Further work will focus on the exploration of the Deep learning approach to characterise human positions from two different datasets.

Acknowledgement

The authors would like to acknowledge the support from the Ministry of Education Malaysia for SLAI scholarship scheme. Also, This work was supported by the Japan Science and Technology Agency under " the Sakura Exchange Program in Science" organised by University of Yamanashi and Center of Excellence for Advanced Sensor Technology (CEASTech), Universiti Malaysia Perlis. Furthermore, gratitude to postgraduate members of the Department of Mechatronics, University of Yamanashi, and postgraduate members of the CEASTech for their help and support toward this project.

References

[1] J. Wilson, N. Patwari and F. G. Vasquez 2009 *Virginia Tech Symp. Wirel. Pers. Commun.*

[2] O. Kaltiokallio, M. Bocca and N. Patwari 2014 *IEEE Trans. Mob. Comput.* **13** 6 p 1159–1172 A fade level-based spatial model for radio tomographic imaging

[3] S. Shukri and L. M. Kamarudin 2017 *J. Netw. Comput. Appl.* vol. **97** pp 157–174.

[4] Joao Figueiras and Simone Frattasi 2010 A John Wiley and Sons Ltd Publication. p 33 - 40

[5] J. Wilson and N. Patwari 2010 *IEEE Trans. Mob. Comput.* vol. **9** no. 5 pp. 621–632

[6] A. J. Wilson 2010 *Comput. Eng.*

[7] L. Heng W. Zheng-human B. Xiang-yuan and A. Jian-ping 2012 *IEEE Int. Conf.*, no. 1, pp. 48–53

[8] B. R. Hamilton, X. Ma, R. J. Baxley and S. M. Matechik, 2014 *IEEE J. Sel. Top. Signal Process* vol. **8** no. 1 pp. 55–65

[9] B. Wagner and D. Timmermann 2012 *Device-Free User Localization Utilizing Artificial Neural Networks and Passive RFID*

[10] L. Zhao, C. Su, H. Huang, Z. Han, S. Ding and X. Li 2019 *Symmetry (Basel)* vol. **11** no. 5 p. 630

[11] Qian Lei, Haijian Zhang, Hong Sun and Linling Tang 2016 *MDPI Sensor Journal*

[12] Mamdouh Refaat 2007 *MDPI Sensor Journal Elsevier*

IOP Conf. Series: Materials Science and Engineering **705** (2019) 012048 doi:10.1088/1757-899X/705/1/012048

Developing an IoT-Based Control System for Existing Air Conditioner using MEMS

Wajdi Araar, Tom Hofacker and Karl Kohlhof*

Cologne University of Applied Science, Cologne, Germany

*E-mail: karl.kohlhof@th-koeln.de

Abstract. Caused by the energy revolution, energy saving becomes a bigger topic than ever. This paper is about the developing of a new controlling system for air conditioning, which could save up to 66% of the energy consumed.

1. Introduction

The air conditioning branch is one of the biggest energy consumers of our planet. Developing a new controlling system based on in ISO 7730 [1] described Predictive Mean Vote (PMV) procedure, could help the existing air conditioners to become more energy efficient. Human thermal comfort as also described in ISO 7730 cannot be expressed by the temperature only. Humidity, air velocity and radiation heat affect the zone of comfort as well as the temperature does. The developed controlling circuit realized on a breadboard ensures to measure these influences by using microelectromechanical sensors (MEMS) and to calculate the measured parameters into simple data which can be sent wireless to the air conditioner itself. Being managed using a smartphone application and simply telling the phone how you feel e.g. cool or normal, the air conditioner would not simply cool till you freeze, it would perfectly stay in your zone of comfort, meanwhile saving up a lot of energy.

2. Basics

Air-conditioning systems as they are used today, mostly regulate room climate only by setting the temperature. Human beings sense for temperature is actually a combination of air velocity, humidity and temperature. So the area of wellbeing can be reached by regulating either of those three. This area of comfort is described in the ISO 7730; measurement regulations are set in the ISO 7726 [1] [2].

2.1. ISO 7730

ISO 7730 presents methods predicting the general thermal sensation and degree of discomfort of people exposed to moderate thermal environments. A human being's comfort zone is mainly related to the thermal balance of the whole body. This balance is influenced by several parameters that could contribute towards human thermal comfort, which includes physiological condition like clothing and activity of human, as well as the environmental parameters like temperature, humidity and air velocity [3]. Once these factors have been measured, the local thermal comfort could be predicted through ISO 7730 by calculating the Predicted Mean vote (PMV). The PMV predicts the mean value of the votes of a large group of people on the ISO thermal sensation scale (+3 = hot; +2 = warm; +1 = slightly warm; 0 = neutral; -1 = slightly cool; -2 = cool; -3 = cold).

IOP Conf. Series: Materials Science and Engineering **705** (2019) 012048 doi:10.1088/1757-899X/705/1/012048

The ISO 7730 also establishes the time interval the sensors need to measure data. Table 1 shows temperature, humidity and air velocity need to get measured every second. To get an average air velocity the mean over three minutes needs to get calculated. This is needed to calculate also the standard deviation of the velocity to calculate the turbulence value, necessary for further calculating the PMV.

Table 1. Measurement frequencies

Value	Type	Time interval
Temperature	Measured	As fast as possible *
Relative Humidity	Measured	As fast as possible *
Air velocity	Measured	Minimum 0.5 s Ideal 0.2 s *
Mean air velocity	Calculated	3 min +
Turbulence	Calculated	3 min +

+ISO 7730 [1], *ISO 7726[2]

2.2. ISO 7726
This standard specifies the characteristics of instruments for measuring physical quantities characterizing an environment as well as the methods for measuring these physical quantities, as in Table 2. The sensors selected needs to for fill the requirements set in ISO 7726.

Table 2. Range and Error of measuring parameters according to ISO 7726 [2]

Value	Range	Error
Air temperature	10 - 40 °C	+/- 0.5 °C
Air velocity	0.05 – 1 m/s	+/- 0.05 m/s
Humidity	0.5 – 3 kPa	+/- 0.15 kPa

2.3. Energy
Comparing the usual operating point of an air conditioner, which means high cooling and high ventilation, to the operating point using the knowledge of ISO 7730[1], a maximum of 2/3 of the power consumed can be saved. By taking the ISO 7730 guidelines and the power consumption of the fan and refrigerator compressor into a graph a new operating field is generated, as shown in Figure 1. As can be seen here, the same comfort can be generated by only 700W (600W compressor + 100W fan) compared to originally 2100W (2000W compressor + 100W fan) which is a massive save of energy. This resolves in ventilating as much as possible and cooling only if necessary, as the fan only takes 1/20 of the power the compressor needs in full power mode. Ventilating the air, depending on the relative humidity and comfortability, can be achieved at higher temperatures and less energy will be consumed. The worst operating point using the IS0 7730 is at 2050W (2000W compressor + 50W fan) but still 50W less than the usual operating point.

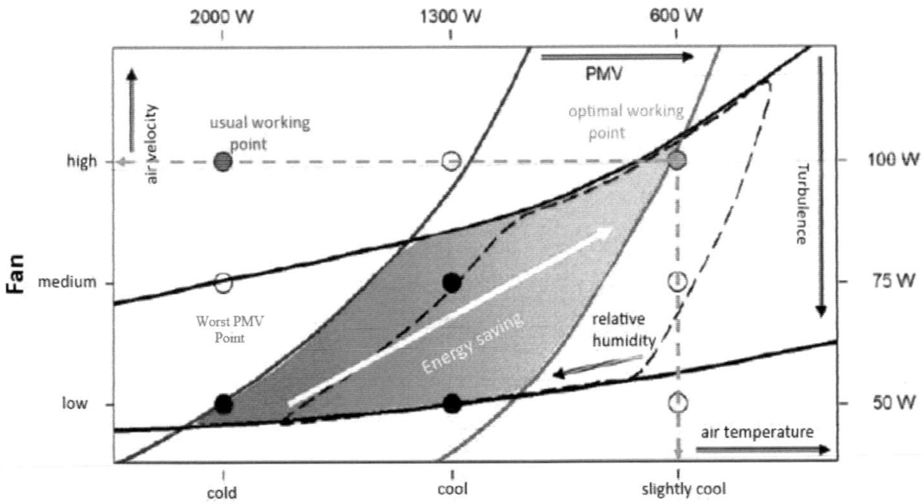

Figure 1. Operating field of an air condition hardware

2.4. Air velocity

To measure the air velocity vector, in principle a 3-dimensinal sensor is needed. As the device will be placed right on the wall, all 3-dimension need not to be measured, because air flowing straight onto the wall will be slowed down by the wall itself to zero. Therefore, a 2-dimensional sensor will do. Furthermore, the dimensional aspect the sensor needs to fit into the guidelines of ISO 7730 as well as ISO 7726.

The selected sensor is the FS7 manufactured in Switzerland by a company specialized in MEMS technology, named IST-AG as seen in Figure 2.

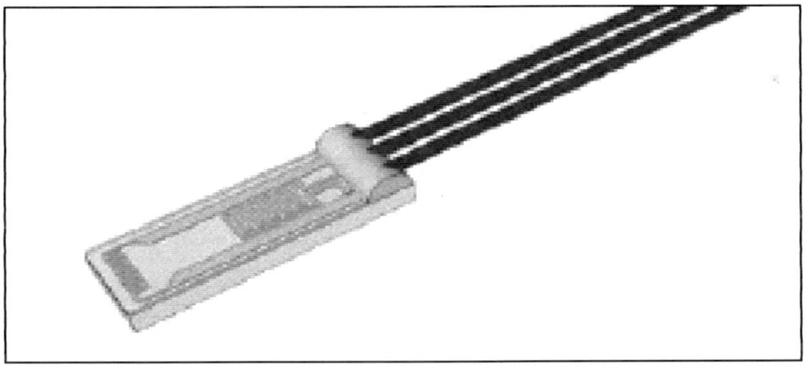

Figure 2. FS7 sensor

Comparing the technical dates of the FS7 (Table 3) to the guidelines set in ISO 7726 (Table 1), the sensor fits right into the ISO. As the sensor is part of the MEMS (micro electro mechanic system) family the entire sensor is very small, which is also required since the plan is to keep the device as small as possible. The biggest advantage of using a MEMS sensor is that the small sized materials got a very small thermal mass and therefore reacts quickly to changes in temperature.

The ground plate of the FS7 is made out of a ceramic substrate, and then a thin platinum layer is added. Using a digital Microscope (refer to Figure 3), the platinum heater resistor (RH = 45 Ohm) that actually senses the airflow shows up not to be squared as it would be preferred. It measures 1.27 x 1.06 mm². That is why airflow from different angles is expected to vary in the result.

Table 3. Technical dates FS7

Range	0m/s – 100m/s
Sensitivity	0.01m/s
Accuracy	3% of measured value
Temperature	-20 C° - 150C°
Temperature sensitivity	< 0.1 %/K Voltage: 2-5 V
Heater: RH	45 Ohm
Temperature sensor RS	1200 Ohm
Size	0.2 x 2.4 x 6.9 mm

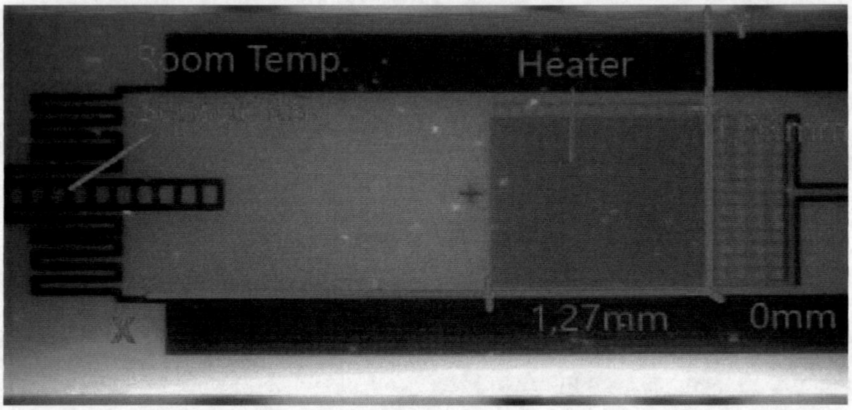

Figure 3. Microscope picture of FS7

Flow sensors are based on the relation of the flow speed to the heat transferred by convection. As flow passes across the heated sensor area, heat is carried away from the heater area by the medium. As flow increases, so does the amount of heat that is transferred. In the Constant Temperature Anemometer (CTA)-Mode a constant temperature difference between the heater and the temperature sensor can be achieved by adapting a Wheatstone-bridge, as seen in Figure 4.

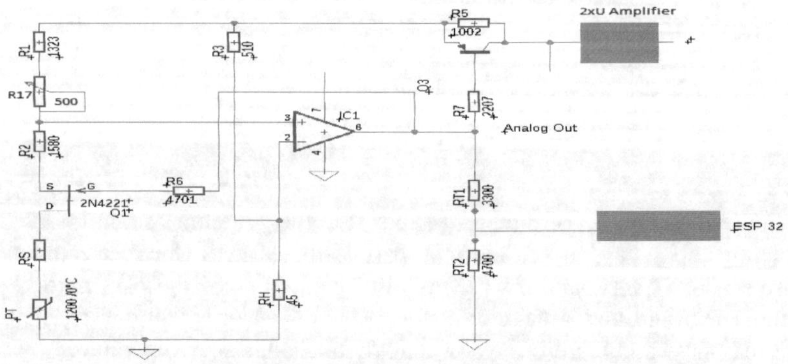

Figure 4. Circuit of FS7 evaluation board for FS5

In that bridge circuit the supplied electrical power controlling the temperature difference is a function of the flow speed. That power is converted into a voltage output signal to be readout easily. The medium temperature variation is measured by the temperature sensor on chip (RS = Pt1200). The resistors R1 to R6 can be chosen as shown in Figure 4. By resistor R1 the temperature difference between heater (RH) and medium (RS) is set to DT=30 K for air. The resistor R2 should be adjustable within ±10 % for calibration [4].

2.5. FS5 Evaluations board

Originally designed for the FS5 sensor, which is the predecessor of the FS7, the flow module also works with the FS7, because the resistors RH and RS of the FS7 and FS5 got the same values (Table 3, Table 4).

Table 4. Technical data of the FS5 evaluation board

Time response	< 100ms
Temperature range	0 – 80°C
Heater RH	45 Ohm +/- 1%
Temperature sensor RS	1200 Ohm +/- 1%
Power supply	5 VDC +/- 5%
Supply current	90 mA
Analogue output, nonlinear	0 – 10 V

2.6. Relative humidity and temperature

In this work, a DHT22 sensor is used to measure the temperature and the relative humidity (Figure 5). It is calibrated und uses digital signal technology to ensure reliability and stability while consuming less power [5]. Features such as high precision, small size, speedy response time und low-costs make the DHT22 sensor as the best choice in comparison to DHT11 and SHT75 (Table 5).

Table 5. Different temperature / humidity sensors

Model	DHT22	DHT11	SHT75
Humidity Range	0 – 100 %	20 – 80 %	0 – 100 %
Temperature Range	-40 – 80 °C	0 – 50 °C	-40 – 123.8 °C
Accuracy	± 2 – 5 % (Humidity) ± 0.5 °C (Temp)	± 5 % (Humidity) ± 2 °C (Temp)	± 1.8 % (Humidity) ± 0.3 °C (Temp)
Repeatability	± 0.3 % (Humidity) ± 0.2 °C (Temp)	± 1 % (Humidity) ± 1 °C (Temp)	± 0.1 % (Humidity) ± 0.1 °C (Temp)
Operating Voltage	3 - 5 V	3 - 5 V	2.4 – 5.5 V
Response time	Average 2 s	Average 2 s	8 s (Humidity) 5 to 30 s (Temp)
Output signal	digital signal	digital signal	digital signal
Price	3 - 9 €	1.5 - 5 €	30 - 45 €

The DHT22 consist of a capacitor-based humidity sensor and a NTC temperature sensor (thermistor). The capacitive sensor has two electrodes separated by a dielectric material.

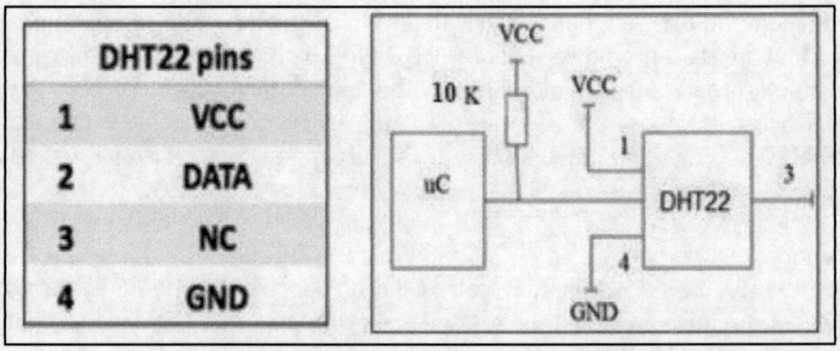

Figure 5. Electrical connection diagram of DHT22

Its dielectric constant changes with humidity, and accordingly the capacitance does so. This change is measured and processed by the chip so that it can be read out by a microcontroller. The thermistor is made by sintering semiconductor materials such as ceramics or polymers to achieve greater resistance change with low temperature change.

The communication between DHT22 and microcontroller is made through a one-wire bus from the company MaxDetect. The microcontroller receives from the sensor 40 bits: 16 bits of 10 times relative humidity value + 16 bits of 10 times temperature value + 8 check-sum bits [6].

If the data is transmitted correctly, then check-sum should be the binary sum of the both 16-bit data. The following example demonstrates the conversion from the 40 data bits to temperature and humidity values:

After measuring a temperature of 26.9°C and a relative humidity of 52.1% the sensor sends the following 40 bits string:
0000 0010 0000 1001 0000 0001 0000 1101 0001 1001

- Relative humidity RH calculation:
  binary = 0000 0010 0000 1001 -> hexadecimal = 0209
  -> decimal= 521= 10 × RH in %
  => RH = 521/10 = 52.1%
- Temperature calculation:
  binary = 0000 0001 0000 1101 -> hexadecimal = 010D
  -> decimal = 269 = 10 × T °C
  => T = 269/10 = 26.9°C
- Check-sum calculation:
  Check-sum = 0000 0010 + 0000 1001 + 0000 0001 + 0000 1101 = 0001 1001

2.7. ESP32 Microcontroller

The ESP32 of the company Espressif has been selected as controller of our system [7]. The ESP32 is the updated version of the ESP8266. Bluetooth and Wifi were added to the board. It is low-cost, low-power system on chip series of microcontrollers with WiFi and Bluetooth capabilities. Comparing this to other development boards on the marked e.g. Arduinos, the ESP32 costs only half of them. It is designed for wearable electronics and internet of things applications. The ESP32 turned out as a reliable option in an industrial environment because of the wide operating temperature range and the multiple input pins [8]. One of the main reasons for choosing the ESP32 except the price and the WiFi module was the built in analog digital converter (ADC), knowing that FS7 combined with the evaluation board gives an analog output that needs to be converted into bits. Moreover, the ESP32 has a built in low-power mode, which is useful when the device is later placed into installation housing. Therefore, the device needs to be operated using a battery, where the low power mode promises an extended operating time.

3. Wiring

In the proposed system the temperature and the relative humidity will be measured through DHT22. The voltage supply of the sensor must be between 3.3 and 6 V (recommended 5V). The DHT22 data pin is related to the digital pin GPIO4 of the ESP32 microcontroller board. The data queue between DHT22 and ESP32 must be related by the pullup-resistor of 10kΩ. Figure 6 shows the overall setup.

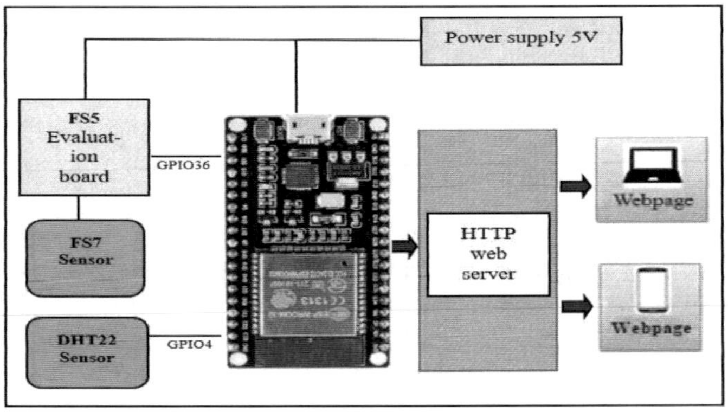

Figure 6a. Block diagram of measurement setup

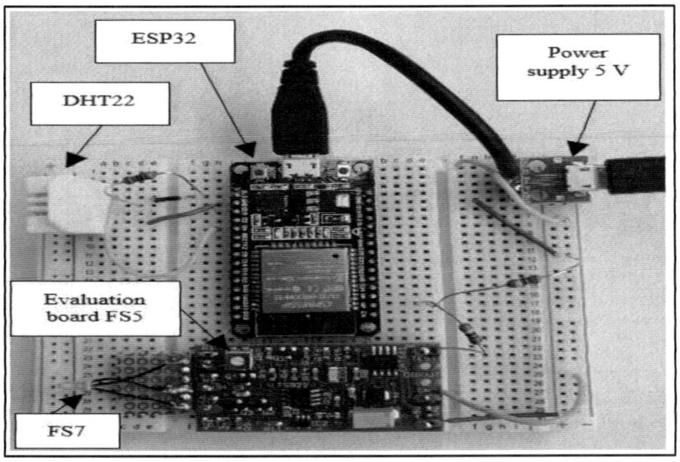

Figure 6b: Device mounted on a bread board

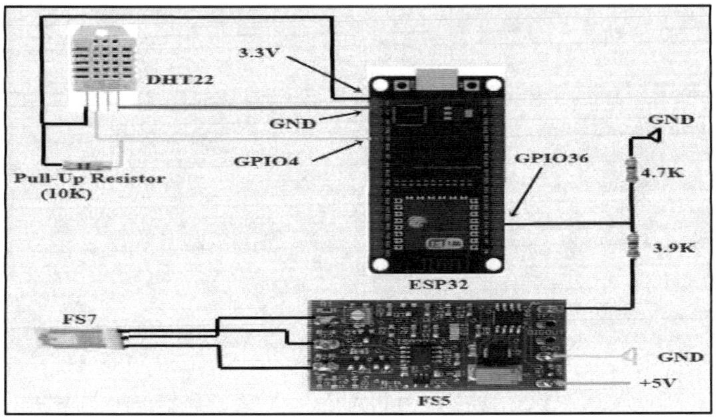

Figure 6c: Measuring components wiring plan

For powering the evaluation board a parallel connection to the USB power supply is needed, which insures a constant 5V power to the evaluation board. Powering the evaluation board through the ESP32, which comes with a built in 5V output-pin VIN, did not work properly, because a pull-up diode in line between USB power in- and 5V output-pins results in a 0.4V loss.

The evaluation board for the FS7-velocity sensor is connected to the ESP32 built in analog-digital-converter input-pin GPIO36. To restrict the input-voltage to the maximum allowed value of 3.3V, a 3.9 k Ohm to 4.7 k Ohm voltage divider is set between them. The ESP32 microcontroller is connected to a webserver that is used to display the results of the measuring.

4. Software

4.1. FS7 Linearization

A challenge occurs programming a sketch (name of an arduino based program) that outputs the right airspeed in m/s because the analog input into the Analog-Digital-Converter of the ESP-32 is not linear, as seen in Figure 7, thus one cannot just scale them up easily.

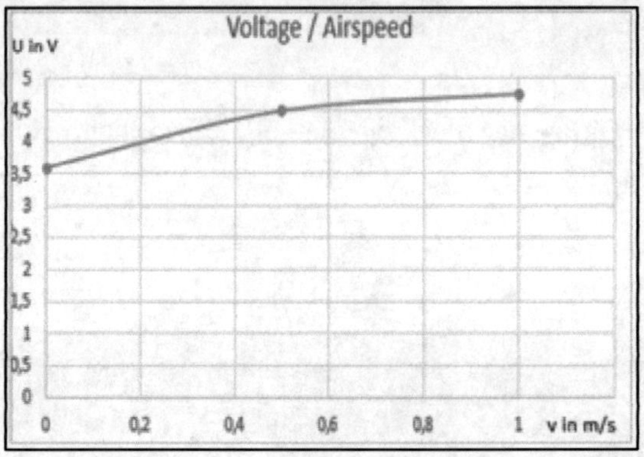

Figure 7. Dependence of FS7 output-voltage on airspeed

For linearization one needs to get three measurement points at 0 m/s airspeed, 0.5 m/s and max airspeed of 1 m/s (Figure 7). By using the CTA-formula for the FS7 anemometer with voltage $U_0 = 3.6$ V at a zero velocity, one can solve the equation system to get the unknown variables k and n:

$$U_V = 3.6V * \sqrt{1 + k * V^n}$$

Subsequently one can apply this formula to convert voltages to the life airspeeds. The variables are simply set fix in the program and only need to be changed if the precise measurement range is changed from 0-1 m/s to another. This process is included into the sketch and is completed every second for each value measured [4].

$$\vec{v} = \frac{[(U - U_0) \cdot (U + U_0)]^{\frac{1}{n}}}{(k^{\frac{1}{n}}) \cdot U_0^{\frac{2}{n}}}$$

4.2. PMV

The Predictive Mean Vote (PMV)-parameter is based on a scientist's study written down in ISO 7730 to show the thermal comfort zone of humans. Every human is different so is the comfort zone, that is why the Predicted Percentage of Dissatisfied (PPD) measured in percent never gets down to zero but stops at 5%. The PPD shows how many people are dissatisfied at the current PMV value. PMV ranges from -3 up to 3 (Figure 8), but the comfort zone described in ISO 7730 is between -0.5 and 0.5, where no more than 10% feel dissatisfied. Further down the scale people start to feel colder and further up more and more people start to feel warmer.

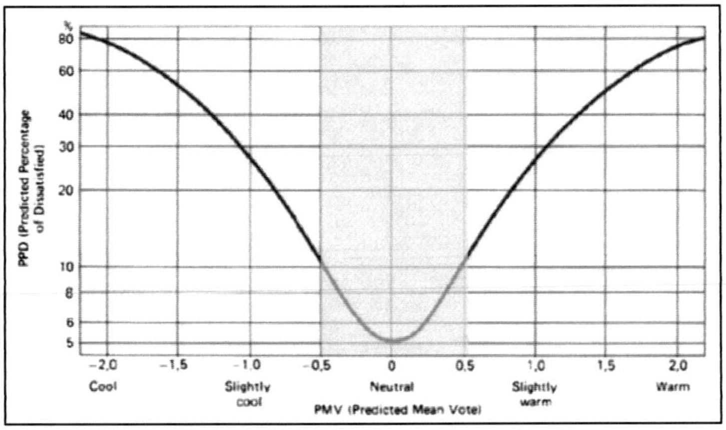

Figure 8.PMV and PPD

For calculation of the PMV value a lot of different variables are needed (Figure 9) to get the right result. Most of them can be measured live with the device designed such as air temperature, air velocity and relative humidity. Turbulence must be calculated from mean air velocity v_m and standard deviation σ_v of air velocity:

$$Turbulence = \sigma_v / v_m$$

Other parameters can't easily be measured and must be preset depending on the level of activity of the person or the insulation by the clothes the person is wearing (Table 6, Table 7).

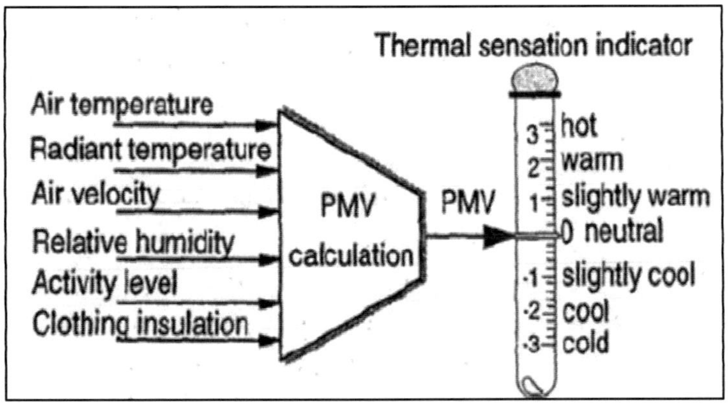

Figure 9. Parameters influencing PMV

ISO 7730 describes the energy consumption of a human depending on different activities. Basically, one has to change the value of energy consumption for the PMV calculation every time depending on people's activity. In the program described here its value is fixed to 58 W/m^2 for a human sitting relaxed. This equals 1met (Metabolic equivalent).

Table 6. Power emission for a human and its metabolic equivalent (ISO 7730) [1]

Activity	W/m²	met	W*
Ajar	46	0,8	83
Sitting, relaxed	58	1,0	104
Sitting, working	70	1,2	126
Standing, easy work	93	1,6	167
Standing, hard work	116	2,0	209
Walking, flat ground			
2 km/h	110	1,9	198
4km/h	165	2,8	297

*heat emission calculated with average skin area of 1,8m² for human

Not only activities do make one feel warm, different heat resistances of clothing also do. The clothing isolation factor measured in [clo] needs to be preset for the PMV calculation. Like the power consumption this factor needs to be adopted depending on the clothes wearing. In this work it is preset to 1 clo = 0,155 m²*K/W [10] which equals wearing t-shirt and trousers, socks, shoes and underwear (Table 7).

Table 7. Clothing-Isolation [9]

Clothing	Isolation in clo	Change of operating temperature in K
Slip	0.03	0.2
T-Shirt	0.09	0.6
Shirt	0.25	1.6
Shorts	0.06	0.4
Pullover	0.28	1.7
Jacket	0.35	2.2
Trousers	0.25	1.6
Skirt	0.15	0.9

5. Calibration

For the first operation test after wiring and programming a wind tunnel was used. Important before getting the values for airspeed is to set the analog output voltage of the evaluation board to U_0 = 3.6V at 0 m/s airspeed, by the potentiometer (R17 = 500 Ohm) built on the evaluation board itself (Figure 4). To ensure 0 m/s the FS7 was covered by a small box turned upside down and the voltage was measured by a Fluke F15B multimeter. After setting the zero voltage the device was placed into the air tunnel and the airspeed was set up to the two other points needed for the linearization, described in Section 4. For reference measurements an air velocity meter VelociCalc 8384[10] was used. During the calibration the device was connected to the computer via micro USB-connector and the digital values were read out by the com window, a tool to display the microcontroller doings.

6. Function

6.1. Test

After calibrating the system, the entire device was placed in thermally different areas to test the system on the edges of the operating field, such in a refrigerator, outside on a windy day and directly in the sun. As to see on the results (Figure 10) given the device works just fine.

```
Airspeed =  0.4035 m/s
Humidity: 46.20 %        Temperature: 25.10 *C

Mean is    0.05972 m/s
Standard deviation from the last 3 min seconds :  0.349 m/s
Turbulence from last 3 min second  : 17.09 %

PMV =    -0.49
```

Figure 10. Output shown in com window device placed inside a lab

The challenge was that the exact PMV value can only be checked using an expensive reference room comfort measurement device such as testo 400 Comfort-set [11]. In this work PMV values could only be estimated roughly by the team's personal feelings. To test the 3-dimensional behavior the bread board was again placed into the wind tunnel. The orientation of the sensor was changed during the different tests, so that all three dimensions could be checked. The result of the 3D test was as expected, that the orientation does only affect the air-velocity measuring neither the humidity nor the temperature.

As the FS7 measures airspeed using a rectangle heater plate to measure heat loss, air coming towards it from the X or Y axis should take different amounts of heat from it. In reality the impact of the rectangular heater can't be confirmed. Both directions got the same effect. That might be understood because the surface area is equal in size in either direction and as it is for the convected thermal mass. Wind coming from the Z axis does have a different behavior, it can't just flow across the sensor, but gets stopped to zero by it and pushed into X and Y axis.

6.2. *Displaying data*

The ESP32 was connected to an integrated webserver running on it, responsible for returning a webpage to the client when the client is connected to it through HTTP (Hypertext Transfer Protocol). The sensor data could be shown on the webpage using a mobile device by connecting it to the local webserver by a local Wi-Fi. In detail, after creating the webserver with the microcontroller, the ESP32 must be connected to the Wi-Fi network, and then an IP address of the ESP32 in the local network will be shown on the Arduino serial monitor. When this IP address will be typed in, the browser will send a request to the webserver for connection. After the connection is established, the webserver will return a webpage which sends requests automatically to the webserver to update the sensors values in their placeholder every 2 seconds.

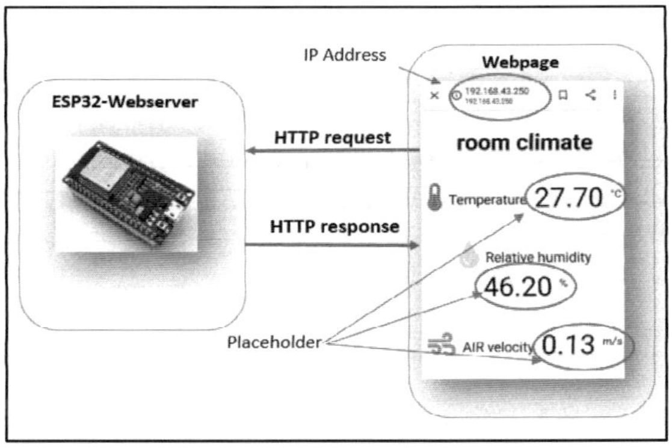

Figure 11. Communication between the webserver and the webpage

The temperature, humidity and air velocity reading from DHT22 and FS7 sensors are displayed in two forms on the webpage. In the first part of the webpage the actual data reading of the sensors are

IOP Conf. Series: Materials Science and Engineering **705** (2019) 012048 doi:10.1088/1757-899X/705/1/012048

displayed (Figure 11) and in the second part three charts are created which could display a maximum of 40 data points (Figures 12). In this work an asynchronous HTTP webserver has been created allowing updating the sensors data reading automatically without the need of refreshing the webpage.

7. Outlook

A prototype device is designed and built on a bread board to measure and control a comfort climate according to ISO 7730 collecting temperature, relative humidity and air velocity data. Being able to display the room climate on any internet capable device the first steps towards the goal of fully functional air condition control are made. In future work and cooperation with the Universiti Malaysia Perlis (UniMAP) the program and the sensor system will be further customized so that a connection to an existing air-conditioner can be made.

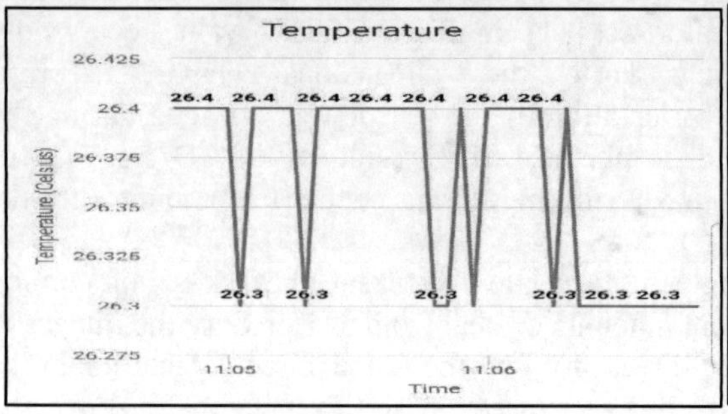

Figure 12a. Temperature chart

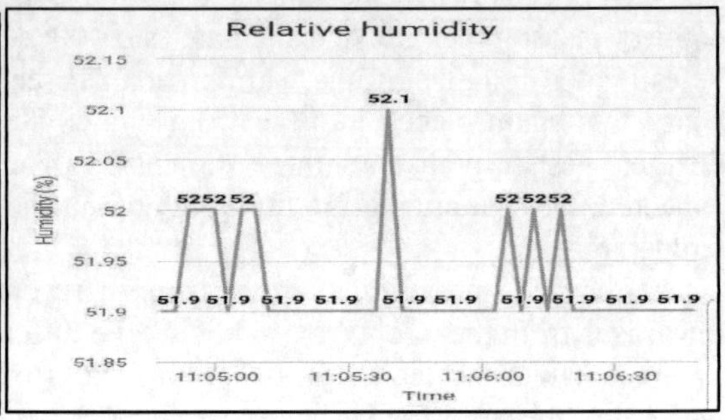

Figure 12b. Relative humidity chart

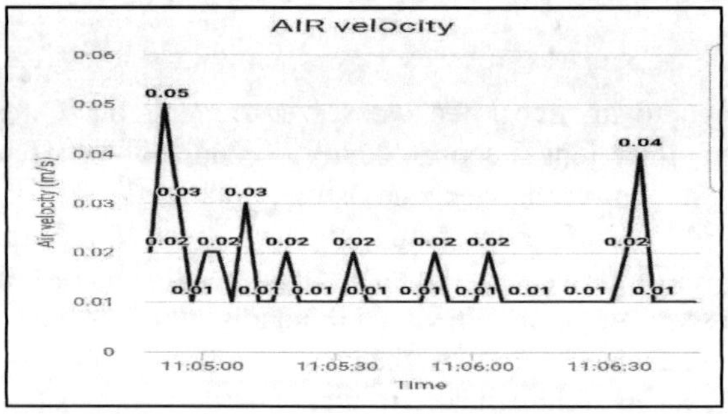

Figure 12c. Air velocity chart

Further via webserver one cannot only display data but also will be able to control the system over internet. At a final step the entire circuit will be redesigned to a smaller and more compact solution, all built together in an installation housing. At this point one will be able to connect multiple devices to the cloud to collect climate data from different locations in the rooms and perfectly operate an air conditioner (Figure 13). Multiple air conditioners can be controlled by only using a mobile or laptop. In the latest step of the project the cloud could be equipped with an artificial intelligence system to get one's thermal comfort self-controlled. The artificial intelligence would also make the most efficient use of the energy available. Trying to set new standards in efficiency of air conditioners the team aims to design a low budget version of the final device to spread it all over the world to take over responsibility in saving energy.

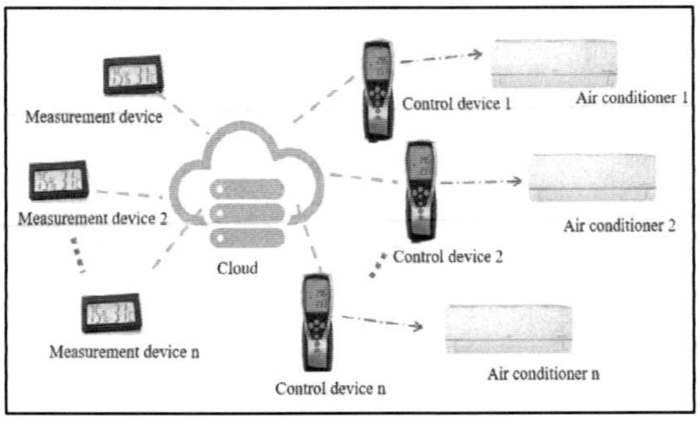

Figure 13. An IoT-Based Smart Co

References

[1] International Standard ISO 7730 2005 *Ergonomics of the thermal environment — Analytical determination and interpretation of thermal comfort using calculation of the PMV and PPD indices and local thermal comfort criterial*

[2] International Standard ISO 7726 1998 *Ergonomics of the thermal environment — Instruments for measuring physical quantities* vol 1998

[3] Ahmad A S, Shukor S A A, Kohlhof K, Jama Z A Z and Mohamad A 2015 Analyzing and evaluation a PMV-based model thermal comfort model in controlling air conditiong system *Simulation* **9** 47–50

[4] Innovative Sensor Technology 2008 *Application Note Thermal Mass Flow Sensor FS7* vol 06

[5] Kodali R K and Gorantla V S K 2018 Weather tracking system using MQTT and SQLite *Proc. 2017 3rd Int. Conf. Appl. Theor. Comput. Commun. Technol. iCATccT 2017* 205–8

[6] Liu T 2015 *Digital-output relative humidity & temperature sensor/module DHT22 (DHT22 also named as AM2302)* vol 22 (Aosong ELectronics Co.,ltd)

[7] Espressif 2019 *ESP32 Series Datasheet*

[8] Rajalakshmi A and Shahnasser H 2018 Internet of things using node-red and alexa *2017 17th Int. Symp. Commun. Inf. Technol. Isc. 2017* 1–4

[9] Wiegel C 2017 *Thermal Comfort in Sun Spaces: To what extend can energy collectors and seasonal energy storages provide thermal comfort in sun spaces?* (Architecture and the Built Environment. https://doi.org/10.7480/abe.2017.11)

[10] TSI Incorporated 2016 *Air Velocity Meter Model 9515-A*

[11] KGaA T S & C testo 400 Behaglichkeits-Set mit Stativ | Raumluftqualität | Gebäude & Bauen | Anwendungen | Testo SE & Co. KGaA

Discovering the Implementation Success Factors for IoT and Big Data Analytics in Transportation System

W N Hussein[1, 2, a], L M Kamarudin[3, 4, b], H N.Hussain[5, c], N A Ishak[1, d], A. Zakaria[3, e] and K J Jadaa[4, f]

[1]School of Human Development and Technocommunication, Universiti Malaysia Perlis (UNIMAP), 02600 Arau, Perlis, Malaysia
[2]AL-Zharaa Medical College, University of Basrah, Basrah, Iraq
[3]Centre of Excellence for Advanced Sensor Technology (CEASTech)
[4]School of Computer and Communication Engineering, Universiti Malaysia Perlis (UNIMAP), 02600 Arau, Perlis, Malaysia
[5]Faculty of Science, University of Basrah, Basrah, Iraq
[a]waleedn9@yahoo.com, [b]latifahmunirah@unimap.edu.my, [c]alhashimy@gmail.com
[d]maizatul@unimap.edu.my, [e]ammarzakaria@unimap.edu.my,
[f]khalid.jamal.jadaa@gmail.com

Abstract. The implementation of intelligent transportation systems, or what is otherwise referred to as IoT and big data analytics in transportation system are affected by several factors. Considering the overall impact of a definite and comprehensive plan that involves business, infrastructure, and organizational administration, improved customers satisfaction can better be achieved when all these antecedent factors are investigated, and their elements are determined. There is no recorded study on multidimensional implementation model where more than one of them is considered. This paper, therefore, stems from the general background of implementation success factors but aims to discover the generality of the definite plan and procedure that must be executed for a successful implementation of big data analytics and IoT-oriented transportation system. This paper employs a systematic literature review and an in-depth interview to address the obvious limitation. This paper also discussed the qualitative approach that was used to gather data on the successful implementation factors in the transportation system. The deliverable from this research will be the implementation success factors for the big data analytics and IoT-oriented transportation system.

Keywords: Internet of Things; Big Data Analytics; Success factors

1. Introduction

Big data, to start with, is characterized by its speed (Velocity), its heterogeneous nature; structured, unstructured and semi-structured (Varieties) and its size (Volume). It is the much-needed themes and pattern of events in data available in the form of unstructured and semi-structured, while the relational databases are exclusive custodians of structured data. This heterogeneity essentially describes big data, and its volumes start from terabytes (TB) to zettabytes (ZB) [1]. Many business corporations in their daily business activities rely on data for executing many of their business decisions [2, 3]. The analysis and transformation of data through varying techniques such as clustering and association rules, therefore, inform their business intelligent decisions [4]. One of the sources of data used in intelligent transportation systems is Internet of Things (IoT) [5, 6].

Internet of Things (IoT), on the other hand, is the ever-growing physical objects in the internet protocol (IP) address [7]. It addresses internet connectivity and ensures communication between internet-enabled objects, devices, and systems. These objects can be connected security systems, thermostats, electronic appliances, or alarm clocks [7]. Internet of things (IoT) is rapidly gaining attention among businesses because it produces sufficient data (through the communication of these objects) that can be used for analytics. These objects are estimated to be up to 12 billion devices and have aided how people can be more connected in 26 times [8]. Table 1 shows the characteristics of big data and the descriptive distribution of Internet of Things installation across domains, respectively.

Table 1. Internet of Things Units Installed Base by Category [9]

Internet of Things Units Installed Base By Category				
Category	2013	2014	2915	2020
Automotive	96.0	189.6	372.3	3,511.1
Consumer	1,842.1	2,244.5	2.874.9	13,172.5
Generic Business	395.2	479.4	623.9	5,158.6
Vertical Business	698.7	836.5	1,009.4	3,164.4
Grand Total	3,032.0	3,750.0	4,880.6	25,006.6

This research main objective is to discover and identify the implementation success factors for big data analytics and IoT-oriented transportation system. Therefore, this paper contributes to the empirical studies that investigated the antecedent factors to the implementation success of big data analytics and IoT-oriented technologies generally, and specifically for the transportation system. It focuses on multiple aspects of the implementation model and presents a holistic and comprehensive empirical investigation in what is called a unified implementation model. In doing this, the next section of this paper review previous studies of the implementation success factors of big data and IoT based transportation system. The third section presents the methodology used and the fourth section presents the results and discussion. The fifth section the last and concluding part of this paper.

2. Review of previous related studies

Studies on implementation models are studies that investigated or/and proposed models that prescribe and describe factors that must be considered in implementing IoT and big data in transportation systems. Business factor, which describes elements like value-driven customers service [10], and competitiveness [11], and adaptable business strategies [12], influences successful implementation of big data and IoT in the intelligent transportation system. Studies such as [10, 11, 13, 14], reported that the challenges of uncertain markets and business process reengineering are better solved by a tailored-business models.

The infrastructure models attend to both hardware and software technical needs in implementing IoT and big data analytics in the intelligent transportation system [13, 15]. The infrastructure implementation models address the challenges of heterogeneous technologies and interdependency. It also highlights how data mining algorithms and statistical models, such as naïve Bayes and logistic regression models, are used in detecting patterns and predicting events like bus arrivals and passengers preferences in the infrastructure model for IoT implementation in transportation sector [16].

The management and administration implementation models for big data analytics and IoT, from past related studies, prescribed customers privacy emphasis [17]. Proper management and strategic administration of the human and infrastructural resources of the big data analytics and IoT-oriented transportation organizations in the sector influence its implementation success [17, 18]. This signifies the management and administration models. This category presents the administrative factor's elements, such as team coordination and workplace culture [18], data adaptation and management policies to IoT implementation [17].

IOP Conf. Series: Materials Science and Engineering **705** (2019) 012049 doi:10.1088/1757-899X/705/1/012049

3. Methodology

This study, being a qualitative research method, employs elements of exploratory and case study analysis. The exploratory approach is to generate a guiding working framework for the study, especially in instances of little or no previous related studies. Therefore, the preliminary study conducted by this study is to identify the state and progress of studies on implementation models of big data analytics and IoT-oriented transportation system. The preliminary study employed a systematic literature review (SLR), a systematic literature review is used in identifying the elements of the implementation success factors for the big data analytics and IoT-oriented transportation system. At this stage, search terms, such as "business strategy/need for big data-based (and IoT) transportation systems", "technology for big data in transportation system", and "management for the implementation of big data and IoT-oriented transportation system", among others, were used. The databases used were from IEEE Xplore, Science Direct, Springer, Web of Science, and Google Scholar. These studies are published, either as conference proceedings or journal articles within the year 2008 and 2018. A total of 468 articles were identified, but only 124 met the inclusion criteria set for this study, and just 21 addressed the research focus of this study. Duplicate and irrelevant studies based on the theme and the search strings are excluded. In addition, a single case study as a qualitative method of research is used to identify the implementation success factors within Konsortium Bas Ekspres Semenanjung (KBES) public transportation bus in Malaysia, being a transportation service company that leverages on IoT and big data analytics. It serves as a single case for in-depth study. With this, the in-depth interview is used to identify the core elements of the implementation factors of big data and IoT-oriented transportation system, which are (a) business, (b) infrastructure/technology, and (c) administrative/managerial due to observed insufficiency of empirical studies on the subject matter for big data analytics and IoT-oriented transportation system. It will be used to validate the elements of the success factors that are identified from the SLR.

In this study, collected data from the in-depth interviews are the technical perspectives of the experts in the big data and IoT-oriented transportation sector. For qualitative approach, an open-ended question was used to collect detailed and unfiltered information about the opinions of the interviewees regarding the core details of the elements of the respective niche-specific implementation success factor of big data and IoT-oriented transportation system. The questions were designed by checking whether the factors and elements in table 2 are significant for the implementation of big data and IoT based transportation system. Investigating the phenomenon of big data analytics and IoT in transportation system in order to identify the implementations success factors, the interview respondents were probed about their perceptions on: (a) the elements of the business factor, (b) the elements of the infrastructure/technology factor, and (c) the elements of the administrative/managerial factor that influence the implementation success of a big data analytics and IoT-oriented transportation system. In-depth interview was held with engineering, subject matter experts in management, hardware, and software. In the same vein, interview and subsequent data were analyzed using NVivo software to identify themes and sub-themes from the transcription of the responses.

Table 2. Big data analytics and IoT success factors

Factor	Element
Business	Technology support
	Innovation and Competitiveness
Infrastructure	Software
	Hardware
Management and Administration	Organizational culture
	Data security and privacy governance

4. Results and Discussion

The preliminary review showed that there are insufficient recorded empirical studies on the implementation of big data and IoT-based transportation system. Previous studies have only investigated business [13, 19] infrastructure/technology [16, 20], and administration/management [21] as factors or/and models separately. Notably, none of these studies investigated multiple factors that contribute to the implementation success of big data analytics and IoT-oriented transportation system. Table 3 shows a summary of the factors and the corresponding elements and sub-themes from the findings of the systematic literature review.

Table 3. Findings of the systematic literature review

Factor	Element	Sub-themes
Business	Technology support	Electronic system adoption
		Data analytics
	Innovation and Competitiveness	Cost strategy
		Standardization and quality assurance
		Business model
Infrastructure	Software	Augmented/Virtual reality
	Hardware	IoT/cloud computing software architecture
		Autonomous vehicles
Management and Administration	Organizational culture	Employee reward system
	Data security and privacy governance	Customers' data privacy and security
		Cyber-attack counter measures and strategy
		Research and development initiative
	Data governance and legal framework	Data reusability
		Data interoperability

Therefore, the interview was done to identify the details of the implementation success factors and their corresponding elements for big data analytics and Internet of Thing-based (IoT) transportation system. The results showed that the business factor binds the shared technologies in the intelligent transportation system. The perspectives of the critical stakeholders in the transportation system revealed that (a) business-related actions for consumer value, (b) organizational tech delivering services, and (c) level and impact of business strategy are the emerged factors. Figure 1, therefore, shows the emerged factors.

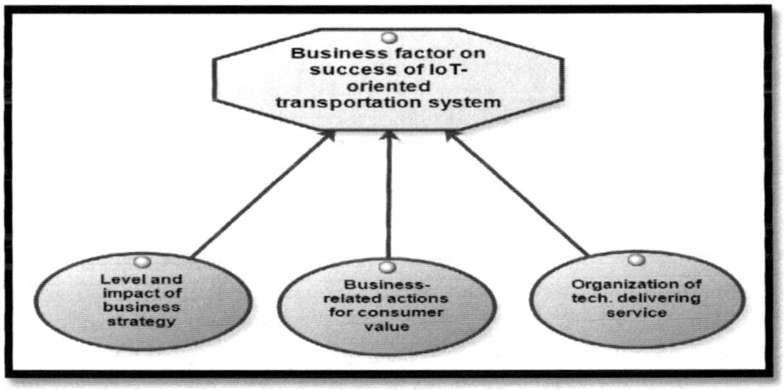

Figure 1. Perceptions of business factors on the success of IoT-oriented transportation system

The results also indicate that technology has been widely adopted because of the suitability, effectiveness, and swiftness of values it brings to the business. A nexus between technology and growth in IoT-oriented transportation system is supported by the emerged response that attests to the positive influence transport technology. Technical skills and expertise for technology service are guaranteed and when there is sufficiency of software for vehicles handling there is a high probability for success of IoT-oriented transportation system. Figure 2, therefore, shows the emerged factors.

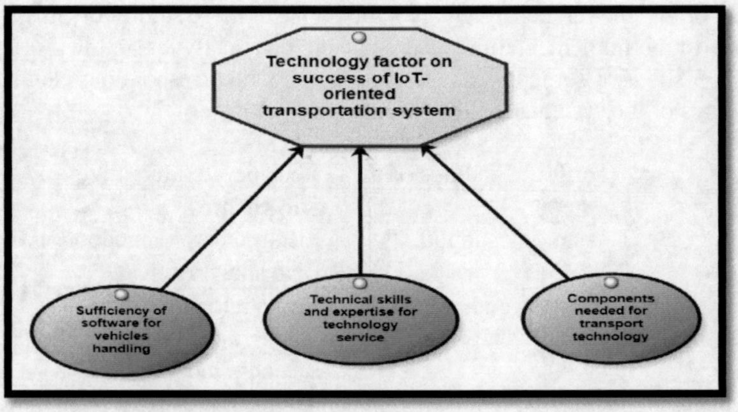

Figure 2. Technology factor on the success of IoT-oriented transportation system

The managerial factor, which addressed the management-related elements of the implementation success factor of Big Data analytics and IoT-oriented transportation system showed the organizational model and best training service as success elements. Figure 3 shows the managerial factor on the success of Big Data and IoT based transportation system analytics of the managerial factor theme.

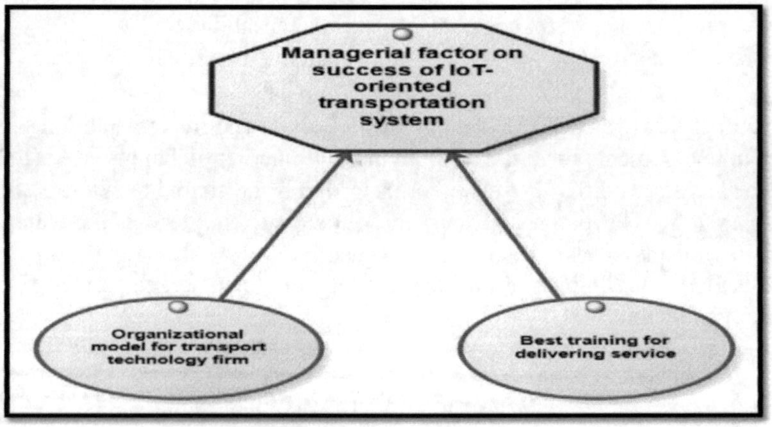

Figure 3. Managerial factor on the success of Big Data and IoT-oriented transportation system

5. Conclusion

Most of the previous studies separately investigated the implementation success of big data and IoT based transportation system. Aside from the significance of a comprehensive study that investigates these factors to identify their interplay in the implementation success of big data analytics and IoT-oriented transportation system, there are also self-acknowledged limitations of these studies. In this paper, the definite plan and procedure that must be executed for a successful implementation of big data analytics and IoT-oriented transportation system have been investigated through identifying the implementation success factors. The success factors highlight business, management, and technology factors. The business factor has three elements which are level and impact of business strategy, business-related actions for consumers' values, and organizational technology for service delivery. The technology factor has software sufficiency, technical skills and expertise, and transport technology as its elements. The managerial has an organizational model for a transport technology firm and the best training for delivering service.

References

[1] Zikopoulos P and C Eaton 2011 *Understanding big data: Analytics for enterprise class hadoop and streaming data.* (Electronic Materials vol 2) ed IBM (New York: McGraw-Hill Osborne Media) chapter 6 pp 160–173

[2] Kopáčková H and Škrobáčková M 2005 *Decision support systems or business intelligence: what can help in decision making?* Scientific papers of the University of Pardubice. Series D, Faculty of Economics and Administration pp 93–100

[3] Hussein W, Kamarudin L and Hussein H 2018 *The Prospect of Internet of Things and Big Data Analytics in Transportation System. Int. Conf. on Big data and Cloud Computing(Sarawak).* vol 1. (journal of physics: Conferrence series) pp 2–4

[4] Zadrozny P and Kodali R 2013 *Big data analytics using Splunk: Deriving operational intelligence from social media, machine data, existing data warehouses, and other real-time streaming sources.* (Electronic Materials vol 1) ed springer (New York: Apress) chapter 3 pp 9–31

[5] Waller M and Fawcett S 2013 D*ata science, predictive analytics, and big data: a revolution that will transform supply chain design and management.* Journal of Business Logistics. 34(2): p 77–84

[6] Grimmer J 2015 *We are all social scientists now: How big data, machine learning, and causal inference work together.* PS: Political Science & Politics. 48(1): p. 80-83

[7] Ezell S 2010 *Explaining international it application leadership: Intelligent transportation systems*.34:p12–16

[8] Atkinson R and Castro D 2008 *Digital quality of life: Understanding the personal and social benefits of the information technology revolution*.p17–20

[9] Gartner 2014 *Gartner says the Internet of Things will transform the data center.* http://www.gartner.com/newsroom/id/2684616 Acessed on 04 May 2019.

[10] Ju J, Kim M and Ahn J 2016 *Prototyping business models for iot service.* Procedia Computer Science 91: pp882–890

[11] Ogbuokiri B, Udanor C and Agu M 2015 I*mplementing bigdata analytics for small and medium enterprise (SME) regional growth.* Department of Computer science, University of Nigeria, Nsukka, Enugu state vol(17): PP 35–43

[12] Fleisch E, Weinberger M and Wortmann F 2015 *Business models and the internet of things, in Interoperability and Open-Source Solutions for the Internet of Things.* Springer. p. 6-10

[13] Drăgan I 2017 *Applying Self-Principles in Heterogeneous Cloud Environments, in Cloud Computing.* Springer. pp 255–274

[14] Dijkman R 2015 *Business models for the Internet of Things.* International Journal of Information Management 35(6): pp 672–678

[15] Guerrero-Ibanez, J, Zeadally S and Contreras J 2015 *Integration challenges of intelligent transportation systems with connected vehicle, cloud computing, and internet of things technologies.* IEEE Wireless Communications 22(6): pp 122–128

[16] He W, Yan G and Da L 2014 *Developing vehicular data cloud services in the IoT environment.* IEEE Transactions on Industrial Informatics. 10(2): pp 1587–1595

[17] Yao L, Sheng Q and Dustdar S 2015 *Web-based management of the internet of things.* IEEE Internet Computing 19(4): pp 60–67

[18] Algan T and Zeeshan A 2017 *Benefits of visual management in construction: cases from the transportation sector in England,* Construction Innovation, 17 (2) pp125–157.

[19] Fugl K 2015 *Business Model Framework Proposal for Internet of Things.*Copenhaguen Business School: Frederiksberg, Denmark, pp13– 35

[20] Liu W and Gao Z 2014 *Study on IoT based architecture of logistics service supply chain.* International Journal of Grid and Distributed Computing 7(1): pp 169–178

[21] Calado J, Osório M, and Prata R 2015 A*n Adaptive IoT Management Infrastructure for EcoTransport Networks.* in Working Conference on Virtual Enterprises Springer p 285

IOP Conf. Series: Materials Science and Engineering **705** (2019) 012050 doi:10.1088/1757-899X/705/1/012050

A tailor-made learning environment workspace for generation y graduates using product service system concept

S Akmal[1], R H Hambali[2], S Maidin[2], M R Shahmir[3], S A Abdul Shukor[4]

[1] Center for Advanced Computing Technology, Institut Pengurusan Teknologi dan Keusahawanan, Universiti Teknikal Malaysia Melaka, Hang Tuah Jaya, 76100 Durian Tunggal, Melaka, Malaysia.
[2]Centre of Smart System and Innovative Design, Fakulti Kejuruteraan Pembuatan, Universiti Teknikal Malaysia Melaka, Hang Tuah Jaya, 76100 Durian Tunggal, Melaka, Malaysia.
[3]Perodua Manufacturing Sdn. Bhd., Lot 1896 Sungei Choh, 48009 Rawang, Selangor Malaysia.
[4]School of Mechatronic Engineering, Universiti Malaysia Perlis, 02600 Perlis, Malaysia

suriatiakmal@utem.edu.my

Abstract. This paper presents a workspace design of a tailor-made learning environment for students using the product-service systems (PSS) concept. Existing research regarding workspace design focuses on the design of the interior of classrooms to ensure that the learning process proceeds with minimum stress and maximum effectiveness. None of the current research had discussed the workspace outside of the classroom. Therefore, the main objective of this study is to propose a learning environment workspace in which students can gather, study, relax, socialize, eat, and work during class gaps. For this purpose, a questionnaire was distributed to identify the behavior of the users to generate a list of requirements in the development of the learning workspace. This paper focused on students at the Engineering Faculty and the learning environment workspace was called the OneForAll Center. Out of 10 functions of the proposed OneForAll Center, only four functions had the highest priority: a space to use a laptop and internet connection, a television, a vending machine and a printing section. The OneForAll Center has been illustrated in a 3D model and shown to the students, in which most of them agreed with the proposed services and facilities such as a space to use the laptop and internet connection and a printing section with a printing service. The OneForAll Center has been identified as a solution to improve learning outside the classroom, specifically to the provision of services to students.

1. Introduction

Generally, 'workspace' is a term which refers to the working process taking place in a room. The term 'workspace design' is inspired by a research at MIT School of Architecture and Planning which proposed to overcome the architectural issues of the workspace and layout of buildings [1]. As a result of the growth of the telecommunication technology, today's workspace designers should consider the impact of their designs on the human population, specifically the generation Y born between 1981 to 1999. The majority of the generation Y are now pursuing their education at university level. This generation is identified as confident and technologically advanced, and they come with a sense of entitlement [2]. Therefore, today's university students are pushing academicians and workspace designers to address the learning spaces beyond classroom walls. They also require a flexible space that allows social interactions in which students and faculty members can meet informally to discuss the current news and study [3].

In recent years, there has been an increased interest in the design of a flexible and smart workspace to improve the diversity in learning activities. For example, Ellis and Holm [4] promoted the integration between a virtual and physical learning space. The research was conducted by assimilating services and support that line up to the results shaped by curriculum requirements, particularly technologically-mediated to offer personalized services and learning support to the students. The results of the study showed that the relationship between the virtual learning space and the physical learning space was not straightforward, but was rather complex and was best understood from the perspective of learning by the students.

Another study by Martinez-Maldonado et al. [5] looked into the approach of exploiting the collaboration between teachers and students by providing a tool called a multi-tabletop classroom system as a way to monitor classroom activities. The work was aimed towards helping students to work in a group in a classroom. For this purpose, a software system was developed, and it was monitored by the teacher who conducted the class. As a result, the multi-tabletop classroom was a method for a teacher to monitor collaborative learning activities. Also, the workspace design focused on solving communication issues to support information exchange during learning activities.

Research to date tended to focus on workspace design for improving learning activities, but none of the existing approaches provided the method to support the customer's activities related to the use of the physical workspace. In one of a public university, the gen-Y engineering students are facing of workspace that has been designed without supporting their activities throughout a day. New age university should provide facilities where students not only gain knowledge but also develop soft skills to succeed both academically and professionally. This indicates the importance and necessity of developing a workspace design that support both academic and professional activities of the gen-Y engineering students.

Therefore, the aim of this paper is to use the product-service systems (PSS) concept to support innovative ways to integrate the physical space and offer services that are based on student activities that have high use of information and communication technology. PSS is an innovation strategy that produces an added value through the integration of products and services by adopting customer-centered activities [6], [7], [8]. The systems focus on fulfilling customers' needs than on product purchases i.e. "I need a printer to print my paper". Value creation during the product and service innovation can be achieved by knowing how the customer recognizes value [7]. Therefore, the role of a designer is challenged to move into a paradigm shift to synthesize different concurrent perspectives such as culture and society needs with respect to certain technological needs.

2. Literature review

2.1. PSS structure

1.1 PSS Structure

According to Morelli [9], the design of a PSS can be outlined by the following:

- The design of a PSS is a suggestion of a new combination of technological artefacts on the basis of functional factors designated by the designer.
- The contact between the designer of the service and the consumer or user is not refereed by the industrial artefact.

To create a PSS, the designer would have little control over the product life cycle because the product is usually a combination of an existing product. The product life cycle element which the designer can control is the stage of use [10]. Another challenge for the designer is the integration of intangible value to promote a long-lasting relationship with the customer. The word 'intangible' based on Webster's New

World Dictionary is defined as something that cannot be touched, incorporeal, impalpable; that represents value but has no intrinsic value or material being; that cannot be easily defined, formulated or grasped [11].

There are seven main design challenges regarding value proposition. The first one is defining the value proposition. This is because of the challenge in obtaining a clear definition of value proposition for the users. Most companies tend to create long a relation with users, thus making excellent value proposition becomes a key element to make the relation last longer. The second challenge is maintaining the value proposition over time. This means maintaining the interaction with the user. The third challenge is creating meaningful high-quality interactions which is by understanding the component of the human. The fourth is creating coherence in a smart PSS. Coherence is important because it has multiple touch points in the system. The fifth challenge is stakeholder management.

This is because multiple stakeholders will be involved, thus management of the stakeholder is an important factor. The sixth challenge is clear communication tools to create a shared vision. The last challenge is the selection of the means and tools for the design process. This is because designing a smart PSS requires good tools and methods to support the design [12].

The development of a PSS needs to consider the procedure and tools used to create the product. PSS can be defined as systems which make up service units and physical objects. Therefore, the design of PSS requires an approach that can link physical objects and services. The physical objects are functional entities that convey the elementary functions of the system, while the service units are entities which are mainly technical that determine the smooth functioning of the whole system [13]. One of the approach is User Centred Design (UCD), which is considered as the key to the usefulness and usability of the product as an effective approach to overcome the limitations of an out-dated system centred design [14].

2.2. User-centered design as a catalyst for innovation

User-centered design (UCD) can be described as a design process in which the user has an impact on the design process for building better and more practical space [15]. UCD is a multiple design approach based on the involvement of users to increase consumer understanding and service requirements and evaluation.

The product engineer should consider the consumer as an individual, possessing individual aptitudes, experiences, and characteristics while taking note the skills and limitations of all potential consumers [16]. Products designed in this way will be suitable to be used by people with the broadest possible range of skills, within the broadest range of conditions, reaching most potential end users. Therefore, UCD is an approach that designs a system in such a way that aims at consumers' attitudes and behaviours as they relate to the tasks that the system is being developed to support. As a result, UCD offers a product that is more efficient, satisfying, and user-friendly for consumers, which may increase sales and customer loyalty.

3. Designing learning environment method

The method that is used to design the OneForAll Center is based on the integration of product and service as discussed in the PSS concept proposed by Morelli [10]. However, the method to identify customer needs accurately and extensively is not mentioned in his method. Therefore, this paper extended his work by integrating the statistical analysis for designing a learning environment workspace for Gen-Y (figure 1). Survey and interview were used to elicit the user requirements in this study.

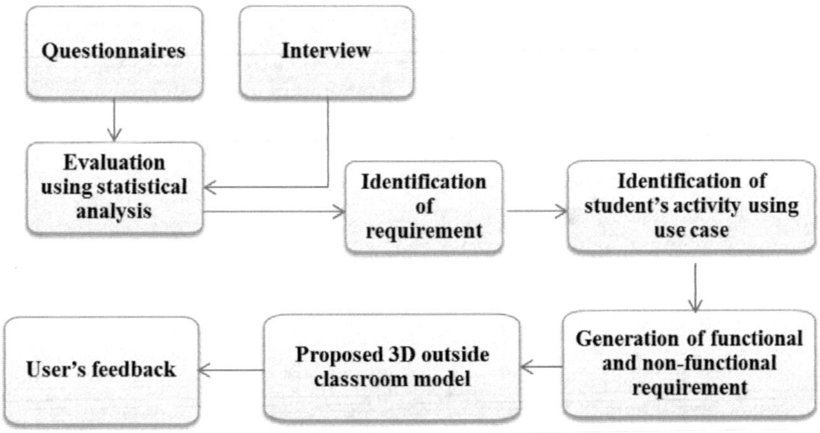

Figure 1. The proposed approach.

The focus of this case study was limited to students at the Engineering Faculty at a public university. The students were provided with good facilities on teaching and learning. However, the learning environment outside the classroom had some deficiencies. Therefore, the OneForAll Center referred to the learning environment outside the classroom. The design of the OneForAll Center was not based towards the learning process, but on the environment of the workspace design of the OneForAll Center which will then affect the learning process.

4. Results and Discussion

The learning environment workspace is called the OneForAll Center. The proposed workspace is a PSS-based design that aims to provide a comfortable area by fulfilling the customer's intangible value in a physical workspace design. The customer's intangible value in this case study referred to the basis of this work which suggested that outdoor activity increases motivation and self-esteem [17].

4.1. Validity of questionnaire

The participants included diploma students, undergraduate students and master students of the Manufacturing Engineering Faculty who attended classes or discussions at UTeM. A total of 389 participants were involved in this survey as shown in Table 1. The Cronbach alpha value for the overall questionnaire was 0.841.

4.2. Demographic Result

The first part of the questionnaire aimed to identify and obtain a picture of the background of the potential users. The questions were formulated to picture the students' lifestyle and their expectations of the OCS development. For this purpose, six questions on demographics were asked including gender, the year of study, whether the student owns a transportation, their current place to stay during class gaps and his/her activity during class gaps.

Table 1 shows the places that students prefer to stay at during a class gap. Most of the students preferred the cafe and the faculty's lobby as the place to stay during a class gap. Based on the feedback from 389 respondents, a significant correlation was found between the year of study and the place to stay during the year of study. In this paper, the authors used the correlation that was proposed by Kotrlik et al. [18]. The more surprising correlation was between transportation and gender, year of study, and the place to stay during the year of study. The correlation was interesting because it showed the mobility of the student on campus. As the students change the year of study, and due to the University's policy, they must stay in a rental house nearby campus. Therefore, there is a need of a motorcycle or car for mobility.

Table 1. The respondents

Respondents	Male	Female	Place to stay during class gap (n=389)					
			Library	Hostel	Rented house	Café and Faculty lobby	Prayer room	General computer room
Master's degree	26	30	17	-	14	9	4	1
First year degree	34	26	17	18	-	23	2	-
Second year degree	31	29	16	-	24	13	7	-
Third year degree	26	35	13	-	6	26	9	7
Fourth year degree	24	39	16	-	18	23	5	1
First year diploma	27	23	5	15	-	24	6	-
Second year diploma	26	24	6	-	10	21	7	6

4.3. Characteristics of learning environment workspace

The second and third parts of the questionnaire aimed to elicit the characteristics of the proposed learning environment workspace. The characterization of the OneForAll Center was based on the requirements and the identified facilities. Table 2 shows the 12 statements on the feedback on requirements and facilities demanded. Due to limited space only a few statements are shown.

Table 2. Statement for requirement identification

Notation	Functional Components
B1	I am certain that students need a place to stay for a while waiting for the next class session.
B2	I really would like to have a place that I can use laptops and connect to the internet
B3	It is good to have a place for discussions.
B9	Is it good to have washroom in the OneForAll Center area
B10	I find it helpful if there is a gym in the OneForAll Center
B11	I would like if a café is placed inside the OneForAll Center
B12	It is good to have a game corners which provide board and card games for the students

The top half of Figure 2 presents the intercorrelation among the six measures of demographics. Interestingly, there is a significant positive correlation (0.829, $p<0.01$) between the year of study and place to stay during the year of study, either at a hostel or rental house. Most of the students stayed outside the campus as hostels accommodated first-year students only. From this data, it can be seen that the transportation used in campus also has a correlation with gender (0.261, $p<0.01$), year of study (-0.227, $p<0.01$) and place to stay during the year of study (-0.193, $p<0.01$). Although the correlation is low, this shows that many factors influence a student to have his/her own transportation. Most of the students who did not have their own transportation liked to stay in campus due to the hot weather. Another factor was that they may not like to use the scheduled buses provided by the university to move between the rented house and campus or around the campus. Also, a carpool concept seemed popular among female students to cater to their student activities as well as reduce their expenses.

The bottom half of the table shows the intercorrelation among the 12 measures of customers' requirements. It is apparent from this table that requirement on printing services (B5) has a high correlation with the place to stay while waiting for the next class session (B1) (0.519, p<0.01), as well as a place to hold discussions (B3) (0.506, p<0.01).

	Gender	Year	Stay	Transportation	Place	Activity	B1	B2	B3	B4	B5	B6	B7	B8	B9	B10	B11	B12
Gender	1			.261**		-.138**			.100*									
Year		1	.829**	-.227**													.105*	.121*
Stay			1	-.193**					.135**		.110*	.133**	.100*	.108*	0.091	.144**	.109*	
Transportation				1		-.115*												
Place					1													
Activity						1	.110*	.121*						.145**				.146**
B1							1	.617**	.512**	.385**	.519**	.302**	.311**	.322**	.455**	.278**	.380**	.202**
B2								1	.464**	.340**	.406**	.285**	.328**	.406**	.444**	.291**	.329**	.275**
B3									1	.494**	.506**	.337**	.390**	.369**	.438**	.298**	.369**	.262**
B4										1	.521**	.409**	.347**	.281**	.347**	.307**	.496**	.327**
B5											1	.397**	.340**	.337**	.385**	.251**	.409**	.244**
B6												1	.378**	.216**	.184**	.328**	.345**	.310**
B7													1	.425**	.441**	.339**	.354**	.233**
B8														1	.545**	.328**	.319**	.272**
B9															1	.376**	.380**	.212**
B10																1	.436**	.263**
B11																	1	.381**
B12																		1

**. Correlation is significant at the 0.01 level (2-tailed).
*. Correlation is significant at the 0.05 level (2-tailed).

Figure 2. The correlation between demography and functional requirements.

The more surprising correlation is for the OneForAll Center to have a place for discussion (B3) and a vending machine (B4) (0.494, p<0.01). The results showed the importance of a vending machine in the OneForAll Center. Similarly, the requirement of the vending machine (B4) has a positive relation with the requirement on printing services (B5) which is 0.521, p<0.01. Nowadays, many companies equip vending machines with cookies or chocolates. Having a short break or some refreshment during discussions or when printing assignments shows the need of a vending machine at the OneForAll Center. Also, there is a positive correlation of 0.545, p<0.01 between students requiring a green environment (B8) and a washroom.

4.4. The facilities and its services

A mean perception score was considered for assessing the functional requirement of the facilities and its services. The functional requirement refers to the area that was required in the OneForAll Center. The respondents were asked to determine the activities they would do if the OneForAll Center exists. Also, a rough design of the OneForAll Center was shown to the respondents during the interview to gain their trust to use the offered services in the future. The evaluation of each functional requirement varied from 1 (the functional requirement is not very important) to 5 (the functional requirement is very important). The results shown in figure 3.

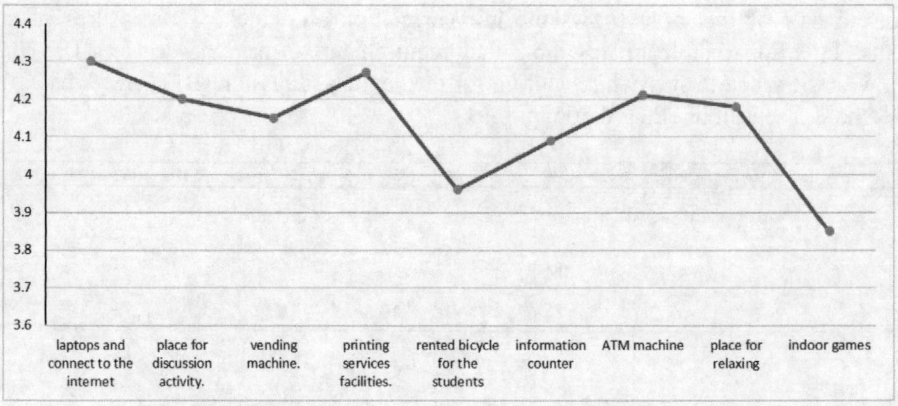

Figure 3. The mean perception score for facilities requirement

The functional requirement area with the highest mean perception scores in the OneForAll Center was the area that has a place for the student to use their laptop and be connected to the network. This finding was consistent with Sudiran [19], who found the importance of the internet among students at the University of Muhammadiyah Malang with a mean perception score of 4.61. The mean perception scores showed that students were heavily dependent on the internet to find reading materials to complete their assignments and communicate with friends on social media. The internet had replaced the textbook in the library and became a way for students to solve problems for their assignments.

To ensure the students would frequently use the functional requirement, it is important to offer a suitable service that can be provided to them. Thus, the next important step is the identification of services to each area that was identified in previous steps through desktop study and the results of the interview. As a result, table 3 shows the tabulated data which consists of the functional components, attributes, and importance of a tentative architecture of the OneForAll Center.

Table 3. The specification and related services of OneForAll Center.

Functional Requirement	Priority	Attributes	Services
Area for discussion	High	Wide space Can be used for 6 - 8 students Comfortable	Equipped with internet access Well-maintenance of facilities Provides rental projector for easy discussion Provides a way to easy rental projector
Area to use laptop/internet	High	Equipped with internet access Equipped with electrical sockets Comfortable Soundproof for privacy	Well-maintenance of facilities
Area for printing and photocopy	High	Easy to access Easy to use Equipped at least 2 photocopy machines Equipped with 2-4 printers.	Provides other services such as binding, paper, poster printing, and colour printing. Well-maintenance of facilities Provides a way to easy payment
Area for vending machine	High	Easy to access Easy to use	Well-maintenance of facilities Provides a way to easy payment Provides coupon to allow refill water

Area to play indoor games	Medium	Variety of indoor games Can be used for 2-5 students	Provide easy rental indoor games
Area for rented bicycle	Low	Easy to access Can be used for 2-5 students	Provide easy rental bicycle Well-maintenance of facilities
Area which provides information/information center	Medium	Easy to access past examination questions and research paper related to engineering.	Helpdesk service Provides rental service Provide online service for easy reading at home

5. Conclusions

This paper introduced a new method to develop a learning environment that promoted out-of-class activities based on the preferences of students. This study was innovative by integrating the PSS concept in producing a series of services in relation to the functional requirement of the proposed area. This study proposed a place whereby the user can have access to a variety of services in one place. However, this study was unable to analyse the cost that would be required to develop the real building. The center was focused on services built in a single location which would lead to greater overall satisfaction. For a designer, PSS facilitates the users to interpret technological and cultural frames embedded in existing products, services and infrastructures, and to search for convergences between the cultural frames expressed by all the actors (people and technologies) involved in the development of the project.

Another advantage of the OneForAll Center is that indirect learning may happen inside the proposed learning environment which eventually may strengthen the social connection outside the classroom. This may happen as students who come and stay inside the proposed OneForAll Center during their class gaps may eventually start to socialize with each other inside the proposed learning environment.

Other than the social aspect, the OneForAll Center is also a sustainable environment, such as by reducing the need for many buildings. This is because the OneForAll Center is a building which has many facilities inside, thus making it much more efficient. With the OneForAll Center, it could reduce the amount of building material used. This will also help the environment in the long run because there would be less construction needed. Other than that, one of the OneForAll Center services and facilities introduces a bicycle renting service. This could establish a green environment lifestyle. With bicycles, the amount of toxic gas emitted by vehicles such as cars or motorcycles could be lessened. This would end up in creating a clean atmosphere around the campus in the future.

References

[1] Seim R and Broberg O 2010 Participatory workspace design: A new approach for ergonomists?. *International Journal of Industrial Ergonomics,* **40** (1) 25

[2] Erickson T 2008 *Plugged in: The Generation Y guide to thriving at work* (Boston: Harvard Business Press)

[3] Cheryan S, Ziegler S A, Plaut, V C, and Meltzoff A N 2014 Designing Classrooms to Maximize Student Achievement. *Policy Insights from the Behavioral and Brain Sciences* **1** (1) 4

[4] Ellis R and Holm J 2013 Towards a smart campus: Integrating virtual and physical learning space. *Association for Tertiary Education Management and Tertiary Education Facilities Managers' Association,* **1** 35

[5] Martinez-Maldonado R, Kay J, Yacef K, Edbauer M T and Dimitriadis Y 2012 *ACM Conference on Interactive Tabletops and Surfaces* **1** 119

[6] Baines T S, Lightfoot H W, Evans S, Neely A, Greenough R, Peppard J, Roy R, Shehab E, Braganza A, Tiwari A, Alcock J R, Angus J P, Bastl M, Cousens A, Irving P, Johnson M, Kingston J, Lockett H, Martinez V, Michele P, Tranfield D, Walton I M and Wilson H 2007 State-of-the-art in product-service systems *Journal of Engineering Manufacture* **221** (10) 1543

[7] Marques P, Cunha P F, Valente F and Leitao A 2013 A methodology for product-service systems development. *Procedia CIRP* **7** 371

[8] Muto K, Kimita K and Shimomura Y 2015 A Guideline for Product-Service-Systems Design

Process *Procedia CIRP* **30** 60-65.

[9] Morelli N 2006 Developing new PSS, methodologies and operational tools *Journal of Cleaner Production* **14** (17) 1495-501
[10] Morelli N 2003 Product-service systems, a perspective shift for designers: A case study: the design of a telecentre *Design Studies* **24** (1) 73-99
[11] Panarotto M, Ericson Å and Larsson T 2012 Intangibles in design of PSS value propositions *The Philosopher's Stone for Sustainability* ed Y Shimomura and K Kimita (Berlin-Heidelberg: Springer) pp 85-90
[12] Diehl J C and Christiaans H H C M 2015 *The International Design Congress*, vol 1 (Gwangju, South Korea) p 476-483
[13] Maussang N, Zwolinski P and Brissaud D 2009 Product-service system design methodology: from the PSS architecture design to the products specifications *Journal of Engineering Design* **20** (4) 349-366
[14] Mao J, Vredenburg K, Smith P and Carey T 2005) The state of user-centered design practice. *Communications of the ACM* **48** (3) 105-109
[15] McKenna S, Staheli D, and Meyer M 2015 *IEEE Symposium visualiztion for cyber security*, 1-8.
[16] Wilkinson R C and De Angeli A 2014 Applying user centred and participatory design approaches to commercial product development *Design Studies* **35** (6) 614-631
[17] Kondric M, Sindik J, Furjan-Mandic G and Schiefler B 2010 Participation Motivation and Student's Physical Activity among Sport Students in Three Countries *Journal of Sports Science & Medicine* **12** (1) 10-18
[18] Kotrlik J W, Williams H A and Jabor M K 2011 Reporting and Interpreting Effect Size in Quantitative Agricultural Education Research *Journal of Agricultural Education* **52** (1) 132-142.
[19] Sudiran 2015 Students' Perception Towards the Use of Internet as Learning Media to Promote Reading Comprehension Skill *Sino-US English Teaching* **12** (9) 684-692.

IOP Conf. Series: Materials Science and Engineering **705** (2019) 012051 doi:10.1088/1757-899X/705/1/012051

Development of Dynamometer for Cutting Force Measurement in Turning Operation

K. Umurani and R.A. Siregar

University of Muhammadiyah Sumatera Utara, Jalan Kapten Muchtar Basri No. 3 Medan, Indonesia

khairulumurani@umsu.ac.id

Abstract. This paper presents the development of dynamometer to measure cutting force in turning operation in particular the tangential force (F_t). The dynamometer utilises a straight bar load cell that connected to holder tool to measure at least one axis of force. The designed dynamometer is capable of computing the forces acting on the workpiece in turning operation using any data acquisition system. Two materials of low elastic modulus, Polyoxymethylene (POM) and Polytetrafluoroethylene (PTFE), were employed as workpieces to test the performance of dynamometer. The performance of the dynamometer to measure one axis of cutting forces is considered satisfactory mainly because of the simplicity and low cost are the main requirements in design and construction.

1. Introduction

Machining is one of the most important manufacturing processes. The industrial revolution and the growth of the manufacturing-based economies of the world can be traced largely to the development of the various machining operations [1]. Industrial products that produce components from the machinery process always emphasize quality to win the competition. For many years the innovation process was needed to produce better quality industrial products to meet the requirements and specifications arising from the demands of the times. The common machinery processes are turning, grinding, milling and drilling in which a sharp cutting tool is used to remove metal from the work piece to produce the desired shape and features. In the turning operation, understanding the cutting force mechanism is very influential on the final result. The cutting forces generated during turning operation have a direct effect on the stability of the system, tool wear, quality of machined surface and accuracy of the workpiece. These forces can be measured theoretically, numerically or experimentally. In the experimental work, the cutting forces can be measured by using a dynamometer that attach to the toolpost. Although it is commercially available, dynamometer is still attractive to be designed to suit their individual needs. The most obvious background is certain price and quality [2-6]. This study was conducted to enrich the design of the current dynamometer to measure low modulus of elasticity materials. Design requirements are dynamometer that can be used to measure at least one directional force simultaneously for low elasticity materials at affordable prices.

IOP Conf. Series: Materials Science and Engineering **705** (2019) 012051 doi:10.1088/1757-899X/705/1/012051

2. Design and Testing of Dynamometer

The design of dynamometer must be simple and can be mounted on a conventional lathe machine. For this objective, the final design of frame that mounted the load cell is shown in Fig.1(a) and (b), while Fig. 1(c) showed the dynamometer apparatus. Inside the frame, a straight bar load cell is used that connected to tool holder. The choice of straight bar load cell, limits only 2 axes can be evaluated simultaneously. For current study, conventional lathe machine and one axis of load cell are employed to measure the cutting force (F_t) as shown in Fig. 2. If three axis are used, three forces can be recorded simultaneously, e.g. F_t = tangential force (cutting), F_r = radial force (thrust) F_f = feed force (axial), however in this case only two forces are sufficient. The load cell is connected to Arduino micro controller board, displayed in LCD board and analysed in MS. Excel (Fig. 3). Static calibration for cutting force was performed in a cantilever beam condition with loads ranging from 0.05 to 0.5 N with linear regression models [6].

2.1. Materials

Two materials with low elastic modulus, Polyoxymethylene (POM) and Polytetrafluoroethylene (PTFE), were employed as workpieces to test the performance of dynamometer. The photograph of both materials can be seen in Fig. 4. The workpiece have dimension of 25 mm diameter and 200 mm long which will be evaluated with a carbide insert tool of DCMT. Mechanical properties of both materials are shown in Table 1 [7].

2.2. Testing method

The tangential force (F_t) is evaluated using one axis of load cell. The first test was performed by changing the spindle speeds of 100, 110, 155 and 190 rpm while the depth of cut was maintained at 0.2 mm. For second test, the depth of cut varies from 0.1, 0.2, 0.3, 0.4 and 0.5 mm while the spindle speed is maintained at 155 rpm. Both tests were carried out in dry conditions. The outer surfaces of the workpiece were machined to minimize the roll forming effects such as the outer surface hardness.

Table 1. Mechanical properties of workpiece.

Work piece	Density (kg/m^3)	Yield Strength (MPa)	Modulus of Elasticity (MPa)
Polyoxymethylene (POM)	1410	65	2760
Polytetrafluoroethylene (PTFE)	2200	23	500

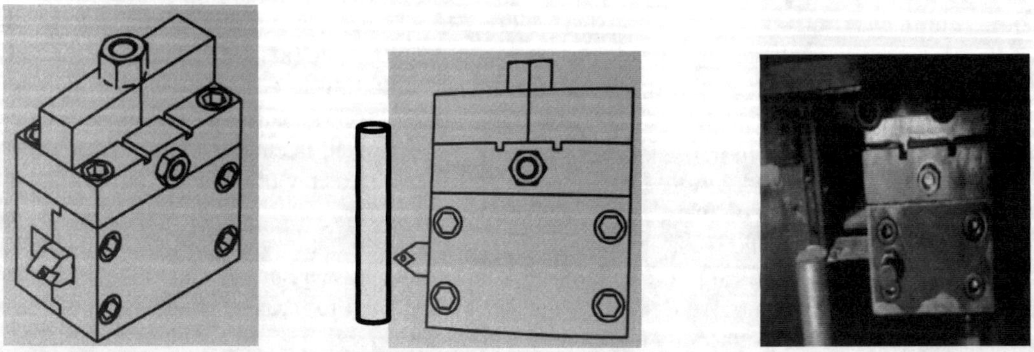

(a) Assembly of dynamometer (b) Top view of dynamometer apparatus (c) Dynamometer apparatus

Figure 1. Design of dynamometer.

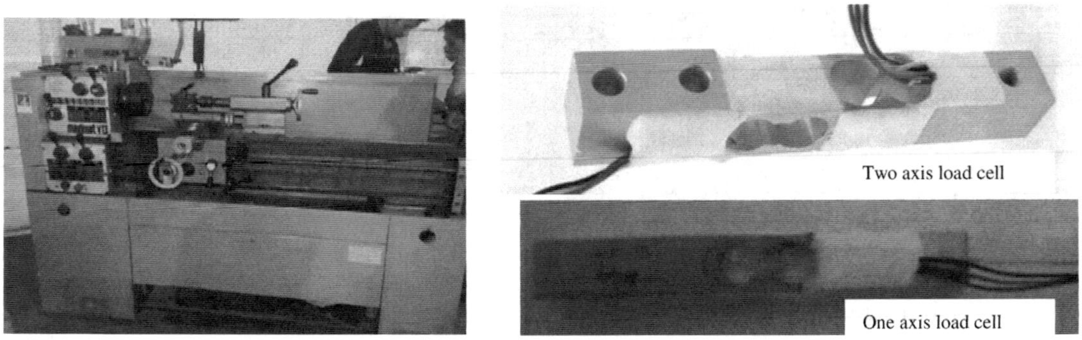

(a) Lathe machine (b) Load cells

Figure 2. Ordinary lathe machine used in this study.

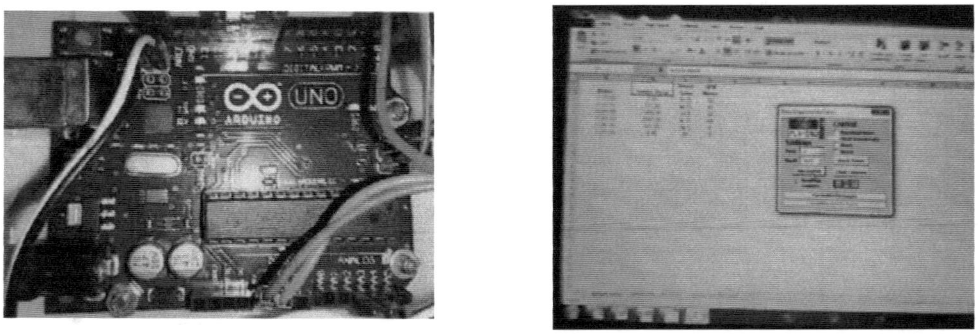

(a) Arduino microcontroller (b) Data analysing in MS. Excel

Figure 3. Materials and force measurement parts.

3. Results and Discussion

Performance current dynamometer to measure the force in one direction that is tangential force can be considered satisfactory especially when simplicity and low cost are key requirements in the design and construction. It can be seen in Fig. 5 that the tangential force (F_t) is not drastically affected by spindle speed, despite a slight increasing when the spindle speed is elevated. For this test, the depth of cut is maintained at 0.2 mm. The response of tangential force for POM material is slightly high compared to PTFE. Both responses also observed when the depth of cut increase gradually as in Fig. 6. POM, which has higher elastic modulus compared to PTFE, shows higher response. It can be noticed that for a depth of cut of 0.5 mm, the response of the cutting force showed a significant different between POM and PTFE materials. This behaviour can be explained by the fact that with the increase in the depth of cut, the tool indents more into the workpiece and the width of cut becomes reasonably high. This effect increases the chip section and material response in stiffness when it contacts with the tool. The correlation between the increases of cutting force responds to material properties, by the current design, considered well explained.

IOP Conf. Series: Materials Science and Engineering **705** (2019) 012051 doi:10.1088/1757-899X/705/1/012051

(a) Polyoxymethylene (POM)

(b) Polytetrafluoroethylene (PTFE)

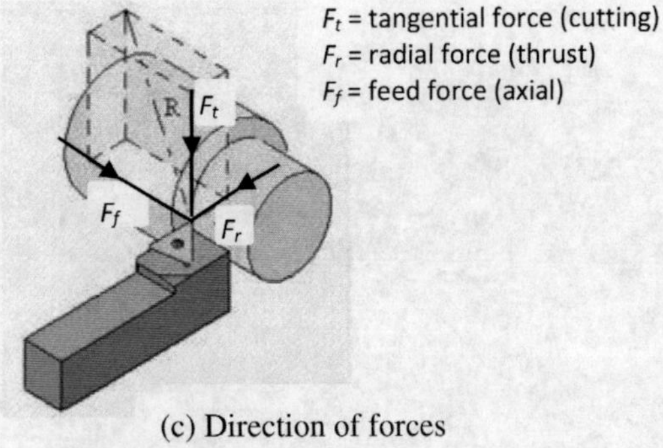

(c) Direction of forces

Figure 4. Materials and forces in turning operation.

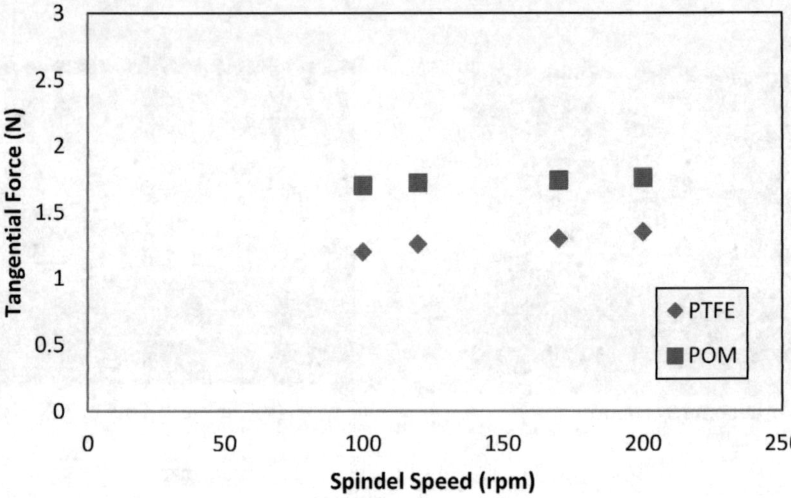

Figure 5. Spindle speed versus tangential force

4. Conclusion

In this paper, lathe tool dynamometer for turning operation has been developed. The dynamometer can only be used in two axes of force elements at the same time during turning. The performance of the dynamometer developed was considered satisfactory, especially when simplicity and low cost are taken into account as requirements for the design and construction of the device. The dynamometer was built at the total cost of USD 120 which can be concluded that it was a cost effective manufacture.

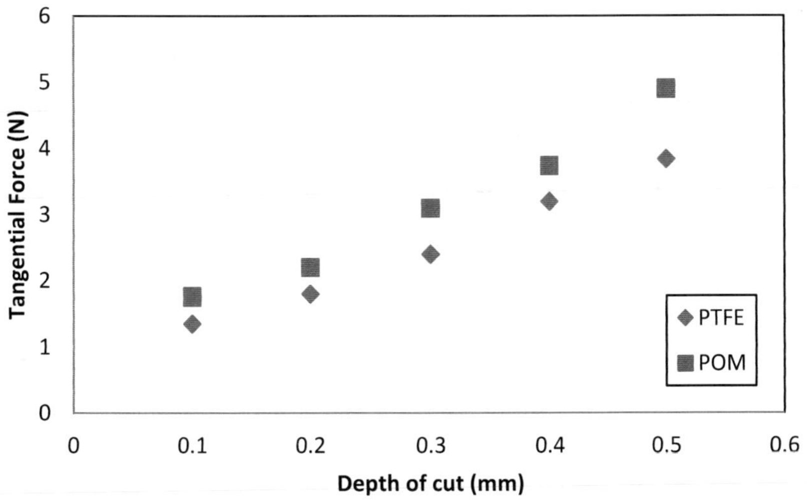

Figure 6. Effect of cut depth to tangential force.

References

[1] Groover, M. 2011 *Principles of modern manufacturing*, John Wiley & Sons Inc

[2] Tulio Hallak Panzera, Paulo Roberto Souza, Juan Carlos Campos Rubio, Alexandre Mendes Abrão, Tanius Rodrigues Mansur 2011 *Int. J. Adv Manuf Technol*

[3] Jain R., Rathore J.K., Gorana V.K. 2016 Design, Development and Testing of a Three Component Lathe Tool Dynamometer Using Resistance Strain Gauges. In: Mandal D.K., Syan C.S. (eds) CAD/CAM, Robotics and Factories of the Future. Lecture Notes in Mechanical Engineering. Springer, New Delhi p 13-21

[4] Muhammad Imran Hanif, Muhammad Aamir, Riaz Muhammad, Naseer Ahmed, Shahid Maqsood 2016 *IJISET*, **3** 9

[5] Yashkumar A. Lad and Ankitkumar A. Patel 2019 *IJSTE* **5**, 8

[6] Hari Dermawan, 2016, *Rancang bangun instrumen pada mata pahat untuk mengukur gaya, kecepatan dan temperature specimen pada mesin bubut* (Unpublished thesis) University of Muhammadiyah Sumatera Utara

[7] Hadi Sucipto, 2016, *Pengaruh bahan mata pahat terhadap gaya potong pada variasi putaran dan specimen pembubutan* (Unpublished thesis) University of Muhammadiyah Sumatera Utara

IOP Conf. Series: Materials Science and Engineering **705** (2019) 012052 doi:10.1088/1757-899X/705/1/012052

Case study on the soil physical parameters disparity and NPK concentrations in regions found in and around Pachapalayam, Coimbatore, Tamil Nadu.

#1Chris Sheba M., #2R. Devaki., #3R.N.Uma

1,2Assistant Professor, 3Professor and Head, #Civil Engineering Department, Sri Ramakrishna Institute of Technology, Coimbatore – 641 010, Tamil Nadu, India.

1chrissheba@gmail.com, 2devakiramasamycivil@gmail.com

Abstract. Deficient soil fertility is a major constraint in crop productivity. Soil characteristics such as pH, bulk density, electrical conductivity, moisture content, etc. play a major role in the health and yield of crops. Major nutrients such as Nitrogen (N), Phosphorus (P) and Potassium (K), are essential for the growth and yield of the crops as well. On account of the importance of soil physical properties and NPK concentration on the soil dynamics and plant development respectively, the study area was spatial distributed and geo-statistically analyzed to produce contour maps that can be used by decision makers in farm management for increased crop productivity.

1. Introduction

Soil is characterized by its physical properties which play a vital role in determining its suitability for environmental, agricultural and engineering uses. Penetration of roots; movement, retention of water and nutrients with its availability to plants; and flow of air and heat are directly related to the physical properties of the soil. Physical properties influence the chemical and biological properties as well. According to World Bank collection of development indicators, it was reported that India uses 60.45% of land area [1] for agriculture and its related activities. However, agriculture in India is highly conditioned by poor fertility of the soil.

Soil fertility refers to the soil's ability to contribute essential plant nutrients and water in sufficient amounts for sustained plant growth and reproduction. In nature, the properties of soil are highly variable. Soils' characteristics and parameters can significantly differentiate the character of the country's soil structure and its utilization. Variability in soil physical properties is inherent in nature due to the geologic soil forming factors and the continuous changes on the earth's surface. It may be induced by agricultural management practices as well. Also, soil susceptibility to erosion is influenced by the spatial variability of soil properties and is related to anthropogenic activities in cultivated lands. [2] The knowledge of the spatial variability of soil properties is important for decision making in farm management.

Soil reaction or soil pH is an indication of alkalinity or acidity of soil and is measured by pH units. Plants take up anions and cations, through their roots from the soil solution to meet their growth requirements. Specific plant requirements and fluid composition in soil determines the relative concentration of the ions. The addition of plant materials during decomposition increases the soil pH by ammonification and decarboxylation processes. [3] As soil nutrients' solubility is different at different pH, the properties of soil such as rate of nutrients leaching out of soil can be affected. Interactions between soil pH and nutrients are agronomically important as different decisions on the rate of fertilizer application must be made. The ideal soil pH is close to neutral and most nutrients are available optimally in 6.5 to 7.5 pH, which is generally compatible for plant root growth. [4] However for normal plant growth and development, moderately acidic to neutral soil pH (5.5 – 7.0) is required for most Indian vegetable crops [5]. Soil pH range for optimum growth of Indian vegetable crops is compiled by [5] and specificity will help farmers decide on the crops to be cultivated.

Saline soil is defined to contain sufficient soluble salts to adversely affect the growth of a plant [6]. Soil salinity reduces crop growth and affects the water uptake by the plant while causing specific ion toxicity and imbalance of nutrients [7]. It is measured by Electrical Conductivity (EC) and if the EC of the soil extract exceeds 4dS/m at 25oC, the soil is considered to be saline [6].

An indicator of soil health and compaction is bulk density (BD) which affects infiltration, water availability, soil porosity, plant rooting depth / restrictions [8], nutrient availability and microbial activity which influence significant processes productivity of soil [9]. BD depends on soil texture, organic matter, mineral density and their packing arrangement [10]. Also, available water capacity is influenced by the soil texture, depth and its restrictive layers. High bulk density affects available water capacity, movement of air and water through soil and root growth. In agricultural practices, unnecessary tillage destroys organic matter and the natural stability of soil aggregates declines, making them vulnerable to erosion. Water capacity of the soil is also affected by soil compaction and organic content [9].

Nitrogen is a key soil nutrient that is required by all plants in balanced amount for growth and development process. Nitrogen supplied by fertilizers can be lost in the soil system due to leaching, denitrification, volatilization, soil erosion and runoff, etc. [11], and the remaining is available for plant uptake as ammonium (NH_4^+) or nitrate (NO_3^-) form [12, 13] and is less influenced by soil pH [4]. At near neutral pH (pH 7), the microbial conversion of NH4+ to nitrate (nitrification) is rapid, and crops generally take up nitrate. In acid soils (pH < 6), nitrification is slow [14], and plants with the ability to take up NH_4^+ may have an advantage. Also, the incorporation of high N containing plant material during decomposition may modify the nitrogen cycle in the soil and thus produce soil pH variations.

Soil pH is also an important factor in the nitrogen (N) nutrition of legumes. Legumes live in symbiosis with Rhizobium which is the bacteria responsible for N fixation in legumes. [15] The survival and activity of Rhizobium declines as soil acidity increases. Soil pH affects the nodule numbers in legumes [16, 17, 18] and certain legume species such as Lupins spp. and Mimosa spp., etc., exhibit acid tolerant nodulation. [19, 20] Soil pH also plays an important critical role in ammonia volatilization losses [21] although soil temperature, soil water content, soil texture and cation exchange capacity and nitrification and hydrolysis inhibitors [22] can affect volatilization.

The form of phosphorus (P), the crucial factor for plant growth & grain production, and its availability in soil is also pH dependent [23]. Therefore, the modification of soil pH to soil type-specific optimal level is required to enhance the efficacy of P fertilizers [24]. As the P uptakes by the crops are from the soil solution, the rate of replenishment of P from other forms is crucial for P availability. Replenishment rate is dependent on soil pH, phosphorus levels in soil, its fixation by the soil, and placement of added phosphorus [25]. Phosphate ions tend to form less soluble compounds at alkaline pH (>7.5) by reacting rapidly with calcium (Ca) and magnesium (Mg) and at acidic pH by reacting with iron (Fe) and aluminum (Al) [4]. But phosphorus does not leach easily from soil as that of N, and removal of P from soil by plants is much less when compared to N or K [26].

Availability of potassium (K) in soil is influenced by the soil properties such as texture [27], pH [28, 29, 30], cation exchange capacity (CEC) [31, 32], moisture content [33] and concentrations of other ions [34] in soil. It plays an important role in improving crop quality and disease resistance [35]. Addition of lime increases K availability, possibly through the K displacement by Ca [36].

Changes in soil pH may affect the balance of the various forms of certain nutrients, as well as their availability to plants. [37] The availability of the micronutrients such as manganese (Mn), iron (Fe), copper (Cu), zinc (Zn), and boron (B) tend to decrease as soil pH increases. The exact mechanisms

IOP Conf. Series: Materials Science and Engineering **705** (2019) 012052 doi:10.1088/1757-899X/705/1/012052

responsible for reducing availability differ for each nutrient, but can include formation of low solubility compounds, greater retention by soil colloids (clays and organic matter) and conversion of soluble forms to ions that plants cannot absorb. In order to identify nutrient deficiencies in the prevailing farmlands, large scale mapping of soil properties was recommended. Variation in soil parameters is a result of interactions of various ongoing soil processes and farm management practices [38]. Therefore, in this study, the spatial variability of soil physical properties and macro nutrients such as N, P and K, that promotes plant growth, was studied in Pachapalayam, Coimbatore, India, in order to identify areas of nutrient deficiency. The village of Pachapalayam, was chosen for study due to its avid farming practices and the people's dependence on soil for their livelihood. An informed guidance is required for economic and strategic crop management.

2. Methodology

2.1. Study Area

Pachapalayam (Figure 1) is a village, between 10°58'06.9"N 76°53'56.1"E and 10°57'12.9"N 76°54'16.0"E located just outside Coimbatore town, Tamil Nadu, India, between Perur and Kovaipudur, at an elevation of 431m above mean sea level. It is a gram panchayat in a total geographical area of 731.97 hectares. This agricultural village has extensive cultivation of various varieties of banana (*Musa sp.*), brinjal (*Solanum melongena*), lady's finger (*Abelmoschus esculentus*), tomato (*Solanum lycopersicum*), etc. and the people dependent on these cultivations for income for many generations. Soil in the study area is cohesive in nature.

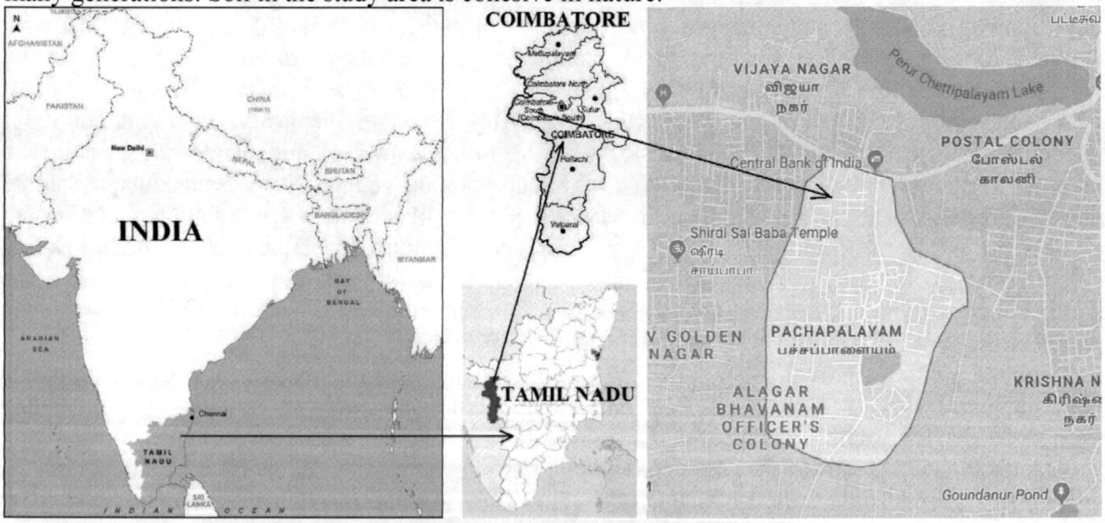

Figure 1. Location Map of the Study Area

2.2. Soil sampling and pre-treatment

The study area was divided into 117 grids of 10,000 sq. m area. From each grid, three soil samples were taken as duplicates at 20 – 40m depths and the sampling points were established and maintained using a Global Positioning System (GPS) device. Samples were collected during January to March 2018 and were tested for soil physical properties such as soil pH, moisture content (MC), bulk density (BD) & electrical conductivity (EC) and for concentration of macro nutrients such as nitrogen, phosphorus and potassium. Soil pH and EC was measured in 2:1 water-to-soil (20cc, ~25g) mixture and using pH meter & conductivity meter respectively. BD was measured on duplicate undisturbed samples in 100 cm3 cylinders [39], after oven drying for 24 h at 105°C and soil MC was determined by gravimetric method [40]. Soil Measuring Kit was used to determine the concentration of N, P & K (turbidometric method).

2.3. Data Processing

The obtained physical properties and concentration of NPK in the soil was subjected to statistical analysis descriptively and geographically. The descriptive analysis of the data was obtained by Microsoft Excel software. The remarkable feature of Geographic Information System (GIS) is its ability to incorporate many forms of dataset with spatial reference (coordinates). The spatial reference (coordinates) of the acquired soil physical properties and NPK were converted to decimal degrees in Excel (CSV format) and imported to ArcMap 10.3 software for further processing for the soil properties conversion to raster format with World Geodetic Survey 1984 (WGS84) as the coordinate reference system.

2.4. Spatial Interpolation

Spatial interpolation is the process of using geographical points with known data values and coordinates to assess the values and coordinates of unknown points within the GIS environment [41]. Hence, it serves as a tool for creating data from sample point for geographical analysis. Many methods of interpolation are prevalent, but for this study, the inverse distance weighted (IDW) ((**1**) was used, which operates under the premise that the closest points has more influence on the unknown point than the points farthest away from it [42].

$$\boldsymbol{Z}\,(\boldsymbol{S_0}) = \frac{\sum_{i=1}^{n} \frac{z(s_i)}{d_i^p}}{\sum_{i=1}^{n} \frac{1}{d_i^p}} \quad \ldots\ldots\ldots\ldots\ldots\ldots\ldots\ldots \text{(1)}$$

where $Z\,(S_0)$ = value at the interpolated point
N = total number of samples in the dataset
S_i = i^{th} data value
d_i = distance between the known data value and the interpolated value
p = weighting power

The x and y coordinates of the acquired sample points are used to interpolate the unknown points and the soil properties are taken as z values for determination at the unknown points.

3. Results And Discussion

The descriptive analysis of the collected sample points for the soil properties (pH, EC, MC and BD) and the concentration of NPK was achieved. From the result of the statistical analysis (Table 1), the observations of pH, EC, BD, P and K are normally distributed whereas the distribution of MC and N indicated positively skewed.

Table 1. Summary of the descriptive analysis of soil physical properties and NPK

Description	pH	EC	MC (%)	BD (g/cm³)	N (Kg/ha)	P (Kg/ha)	K (Kg/ha)
Mean	7.24	0.66	2.56	1.11	41.71	9.90	54.74
Standard Error	0.08	0.04	0.62	0.05	5.02	1.12	6.66
Median	7.22	0.53	0.11	1.39	16.40	3.55	17.00
Mode	8.09	1.00	0.01	1.32	15.60	1.20	17.00
Standard Deviation	0.83	0.41	6.68	0.59	54.26	12.11	72.08
Range	4.66	1.84	47.96	1.73	301.65	61.76	291.31
Minimum	5.20	0.13	0	0.03	2.40	0.26	7.59
Maximum	9.86	1.97	47.96	1.77	304.05	62.02	298.90
Count	117	117	117	117	117	117	117

3.1. Spatial Analysis

All data were converted to raster format according to their geospatial reference and were subjected to spatial interpolation. IDW technique was used to analyze and study the spatial distribution across the area of study. The thematic interpolated layers of the soil physical parameters and concentration of NPK are illustrated in Figure 2 (a – g).

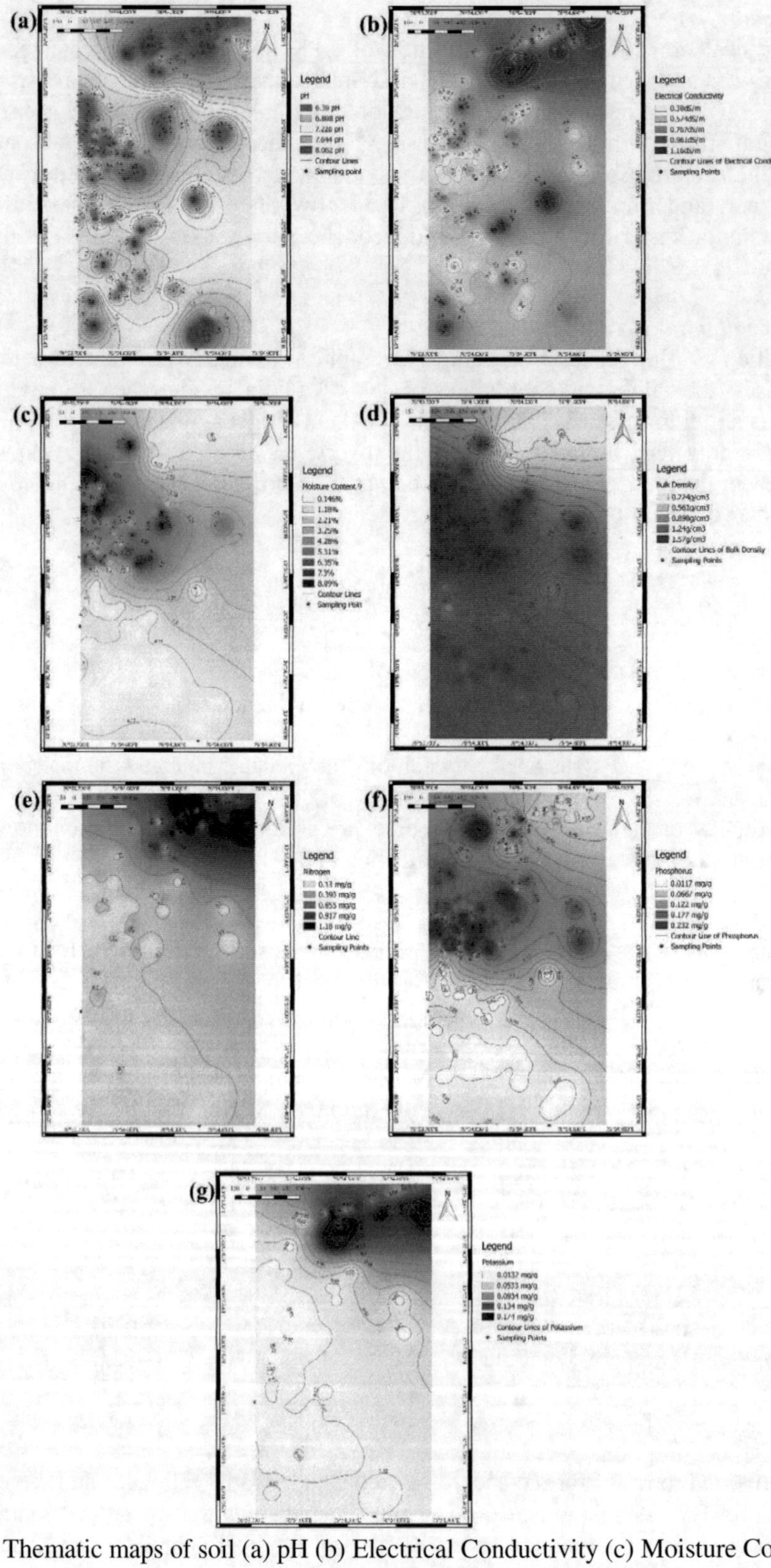

Figure 2. Thematic maps of soil (a) pH (b) Electrical Conductivity (c) Moisture Content (d) Bulk Density (e) Nitrogen (f) Phosphorus (g) Potassium

From classification of the interpolated data (Table 2), 57.2% of the area under study showed soil pH of 7.1 to 10 while most Indian vegetable crops require 5.5 – 7.0 soil pH for normal plant growth and development [5]. The EC of the soil extract is low and does not exceed 4dS/m at 25oC and therefore the soil in the study area is not saline. The study area has clayey to loamy soil with only 10.3% of the area having BD of < 2g/cm3. According to [8], BD < 1.1 – 1.9 g/cm3 is ideal for plant growth and BD > 1.47 – 1.8 g/cm3 restrict root growth. Since 89.7% of the study area has exceeding BD (Table 2), the agricultural practices have to be considered. The moisture content is also low in the study area for cultivation of crops which could be due to the time of sampling. During the sampling period the monsoon rain is low to moderate. Therefore, additional watering of the crops is required.

The tested soil samples were classified according to the soil rating chart of NPK [43] and it is found that a majority of the study area has low concentration of nitrogen available to plants (Table 3) which could be attributed to changes in cultivation [44]. Also, the study area does not contain high levels of N and K, however 36% of the study area shows high level of phosphorus which does not leach through the soil. As the study area contains more of vegetables and fruit plantations, it is recommended to cultivate pulses or oil seed crops which require higher P and to consider phyto-remediation for P removal. Plant waste, vermicompost, yard manure, etc could be used to correct the NPK deficiency with the help of precision farming [45, 46, 47, 48]. Contour maps (Figure 2) generated in this study is GPS linked and hence the area to be managed accurately for the nutrient levels.

Table 2. Classification of Study Area according to the Interpolated Data (Soil Physical Parameters)

pH		**EC (dS/m)**		**Bulk Density (g/cm³)**		**Moisture Content**	
Range	**Percent Area**	**Range**	**Percent Area**	**Range**	**Percent Area**	**Range**	**Percent Area**
5.3 - 6	7.9%	0 - 0.5	36.0%	0 - 10	51.5%	0 - 10	50.2%
6.1 - 7	34.9%	0.6 - 1	39.0%	10.1 - 20	15.2%	10.1 - 20	15.3%
7.1 - 8	44.8%	1.1 - 1.5	19.2%	20.1 - 30	12.1%	20.1 - 30	13.0%
8.1 - 9	10.1%	1.6 - 2	5.8%	30.1 - 40	12.1%	30.1 - 40	12.1%
9.1 - 10	2.3%			40.1 - 50	9.1%	40.1 - 50	9.3%

Table 3. Classification of Study Area according to the Interpolated Data (NPK)

Classification	**Nitrogen (Kg/ha)**		**Phosphorus (Kg/ha)**		**Potassium (Kg/ha)**	
Low	< 240	92.97%	< 11.0	34.62%	< 110	48.28%
Medium	240- 480	7.03%	11 – 22	29.33%	110-280	51.72%
High	> 480	0%	> 22	36.06%	> 280	0%

4. Conclusion

The study area is an agricultural dependent village. Soil nutrients are important for the yield of crops cultivated here and the use of geostatistical tools has made it possible to generate contour maps by which decision makers and farmers can identify easily areas of low or high soil properties and NPK concentrations. Soil physical properties and macro-nutrients content influence greatly the plant growth and development. The results demonstrate the spatial variability of the study area with regard to pH, electrical conductivity, moisture content, bulk density and NPK. Considering the derived classification of the properties, eco-friendly, precise management of farms is required for improvement of the current soil condition and crop productivity. Spatial distribution of micro-nutrients such as Cu, Ni, Mn, Cr, Fe, etc could be studied for a holistic understanding of the soil dynamics in the cultivated area.

Acknowledgement
The authors are thankful to SNR Sons Charitable Trust, Coimbatore, India, for the financial support and also a number of people, who thoughtfully helped during the course of the study.

References

[1] "Agricultural land (% of land area)," Food and Agriculture Organization, [Online]. Available: https://data.worldbank.org/indicator/AG.LND.AGRI.ZS. [Accessed 23 July 2018]

[2] Y. Q. Wang and M. A. Shao 2013, "Spatial variability of soil physical properties in a region of the loess plateau of PR China subject to wind and water erosion," *Land Degrad. Dev.,* vol. 24, pp. 29 - 3

[3] F. Yan, B. Hütsch and S. Schubert 2006, "Soil-pH dynamics after incorporation of fresh and oven-dried plant shoot materials," *J Plant Nutr Soil Sci,* vol. 169, p. 506–508

[4] T. L. Jensen 2010, "Soil pH and the availability of Plant Nutrients," *IPNI Plant Nutrition TODAY,* no. 2, Fall

[5] A. N. Ganeshamurthy, D. Kalaivanan and G. C. Satisha 2016, "Management of Vegetable Crops in Acid Soils of India," *Innovations in Horticultural Sciences,* pp. 559 - 584

[6] S. S. S. o. A. 2001 "Glossary of Soil Science Terms," Soil Science Society of America, Madison, WI

[7] D. L. Corwin and S. M. Lesch 2003, "Application of soil electrical conductivity to precision agriculture: theory, principles and guidelines," *Agron. J.,* vol. 95, pp. 455 - 471

[8] USDA 1987, Soil Mechanics Level I. Module 3 – USDA Textural Soil Classification. Study Guide., Stillwater, OK, USA: USDA, Soil Conservation Service

[9] A. Nyéki, G. Milics, A. J. Kovács and M. Neményi 2017, "Effects of Soil Compaction on Cereal Yield," *Cereal Research Communications,* vol. 45, no. 1, pp. 1 - 22

[10] N. C. Brady 1990, The nature and properties of soils, Macmillan Publishing Company Incorporated

[11] J. A. Lamb, F. G. Fernandez and D. E. Kaiser 2014, "Understanding Nitrogen in Soils," UMN Extension Specialists in Nutrient Management, Minnesota. AG-FO-3770-B

[12] S. A. Barber 1995, Soil Nutrient Bioavailability: A Mechanistic Approach, John Wiley & Sons

[13] C. Ste-Marie and D. Paré 1999, "Soil, pH and N availability effects on net nitrification in the forest floors of a range of boreal forest stands," *Soil Biology and Biochemistry,* vol. 31, no. 11, pp. 1579-1589

[14] D. M. Sylvia, J. J. Fuhrmann 2004, P. G. Hartel and D. A. Zuberer, Principles and Applications of Soil Microbiology, Second ed., Prentice Hall

[15] H. H. Zahran 1999, "Rhizobium-Legume Symbiosis and Nitrogen Fixation under Severe Conditions and in an Arid Climate," *Microbiological and Molecular Biology Reviews,* vol. 63, no. 4, pp. 968 - 989

[16] T. A. Lie 1969, "The effect of low pH on different phases of nodule formation in pea plants," *Plant and Soil,* vol. 31, no. 3, p. 391–406

[17] S. Mohebbi and R. L. Mahler 1989, "The effect of soil pH on wheat and lentils grown on an agriculturally acidified northern Idaho soil under greenhouse conditions," *Communications in Soil Science and Plant Analysis,* vol. 20, no. 3-4, pp. 359-381

[18] A. A. T. Vargas and P. H. Graham 1989, "Cultivar and pH effects on competition for nodule sites between isolates of Rhizobium in beans.," *Plant and Soil,* vol. 117, no. 2, p. 195–200

[19] F. B. J. dos Reis, M. F. Simon, E. Gross, R. M. Boddey, G. N. Elliott, N. E. Neto, M. F. Loureiro, L. P. de Queiroz, M. R. Scotti, W. M. Chen, A. Norén, M. C. Rubio, S. M. de Faria, C. Bontemps, S. R. Goi, J. P. Young, J. I. Sprent and E. K. James 2010, "Nodulation and nitrogen fixation by Mimosa spp. in the Cerrado and Caatinga biomes of Brazil.," *The New Phytologist,* vol. 186, no. 4, pp. 934-46

[20] J. I. Sprent 2009, Legume Nodulation: A Global Perspective., Chichester, UK: Wiley-Blackwell

[21] P. Rochette, D. A. Angers, M. H. Chantigny, M.-O. Gasser, J. D. MacDonald, D. E. Pelster and N. Bertrand 2013, "NH3 volatilization, soil NH4+ concentration and soil pH following subsurface banding of urea at

increasing rates," *Canadian Journal of Soil Science,* vol. 93, no. 2, pp. 261 - 268

[22] D. E. Clay, G. L. Malzer and J. L. Anderson 1989, "Ammonia Volatilization from Urea as Influenced by Soil Temperature, Soil Water Content, and Nitrification and Hydrolysis Inhibitors," *Soil Science Society of America Journal,* vol. 54, no. 1, pp. 263-266

[23] N. Devau, E. L. Cadre, P. Hinsinger, B. Jaillard and F. Gérard 2009, "Soil pH controls the environmental availability of phosphorus: Experimental and mechanistic modelling approaches," *Applied Geochemistry,* vol. 24, no. 11, pp. 2163-2174

[24] S. von Tucher, D. Hörndl and U. Schmidhalter 2018, "Interaction of soil pH and phosphorus efficacy: Long-term effects of P fertilizer and lime applications on wheat, barley, and sugar beet," *Ambio,* vol. 47, no. (Suppl. 1), pp. S41-S49

[25] D. B. Beegle and P. T. Durst, "Managing Phosphorus for Crop Production," [Online]. Available: https://extension.psu.edu/programs/nutrient-management/educational/soil-fertility/managing-phosphorus-for-crop-production. [Accessed 02 November 2018]

[26] M. Sillanpää 1982, "Micro nutrients and the nutritent status of soils: a global study," Food and Agriculture Organization of the United Nations, Rome

[27] Y. Pal, R. J. Gilkes and M. T. F. Wong 2001, "Mineralogy and potassium release from some Western Australian soils and their size fractions," *Austr. J. Soil Res.,* vol. 39, pp. 813 - 822

[28] S. Sahu and S. K. Gupta 1987, "Fixation and release of potassium in some alluvial soils," *J. Indian Soc. Soil Sci.,* vol. 35, pp. 35 - 40

[29] E. Uribe and F. R. Cox 1988, "Soil properties affecting the availability of potassium in highly weathred soils," *Soil Sci.Soc. Am. J.,* vol. 52, pp. 148 -152

[30] J. Zhao, Y. Dong, X. Xie, X. Li, X. Zhang and X. Shen 2011, "Effect of annual variation in soil pH on available soil nutrients in pear orchards," *Acta Ecologica Sinica,* vol. 31, pp. 212 - 216

[31] A. N. Sharpley 1990, "Reaction of fertilizer potassium in soils of different mineralogy," *Soil Sci.,* vol. 149, pp. 44 - 51

[32] K. Sardi and G. Csitari 1998, "Potassium fixation of different soil types and nutrient levels," *Commun. Soil Sci. Plant Anal,* vol. 29, p. 1843 – 1850

[33] Q. Zeng and P. H. Brown 2000, "Soil potassium mobility and uptake by corn under differential soil moisture regimes," *Plant and Soil,* vol. 221, pp. 121 - 134

[34] L. Zawartka, G. Huszcza-Ciolkowska and E. Szumska 1999, "Effects of Poly and Orthophosphates on the Dynamics of some Macro and Micronutrient Elements in Soil Material of Varied pH. III. Potassium Communications in Soil Science and Plant Analysis," *Commun. Soil Sci. Plant Anal.,* vol. 30, pp. 655 - 661

[35] R. Mouhamad, A. Alsaede and M. Iqbal 2016, "Behavior of Potassium in Soil: A mini review.," *Chemistry International,* vol. 2, no. 1, pp. 58 - 69

[36] F. R. Magdoff and R. J. Bartlett 1980, "Effect of Liming acid soils on Potassium availability," *Soil Science,* vol. 129, no. 1, pp. 12 - 14

[37] W. L. Lindsay 1979, Chemical equilibria in soils, New York (NY): Wiley, p. 449

[38] T. B. Parkin 1993, "Spatial variability of microbial processes in soil — a review," *Journal of Environment Quality,* vol. 22, p. 409 – 417

[39] N. J. McKenzie, D. J. Jacquier, R. F. Isbell and K. L. Brown 2004, Australian Soils and Landscapes An Illustrated Compendium., Collingwood, Victoria.: CSIRO Publishing

[40] I. S. McQueen and R. F. Miller 1968, "Calibration and Evaluation of a Wide-Range Gravimetric Method for measuring Water Stress," *Soil Science,* vol. 106, no. 3, pp. 225 - 231

[41] J. Li and A. D. Heap 2011, "“A review of comparative studies of spatial interpolation methods in environmental sciences: Performance and impact factors," *Ecol. Inform.,* vol. 6, no. 3 - 4, p. 228 – 241

[42] C. Childs 2011, "Interpolating Surfaces in ArcGIS Spatial Analyst," *ArcGIS User, ESRI Educ. Serv.,* p. 32 – 35

[43] TNAU, "Resource Management :: Soil :: Soil Rating Chart," Tamil Nadu Agricultural University, 2016. [Online]. Available: http://agritech.tnau.ac.in/agriculture/agri_soil_soilratingchart.html. [Accessed 27 May

2019]

[44] W. M. Post and L. K. Mann, "Changes in Soil Organic Carbon and Nitrogen as a Result of Cultivation," 2005. [Online]. Available: https://doi.org/10.3334/CDIAC/TCM.006. [Accessed 27 May 2019]

[45] W. Whiteley 2002, "The Use of Precision Farming Technologies in Crop Decision Making," in *7003, 13th Congress International Farm Management Association.*, Wageningen, The Netherlands

[46] M. Pandit, K. P. Paudel, A. K. Mishra and E. Segarra 2012, "Adoption and Nonadoption of Precision Farming Technologies by Cotton Farmers," in *Agricultural & Applied Economics Association's 2012 AAEA Annual Meeting,* , Seattle, Washington

[47] D. R. Krindred and R. Sylvester - Bradley 2014, "Using Precision Farming technologies to improve nitrogen management and empower on-farm learning," *Aspects of Applied Biology,* vol. 127, pp. 1 - 8

[48] A. M. Blackmer and S. E. White 1998, "Using precision farming technologies to improve management of soil and fertiliser nitrogen," *Australian Journal of Agricultural Research* , vol. 49, no. 3, pp. 555 - 564

IOP Conf. Series: Materials Science and Engineering **705** (2019) 012053 doi:10.1088/1757-899X/705/1/012053

Modeling Effect of Road Topology on Signal Analysis in Connected Autonomous Vehicle Communication

J S C Turner[1, 3], A B Shahriman[1, 2], A Harun[1, 3, 4], Z M Razlan[1, 2], M S M Hashim[1, 2], W K Wan[1, 2], M A Fadzilla[1, 3], S A Z Murad[3], N A M A Hambali[3] and C Y Phoon[3]

[1]Motorsports Technology Research Unit (MoTECH), Universiti Malaysia Perlis, Pauh Putra Campus, 02600 Arau, Perlis, Malaysia
[2]School of Mechatronic Engineering, Universiti Malaysia Perlis, Pauh Putra Campus, 02600 Arau, Perlis, Malaysia
[3]School of Microelectronic Engineering, Universiti Malaysia Perlis, Pauh Putra Campus, 02600 Arau, Perlis, Malaysia
[4] Corresponding author: aziziharun@unimap.edu.my

Abstract. Connected car has become one of emerging technology in the automotive industries and vehicular communication studies on the other hand, has become one heavily researched topic in academia which focuses on the performance of various parameters upon vehicle-to-vehicle (V2V) and vehicle-to-infrastructure (V2I) communication-based in the IEEE 802.11p vehicular communication studies. In urban environment, obstructions from existence of building, trees, street's layout and other obstruction or objects can strongly influence the receive signal levels at 2.4 GHz frequency band. To ensure reliable and stable communication link for efficient future deployments of cooperative RSUs deployments, will require analyses on the signal strength upon environments that generally characterized by challenging propagation conditions. This article presents the results of experimental field testing of signal characteristics of different types of road topology such as grass, gravel and tar surface on open field environment. The results show that road topology or streets layout have an effect on V2I communication and the attenuation resulted from certain types of road's surface can give significant impact on signal propagation.

1. Introduction

Connected car and autonomous driving or Connected Autonomous Vehicle (CAV) has become one of emerging technology and a heavily researched topic in academic and automotive industry. The evolutions and drastic change in autonomous vehicles technology has been happening since the first radio controlled vehicles was designed in 1920s. In 1960s, similar electronic guide system was used in autonomous cars and this feature is improvised in 1980s [1]. Today, there have been numerous projects, innovations and developments towards autonomous and connected car technology using similar and modified technologies [2]. Most of the projects are vehicular communication (VC) based applications for semi-autonomous cars target to enhance the transportation safety, efficiency and offer services to users or drivers and passengers. One of the famous concepts of wireless based technology in connected cars is such as the vehicle-to-infrastructure (V2I) and vehicle-to-vehicle (V2V) communication. As such, there are other classes of non-automotive applications that use similar concept which existed some time ago namely autonomous underwater vehicles (AUVs), swarm robots and mobile crowd sensing [3][4]. These applications achieve connectivity by connecting nodes directly and come to an agreement by handshaking with each other's message before proceeding to other task. There are often used in military which are mostly deployed on the ground and sea and the system can be form of robots, automation and human. On the contrary, mobile crowd sensing is often used for environmental monitoring while there are also other example of methods that can be used to control swarm robot such as cooperative and predictive control. The system can provide information by allowing other system to read its mind so that the other system is able to make decision towards desired goal. All of these requires wireless system and control algorithm.

In automotive industries the use of both V2I and V2V applications are very wide. V2V and V2I communication interact with other cars and objects around the environment in order to communicate, collect and share data among themselves with aims of increasing safety by reducing accidents and increasing efficiency by managing traffic flow. Each application has its own characteristics with assigned latency and beaconing rates that are based on types of applications and its priority. Several examples of semi-autonomous features introduced in modern vehicle that target safety are such as emergency brake, slow warning, hazard warnings, pre-crash sensing, lane change and collision warning. These types of application are very high in priority and must be assigned with high beacon rates and very low latency for accurate positioning and broadcasting. While transportation efficiency is as crucial as safety, it is assigned as medium priority and requires low beaconing rates and high in latency for unicast V2V or cellular communications purposes. Some examples of applications that target on efficiency are such as adaptive cruise control, electronic toll collection, and so on. These projects and application including the service for users are mostly using ad hoc based, vehicle-to-vehicle (V2V) and vehicle to infrastructure (V2I) communication [5].

In transportation safety and traffic efficiency, the 802.11p or Dedicated Short Range Communication (DSRC), LTE, VANET and Wi-Fi such as WAVE are several examples of available wireless technologies that support V2V and V2I communication. In terms of performance, major crucial aspects that need to be considered to ensure an efficient cooperative control are reliability of communication performance as discussed in [6]. Moreover, a few studies in [7] have explored the impact of terrain elevation and street's topology obstruction in an urban environment where the physical obstruction from the surrounding environment give an impact on the signal strength. In this paper, we aim to investigate the effect of surface attenuation on signal characteristics.

2. Research Background

Empirical study and channel measurement have been used extensively by means to provide understanding of deterministic models for modeling signal propagation thorough surface attenuation. There are a number of research work performed in areas such as agricultural field, environmental monitoring, land and seas for the purpose of future planning and efficient sensor deployment to maintain and sustain reliable communication [8] - [12]. These authors performed indoor and outdoor analysis through wireless sensor network applications and provide RSSI mapping, coverage measurement, antenna positioning impact, data rate measurement and more including modeling signal propagation thorough object, surface and vegetation attenuation. Several data prediction and methods available for modeling signal propagation as recommended by standard ITU-R recommendation for vegetation attenuation [13] and outdoor radio wave propagation [14] are as crucial as to find the right prediction methods through specific objects and types of environment. Recent works have been performed in [15] to model attenuation profile on roadside areas with tall, dense trees and plantation along the roadside area in sub urban environment. The authors analyzed both trees and shrub using different antenna heights. The RMSE and attenuation profiles are compared to existing vegetation attenuation models to find the most suitable models for that particular environment. In this paper, the research work is extended on exploring surface attenuation profile based on several types on surfaces especially flat terrains. Theoretically, there are a number of outdoor propagation models to discuss thorough surface attenuation models. One of the common outdoor propagation models in free space radio-wave propagation is the free space path loss, FSL. FSL model can be used to predict received signal strength when both transmitter and receiver have clear line of sight (LOS) environment. This model is applicable where one single path between the transmitter and receiver with its received signal as a function of transmitted power, antenna gain and distance. The received signal power, L_{FSL} expressed in dB is summarized in equation (1) [11] [15].

$$L_{FSL}(dB) = -27.56 + 20\, log_{10}(f) + 20\, log_{10}(d) \tag{1}$$

Where f is the frequency in MHz, d is the distance between the isotropic transmitting and receiving antennas in meters. Equation (1) can be calculated originally using Friis free space equation as given in equation (2) [16] which includes antenna gains as unity.

$$P_R = P_T G_T G_R \left(\frac{\lambda}{4\pi d}\right)^2 \tag{2}$$

Where P_R is received signal power, P_T is transmitted signal power, G_T and G_R are antenna gains of both the transmitter and receiver, respectively, with G_T=1, $G_R = 1$ (unity gain antenna), $\lambda = c/f$ is the wavelength in meters.

During field measurement the transmitting and receiving antenna heights are being placed close to the ground. Plane Earth (PE) propagation model can be applied when considering transmitter and receiver which are placed close to the surface within 0 meter to 3 meters range. Thus, PE model can be used to describe path loss of ground wave propagation more accurately. The received power can be described using PE propagation model by calculating cross-over distance value as in equation (3). The relationship between received power, P_R and distance between communicating vehicle nodes based on values of d_{CROSS} is given in equation (4).

$$d_{CROSS} = \frac{4\pi h_T h_R}{\lambda} \tag{3}$$

$$P_R = \begin{cases} P_T G_T G_R \left(\frac{\lambda}{4\pi d}\right)^2, d \le d_{CROSS} \\ \frac{P_T G_T G_R h_T^2 h_R^2}{d^4}, d > d_{CROSS} \end{cases} \tag{4}$$

Where P_R is the received power in dB, P_T is transmitted signal power h_T and h_R are transmitting and receiving antenna heights in meter, d is distance between transmitter and receiver in meter, λ is wavelength in meter and G_T and G_R are antenna gains in unity respectively.

In non-line-of-sight (NLOS) environment, shadowing effect occurs as obstruction blocks the signal path between transmitter and receiver. The average received signal strength which decreases logarithmically can be estimated using Log-Normal shadowing model. This is the basic probabilistic propagation model that distributes the reception power in logarithmic domain by applying variance to a normal distribution. This model can be used to compare with the deterministic model as random variable whereby the received power of this model will follow the Log-normal distribution. Log-normal shadowing model is given in equation (5) [11] [15].

$$P_R(d) = P_{R_0} - 10n log(d) + X_\sigma \tag{5}$$

Where $P_R(d)$ is received power in dBm, d is distance in meters from the transmitter, P_{R_0} is signal strength at 1 meter antenna separation, n is path loss exponent and X_σ represents Gaussian random variable with zero mean and standard deviation of σ dB.

Weissberger's model [10][15] as given in equation (6) is applicable for measuring the transmitter and receiver propagation path that are composed by dense, dry and in-leaf trees in temperate climates. This model plays an important role when modeling propagation path through a groves of trees or vegetation.

$$L_W(dB) = \begin{cases} 1.33 \times f^{0.284} d^{0.588}, 14m < d \leq 400m \\ 0.45xf^{0.284} d_f, 0m \leq d_f \leq 14m \end{cases} \quad (6)$$

Where L_W is path loss in dB, f is transmission frequency in GHz, d is foliage depth along the path not less than 14 meters and not exceeding than 400 meters and d_f is foliage depth along the path not exceeding 14 meters.

Two-ray ground reflection model is applicable for predicting large-scale signal strength over LOS in urban environment. This model is applicable when modeling propagation path when vegetation is present outside the transmitter and receiver's path with antenna measurement at significant heights. Two-ray ground reflection model is given in detail in [15] and path loss for Two-ray model can be calculated using equation (7) [16][17].

$$L_{RAY}(dB) = 40logd - (10logG_T + 10logG_R + 20logH_T + 20logH_R) \quad (7)$$

Where, d is separation distance of transmitter and receiver in meters, G_T is transmitter antenna gain, G_R is receiver antenna gain, H_T is transmitting antenna heights and H_R receiving antenna heights in meters.

3. Measurement Location and Equipment

The channel measurement has been conducted in Universiti Malaysia Perlis campus with three different LOS environments such as grass surface (A), tar surface (B) and gravel surface (C). These areas are mainly surrounded by foliage, vegetation and buildings with clear LOS and unobstructed propagation path between the transmitting and receiving antennas. The location of these surfaces based on satellite view is as depicted in Figure1. The view of open field measurement and data collection on grass, tar and gravel surfaces is as depicted in Figure 2, 3 and 4, respectively. Based on Figure 2, the surface is covered entirely with a carpet of grass and only a few crests, hills and trees surrounded the area some distance away. Tar surface in Figure 3 shows a smooth road covered with tar along the way with huge building, trees and hills situated a distance away. Gravel surface as shown in Figure 4 consists of small stones, pebbles and sand but there are vegetation surrounded the compound area quite a distance away. All three medium of surfaces represents similar road surface that can be found along roadways, streets, highways in Malaysia, as such there were chosen for channel measurement and data collection in this paper. This is to ensure that propagation path on different types of surface medium can be obtained to model through surface attenuation specifically for future implementation and effective CAV deployment program on Malaysia's roadway.

The hardware used in this experiment for channel measurement and data collection is low mobility IEEE 802.15.4 XBee S2C compliant devices which operate under free licensed Wi-Fi at 2.4 GHz frequency band. The XBee S2C wireless sensor nodes were programmed as transmitter and receiver using XCTU software [18], respectively as shown in Figure 5. The transmitter (XBee TX) was programmed as the base or local and connected to a laptop with fixed position whereas receiver (XBee RX) was programmed as remote. A real on-field-setup of these wireless sensor nodes is as shown in Figure 6 and the a more detailed specification of XBee S2C sensor nodes is given in [15].

Figure 1. Field measurement held inside Universiti Malaysia Perlis area A, B and C.

Figure 2. Data collection on line of sight environment A, grass surface.

Figure 3. Data collection on line of sight environment B, tar surface.

Figure 4. Data collection on line of sight environment C, gravel surface.

Figure 5. XBee S2C (left) and XBee Explorer (right) connected to laptop.

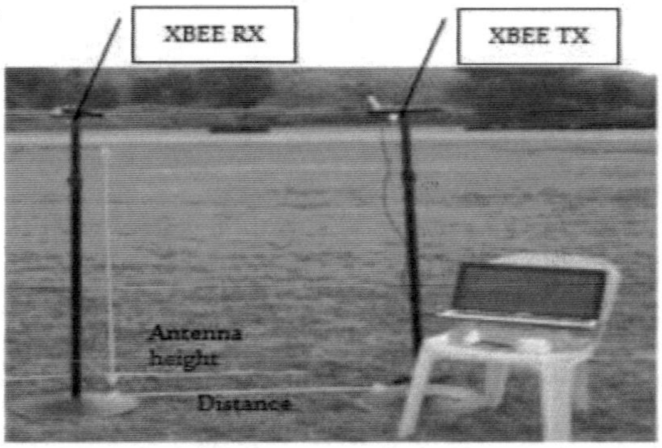

Figure 6. On field measurement setup with XBee receiver (left) and XBee transmitter (right) connected to a laptop.

4. Experimental Design

Before conducting channel measurement, the 1st Fresnel Zone [16] as illustrated in Figure 7 on each surface was calculated based on general equations in [15][19] to measure the maximum distance and allowable heights for obstruction between the transmitting and receiving antennas. This is to ensure that during channel measurement, the 1st Fresnel Zone is free from other source of interference that may affect the performance of data transmission in this particular zone. Knowing the allowable heights of obstruction may prevent or reduce signal interference from unknown source during signal transmission. Measurement setup between transmitting and receiving antenna were estimated to be 100 meters or 0.1 kilometers. The measured clearance radius or maximum allowable height in Fresnel Zone is 1.37 meters. As such when performing channel measurement, objects (e.g. laptop, chair) used in-situ during channel measurement between the transmitting and receiving antennas did not exceed 1.37 meters.

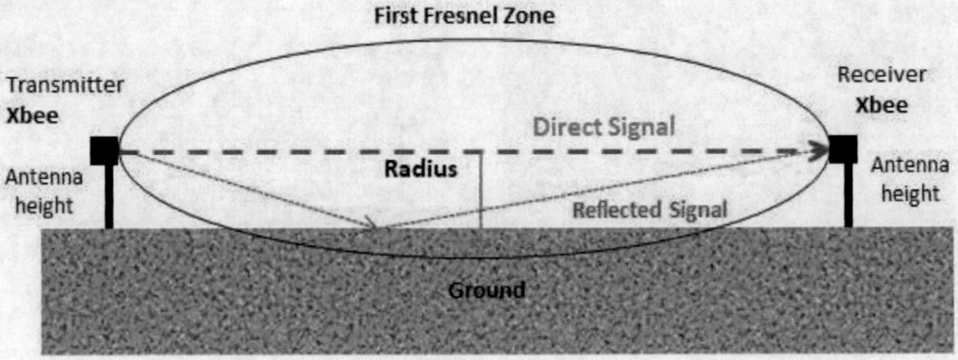

Figure 7. Transmitter and receiver measurement setup and illustration of Fresnel Zone

The channel measurement for all three surfaces was performed by mounting both transmitter and receiver nodes on metal poles. The transmitter (XBee) TX or road-side-unit (RSU) was programmed as base station and connected to a laptop which runs XCTU software. The receiver (XBee RX) or on-board-unit (OBU) was placed and moved every 1 meter increment until the maximum distance of receiver's sensitivity can be reached. At every 1 meter distance, RSSI vales were recorded for two minutes, repeated two times and averaged. The measurements were taken with two different antenna heights, 1.3 meters and 0.15 meters at all three types of surfaces. The XCTU software was used to output the link's quality during data transmission in terms of received signal strength indicator (RSSI) in dBm and packets delivery ratio (PDR). In this paper, only RSSI values were used to model the surface attenuation using root mean square error RMSE to find the closest fit model and applicable in this particular environment. The recorded values were concluded and presented in terms of power level or RSSI (dBm) in the following section.

5. Results and Discussion

First measurement was carried out on grass field environment for two different antenna heights, 0.15 m and 1.3 m. The bed of grass is about 100 m x 50 m in size and situated at Kompleks Sukan Syed Sirajuddin Areeb Putra UniMAP, highlighted as A in Figure 1. Measurement was carried out over the path length starting from 0 m up to 100 m with 1 m increment. As signal propagates along grass field, it is observed that there are significant differences found in signal power variation for both 0.15 m and 1.3 m antenna heights as shown in Figure 8 and Figure 9, respectively. Based on Figure 8, a significant drop from -19 dBm to -40 dBm is observed at 1 m distance for 0.15 m antenna heights. Signal continues to decrease abruptly up to 20 m distance with signal power of -79 dBm. There is a slight signal fluctuation from 21 m onwards until it reaches its maximum range at 64 m distance. The signal only reaches 64 m due to several reasons.

Both transmitting and receiving antenna heights were placed 0.15 m from the ground. As such, within radius of Fresnel Zone the signal has been obstructed by ground effect. This phenomenon creates signal to scatter and disperse hence limits the propagation path. When the antenna height was increased to 1.3 m from the ground, the signal decays gradually starting from 0 m (-23 dBm) to 30 m (-68 dBm) distance as shown in Figure 9. Signal fluctuation is observed between 5 m and 8 m as the signal increases before it decreases abruptly between 9 m and 11 m distance which is from -45 dBm to -57 dBm. Uneven surface of the grass caused the signal to fluctuate starting at 31 m as the signal range is increased further up to 88 m distance. The difference in power loss for first 10 m distance is 30.3 dB compared to antenna height at 1.3 m which is 16.3 dB for grass surface.

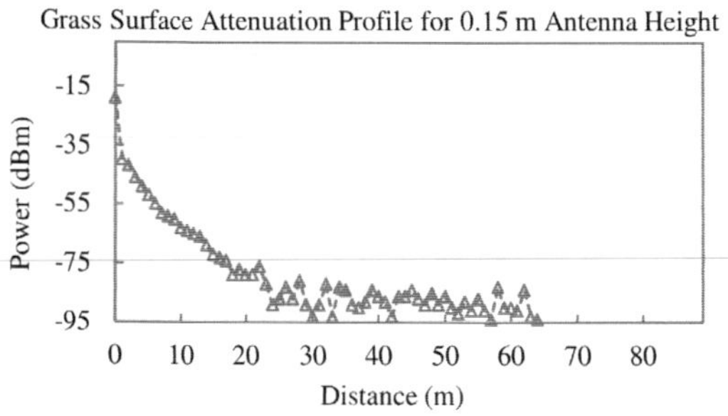

Figure 8. Grass attenuation profile for antenna height 0.15 m

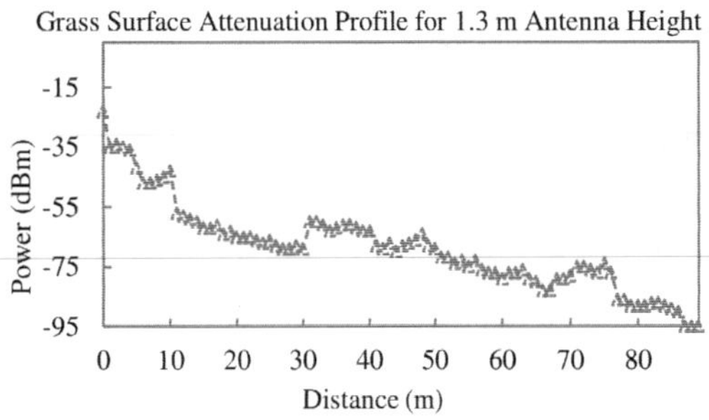

Figure 9. Grass attenuation profile for antenna height 1.3m

Second experiment was carried out on tar surface area along the Universiti Malaysia Perlis Racing Circuit road, highlighted as B in Figure 1. The results of surface attenuation are as shown in Figure 10 and Figure 11 for 0.15 m and 1.3 m antenna heights, respectively. As the signal propagates along the tar surface, signal power drops significantly from -17 dBm to -81 dBm starting from 0 m up to 24 m distance for the 0.15m antenna height as depicted in Figure 10. With antenna heights positioned close to the surface, phenomena known as multipath fading and reflection caused signals to arrive at base station via multiple paths. After decreasing gradually, signal power started fluctuating at 25 m distance for 0.15 m antenna height. After 26 m, signal increased and decreased sharply until it reaches maximum range at 46 m distance due to strong winds during data collection. At 20 m distance, the signal power drop to – 82 dBm because the transmitting and receiving antennas is in clear LOS but strong wind caused the signal to overlap and generate multipath interference. The signal gets distorted and results in signal loss hence the signal only propagated a distance and lost its signal at 47 m.

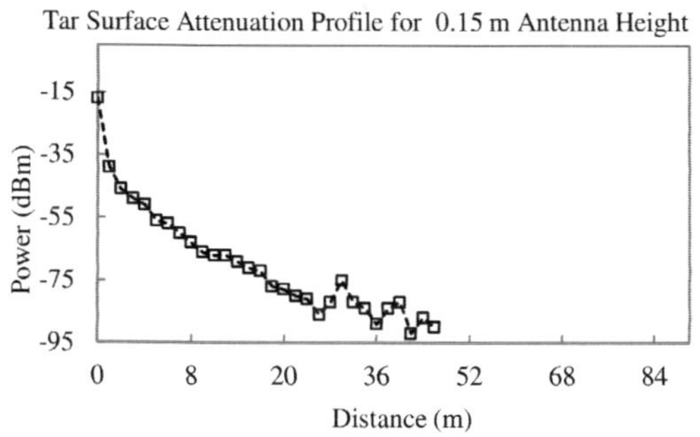

Figure 10. Tar attenuation profile for antenna height 0.15 m

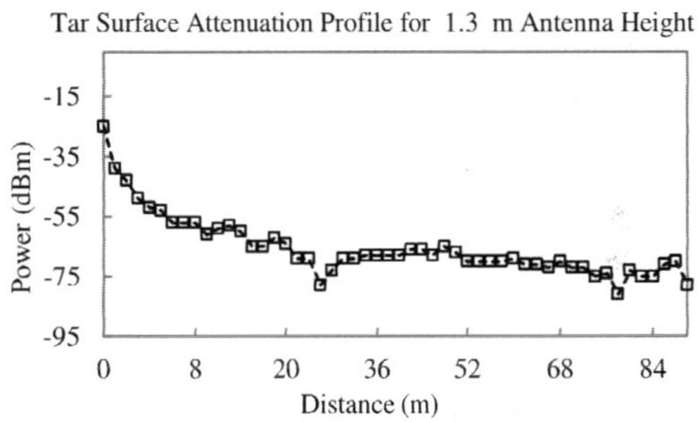

Figure 11. Tar attenuation profile for antenna height 1.3 m

Results for 1.3 m antenna heights tar surface attenuation shows that signal significantly decreases from -25 dBm to -78 dBm starting from 0 dBm to 26 m distance as shown in Figure 11. Significant drop is observed only at 26 m and 90 m which is – 78 dBm for both distances. Less signal fluctuation is observed as propagation along occurs on smooth and even tar surface which contributed to less signal scattering. The difference in power loss between 0.15 m and 1.3 m antenna heights for the first 26 m is 48.6 dB and 34.21 dB, respectively.

Third experiment was carried out on gravel filled surface which consists of sand, small and large pebbles. The test bed is 60 m in length and located at the entrance of Universiti Malaysia Perlis Racing Circuit entrance, as highlighted in C (Figure 1). The results of 0.15 m and 1.3 m antenna height for gravel filled surface attenuation are shown in Figure 12 and Figure 13, respectively.

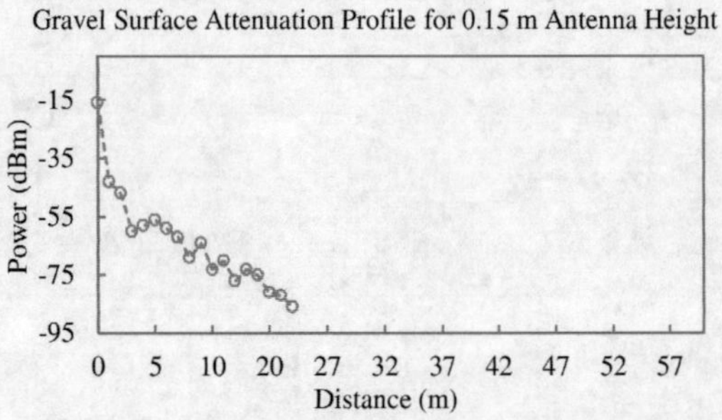

Figure 12. Gravel attenuation profile for antenna height 0.15 m

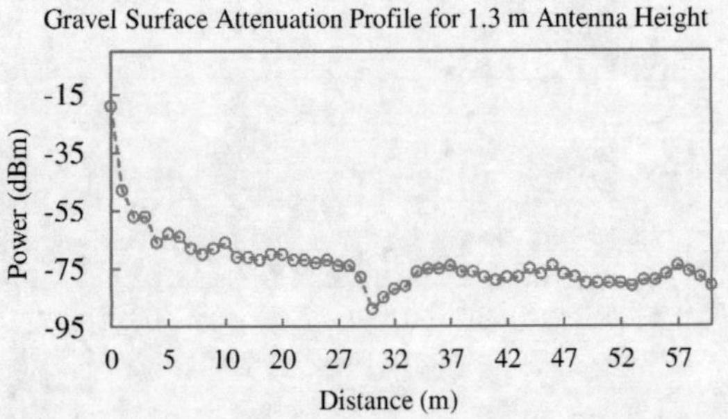

Figure 13. Gravel attenuation profile for antenna height 1.3 m

Signal drop is observed to be significant for both antenna heights at first 10 m distance as signal propagates along gravel surface. The power loss for first 10 m distance is 37.4 dB and 39 dB for 0.15 m and 1.3 m antenna height, respectively. Signal power is observed to reduce from -15 dBm to -73 dBm for 0.15 m antenna height and from -19 dBm to -66 dBm between 0 m and 10 m distance as shown in Figure 12. The propagation range on gravel surface is very limited when the antenna is placed closed to the ground because signal power continues to diminish at 24 m distance. Due to that, signal scatter and divert in various paths and signal are easily absorbed by ground objects (gravel, sand, pebbles) and low antenna height positioned to ground caused signal degradation and limitation in propagation range.

On the contrary with 1.3 m antenna heights, signal is observed to decline abruptly at 30 m distance from -78 dBm to -89 dBm due to existence of larger gravel size and hump along the signal path. At 31 m distance, signal increased to -74 dBm at 37 m distance. The graph show mild signal fluctuations until 57 m distance before decreasing to -81 dBm at 60m. This phenomenon occurred when multipath propagation caused two and more signals to arrive at receiver at different time intervals and from multiple path lengths. Hence, signals that arrived at receiver simultaneously suffers distortion especially signal power having similar signal strength.

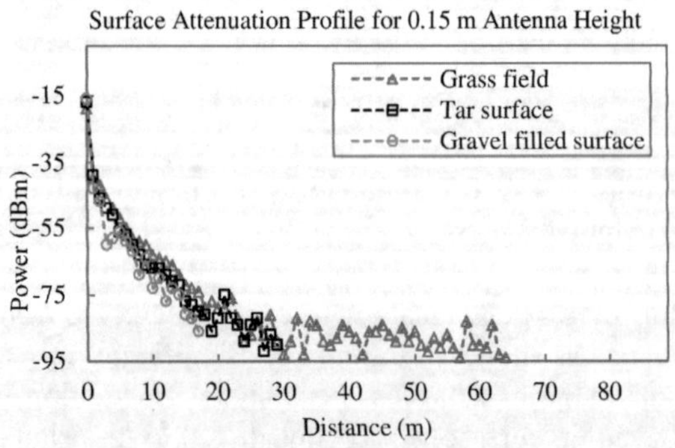

Figure 14. Ground surface attenuation profile for antenna height 0.15 m

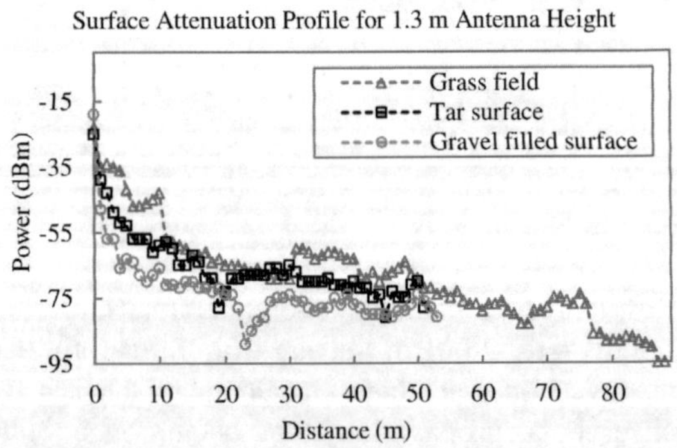

Figure 15. Ground surface attenuation profile for antenna height 1.3 m

Comparisons between three different surfaces of attenuation profile are as shown in Figure 14 and Figure 15 for 0.15 m and 1.3 m antenna height, respectively. Figure 14 shows significant signal attenuation between first 20 m distance before having fair signal fluctuations for all three surfaces. Figure 15 shows gradual significant signal drop between first 10 m distance. It is observed from the graphs that having different medium of surfaces could also affect the characteristics of signal propagation. Grass surface shows the least significant followed by tar and gravel surface. In addition, antenna heights, elevation angle and orientation contribute to variations in signal power. Thus these are several key factors to consider when planning incorporating results and execution of future modeling properly. The signal propagation range for each surface is given in Table 1 below.

Table 1. Signal strength (dBm) and travel distance (m) for grass, tar and gravel surface

	0.15 m Antenna Height	1.3 m Antenna Height
Grass Field	64 m (-94 dBm)	88 m (-95 dBm)
Tar Surface	47 m (-90 dBm)	90 m (-78 dBm)
Gravel Filled Surface	24 m (-81 dBm)	60 m (-86 dBm)

Based on the table above, it can be concluded that signal transmission with clear and unobstructed LOS propagation path can transmit farther and in greater range. With appropriate deployment of antenna heights based on surrounding environment, streets layout and road topology can aid in future planning of sensor nodes deployment between road infrastructure and road side unit.

6. Conclusions

The importance of modeling and investigating effect of different types of surfaces has been demonstrated in this paper and results have shown that certain types of medium on the road can give significant impact of propagation signal. Street's layout and road topology including terrain elevation, surface and many more play a major role in V2I communication as they may contribute to changes in link's connectivity. To implement reliable V2I and V2V communication in order to achieve reliable communication range between vehicle and infrastructure one of the solutions is to investigate the effect of obstruction from the surrounding environment. In our future work, we will model the impact of surface attenuation and develop empirical models to quantify the results.

7. References

[1] Keshav B 2015 A*utonomous Cars: Past, Present and Future - A Review of the Developments in the Last Century, the Present Scenario and the Expected Future of Autonomous Vehicle Technology* pp 191–198

[2] Panos P, Arnaud L F, Knut E, Robert B and Stefano C 2009 *Vehicular systems: Enablingtechnologies, applications, and future outlook on intelligent transportation* (IEEECommun. Mag.) pp 84–95

[3] Matthew A J and Jamshidi M 2010 *Underwater Swarm Robots* (IEEE Syst. J. vol. 4, no. 1) pp 65–73

[4] Jayaraman P P, Perera C, Georgakopoulos D and Zaslavsky A 2013 *Efficient Opportunistic Sensing using Mobile Collaborative Platform MOSDEN* (9th IEEE Int. Conf. Collab.Comput. Networking, Appl. Work) pp 77–86

[5] Goel S and Yu Y 2015 *Emerging Research in Connected Vehicles* (Guest Editorial IEEE Intell. Transp. Syst. Mag., vol. 7, no. 2) pp 6–9

[6] Bai F and Krishnan H 2008 *Reliability Analysis of DSRC Wireless Communication for Vehicle Safety Applications* (IEEE Intell. Transp. Syst. Conf.) pp 355–362

[7] Gozalvez J, Selpruce M and Bauza R 2012 *IEEE 802.11p Vehicle to Infrastructure Communications in Urban Environments* (IEEE Commun. Mag., vol. 50, no. 5) pp 176–183

[8] Turner J S C, Kamarudin L M, Ndzi D L, Harun A, Zakaria A, Shakaff A Y M, Saad A R M, S and Mamduh S M 2014 *Modelling indoor propagation for WSN deployment in smart building* (2nd Int. Conf. Electron. Des. ICED 2014) pp 398–402

[9] Ndzi D L, Harun A, Ramli M F, Kamarudin L M, Zakaria A, Shakaff A Y M, Jaafar M N, Zhou

S and Farook R S 2014*Wireless sensor network coverage measurement and planning in mixed crop farming* (*Comput. Electron. Agric.*, vol. 105) pp 83–94

[10] Harun A, Shakaff A Y M, Zakaria A, Kamarudin L M, Jaafar M N and Ndzi D L 2011 *Wireless sensor networks mapping and deployment in tropical precision farming* (Proc. -CIMSim 2011 3rd Int. Conf. Comput. Intell. Model. Simul.) pp 346–350)

[11] Harun A, Ramli M F, Kamarudin L M, Ndzi D L, Shakaff A Y M, Zakaria A and Jaafar M N 2012 *Comparative performance analysis of wireless RSSI in wireless sensor networks motes in tropical mixed-crop precision farm* (Proc. - 3rd Int. Conf. Intell. Syst. Model. Simulation, ISMS 2012) pp 606–610

[12] Ndzi D L, Arif M A M, Shakaff A Y M, Ahmad M N, Harun A, Kamarudin L M, Zakaria A, Ramli M F and Razalli M S 2012 *Signal Propagation Analysis for Low Data Rate Wireless Sensor Network Applications in Sport Grounds and on Roads* (Prog. Electromagn. Res., vol. 125) pp 1–19

[13] ITU-R 2012 *ITU-R Recommendation P.833-7 Attenuation in vegetation* vol. 7

[14] ITU-R 2017 *Propagation data and prediction methods for the planning of short-range outdoor radiocommunication systems and radio local area networks in the frequency range 300 MHz to 100 GHz* (vol. P, no. 1411–9)

[15] Turner J S C et al. 2019 *Effect of Roadways Plantation on Signal Propagation Analysis in Connected Autonomous Vehicle Communication* (IOP Conf. Ser. Mater. Sci. Eng., vol. 557) p 012056

[16] Rappaport T S 2002 *Wireless Communication: Principles and Practice* (Second Edi. NJ 07458: Prentice Hall PTR)

[17] Jha C K and Jain R 2014 *Literature Survey on Various Outdoor Propagation Model for Fixed Wireless Network* (vol. 3, no. 8) pp 2012–2015

[18] Digi 2015 *Configuration and Test Utility XCTU Software*

[19] Turner J S C, Ramli M F, Kamarudin L M, Zakaria A, Shakaff A Y M, Ndzi D L, Nor C M, Hassan N and Mamduh S M 2013 *The Study of Human Presence Effects on Signal Strength in Buildings for Indoor WSN Deployment* pp 30–35

IOP Conf. Series: Materials Science and Engineering **705** (2019) 012054 doi:10.1088/1757-899X/705/1/012054

A Review on the efficiency and accuracy of localization of moisture distributions sensing in agricultural silos

A A Almaleeh[1], A Zakaria[1, a], S M M S Zakaria[2], L M Kamarudin[3], M H F Rahiman[4,] A S A Sukor[1], Y A Rahim[5] and A H Adom[1]

[1]Centre of Excellence for Advanced Sensor Technology (CEASTech), Universiti Malaysia Perlis, Pauh Putra Campus, 02600 Arau, Perlis, Malaysia.
[2]School of Microelectronic Engineering; Universiti Malaysia Perlis, Pauh Putra Campus, 02600 Arau, Perlis, Malaysia.
[3] School of Computer and Communications Engineering, Universiti Malaysia Perlis, Pauh Putra Campus, 02600 Arau, Perlis, Malaysia.
[4]School of Mechatronic Engineering, Universiti Malaysia Perlis, Pauh Putra Campus, 02600 Arau, Perlis, Malaysia.
[5]School of Manufacturing Engineering, Universiti Malaysia Perlis, Pauh Putra Campus, 02600 Arau, Perlis, Malaysia.

Email: ammarzakaria@unimap.edu.my

Abstract. The moisture distribution in the silos depends upon various seeds parameters such as type and size of seeds, amount of storage, external weather, and storage period as well as structural and environmental factors. It is very difficult to predict moisture distribution in silos effectively while taking all the above aspects into consideration. This study aims to investigate the efficiency and accuracy of localization of moisture distributions sensing in agricultural silo. The work is mainly focussed on three major elements: Radio Frequency (RF), tomographic imaging and classification process using machine learning. In particular, RF-based signal and volume tomographic images are used to predict the moisture distribution. Furthermore, computational intelligence techniques such as artificial neural network (ANN) is applied to develop models based on previous data. The generalization of these models towards new set of data is discussed in making sure a successful application of a model. A detailed study of the relative performance of computational intelligence techniques has been carried out based on different statistical performance criteria.

Keywords. *Radio-frequency, Microwave energy, artificial neural network, Moisture Measurement.*

1. Introduction

Moisture determination has been a major problem in many branches of store grain for many years particularly in the agricultural silos. Accurate moisture determination testing in silos is very important in managing and selling paddy and grain. This is because each grain has a completely different ideal moisture determination. Inaccurate Moisture determination measurements will lead to lower head grain. Once milled at wrong moisture content, it will endure additional drying cost and loss of quality especially when the paddy is dried too much. Furthermore, if the grain is just too wet in storage, harvested paddy will become wetter than necessary and this can contribute to the loss in profit when the paddy is sold. Therefore, it is important to develop moisture measurement systems to determine exact moisture content (MC).

IOP Conf. Series: Materials Science and Engineering **705** (2019) 012054 doi:10.1088/1757-899X/705/1/012054

Radio Frequency (RF) based volume tomography is considerably a quite new and challenging area, especially for non-invasive 3D imaging technologies which the primarily concept idea is adopted from Electrical Capacitance Volume Tomography [1]. In conventional electrical capacitance tomography, the measured capacitance will be placed between plates combinations. Then, it is used to reconstruct 2D images of moisture distribution. Similarly, RF based volume tomography are also constructed to have multiple layers of antenna measuring different attenuation of the transmitted signal. To reconstruct 2D images, the multi-path or diffraction of the signal from the silo walls is considered as a source of distortion to the final reconstructed images. Time of Arrival approaches may be able to mitigate the effect of distortion. However, it may not be true for all conditions. When grain samples have different layers of moisture, 2D image reconstruction can be quite complicated.

3D tomography reconstruction can also be accomplished by stacking frames from a sequence of time intervals of measurements. However, the volumetric profile of the dielectric distribution (in this case will be based on the signal attenuation), may be incoherent due to difference of grain size, shape, and etc. Various challenges remain associated with such reconstruction and need to be handled intelligently. Baseline measurements need to be captured prior to the assessment or handled properly during the tomography imaging. However, when comes to the actual implementation, these baseline information may not be available or difficult to obtain. To counter these issues, the mapping process could incorporate or employ intelligent machine learning approaches.

This study aims to investigate suitable approaches that can measure efficiency and accuracy of localization of moisture distributions sensing in agricultural silo. This is done in three parts, where the first part reviews RF frequency operating at microwave band that gives the best response towards moisture content sensing at different temperature, humidity and grain type. Then, using tomography approach to assess the non-linear image reconstruction accuracy to examine moisture distribution in the silo. Finally, using intelligent machine learning techniques to measure and quantify the distribution of moisture level of rice grain in silo storage using the ANN models. This would benefit the agricultural sector in securing the food storage for the country and help to reduce damages of grain in the silos.

2. Radio Frequency Signal

Radio-frequency (RF) technique is basically principles are described for mistreatment including high frequencies and microwaves insulator properties, or permittivity, of grain for sensing moisture, or permittivity, of grain for sensing moisture through their correlation with moisture content. As a result of high correlations that occur at any frequencies between insulator properties of grain and quantity of water gift within the grain, the radio frequencies are useful in fast and non-destructive sensing of moisture content. Most of those dielectric-type moisture meters use frequencies within the range between about 1-20 MHz. After which the attenuation and phase shift were perceived to be functions of moisture content, grian hulk density, and temperature, illustrated that the determinations of temperature-compensated and density-independent moisture content were acquired with 15.2 GHz measurements on shelled corn [2].

Nearly 50 years ago, the permittivity of grain were obtained through the first quantitative data in the 1–50-MHz frequency range [3], [4]. At both lower and higher frequencies, measurements on grain samples were developed on techniques for dielectric proper-ties [5] [6]. More than 25 years ago, studies explicates the first report regarding data on dielectric properties of grain at microwave frequencies [7] [8]. Several sort of grain and soybeans were developed into models, covering wide frequency ranges [9] [10]. Even in recent times, there have been availability of additional data with respect to the function of moisture content, bulk density, and temperature on the dielectric properties of grain [11] [12]. There have been development of reviews recently publishes covering the RF range below 1 GHz [13] and at microwave frequencies in instrumentation for sensing moisture in cereal grains and other granular materials [14]. There are diverse types of moisture sensing instruments and sources of such moisture sensing equipment are identified in each of these reviews.

A microwave transceiver using a free-space transmission method was designed and implemented to construct the prototype grain moisture meter [15]. The constructed transceiver is illustrated in Fig. 1. A 10.5 GHz microwave signal at 11 Mw generated by an oscillator with a dielectric resonator [11] is transmitted through an isolator and radiated from a transmitting horn antenna into the sample holder. The microwave signal, attenuated by the grain, is collected by a receiving horn antenna and detected by a Schottky diode and RF impedance matching circuit with excellent frequency characteristic [16] [12]. The oscillator consists of a single source biased oscillator with a dielectric resonator, which has a high Q and good temperature stability. The detector converts input RF signal to dc voltage.

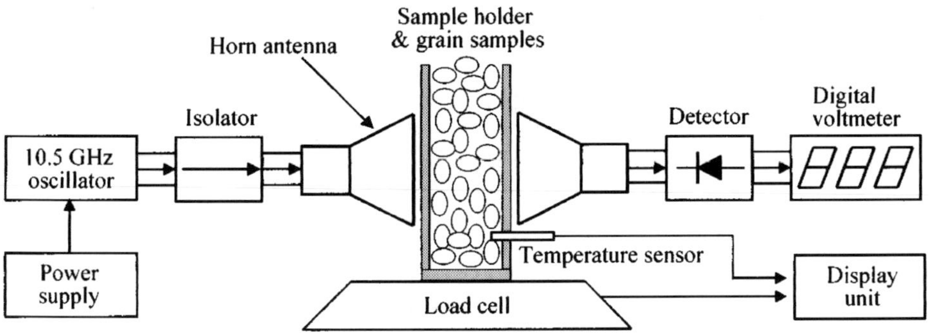

Figure 1. Schematic diagram of the prototype grain moisture meter [17].

3. Tomography Imaging

An investigation on the efficiency and accuracy of localization towards moisture distribution sensing based-on tomography method, the correlation between multiple RF attenuation information can be gathered and analysed using tomography method to locate the moisture distribution in the silo. The optimum number of RF points that act as the sensor in order to provide sufficient accuracy of tomography data. Hence, this enables cross-section images which disclose the information of the location of the moisture in the silo. Uneven nature of the moisture distribution is expected and therefore the tomography image reconstruction requires a non-linear reconstruction approach that results in 2D cross-section images of the silo. This 2D images will be used in the Machine Learning to predicting and assessment of the moisture distribution in the silo and provide moisture volume estimation.

Tomography Imaging Technique has successfully been applied an imaging modality in which internal three-dimensional conductivity distribution of the target is reconstructed on the basis of electrical measurements from the target surface and air voids, moisture [18]. And also this technique has been applied to geophysical monitoring, industrial process tomography and medical imaging. However, Tomography of concrete has been considered only in a few publications [19].

4. Machine Learning

Intelligent Machine Learning Technique for Predicting the Moisture Distribution and Provide Quantification Assessment Using Deep Convolutional Neural Network It has been a big challenge to predict a 3D distribution or performing the quantification of large volumetric datasets derived from computed tomography images. With limited tomography datasets, the information that describes the content of moisture in the silo could be truncated. To provide better quantification of moisture content in the silo, a 3D moisture distribution is proposed. This could be achieved by reinforcing machine learning to provide a better solution with a limited number of training datasets. Using Deep Artificial

Neural Network algorithm, this could effectively predict the 3D moisture distribution and perform the quantification of large volumetric datasets derived from computed tomography images.
In the Radio frequency impedance tomography in the reconstruction of the image, the so-called Generalized Tikhonov regularization [20] is very often used. In a generalized form, this method also represents the Gauss-Newton algorithm.

The application of the least squares method originated from the Gauss-Newton method wherein the matrix $Z^{(l)}$ fulfills the role of matrix X (first partial derivatives relative to fixed approximations $\beta^{(l)}$ and observed values of independent variables), and the role of the vector y (observation of the dependent variable) vector $e^{(l)}$. Hence, the empirical values between the dependent variable and the lth of its approximations that represents the vector of differences.

$$\digamma(X_t,\beta^{(l)}) \tag{1}$$

The study employed Gauss-Newton algorithm to evaluate the structural parameters of non-linear models. The non-linear function in its general form is illustrated below:

$$y_t = \digamma(x_t, b) + \varepsilon_t \tag{2}$$

Also, the artificial neural networks and the electrical tomography method was also used. Thus far, there have not been wide assessment of the silos by using the tomographic and neural networks methods. The low accuracy of mappings and the low resolution of the reconstructed image have been the major reason [21]. Where:

y_t—observations of the explanatory variable,
$x_t = [x_t]$—P vector of observations for explanatory variables,
$\beta_t = [\beta_j]$—K vector of structural parameters,
ε_t—implementations of random elements (we assume that random components are uncorrelated, have an average of zero and equal, positive and finite variance).

The reform of the internal image of the investigated sample is associated with the determination of the global minimum of the fitness function in the Gauss-Newton method. In order to carry out quantitative considerations, we assume that the tested sample is polarized with an alternating low-frequency current. Then, a function with real values describes the electrical material properties. But then, in the generalized Laplace equation, the word relative to the frequency is neglected, and this function can be equated with the electrical conductivity (real isotropic admittivity case).

4.1. Humidity Testing by the Neural Imaging

In solving the moist grain imaging problem, the artificial neural networks as well as the electrical tomography method was also examined. And thus far, there have not been a wide dissemination in the assessment of the grain with regards to tomographic and neural networks methods. Reason been that of low accuracy of mappings and the low resolution of the reconstructed [21]. A set of new method was developed centered on a set of many separately trained neural networks to increase the resolution of tomographic reconstructions depicting the degree of internal humidity of silos. The number of neural networks corresponds to the 3D resolution of the lattice dividing the inside of the silos into individual pixels, which was divided into 8099 points. In a short intervals, by making use of a device termed multiplexer, 192 value of voltage drops were generated by the tomographic system which reads between different electrode pairs. These are the input data for the neural network system. The neural networks are designed in such a way that on the basis of an input vector containing 192 elements, each of the 8099 neural networks generates the value of a single pixel of the output image. Fig. 2 illustrates the neural model employed during the simulations experiments in mathematical form.

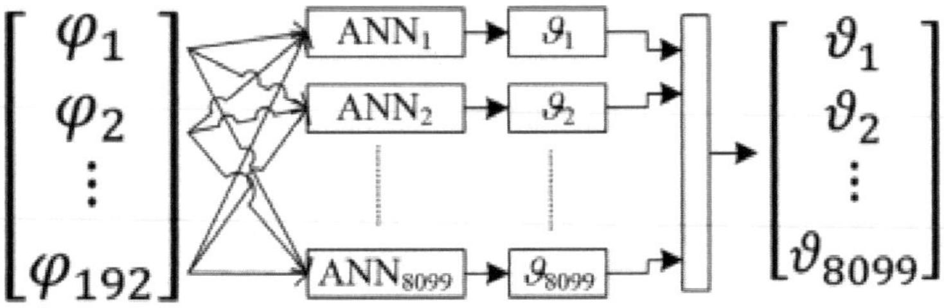

Figure 2. The scheme of converting electrical signals into the image pixels[22].

5. Conclusions

The main goal of this review was to find the best solution of localization of moisture distributions sensing in agricultural silos. Non-destructive methods and algorithms have been studied, which allow estimation of moisture distributions sensing in agricultural silos. A new concept of a non-destructive system based on Radio Frequency (RF) based volume tomography was reviewed. Then, tomographic imaging as well as computational intelligence approaches such as ANN are also have been studied to investigate the efficiency and accuracy of localization of moisture distributions sensing in agricultural silo. The correlation between multiple RF sensor network attenuation information can be gathered and analysed using tomography method to locate the moisture distribution in the silo. From the literatures that have done, it can be concluded that these three approaches have very effective for measuring the moisture content of grain and should be adaptable for fast, accurate, nondestructive moisture measurements on grain.

Acknowledgement

This Research was supported by Ministry of Education (MOE) through Fundamental Research Grant Scheme (Ref: TRGS/1/2018/UNIMAP/02/4/3) with title: Intelligent machine learning technique for predicting the moisture distribution and provide quantification assessment using Deep Convolutional Neural Network. The authors also thank Government of Malaysia and Universiti Malaysia Perlis (UniMAP) for the facilities and technical support, and also the authors would like to thank IPMA Sdn. Bhd. for providing technical knowhow and access to equipments for this research work

IOP Conf. Series: Materials Science and Engineering **705** (2019) 012054 doi:10.1088/1757-899X/705/1/012054

References

[1] Wilson J, Patwari N 2010 *Radio tomographic imaging with wireless networks*. IEEE Transactions on Mobile Computing, 9:621-632

[2] Kraszewski A, Trabelsi S, Nelson S 2000 *Density-independent and Temperature-compensated Moisture Content Determination in Shelled Corn by Microwave Sensing*. Sensors update, 7:51-64

[3] Nelson S, Soderholm L, Yung F 1953 *Determining the dielectric properties of grain* Agricultural Engineering, 34:608-610

[4] Knipper N 1959 *Use of high-frequency currents for grain drying*

[5] Corcoran PT, Nelson SO, Stetson LE, Schlaphoff CW 1970 *Determining dielectric properties of grain and seed in the audiofrequency range* Transactions of the ASAE, 13:348-0351

[6] Stetson LE, Nelson SO 1972 *Audiofrequency dielectric properties of grain and seed.* Transactions of the ASAE, 15:180-0184

[7] Nelson SO 1973 *Microwave dielectric properties of grain and seed*. Transactions of the ASAE, 16:902-0905

[8] Chugh RK 1973 *Dielectric properties of wheat at microwave frequencies*.

[9] Nelson SO 1987 *Models for the dielectric constants of cereal grains and soybeans*. Journal of Microwave Power and Electromagnetic Energy, 22:35-39

[10] Kraszewski A, Nelson S 1989 *Composite model of the complex permittivity of cereal grain* Journal of Agricultural Engineering Research, 43:211-219

[11] Lawrence K, Nelson S, Kraszewski A 1990 *Temperature dependence of the dielectric properties of wheat*. Transactions of the ASAE, 33:535-0540

[12] Nelson S, Trabelsi S, Kraszewski A 1998 *Advances in sensing grain moisture content by microwave measurements*. Transactions of the ASAE, 41:483

[13] Lawrence K, Nelson S 1998 *Sensing moisture content of cereal grains with radio-freequency measurements-a review* In Papers and Abstracts from the Third International Symposium on Humidty and Moisture: 195-203

[14] Kraszewski A 1980 *Microwave aquametry—a review* Journal of Microwave Power, 15:209-220

[15] Kim K-B, Park S, Noh SH, Kim MS, Choi MY 2005 *Moisture measurement in grain by free-space microwave transmission at 10.5 GHz* In 2005 ASAE Annual Meeting. American Society of Agricultural and Biological Engineers;: 1

[16] Kraszewski A, Trabelsi S, Nelson S 1996 *Wheat permittivity measurements in free space* Journal of microwave power and electromagnetic energy, 31:135-141

[17] Kim K-B, Kim J-H, Lee SS, Noh SH 2002 *Measurement of grain moisture content using microwave attenuation at 10.5 GHz and moisture density* IEEE Transactions on Instrumentation and Measurement, 51:72-77

[18] Karhunen K, Seppänen A, Lehikoinen A, Monteiro PJ, Kaipio JP 2010 *Electrical resistance tomography imaging of concrete* Cement and Concrete Research, 40:137-145

[19] Daily W, Ramirez A, Binley A, Henry-Poulter S 1993 *Electrical resistance tomography of concrete structures. Lawrence Livermore National Lab., CA (United States);*

[20] Lechleiter A, Rieder A 2008 *Newton regularizations for impedance tomography convergence by local injectivity* Inverse Problems, 24:065009

[21] Rymarczyk T, Kłosowski G 2018 *Application of neural reconstruction of tomographic images in the problem of reliability of flood protection facilities.* Eksploatacja i Niezawodność, 20

[22] Rymarczyk T, Kłosowski G, Kozłowski E 2018 *A Non-Destructive System Based on Electrical Tomography and Machine Learning to Analyze the Moisture of Buildings*. Sensors, 18:2285

IOP Conf. Series: Materials Science and Engineering **705** (2019) 012055 doi:10.1088/1757-899X/705/1/012055

Comparison of feed forward, Cascade forward and Layer Recurrent Algorithm model for breast cancer prediction

Dr K SheelaSobana Rani[1], Dr S Anila[2]
[1]Associate Professor, Department of Electrical and Electronics Engineering,
[2]Professor, Department of Electronics and Communication Engineering
Sri Ramakrishna Institute of Technology
Coimbatore, Tamil Nadu, India

Abstract. The number of cases of Breast cancer seems to be constantly increasing worldwide in the recent years. According to 2012 world cancer report, the breast cancer incidence rate in Asia appears to be 37.5% and the mortality rate is 13.21 %. As per the recent statistics report of 2018, it has been estimated that the number of new cases of breast cancer is 14% in India. The percentage is higher when compared to lung cancer, oral cancer and other types of cancers. It has been identified that it occurs mostly in the age group between 41-50 years and the distribution of this is found to be 42%, whereas it is 18%, 24% for the age groups of 31-40 and 51-60 respectively. The five stages of survival include the survival rate of 100%, 98%, 88%, 52%, 16% for each stage respectively. Diagnosis in the early stage can reduce the death rate. Even during the stage 3 if the cancer is predicted, it can be treated. Machine learning using Artificial Neural Network(ANN) techniques can be effectively utilised for the prediction of cancer. In the proposed method, networks designed using various ANN Algorithm are used to predict the breast cancer if it is benign or malignant. Three different networks such as feed-forward back prop, Cascade forward and layer recurrent has been implemented. The performance measures such as Accuracy, error, specificity, sensitivity, positive predictive value and negative predictive value are obtained. From the table 2, it is inferred that among these algorithms, feed-forward algorithm has better performance compared to other two algorithms.ANN has been a powerful tool for analyzing the data when there are non-linear interactions between the input and the output to be predicted. The results show that the accuracy of ANN for the prediction was better than other approaches. The Wisconsin breast cancer database has been used for the analysis. The results are very competitive and can be used for diagnosis, prognosis, and treatment.

Index Terms- Breast Cancer, Neural Network, Wisconsin dataset, Feed-forward, cascade, layer recurrent.

1. Introduction

The uncontrollable growth of cells in an organ is called tumors which are termed as cancer. There are two kinds of tumors, benign and malignant as shown in Figure 1. Benign are non-cancerous tumors, non-invasive, slow growing and it has a capsule like structure which forms the outer layer. These type of cancer are non-contagious and are not life threatening whereas, malignant or cancerous tumors are proliferate, highly invasive, grows rapidly and life threatening. Breast cancer is the most frequently occurring disease among women worldwide. It occurs in both men and women. Breast cancer is the leading cause for mortality between ages 40 and 55. Diagnosis in early stage can reduce the death rate.

Breast Cancer screening is the process of checking the presence of Cancer cells in women who has the early symptoms of breast cancer.

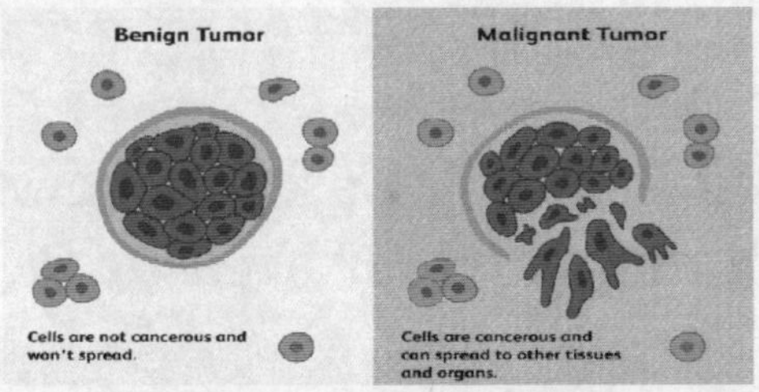

Figure 1 Classification of tumor

At present the widely used screening technique is mammography which is 90% sensitive to cancer cells, which means it cannot detect the 10% cancer cells. With the help of existing technologies such as data mining and machine learning, diagnosis can be made more accurate and medical errors can be minimized.

2.Literature Review

There has been large number of machine learning techniques and algorithms. The utilization of data science and machine learning approaches in medical fields proves to be productive as such approaches may be considered of great assistance in the decision making process. With an unfortunate increasing trend of breast cancer cases, there comes a big deal of data which is of significant use in clinical and medical research and more to data science and machine learning application in this domain. Comprehensive surveys are found to be in existing approaches.

Abien Fred M.Agarap [1] presented an application of different machine learning algorithms for the diagnosis of breast cancer and they exhibited better performance on the binary classification of breast cancer. Venkatesan P and Suresh M L [2] have implemented the ANN model for breast cancer survival prediction. They have presented the ability of the neural network to generalize to new population samples and concluded that the ANN model performs better than logistic model. MihirBorkar et al., [3] have developed an ANN model that helps to determine if patients have breast cancer or not. If patients have a cancerous tumour it can be determined by analysing certain attributes of the cells. Data obtained from 699 patients have been used to train the network. The results were tested using ROC analysis. Chandra et al.,[4] in 2014 have developed ANN with extreme learning techniques for diagnosing breast cancer based on Breast Cancer Wisconsin Dataset. Results showed that Extreme Learning Machine Neural Networks (ELM ANN) have better generalization classifier model than BP ANN. The development of this technique is promising as an intelligent component in medical decision support systems. R.R.Janghel et al.,[5] presented a number of neural network approaches for disease diagnosis. A number of configurations of these networks were trained and tested against the Breast Cancer database. The results showed the ability of the different neural networks to classify the applied input into either of the two classes. It has been suggested that working on better attributes and methods of diagnosis may largely boost the diagnosis accuracy.

KalpanaKaushik and Anil Arora [6] presented the Databases of breast cancer (Wisconsin Breast Cancer (WBC), Wisconsin Diagnosis Breast Cancer (WDBC) and Wisconsin Prognosis Breast Cancer (WPBC) by using classification accuracy and confusion matrix based on 10-fold cross validation method. Also, a fusion at classification level between these classifiers to get the most suitable multi-classifier approach for each data set was introduced. Recently, the neural network has become a popular tool in the classification of cancer datasets. Ahmetmert et al., [7] proposed a method by reducing the dataset of Wisconsin Diagnostic Breast Cancer (WDBC) to one-dimensional feature vector computing an Independent Component (IC). The original data with 30 features and reduced one feature (IC) are used to evaluate diagnostic accuracy of the classifiers such as k-Nearest Neighbor (k-NN), ANN Radial Basis Function Neural Network (RBFNN), and Support Vector Machine (SVM). The comparison of the proposed classification using the IC with original feature set is also tested on different validation (5/10-fold cross-validations) and partitioning (20%–40%) methods. These classifiers are evaluated based on how the tumors are categorized as benign and malignant in terms of specificity, sensitivity, accuracy, F-score, discriminant power, and the Receiver Operating

Characteristics (ROC) curve with its criterion values including Area Under Curve (AUC) and 95% Confidential Interval (CI). This represents an improvement in diagnostic decision support system, while reducing computational complexity. Esraa A. AL-Dreabi et al.,[8] developed a diagnostic system that could classify breast tumors as either malignant or benign to provide a faster and more reliable method for patients. HtetThazinTikeThein and Khin Mo MoTun [9] have proposed an approach for breast cancer to distinguish between different classes of breast cancer. This approach is based on the Wisconsin Diagnostic and Prognostic Breast Cancer and the classification of different types of breast cancer datasets. AutsuoHiga [10] has implemented two powerful classification algorithms namely decision tree and ANN that have been applied for breast cancer prediction.

3. Artificial Neural Network

ANN is one of the artificial intelligence techniques that provide solution for classification and regression problems. It is observed as a best technique for data mining tasks. It has framework for different machine learning algorithms to perform together with data inputs. ANN learns how to predict the output from a set of attributes. The algorithm learns to predict with a training datasets. It has been observed from a lot of research that ANN provided good accuracy in breast cancer prediction. The goal of ANN is to find solution to problems in a similar manner that a human brain does. The network is trained with the training dataset using different algorithms.

3.1Feed-Forward Model

The Network contains feed-forward topology and it will work in one direction, which means the information must flow from input to output in only one direction with no back-loops. The feed-forward algorithm consists of input layer, hidden layers and output layer. While training this network, calculations are carried from input layer to output layer through the hidden layers, and error values were then propagated to previous layers. Feed-forward networks have one or more hidden layers of neurons followed by an output layer of neurons as in Figure 3.1 and its working is shown in Figure 3.2

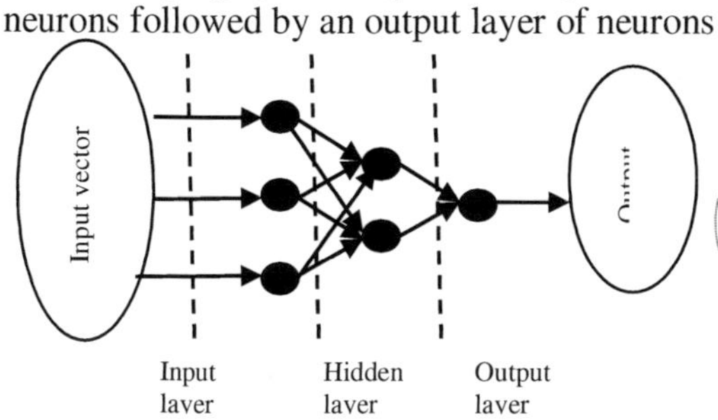

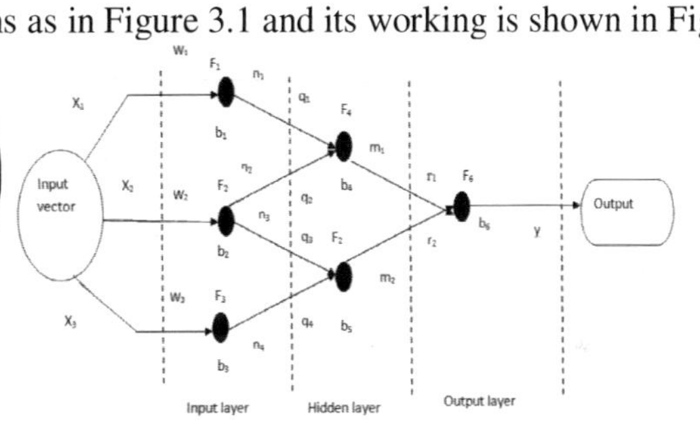

Figure.3.1 Basic Feed-Forward topology **Figure 3.2 Working of Feed-Forward**

The feed-forward artificial neural network can lead to long mathematical descriptions where solving the ANN's parameter optimization problem is complicated and impractical.

3.2 Cascade Forward Network Model

Cascade forward models are similar to the feed-forward networks, but the only difference is, it includes a weight connection from the input to each layer and from each layer to the successive layers. While two-layer feed forward networks can effectively learn any input-output relationship, feed-forward networks with more layers might learn complex relationships more quickly. Cascade forward network has connections from layer 1 to layer 2, layer 2 to layer 3, and layer 1 to layer 3 as shown in Figure 3.3. These networks also have connections from the input to all three layers. The additional connections will improve the speed of the network learning process.

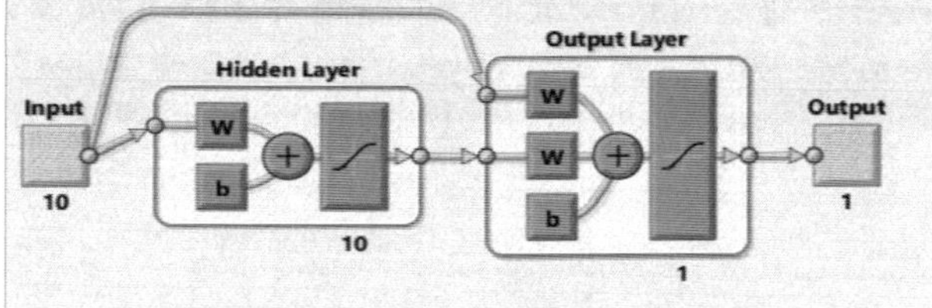

Figure 3.3 Cascade Forward model

3.3 RECURRENT MODEL

In recurrent model, information is transmitted in both directions, forward and also backwards i.e. it has feedbacks as in Figure 3.4. Different recurrent artificial neural networks such as Hopfield, Elman, Jordan, bi-directional and other networks are just special cases of recurrent artificial neural networks.

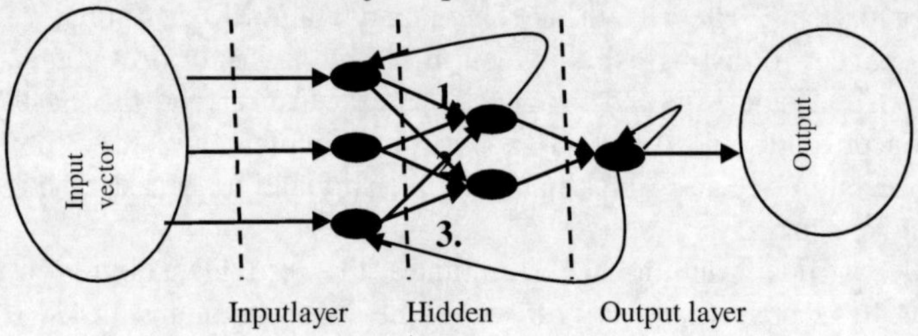

Figure.3.4 Basic recurrent topology model

3.4 Layer Recurrent Model

Layer recurrent neural networks are also similar to feedforward and cascade networks, in which each layer has a recurrent connection with a tap delay associated with it. This allows the network to have an infinite dynamic response to time series input data. This is similar to time delay and distributed delay layer networks which have finite input responses and is shown in Figure 3.5

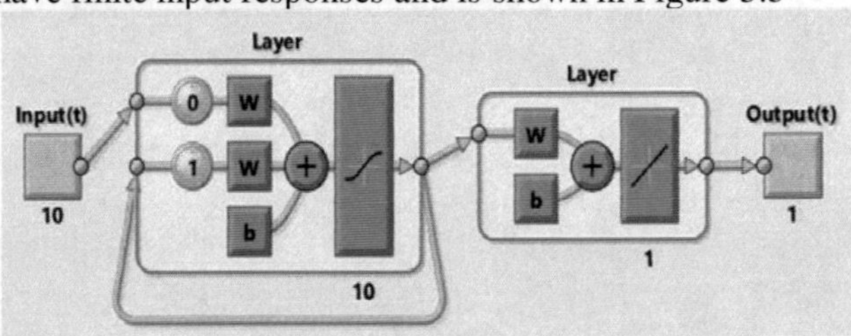

Figure.3.5 Layer recurrent model

4.Methodology

The flow of methodology is shown in Figure 4.1.Wisconsin diagnostic breast cancer (WDBC) dataset is used to implement the machine learning algorithm for the breast cancer diagnosis. There are 569 data entries which contain 32 columns where the first column contains patient ID and the next followed by diagnosis results(benign or malignant) and the rest of the column has the attributes. In the dataset which has 569 entries there are 212 malignant cases and 357 benign cases. The features are the mean, standard deviation and worst of Radius, Texture, Perimeter, Area, Smoothness, Compactness, Concavity, Concave points, Symmetry, Fractal dimension.

The system has to be trained with set of data so that it can give the result depending on past experience. The higher the training rate the greater the accuracy. But training the system will consume a lot of time. The training and testing data should be separated so that the accuracy will be large with minimum number of training data.The ANN is implemented using MATLAB software with Neural Network tool.

Performance evaluation has been done using the various performance measures such as, Accuracy, Error, Sensitivity, Specificity, Positive Predictive Value, Negative Predictive Value and Confusion matrix. A confusion matrix will comprises of actual and predicted results of the cancer. From the confusion matrix the true positive(tp), true negative(tn), false positive(fp) and false negative(fn) values are generated. These values are used to find various performance measures.

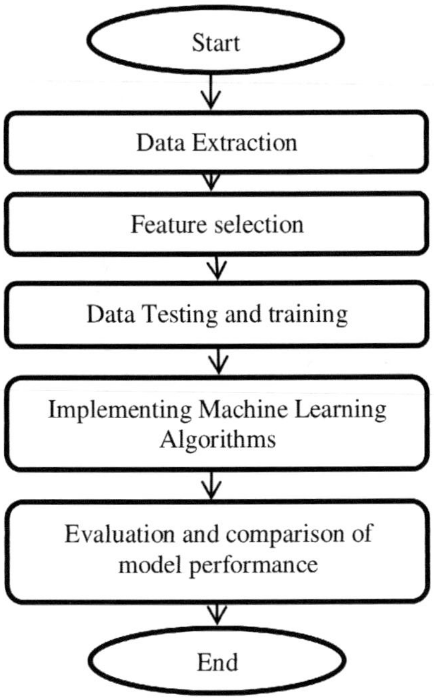

Figure 4.1 Flow of methodology

$$Accuracy = \frac{tp+tn}{all\ entries\ in\ the\ dataset} \quad (1)$$

$$Error = 1 - Accuracy = \frac{fp+fn}{all\ entries\ in\ the\ dataset} \quad (2)$$

$$Sensitivity = \frac{tp}{tp+fn} \quad (3)$$

$$Specificity = \frac{tn}{fp+tn} \quad (4)$$

$$Positive\ Predictive\ Value = \frac{tp}{tp+fp} \quad (5)$$

$$Negative\ Predictie\ Value = \frac{tn}{tn+fn} \quad (6)$$

5. Result And Discussion

The experiments have been conducted using WISCONSIN Breast Cancer Dataset. The data set has been divided as training and testing dataset in the ratio of 70-30% respectively. The three different algorithms such as feed-forward back propagation, Cascade forward and layer recurrent has been implemented. The performance measures such as Accuracy, error, specificity, sensitivity, positive predictive value and negative predictive value are found for the three algorithms. From the table 2, it is inferred that among these algorithms, feed-forward algorithm has better performance compared to other two algorithms.

The training window of the three models, the regression graph and the performance of the three models, FF model, CF model and LR model are shown below.

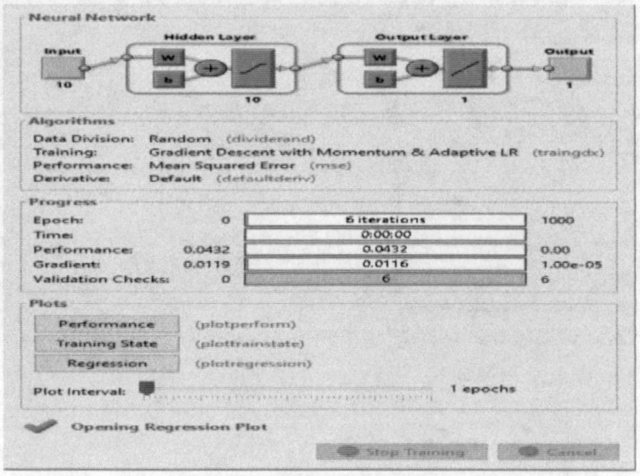

Figure 5.1Training window of FF

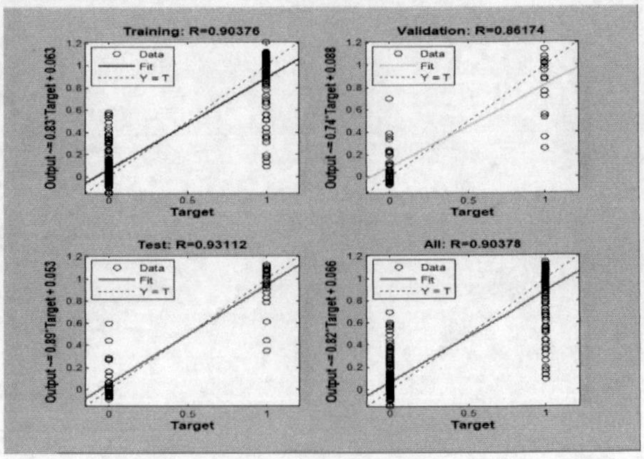

Figure 5.2 Regression graph of FF

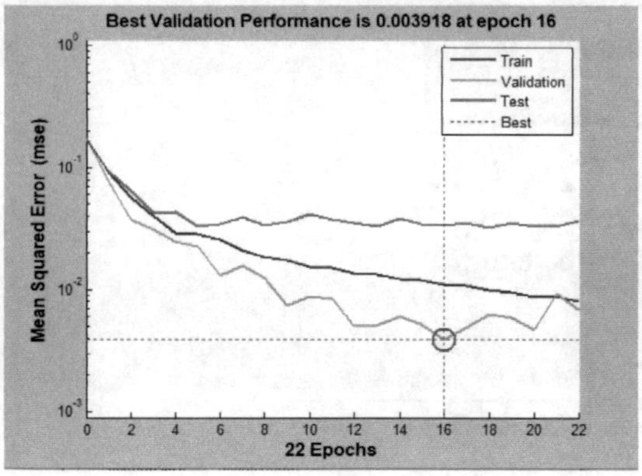

Figure 5.3 Performance graph of FF

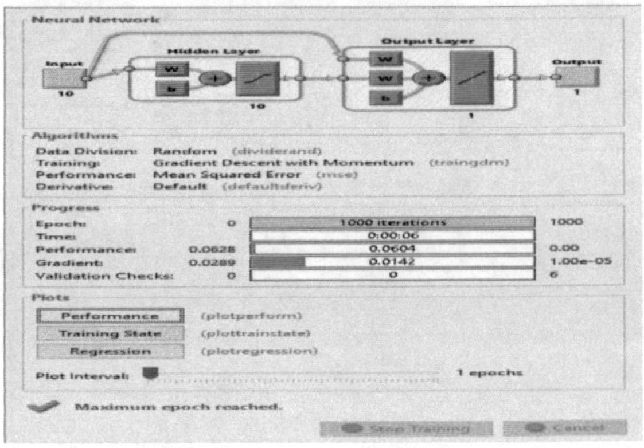

Figure 5.4Training window of CF

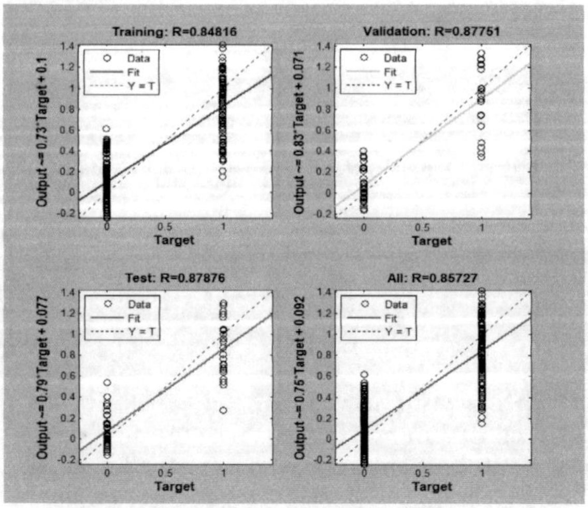

Figure 5.5 Regression graph of CF

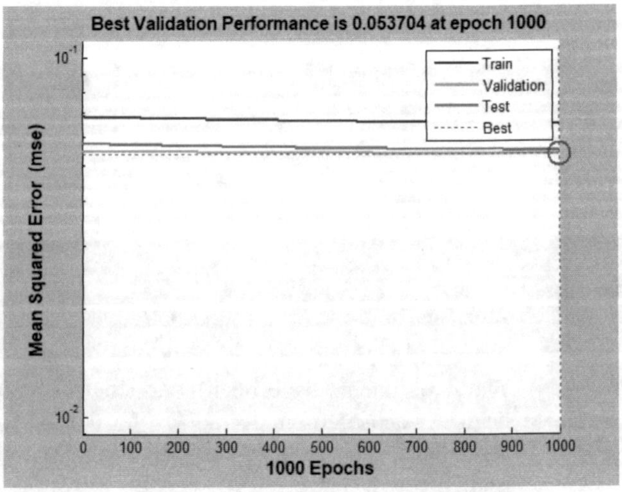

Figure 5.6 Performance graph of CF

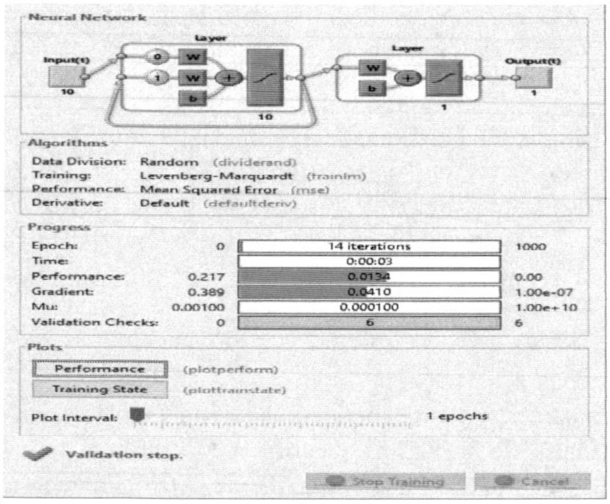

Figure 5.7 Training window of LR

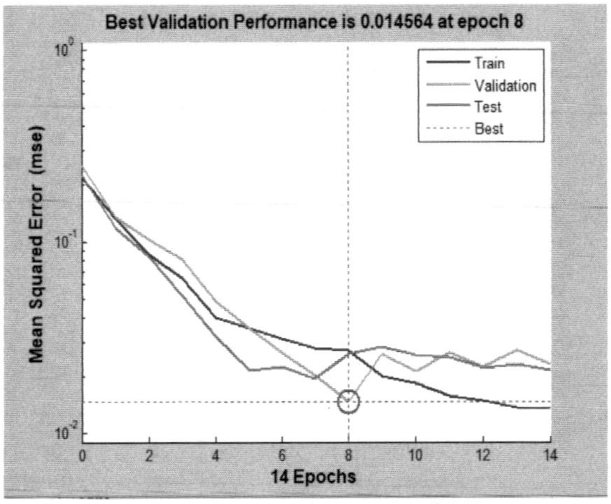

Figure 5.8 Performance graph of LR

Table 5.1 Comparison of performance measures of FF, CF and LR algorithm

Parameters	Feed-Forward	Cascade	Layer Recurrent
Accuracy	0.9534	0.9302	0.927
Error	0.0466	0.0698	0.073
Sensitivity	0.8730	0.8253	0.8666
Specificity	1.000	0.9908	0.9642
Positive Predictive Value	1.000	0.9811	0.9285
Negative Predictive Value	0.9316	0.9075	0.9310

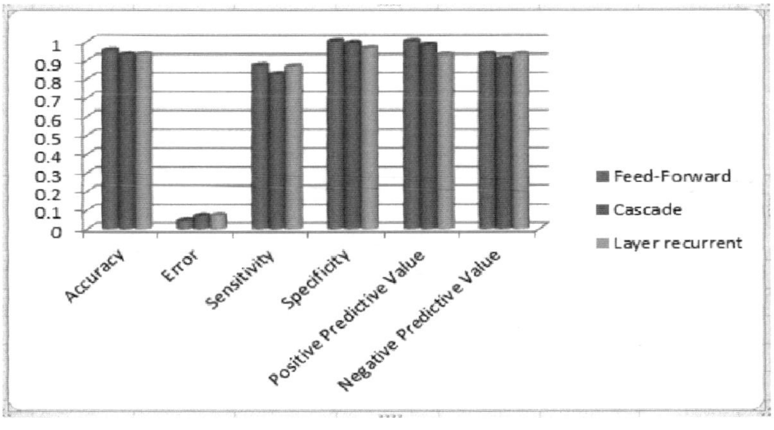

Figure 5.9 Comparison of Performance measures of different models

From figure 5.2, Feed forward algorithm has 95% accuracy with less error of 4% whereas the sensitivity and specificity is 87% and 100% respectively. It can predict the positive values with 100% efficiency and negative value with 93% efficiency. Cascade networks gives 93% accuracy with 6% error. The sensitivity and specificity values obtained were 82% and 99% respectively. The algorithm can predict the positive values with 98% efficiency and negative values to 90%. In Layer recurrent network the accuracy is 92% with the error rate of 7%. Sensitivity and specificity is 86% and 96% respectively. The algorithm predicts the positivevalue to 92% and prediction of negative value is 93%.

It is inferred that among these algorithms, feed-forward algorithm has better performance compared to other two algorithms.

6. Conclusion

The experiment was conducted using WISCONSIN Breast Cancer Dataset. The dataset was divided as training and testing dataset in the ratio of 70-30% respectively. Three different algorithms such as feed-forward back prop, Cascade forward and layer recurrent has been implemented. The performance measures such as Accuracy, error, specificity, sensitivity, positive predictive value and negative predictive value are obtained. From the table 5.1, it is inferred that among these algorithms, feed-forward algorithm has better performance compared to other two algorithms.

References

[1] Abien Fred M.Agarap, 2017 On Breast Cancer Detection: An application of Machine Learning Algorithms on the Wisconsin Diagnostic DatasetarXiv:1711.07831v1 [cs.LG] 20
[2] Venkatesan Pand Suresh M.L 2009 Breast Cancer Survival Prediction using Artificial Neural NetworkIJCSNS International Journal of Computer Science and Network Security 9 5
[3] MihirBorkar, KhushaliDeulkar and Abhinav Garg 2015 Prediction of Breast Cancer using Artificial NeuralNetworksInternational Journal of Engineering Research & Technology 4 09
[4] Chandra PrasetyoUtomo, AanKardiana and Rika Yuliwulandari 2014 Breast Cancer Diagnosis using Artificial Neural Networks with Extreme Learning TechniquesInternational Journal of Advanced Research in Artificial Intelligence 3 7
[5] R.R.Janghel, AnupamShukla, RituTiwari and Rahul Kala 2010 Breast Cancer Diagnosis using Artificial Neural Network Models Source: IEEE Xplore
[6] KalpanaKaushik and Anil Arora 2016 Breast Cancer Diagnosis using Artificial Neural Network International Journal of Latest Trends in Engineering and Technology 7 2
[7] AhmetMert, NiyaziKilic and ErdemBilgili 2015 Breast cancer Detection with Reduced Feature set Computational and Mathematical methods in medicine
[8] Esraa A. AL-Dreabi, Mohammad M. Otoom, Bareqa Salah, Ziad M.Hawamdeh, Mohammad Alshraideh2017Automated Detection of Breast Cancer Using Artificial Neural Networks and Fuzzy Logic International Journal of Sciences: Basic and Applied Research 35 3
[9] HtetThazinTikeThein and Khin Mo MoTun 2015An approach for Breast Cancer diagnosis classification using Neural Network Advanced Computing: An International Journal 6 1
[10] AutsuoHiga 2018 Diagnosis of Breast Cancer using Decision Tree and Artificial Neural Network Algorithms International Journal of Computer Applications Technology and Research 7 01

AUTHOR INDEX